T0276548

Animal Husbandry

Animal Husbandry

Editor: Christian Snider

www.callistoreference.com

Callisto Reference,
118-35 Queens Blvd., Suite 400,
Forest Hills, NY 11375, USA

Visit us on the World Wide Web at:
www.callistoreference.com

© Callisto Reference, 2017

This book contains information obtained from authentic and highly regarded sources. Copyright for all individual chapters remain with the respective authors as indicated. All chapters are published with permission under the Creative Commons Attribution License or equivalent. A wide variety of references are listed. Permission and sources are indicated; for detailed attributions, please refer to the permissions page and list of contributors. Reasonable efforts have been made to publish reliable data and information, but the authors, editors and publisher cannot assume any responsibility for the validity of all materials or the consequences of their use.

ISBN: 978-1-63239-779-9 (Hardback)

The publisher's policy is to use permanent paper from mills that operate a sustainable forestry policy. Furthermore, the publisher ensures that the text paper and cover boards used have met acceptable environmental accreditation standards.

Trademark Notice: Registered trademark of products or corporate names are used only for explanation and identification without intent to infringe.

Printed in the United States of America.

Cataloging-in-publication Data

Animal husbandry / edited by Christian Snider
p. cm.
Includes bibliographical references and index.
ISBN 978-1-63239-779-9
1. Animal culture. 2. Domestic animals. 3. Livestock. 4. Animal genetics. I. Snider, Christian.
SF77 .A55 2017
636--dc23

Table of Contents

Preface

The management and care of farm animals through which the genetic qualities and behavior of the animals are designed to be advantageous to humans is called animal husbandry. Selective breeding is an alternative term used to refer to animal husbandry. This book explores all the important aspects of animal husbandry in the present day scenario focusing on different breeding techniques and advanced husbandry practices. This book will provide interesting topics for research which interested readers can take up. The researchers included herein are of utmost significance and bound to provide incredible insights to readers. This book includes contributions of experts and scientists which will provide innovative insights into this field.

This book is a result of research of several months to collate the most relevant data in the field.

When I was approached with the idea of this book and the proposal to edit it, I was overwhelmed. It gave me an opportunity to reach out to all those who share a common interest with me in this field. I had 3 main parameters for editing this text:

1. Accuracy - The data and information provided in this book should be up-to-date and valuable to the readers.
2. Structure - The data must be presented in a structured format for easy understanding and better grasping of the readers.
3. Universal Approach - This book not only targets students but also experts and innovators in the field, thus my aim was to present topics which are of use to all.

Thus, it took me a couple of months to finish the editing of this book.

I would like to make a special mention of my publisher who considered me worthy of this opportunity and also supported me throughout the editing process. I would also like to thank the editing team at the back-end who extended their help whenever required.

Editor

1

Typing Late Prehistoric Cows and Bulls—Osteology and Genetics of Cattle at the Eketorp Ringfort on the Öland Island in Sweden

Ylva Telldahl[1]*, Emma Svensson[2], Anders Götherström[2], Jan Storå[1]

1 Osteoarchaeological Research Laboratory, Department of Archaeology and Classical Studies, Stockholm University, Stockholm, Sweden, 2 Department of Evolutionary Biology, Uppsala Universitet, Uppsala, Sweden

Abstract

Human management of livestock and the presence of different breeds have been discussed in archaeozoology and animal breeding. Traditionally osteometrics has been the main tool in addressing these questions. We combine osteometrics with molecular sex identifications of 104 of 340 morphometrically analysed bones in order to investigate the use of cattle at the Eketorp ringfort on the Öland island in Sweden. The fort is dated to 300–1220/50 A.D., revealing three different building phases. In order to investigate specific patterns and shifts through time in the use of cattle the genetic data is evaluated in relation to osteometric patterns and occurrence of pathologies on cattle metapodia. Males were genotyped for a Y-chromosomal SNP in *UTY19* that separates the two major haplogroups, Y1 and Y2, in taurine cattle. A subset of the samples were also genotyped for one SNP involved in coat coloration (*MC1R*), one SNP putatively involved in resistance to cattle plague (*TLR4*), and one SNP in intron 5 of the *IGF-1* gene that has been associated to size and reproduction. The results of the molecular analyses confirm that the skeletal assemblage from Eketorp is dominated by skeletal elements from females, which implies that dairying was important. Pathological lesions on the metapodia were classified into two groups; those associated with the use as draught animals and those lesions without a similar aetiology. The results show that while bulls both exhibit draught related lesions and other types of lesions, cows exhibit other types of lesions. Interestingly, a few elements from females exhibit draught related lesions. We conclude that this reflects the different use of adult female and male cattle. Although we note some variation in the use of cattle at Eketorp between Iron Age and Medieval time we have found little evidence for the use of different types of animals for specific purposes. The use of specific (genetic) breeds seems to be a phenomenon that developed later than the Eketorp settlement.

Editor: Vincent Laudet, Ecole Normale Supérieure de Lyon, France

Funding: Financial support to Emma Svensson from Birgit & Gad Rausing and Helge Ax:son Johnsson stiftelsen, P o Lundells, Lars Hiertas Minne. Financial support was received from Berit Wallenbergs Stiftelse and Birgit and Gad Rausings stiftelse to Ylva Telldahl. Anders Götherström was supported by the Royal Swedish Academy of Science. The funders had no role in study design, data collection and analysis, decision to publish, or preparation of the manuscript.

Competing Interests: The authors have declared that no competing interests exist.

* E-mail: ylva.telldahl@ofl.su.se

Introduction

Breeding of cattle has been suggested to have a long tradition in Europe [1]. For example in Italy (Etruria) some contemporary breeds are believed to predate the Roman age [2]. In Northern Europe there may be native breeds, which can be traced some 1000 years back in time [3]. Although farming was introduced to Scandinavia during the Neolithic it is from the Bronze Age and onwards that livestock was used more regularly in the cultivation of land [4] The interaction between farming and animal husbandry became more expressed during the Iron Age [4–7].

During the Viking Age and Medieval Age there were various trade markets in the Baltic region, organized through several ports. The ports also served as hubs for the spread of new agricultural technological innovations and often farms were located close to these ports. The Eketorp ringfort, in the southern parts of Öland, Sweden (Figure 1) offers a unique insight into Iron Age and medieval period husbandry. At an early stage the ringfort was a farming settlement, which over time developed to a garrison. Three settlement phases, I-III, have been identified. Phase I is from Late Roman Iron Age ca. 300–400 A.D. and phase II from Germanic Iron Age ca.400–700 A.D. The ringfort was then abandoned and used again from about 1170–1220 A.D. [8,9]. The animal bones from the ringfort have previously been studied by Boessneck et al (1979). The vast majority of the osteological material is from the second and third phases of the settlement, and thus we have focused on these two periods.

Some of the 53 house foundations dating to phase II at Eketorp are remains of three-aisled houses [10] which were a new type that were built with a byre for stalling animals during winter [11–13]. The remains of 13 byres with stalls for approximately one hundred cattle have been excavated [14]. The faunal assemblage recovered at the Eketorp ringfort is one of the largest in Scandinavia from that time period; 0.5 tons from phase II and 1.3 tons from phase III [8,15]. The recovered bones represent food debris from domesticates such as cattle, sheep and pig, where cattle probably was the most important source of meat. Approximately 75% of the slaughtered adult animals were females [15,16], which illustrate the frequency bias in sex when females are kept to an adult age for milk production. In phase III the bones mainly represent debris of

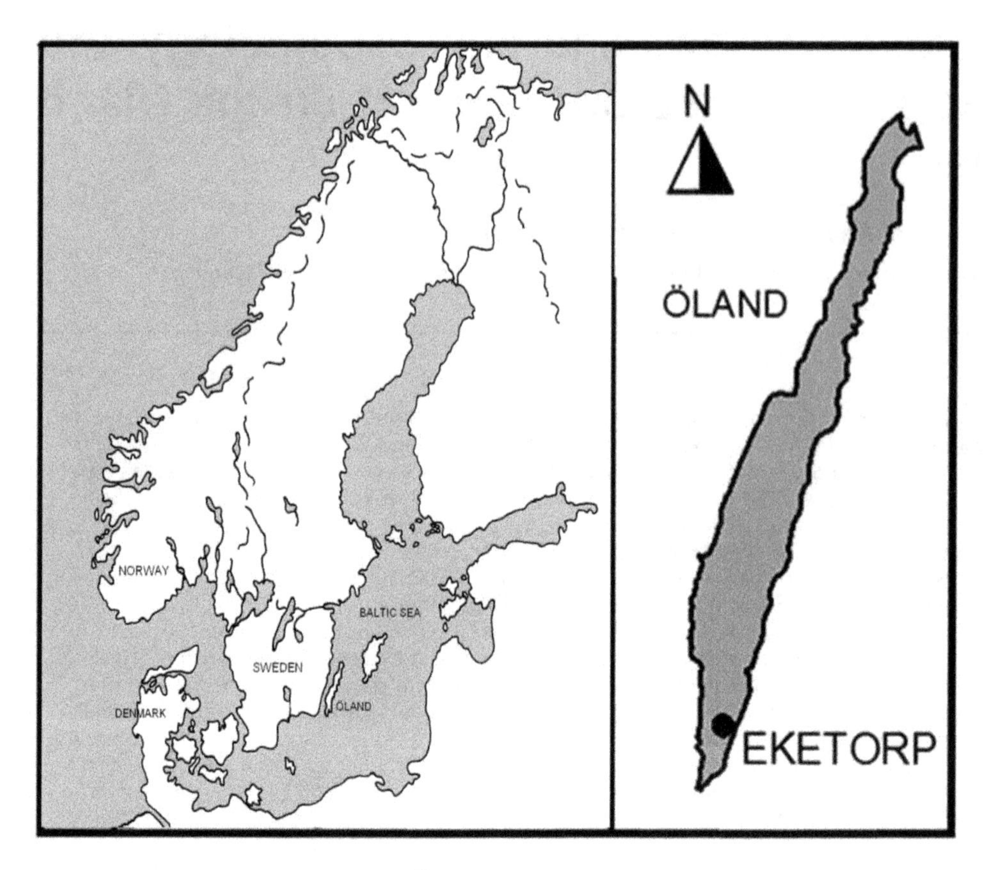

Figure 1. Map showing the location of Öland and Eketorp.

meat that had been brought to the site. Stables have been identified but they do not contain byres for cattle. The presence of long and slender metapodia shows that castration was practiced during both phases (II and III) [15,16]. Fifty-seven bones (23,1%) out of a total of 247 metatarsals and 41 bones (15,2%) of 269 metacarpals exhibit pathological lesions.

The mortality pattern for cattle shows a dominance of bones from sub adult animals in both phases. This is confirmed by Boessneck et al's data on tooth eruption and epiphysial closure on tibia and metapodials where 53.6% in phase II and 59.3% in phase III of the cattle were slaughtered before the age of 2 ½ years old. Boessneck et al also presents a size comparison of unfused first phalanx showing that the majority of sub adult cattle were slaughtered between 6–8 month and 1 ½ – 1 ¾ year of age. The calves were mainly slaughtered in the late autumn and early winter [15]. The average withers height for Eketorp bulls was 111 cm and for cows 109 cm in phase II while it in phase III was approximately two cm higher for both sexes. Compared to cattle from other Swedish and North European sites the Eketorp cattle from period III are among the largest [15,17].

Body size has been used to a wide extent to discuss prehistoric breeding and changes in body size have often been seen as a sign of improvement of livestock [18].Biomolecular analyses may provide complimentary data to the osteological data and holds the possibility to identify genetically different types of animals. Variation in genes coding for specific traits, such as, pigmentation, and muscle mass, may be used to detect breeding [19–21]. However, as breed identification of modern animals demands a panel of some 30 microsatellite markers [22] an even larger panel of SNPs suitable for degraded DNA would be needed [23]. This makes genetic breed identification based on ancient DNA (aDNA) a challenging task given the poor preservation of the DNA and the techniques available at present. We choose to investigate size differences that may be related to genetic characteristics, and if such exist, it would be an indication of advanced animal husbandry. Also, changes in single genetic systems over time can be an indication of selection/breeding.

Here we use a combination of morphological data; sex, physical characteristics and pathological patterns (studied by Telldahl) with molecular data on sex and genetic variation to identify possible shifts in cattle breeding strategies at Eketorp. We consider Eketorp as a model site for northern European farming. If this is a key period in rapid sophisticated specialisation within farming, we expect to find a change in sex proportions and morphology. If, on the other hand, we do not find such change, it can be taken as support for continuity, or a slower rate of specialisation during this period.

Materials and Methods

Our study focuses on metapodia from cattle (*Bos taurus*) recovered at Eketorp. A total of 4470 metapodia have been identified at Eketorp whereof 1879 are from phase II and 2572 from phase III [15]. The bones are highly fragmented and from the assemblage we were able to retrieve 340 specimens of metapodia that offered possibilities for osteometric analyses and/or analyses of specific skeletal lesions. The total sample comprises of 190 metatarsalia and 150 metacarpalia (McIII-IV and Mt III–IV – hereafter abbreviated Mc or Mt). Both complete and fragmentary metapodia from fully-grown and sub adult cattle were analysed. In cattle the distal epiphysis in metacarpals fuses at the age of 2–2 ½ and metatarsals fuses approximately ½ years later

[24–26]. In order to investigate the slaughter patterns of calves at Eketorp bones from sub adults were also selected for molecular sexing.

For size comparisons we use the breadth of the distal epiphysis (Bd) taken according to definitions from von den Driesch [27]. This measurement has proven useful for morphological sexing of male and female metapodia [28,29] and is commonly used in osteoarchaeological analyses. All measurements were documented using a digital calliper to the nearest 0.01 mm.

Seven different types of skeletal lesion on metapodia were identified in the Eketorp assemblage by Telldahl: lipping, new bone formation, eburnation, bone inflammation causing thickening of the diaphysis, depressions in articular facet, carpals/tarsals ankylosis and broadening of trochlea capitis medialis of metapodia. Lipping and broadening of trochlea capitis of metapodia is an overgrowth or bone modification beyond the joint margins. Exostosis is seen as new bone formation near the articular facets. Eburnation is seen as polished bone surface where the cartilage is damaged [30]. Thickening of the diaphysis could be the result of an inflammation where bacteria has access through the connective tissue[31]. The depressions are recorded in both proximal and distal articular facets. The etiology of carpal/tarsal ankylosis is uncertain but research have shown a correlation with age, conformation of the legs and increased load [32].

Two of these lesions, lipping and broadening of trochlea capitis medialis, are probably related to the use as draught animals [33,34,49] (Figure 2). In the present study the lesions are classified into two groups, lesions associated with draught use (workload related) and lesions with an unknown aetiology, i.e. not with certainty related to draught use. Here we report the lesions as present or absent.

A total of 133 metapodia dated to phase II and III were chosen for molecular analyses; 44 metacarpals and 89 metatarsals including 31 metatarsals from juvenile animals with an unfused distal epiphysis too young to be sexed morphometrically; 13 from phase II and 18 from phase III. The selection of the bones was conducted in order to gain a representative sample covering as completely as possible the full morphological size variation and also the presence of skeletal pathologies. A laboratory dedicated to work on aDNA, physically separated from work on modern DNA and PCR products, with positive air pressure and UV lightning was used for all aDNA extractions, a previously described method [21] was used and one extraction blank was included per every six extracts. PCR for sex identification was set up as in [28] using the forward sequencing system (F+R−biotinylated PCR primers and forward sequencing primer) S4 targeting a 63bp fragment. Positive PCR products were genotyped using pyrosequencing technology with a PSQ 96MA following guidelines from the manufacturer.

All samples identified as males were further genotyped for a Y chromosomal SNP in *UTY19*, this SNP has been shown to differentiate North and South European breeds in modern cattle, haplogroup Y1 and Y2 respectively [35]. A primer set targeting 74 base pairs was developed to increase the amplification success of the relatively degraded DNA (Figure S4). A subset of the samples genotyped for the sex identifying SNP were selected for further genotyping based on molecular preservation. Samples from both periods were genotyped in one SNP located in intron 5 of the *Insulin-like Growth Factor 1* (*IGF-1*) gene [38] (Figure S4). *IGF-1* is essential for in vivo follicular development in cattle [39,40] and it is also known to play an important role in various aspects of muscle growth and development [41,42]. The SNP is located in a QTL for twinning rate [38,43], and has been picked up in genome scans for selection in modern cattle [38,44,45]. Further one SNP involved in coat coloration in the *Melanocortin receptor 1* (*MC1R*) gene [36] and one SNP putatively involved in resistance to cattle plague, *Toll-like receptor 4* (*TLR4*) [37] were genotyped. These two

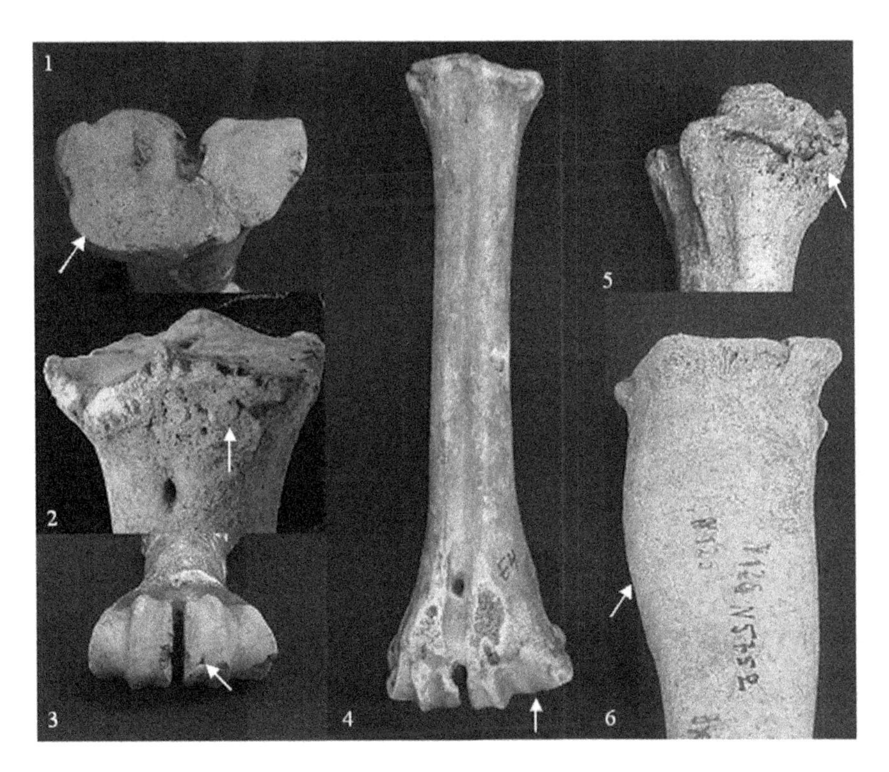

Figure 2. Six skeletal lesions in metapodials from Eketorp ringfort, Öland, Sweden. The lesions comprise of lipping (1), exostosis (2), depression on distal trochlea (3), broadening of the medial trochlea (4), tarsal ankylosis (5) and bone inflammation causing thickening of the diaphysis (6).

SNPs have been suggested to be under selection in northern European cattle (REF 21, Svensson et al, 2007 animal genetics) and the same PCR and sequencing conditions as in (21) was used.

All new primer systems were first blasted to ensure specificity to cattle, and tested on human DNA in the optimization process. The nature of pyrosequencing, where not only the SNP position, but also adjacent nucleotides is given also ensures that correct and specific results are obtained. Allelic dropout was assumed in cases where one or more replicates were homozygous while the other replicates were heterozygous or homozygous for another allele. An estimate of the probability of a false homozygote after (n) replicates was calculated according to [46]: P (false homozygote) = K x $(K/2)^{n-1}$, where K is the observed number of allelic dropouts divided by all heterozygous individuals. Allele frequencies were calculated by hand. Chi 2 test and Fisher's exact test as implemented in STATISTICA 9 were used to test for differentiation in allele frequencies between period II and III. Detailed descriptions information of each element is provided in the supporting information (Figures S1, S2, S3).

Results

DNA was successfully extracted from 104 of the 133 metapodia chosen for analysis, based on a minimum of 4 typings for females and 2 typings for males. However, 4 samples yielded insufficient data for a conclusive result. We were unable to extract DNA from 28 of the bones. The success rate was 78.9 % (Table 1). Allelic dropout was 0.33, providing significance with 4 observations for a female (p false homozygote = 0.001476). No bias was detected in which of the two alleles that was lost in male samples with allelic dropout. The osteological sex identification was confirmed by the molecular result in all cases.

The size distribution confirms that female animals dominate the Eketorp assemblage in both phase II and III. The distal breadth of the epiphysis of the metacarpal and metatarsal bone shows a good separation of the sexes with a small overlap around 53 mm for Mc and around 51 mm for Mt (Figure 3). Mean values and standard deviations of the distal breadth on successfully sexed metapodials confirm a minor overlap (Table 2). DNA analysis of young animals indicate that, more males were slaughtered at young age in period II compared to period III, however the differences is not significant, p = 0.09 (Chi2) (Table 1).

Table 1. Results of molecular sexing.

Adults	Phase	Female	Male	no result
Metacarpals	II	3	3	2
	III	11	14	10
	II/III			1
Total		**14**	**17**	**13**
Metatarsals	II	6	11	4
	III	16	15	4
	II/III	2	1	2
Total		**24**	**27**	**10**
Subadult metatarsals				
	II	2	7	3
	III	8	5	3
Total		**10**	**12**	**6**

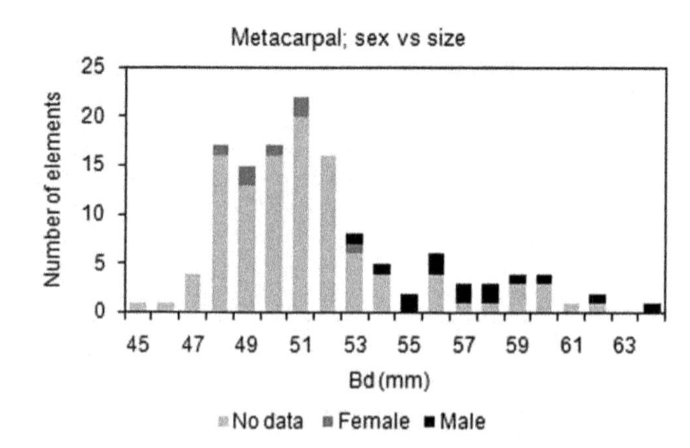

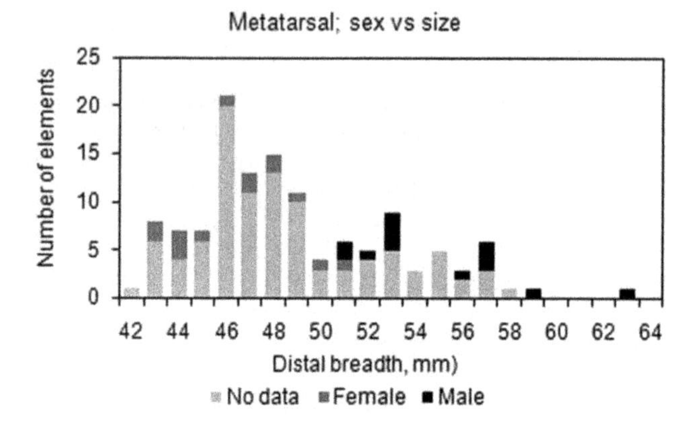

Figure 3. Histograms showing the size of cattle metatarsals and metatarsals according to the breadth of the distal epiphysis (Bd). Sex according to molecular analyses. Included are all available measurements.

The number of different fully grown elements identified was 135 metatarsals and 151 metacarpals. Twenty-three of the metatarsals elements exhibited pathological lesions of an unspecific aetiology while 13 elements (9.6%) exhibited draught related lesions. For metacarpals the corresponding frequencies were 20 unspecific and 8 (5.3%) draught related lesions. There is a slight difference in frequency of draught lesions between Mc and Mt in period II while the frequencies are more even in period III, the observations are too few for a more detailed interpretation, Table 3. The lesions associated to draught use were found mainly on metapodia from male animals while the other types of lesions are found on bones mainly coming from female animals (Figure 4). Noteworthy in

Table 2. Results of the mean value and standard deviation (s.d.) of the distal epiphysis (Bd) in molecular sexed metapodials from Eketorp ringfort, Öland in Sweden.

		Female			Male		
Adults	Phase	n	mean	s.d.	n	mean	s.d.
Metacarpals	II	2	52.61	1.52	3	57.53	1.27
	III	5	49.72	1.19	11	58.58	3.31
Metatarsals	II	2	45.28	2.79	4	57.06	2.50
	III	10	46.67	2.48	8	54.81	4.25

Table 3. Frequency of pathological lesions on cattle metapodia at Eketorp in phase II and III studied by molecular analyses.

		Female		Male		No result	
Element	Phase	Unspecificpat.	Workload related	Unspecific pat.	Workload related	Unspecificpat.	Workload related
Metacarpals	II	2		1	1	4	3
	III	10		4	4		
Metatarsals	II	3		3	5	1	
	III	8	1	7	3		2
	II/III	1			1		1

Phase III is that three metatarsals exhibiting workload related pathologies apparently come from female cattle on the basis of their distal breath measurement. Molecular sexing was inconclusive for two of the elements while the third one was not analysed (Figure S3).

Thirty-three male metapodia were successfully typed for the *UTY19* SNP, all but two have the Y2 defining allele (A). One sub adult from period II and one adult from period III belong to haplogroup Y1(C). The metapodia of the Y2 males exhibit a marked size variation (Figure 5). Only a limited number of animals were genotyped for the coat colour SNP *MC1R* (n = 34), but the result is interesting; no animals indicated the C/C genotype consistent with dominant black coat colour. Instead all animals were either heterozygous (which also results in black pigmentation) or homozygous for the wild type allele, which suggests that the Eketorp cattle mainly were of red or light coat coloration. The *MC1R* results did not yield any significant difference between the two periods, possibly due to limited sample size, or because there was no difference in the coat coloration. When 85 animals were genotyped for the 2021C>T *TLR4* SNP putatively involved in resistance to cattle plague, no significant change in allele frequency over time p = 0.08 was found (Fishers exact two-tailed). The ancestral G allele (G is due to reverse sequencing of the SNP), linked to possible resistance to cattle plague that increases from 0.605 in period II to 0.769 in period III. Finally, 18 animals were typed for a C/T mutation in *IGF-1*, five from Eketorp II and 13 from Eketorp III. The difference between the two periods is on the verge of significant (p = 0.0532) indicating an increase of allele C, from 0.4 to 0.77, in phase III, but it should be noted that the sample set is relatively small. No obvious morphological trait could be assigned to the variation in *IGF-1*. However it should be noted that the frequency of the C allele in modern milk and meat breeds is 0.9 and 0.71 respectively, thus the allele frequency in Eketorp III is more similar to modern beef cattle. No size clusters correlated to *IGF-1* genotype are visible, since all genotyped animals fell within the same size range as the other animals (Figure 6).

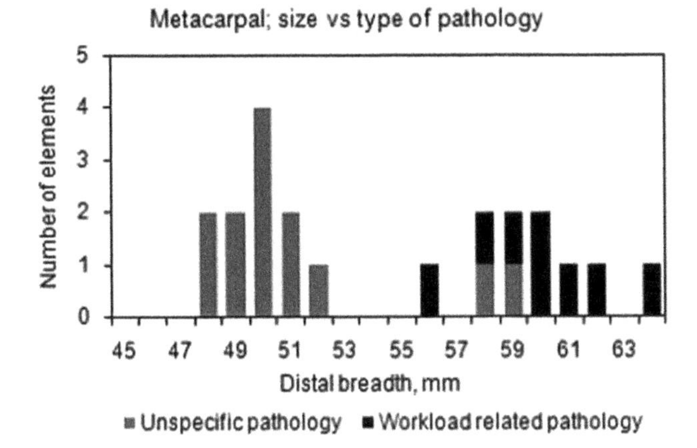

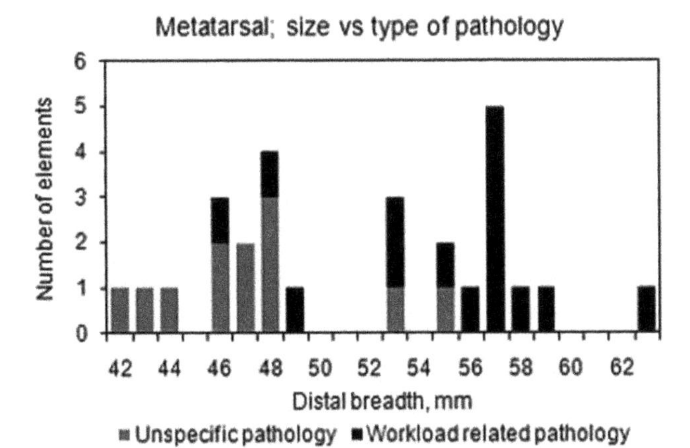

Figure 4. Histograms showing the size variation of metapodia with unspecific or workload related pathological lesions.

Discussion

Our results confirm that the skeletal assemblage of cattle at Eketorp mainly consists of bones from female animals. The predominance of cows is by no means unusual as cows provided both milk and calves for breeding purposes. The need for bulls was not as great and only a handful was most likely enough in order to cover the need for breeding. The use of cattle was not restricted to dairying or meat procurement. At Eketorp the metapodia of males and females exhibit a different pattern of pathological lesions. Females exhibit a dominance of lesions with an unspecific etiology while males also exhibited lesions that may be associated to draught use.

Boessneck [15] observed that calves at Eketorp exhibited a seasonal pattern of slaughter (Figure 7). Phase II exhibits a roughly similar frequency of calves slaughtered in their first or second autumn while phase III shows a clear dominance of calves slaughtered in their first autumn (*ibid.*). The comparison is to some extent affected by the inclusion of both anterior and posterior elements, which may exhibit a slightly different growth pattern. This however is of no major consequence for the comparison. Furthermore, female and male cattle exhibit a different growth pattern. We show that the majority of the sub adult cattle were males, which is in accordance with the (opposite) sex distribution of the adults (Figure 8).

The difference in kill-off patterns between phases II and III is related to an increased reliance on meat, which is in line with the

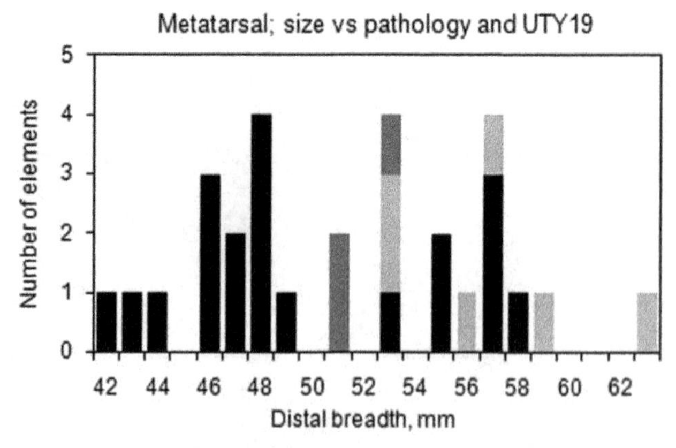

Figure 5. Histogram showing size variation and pathological lesions for males correlated with Y haplogroup.

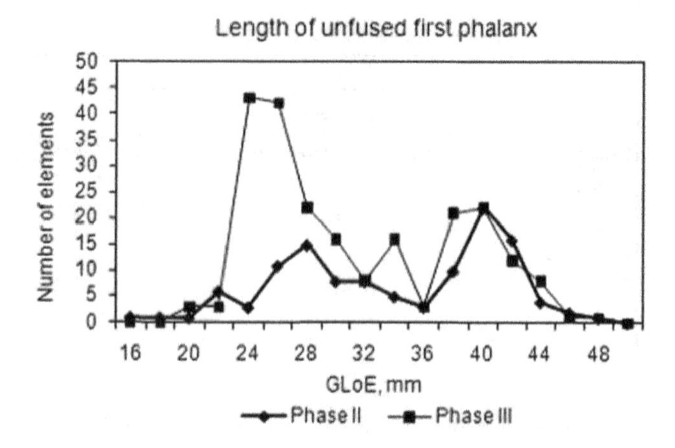

Figure 7. Line graph showing the size variation of unfused first phalanx at Eketorp, phases II and III. The GloE24-30 mm represent calves aged between 6–8 month and GloE 38–43 mm represents calves between 1 ½–1 ¾ years of age. Data are found in Boessneck (et al. 1979:tab 21).

change of the ringfort from a farming settlement to a fortified complex. However, the kill off pattern in phase III was not an ideal one towards husbandry practices focusing on milk production. Instead it indicates a specialization for meat production to the ringfort. The culling of many female calves is not commonly observed and, in fact, this may have led to a depletion of the cattle stock around Eketorp. Linked to this may have been the need to use females for draught purposes during phase III. Earlier studies [15,16] show that castration was most probably conducted on some male animals at Eketorp and thus a link between castration and draught use may be assumed. Systematic breeding of draught cattle is known in written sources in Sweden since the 16th Century when it became a profitable trade, especially in Southern Sweden when farming expanded [47]. We have no indications that the castrated males have been of a selected breed or specific type of animal. Bulls may have been chosen from a common pool of animals. All studied castrates belonged to haplogroup Y2, but note that we only observed two cases of Y1 among all males analysed.

The increase in body size observed by Boessneck et al. [15] might imply an introduction of a different type of livestock during phase III. However, we found no support for the use of different types of animals for specific purposes. The genetic data indicate that the cattle population at Eketorp was homogeneous both within each phase and between phase II and III, the latter also indicates that the population of cattle on Öland wasn't subjected to any major genetic changes for at least 400 years. There is no statistically significant difference in genetic data, but trends indicate possible differences. Taken together, the genetic result and the morphology suggest that there is a small but noticeable difference between the population from period II and that from period III. The resistance to cattle plague was possibly slightly higher in phase III. Only in less than one case out of ten would we expect to see the present difference *TLR4* by pure chance, the trend is even more obvious in *IGF-1*. If the trends are interpreted as true differences, then the animals were probably exposed to natural or artificial selection that is disease or breeding. However, given the time elapsed between the two periods and the relatively

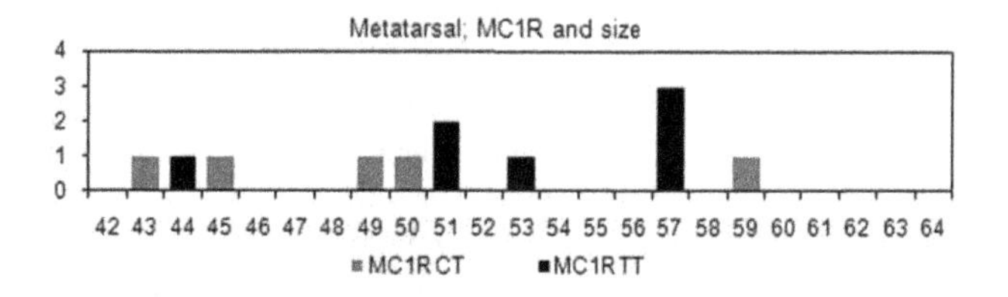

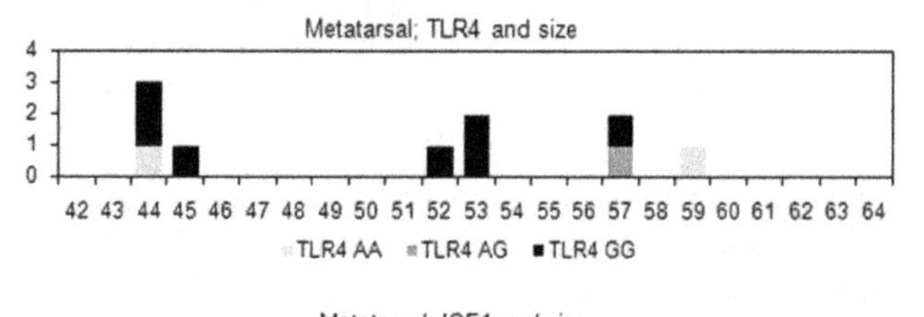

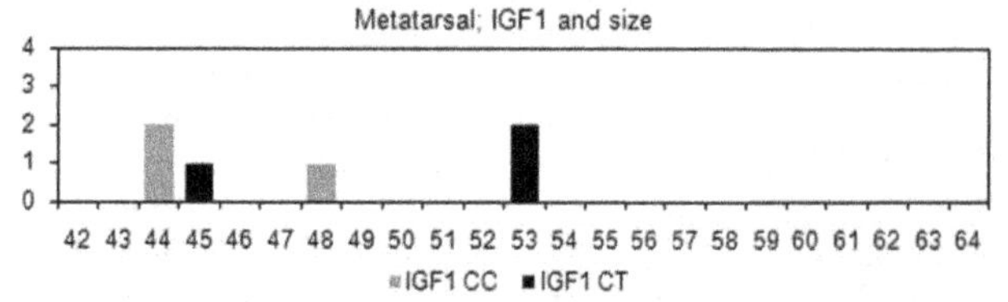

Figure 6. Histograms showing the size variation of the genotyped animals (metatarsals) for the *MC1R*, *TLR4* and *IGF1* genes.

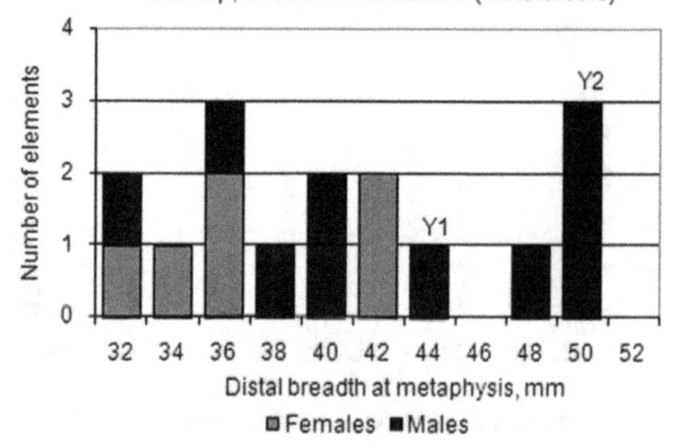

Figure 8. Histogram showing the size distribution of the sub adult metatarsals in phase III.

small sample set genetic drift could also explain the differences in allele frequencies.

Because of the relatively isolated location, the ringfort was probably dependent on local cattle. The military ringfort during phase III probably had to have some kind of organization to secure meat resources. But, it is unlikely that large numbers of cattle of different breeds were imported to Eketorp from other areas. However, it cannot be excluded that some animals or meat from animals that did not belong to the (local) breeding population were brought to the ringfort. Two of the male animals belong to a different Y-chromosomal haplogroup (Y1) than the majority of males, which similarly to other animals from early medieval northern Europe belong to haplogroup Y2 [48]. Although, this SNP is correlated with breeds in modern animals [35], we cannot claim that this is the case for the Eketorp cattle; since analysis of aDNA have shown temporal rather than geographical structure in historic populations [28]. It is therefore more appropriate to state that a minimum of two different male lineages were present at the site during both periods, one in majority and one very rare.

Summarizing our results, we uncovered trends that can be interpreted as a chronological shift, possibly towards a farming economy where cattle gained a new role. We also discovered patterns, compatible with early breeding and an increased level of specialisation in phase III, although these patterns were not obvious enough to be interpreted as undisputable evidence. We see a varied use of cattle at Eketorp and its surroundings, utilization strategies that require breeding efforts and conscious decisions on actions such as castration. In spite of this we have found little if any evidence of the use of genetically specific types of animals for specific purposes. The use of specific (genetic) breeds seems to be a later phenomenon. The usage of the bulk of the cattle seems to have been constant from period II to period III, indicating that the shift we describe was not a fast one.

Ethics statement

The animal bones used in this article are food debris from the excavated Eketorp ringfort on the Öland island in Sweden dated between 300–1200/50 A.D. The Museum of National Antiquities, Stockholm, Sweden, has permitted the analysis.

Acknowledgments

We will thank the anonymous referees for their helpful comments.

Author Contributions

Conceived and designed the experiments: YT ES AG JS. Performed the experiments: YT . Analyzed the data: YT AG JS. Contributed reagents/materials/analysis tools: YT AG JS. Wrote the paper: YT ES AG JS. Performed the DNA analysis: ES.

References

1. Albarella U, Johnstone C, Vickers K (2008) The development of animal husbandry from the Late Iron Age to the end of the Roman period: a case study from South-East Britain. Journal of Archaeological Science 35: 1828–1848.
2. Pellecchia M, Negrini R, Colli L, Patrini M, Milanesi E, et al. (2007) The mystery of Etruscan origins: novel clues from Bos taurus mitochondrial DNA. Proceedings of the Royal Society B: Biological Sciences. 274: 1175.
3. Kantanen J, Olsaker I, LE Holm (2000) Genetic diversity and population structure of 20 North European cattle breeds. Journal of Heredity 91: 446.
4. Cserhalmi N (1998) Fårad mark: handbok för tolkning av historiska kartor och landskap: Sveriges hembygdsförb. 175 p.
5. Myrdal J (1985) Medeltidens åkerbruk: agrarteknik i Sverige ca 1000 till 1520 Nordiska museet. 294 p.
6. Sweeney D, Bailey M (1995) Agriculture in the middle ages: technology, practice, and representation. University of Pennsylvania Press.
7. Pedersen, EA, Widgren M (1998) Fähusdrift, järn och fasta åkrar. In: Welinder S, Pedersen EA, Widgren M, eds. Det svenska jordbrukets historia, Jordbrukets första femtusen år., Natur och Borås: Borås. pp 239–266.
8. Borg K (2000) Eketorp-III: Ett medeltidsarkeologiskt projekt: University of Lund, Institute of Archaeology. 188 p.
9. Borg K, Näsman U, Wegraeus E (1976) Eketorp: Fortification and Settlement on Öland, Sweden. The Setting. 127 p.
10. Näsman U (1976) Introduction to the Descriptions of Eketorp-I,-II &-III. In: Borg K, Näsman U, Wegreaus E, eds. Eketorp. Fortification and Settlement on Oland/Sweden The Monument. 215 p.
11. Zimmermann W (1988) Why was cattle-stalling introduced in prehistory? The significance of byre and stable and of outwintering, In: Fabech C, Ringtved J, eds. Settlement and landscape proceedings of a conference in Århus, Denmark, May 4-7, 1988 Jutland Archaeological Society.
12. Årlin C (1999) Under samma tak- Om "husstallets" uppkomst och betydelse under bronsåldern ur ett sydskandinaviskt perspektiv. In: Olausson, M, eds. Spiralens öga: tjugo artiklar kring aktuell bronsåldersforskning. pp 291–307.
13. Herschend F (2009) The Early Iron Age in South Scandinavia: Social Order in Settlement and Landscape. Uppsala: Institutionen för arkeologi och antik historia, Uppsala universitet. 449 p.
14. Näsman U (1981) Borgenes Ö. Skal. 1: 19–27.
15. Boessneck J (1979) Eketorp: Befestigung und Siedlung auf Öland/Schweden; Die Fauna. Almqvist & Wiksell. 504 p.
16. Telldahl Y (2005) Can palaeopathology be used as evidence for draught animals, in: Diet and health in past animal populations. Oxford: Oxbow Books. pp 63–67.
17. Benecke N (1994) Archäozoologische Studien zur Entwicklung der Haustierhaltung: in Mitteleuropa und Südskandinavien von den Anfängen bis zum ausgehenden Mittelalter. Berlin.
18. Albarella U (1999) The animal economy of rural settlements: a zooarchaeological case study from Northamptonshire. Medieval Settlement Research Group Annual Report, 9: 16–17.
19. Schlumbaum A, Stopp B, Breuer G, Rehazek A, Blatter R, al et (2003) Combining archaeozoology and molecular genetics: the reason behind the changes in cattle size between 150BC and 700AD in Northern Switzerland. Antiquity 77: 298.
20. Schlumbaum A, Turgay M, Schibler J (2006) Near East mtDNA haplotype variants in Roman cattle from Augusta Raurica, Switzerland, and in the Swiss Evolene breed. Animal Genetics 37: 373–375.
21. Svensson E, Anderung C, Baubliene J, Persson P, Malmström H, et al. (2007) Tracing genetic change over time using nuclear SNPs in ancient and modern cattle. Animal Genetics 38: 378–383.
22. Wiener P, Burton D, Williams J (2004) Breed relationships and definition in British cattle: a genetic analysis. Heredity 93: 597–602.
23. McKay S, Schnabel RD, Murdoch BM, Matukumalli LK, Aerts J, et al. (2008) An assessment of population structure in eight breeds of cattle using a whole genome SNP panel. Bmc Genetics 9: 37.
24. Silver I (1969) The ageing of domestic animals. Science in archaeology 26: 283–302.
25. Habermehl KH (1961) Die Altersbestimmung bei Haustieren, Pelztieren und beim jagdbaren Wild Paul Parey.
26. Schmid E (1972) Atlas of animal bones: For prehistorians, archaeologists and Quaternary geologists. Knochenatlas. Für Prähistoriker, Archäologen und Quartärgeologen Elsevier Science Ltd. 159 p.
27. Von Den Driesch A (1976) A guide to the measurement of animal bones from archaeological sites. Cambridge MA. 101 p.
28. Svensson E, Götherström A, Vretemark M (2008) A DNA test for sex identification in cattle confirms osteometric results. Journal of Archaeological Science 35: 942–946.
29. Mennerich G (1968) Römerzeitliche Tierknochen aus drei Fundorten des Niederrheingebiets Universitat Munchen. 176 p.

30. Roberts C, Manchester K (1997) The Archaeology of Disease. 243 p.
31. Hoerr NL, Osol A (1956) Blakiston's new Gould medical dictionary. McGraw-Hill. 1528 p.
32. Axelsson M (2000) Bone spavin. Clinical and epidemiological aspects of degenerative joint disease in the distal tarsus in Icelandic horses .Acta Universitatis Agriculturae Sueciae. Veterinaria (Sweden).
33. Bartosiewicz L, Van Neer W, Lentacker A (1997) Draught cattle: their osteological identification and history. Annales-Musee Royal de l'Afrique Centrale Sciences Zoologiques (Belgium). 147 p.
34. Cupere BDe, Lentacker A, Neer WVan, Waelkens M, et al. (2000) Osteological evidence for the draught exploitation of cattle: first applications of a new methodology. International Journal of Osteoarchaeology 10: 254–267.
35. Götherström A, Anderung C, Hellborg L, Elburg R, Smith C, al et (2005) Cattle domestication in the Near East was followed by hybridization with aurochs bulls in Europe. Proceedings of the Royal Society B: Biological Sciences 272: 2345.
36. Klungland H, Vage DI, Gomez-Raya L, Adalsteinsson S, Lien S (1995) The role of melanocyte-stimulating hormone (MSH) receptor in bovine coat color determination. Mammalian Genome 6: 636–639.
37. White S, Kata S, Womack J (2003) Comparative fine maps of bovine toll-like receptor 4 and toll-like receptor 2 regions. Mammalian Genome 14: 149–155.
38. Lien S, Karlsen A, Klemetsdal G, Våge DI, Olsaker, et al (2000) A primary screen of the bovine genome for quantitative trait loci affecting twinning rate. Mammalian Genome 11: 877–882.
39. Beg M, Bergfelt DR, Kot K, Ginther OJ (2002) Follicle selection in cattle: dynamics of follicular fluid factors during development of follicle dominance. Biology of reproduction 66: 120.
40. Ginther O, Bergfelt DR, Beg MA, Meira C, Kot K (2004) In vivo effects of an intrafollicular injection of insulin-like growth factor 1 on the mechanism of follicle deviation in heifers and mares. Biology of reproduction 70: 99.
41. Bunter K, Hermesch S, Luxgford BG, Graser HU, Crump RE (2005) Insulin-like growth factor-I measured in juvenile pigs is genetically correlated with economically important performance traits. Australian Journal of Experimental Agriculture 45: 783–792.
42. Davis M, Simmen R (2006) Genetic parameter estimates for serum insulin-like growth factor I concentrations, and body weight and weight gains in Angus beef cattle divergently selected for serum insulin-like growth factor I concentration. Journal of animal science 84: 2299.
43. Meuwissen T, Karlsen, A, Lien S, Olsaker, I, Goddard ME (2002) Fine mapping of a quantitative trait locus for twinning rate using combined linkage and linkage disequilibrium mapping. Genetics 161: 373.
44. Flori L, Fritz, S, Jaffrézic F, Boussaha M, Gut I, al et (2009) The genome response to artificial selection: a case study in dairy cattle. PLoS One 4: 6595.
45. Qanbari S, Pimentel ECG, Tetens J, Thaller G, Lichtner P, et al. (2009) A genome-wide scan for signatures of recent selection in Holstein cattle. Animal Genetics;doi:10.1111/j.1365-2052.2009.02016.x.
46. Gagneux P, C Boesch, Woodruff D (1997) Microsatellite scoring errors associated with noninvasive genotyping based on nuclear DNA amplified from shed hair. Molecular Ecology 6: 861–868.
47. Myrdal J (1999) Jordbruket under feodalismen 1000-1700. Natur och kultur/ LTs förlag. 407 p.
48. Svensson E, Götherström A (2008) Temporal fluctuations of Y-chromosomal variation in Bos taurus. Biology Letter 4: 752–754.
49. Johannsen Nørkjær N (2006) Draught cattle and the South Scandinavian economies of the 4th millennium BC. Environmental archaeology 11: 35–48.

2

Congenic Mice Provide Evidence for a Genetic Locus That Modulates Spontaneous Arthritis Caused by Deficiency of IL-1RA

Yanhong Cao[1,2], Xiaoyun Liu[2], Nan Deng[3], Yan Jiao[2], Yonghui Ma[2], Karen A. Hasty[2], John M. Stuart[3,4]*, Weikuan Gu[2]*

1 Institute of Kaschin-beck Disease, Center for Endemic Disease Control, Chinese Center for Disease Control and Prevention, Harbin Medical University; Key Laboratory of Etiologic Epidemiology, Education Bureau of Heilongjiang Province and Ministry of Health (23618104), Harbin, China, 2 Departments of Orthopaedic Surgery and Biomedical Engineering, University of Tennessee Health Science Center, Memphis, Tennessee, United States of America, 3 Department of Medicine, University of Tennessee Health Science Center, Memphis, Tennessee, United States of America, 4 Research Service, Veterans Affairs Medical Center, 1030 Jefferson Avenue, Memphis Tennessee, United States of America

Abstract

To understand the role of genetic factors involved in the development of spontaneous arthritis in mice deficient in IL-1 receptor antagonist protein (IL_1RA), we have identified a genomic region containing a major quantitative trait locus (QTL) for this disease. The QTL is on chromosome 1 and appears to be the strongest genetic region regulating arthritis. To confirm the importance of the QTL and to identify potential candidate genes within it, we conducted speed congenic breeding to transfer the QTL region from DBA/1 mice that are resistant to spontaneous arthritis into BALB/c$^{-/-}$ which are susceptible. Genetic markers along every chromosome were used to assist in the selection of progeny in each generation to backcross to BALB/c$^{-/-}$. By the 6th generation we determined that all of the chromosomes in the progeny were of BALB/c origin with the exception of portions of chromosome 1. At this stage we intercrossed selected mice to produce homozygous strains containing the genomic background of BALB/c$^{-/-}$ except for the QTL region on chromosome 1, which was from DBA/1. We were able to establish two congenic strains with overlapping DBA/1 DNA segments. These strains were observed for the development of spontaneous arthritis. Both congenic strains were relatively resistant to spontaneous arthritis and had delayed onset and reduced severity of disease. The gene/s that regulates this major QTL would appear to be located in the region of the QTL that is shared by both strains. The common transferred region is between D1Mit110 and D1Mit209 on chromosome 1. We evaluated this region for candidate genes and have identified a limited number of candidates. Confirmation of the identity and precise role of the candidates will require additional study.

Editor: Pierre Bobé, INSERM-Université Paris-Sud, France

Funding: The study was supported by grants from the National Institute of Arthritis and Musculoskeletal and Skin Diseases, National Institutes of Health (R01 AR51190 to WG), National Natural Science Foundation of China (Project 81171679 to YHC) and program directed funds from the Department of Veterans Affairs. The funders had no role in study design, data collection and analysis, decision to publish, or preparation of the manuscript.

Competing Interests: The authors have declared that no competing interests exist.

* E-mail: jstuart1@uthsc.edu (JMS); wgu@utusc.edu (WG)

Introduction

Interleukin 1 (IL-1) is a major contributor to the development of immune mediated arthritis. This cytokine is expressed by macrophages, monocytes and synovial fibroblasts. Its action is in part regulated by the IL-1 receptor antagonist protein (IL-1RA) which is the product of the Il1rn gene. The importance of IL-1RA in regulation of IL-1 activity has been established by generating mice deficient in IL-1RA. These IL-1RA deficient mice develop spontaneous arthritis as first described by Horai and colleagues [1]. BALB/c mice that are homozygous for the deficiency (BALB/c$^{-/-}$) develop inflammation in the hind limbs beginning at about 6 weeks of age and achieving an incidence approaching 100% by 3 months of age. Histopathologic examination of the joints of these mice shows infiltration of inflammatory cells and synovial proliferation. The development of disease is in part dependent upon genetic background since DBA/1 mice with a similar deficiency in IL-1RA do not develop spontaneous arthritis [2].

Deficiency of IL-1RA (DIRA) as a result of IL1RN mutation has also been identified in humans and results in a rare autosomal recessive autoinflammatory syndrome. DIRA is manifested by systemic inflammation including rash, painful movement, joint swelling and bone involvement [3]. Polymorphism of this gene in humans has also been associated with increased risk of osteoporotic fractures [4] and with gastric cancer [5]. Although the manifestations of IL-1RA deficiency in humans are somewhat different from those observed in mice, it is clear that IL-1RA is involved in human arthritis. Recombinant human IL-1RA has been developed as the therapeutic product Anakinra [6]. Administration of Anakinra has been shown to alleviate rheumatoid arthritis [7] and several other inflammatory disorders including systemic-onset juvenile idiopathic arthritis, familial Mediterranean fever and others. Because of its involvement in human disease there has been substantial interest in the mechanisms by which IL-1RA modulates arthritis.

Spontaneous arthritis is dependent not only on IL-1RA deficiency but also other as yet unidentified genetic factors. In order to identify those factors we used classical genetic techniques and bred susceptible and resistant mice to obtain an F2 generation and identified QTL associated with arthritis susceptibility [8]. After we conducted QTL analysis with phenotypic and genotypic determination of 191 F2 progeny, we obtained evidence for potential QTL on chromosomes 1, 6, 11, 12, and 14 [8]. The data suggested that the QTL on chromosomes 1 and 6 had the greatest influence on disease whereas there was weaker evidence for the involvement of potential QTL on chromosomes 11, 12, and 14 [8]. The QTL on chromosome 1 covers a large region at the distal end of the chromosome. Because of the strength of the association of spontaneous arthritis and this QTL we undertook additional studies to further characterize it. We hypothesized that one gene or a few genes within the QTL region regulate spontaneous arthritis. Accordingly, if a genomic fragment that contains the gene/s responsible for regulation of IL-1RA and development of spontaneous arthritis in BALB/c was replaced by the analogous fragment from DBA/1 mice which are resistant to spontaneous arthritis, then disease in the new strain would be reduced in incidence and/or severity. To test our hypothesis, we have developed congenic strains that contain the genomic fragments in the region of the QTL from DBA/1 mice on a BALB/c background.

Congenic strains are animals in which a genetic locus (often containing a QTL of interest) has been moved from one strain/line (donor) to the background of another strain/line (recipient) by back-crossing. Polymorphism of molecular markers is used to detect the source strain of the genome components of a congenic strain. For these experiments, we used speed or marker-assisted congenic breeding. Theoretically, the classical protocol of congenic breeding needs about 10 generations, at which 99+% of the genetic background of the progeny is of recipient origin while still retaining heterozygosity at the region of interest [9–10]. However, the availability of dense genetic maps of the mouse genome has allowed the development of marker-assisted breeding strategies [10], which reduce the number of generations required to eliminate donor strain-derived alleles outside the genetic region of interest. By employing this strategy of "speed congenics" we were able to rapidly produce the mouse strains used in this study.

Materials and Methods

Mice

All mice have been maintained in the animal facility of the Department of 'Veterans Affairs Medical Center, Memphis. Experimental animal procedures and mouse husbandry were performed in accordance with the National Institutes of Health's Guide for the Care and Use of Laboratory Animals and approved by the VAMC Institutional Animal Care and Use Committee.

Microsatellite Markers

A total of 123 microsatellite markers were selected for genotyping of progeny to assist with identifying the most informative backcrosses for breeding (Table S1). Those 123 markers are polymorphic between the two parental strains DBA/1 and BALB/c, and are evenly distributed along the 19 autosomal chromosomes, with distances of less than 20 cM from each other. The number of markers on each of chromosomes is 19, 8, 9, 6, 6, 6, 6, 6, 6, 7, 5, 4, 8, 5, 5, 4, 3, 5, 5 from chromosome 1 to 19, respectively.

Genotyping

Genomic DNA was extracted from tissues obtained by ear punch. The procedure used has been previously described [8]. Briefly, DNA was extracted from the tissue and amplification of microsatellite markers conducted by polymerase chain reaction (PCR). PCR products were analyzed using poly-acrylamide gel electrophoresis using the Mega-Gel Dual High-Throughput Vertical Electrophoresis System (C.B.S. Scientific, Del Mar, CA).

Breeding Procedure

The following procedure was used for the congenic breeding (Figure S1): 1) Mice with IL-1RA deficiency on the BALB/c background were crossed with DBA/1 mice which were also deficient in IL-1RA to produce heterozygous (F1) mice; 2) The F1 progeny were backcrossed to BALB/c$^{-/-}$ to produce N1 mice. The N1 mice were genotyped with 123 microsatellite markers. The individual mice with the fewest genomic markers for DBA/1 background but with the QTL region from the DBA/1 on chromosome 1 were selected for the next generation; 3) The selected N1 mice back cross to BALB/c$^{-/-}$ to produce an N2 generation. The N2 mice were then genotyped for the same 123 microsatellite markers. The individual mice with the most BALB/c$^{-/-}$ genetic background but possessing DBA/1 genomic DNA within the heterozygous QTL region on chromosome 1 were selected for the next generation; 4) The selected N2 mice were backcrossed to BALB/c$^{-/-}$ to produce the N3 generation. This process was repeated with selection of individual mice with the most BALB/c$^{-/-}$ background as well as the heterozygosity for DBA/1 at the QTL region for six generations; 5) At the end of the sixth generation it was determined that selected mice were homozygous for BALB/c$^{-/-}$ background but contained heterogygous DBA/1 in the region of the QTL; 6) These mice were then interbred to generate BALB/c$^{-/-}$ which were homozygous congenic for DBA/1 within the QTL;7) Ultimately we were able to establish 2 congenic strains. We first obtained the congenic strain BALB.D1-1$^{-/-}$. The transferred region from DBA/1 is from marker D1Mit55 to D1Mit209, which are located at 155166854–155167004 bp and 191493187–191493284 bp, respectively. The second congenic strain is BALB.D1-2$^{-/-}$. The transferred genomic region from DBA/1 is from D1Mit359 to the distal end of chromosome 1; and 8) The new congenic strains were then observed for the development of spontaneous arthritis.

Phenotype Evaluation

Mice from the BALB/c.D1$^{-/-}$ congenic strain were observed for the development of spontaneous arthritis. Individual mice were visually inspected for the presence of arthritis at least three times weekly. Each limb was scored on a scale of 0–4. Statistics used for the analysis of arthritis severity/incidence were as same as we have previously described ((0-no evidence of erythema and swelling, 1-mild redness and swelling of joint and ankle, 2-definite swelling, 3-severe swelling of entire limb, and 4-limb burned out and deformed). A severity score was calculated for the 4 limbs. The maximum score for an individual mouse is 16 [2,8]. The disease severity of a mouse is calculated with total scores of all limbs divided by the maximal score possible (in this case, 16) and multiplied by 100. The disease severity (total score of arthritis) of a strain is calculated with total scores of all mice divided by the total number of mice. The incidence was calculated with the number of affected mice divided by the total number of mice [2,8]. Differences were analyzed by Fisher's exact test for incidence and by ANOVA for severity.

Cytokine Assays

Cells were harvested from mouse spleens at 4 months of age. The spleens were dissociated, washed and cultured in HL-1 medium. Cells were cultured in 48 well plates at a density of 2×10^6 per well and CD3CD28 stimulation beads (Life Technology) added at the initiation of culture. Supernatant fluids were removed after 48 hours and cytokine levels were measured using Milliplex kits by Multianalyte Technology (Millipore, MA) according to the manufacturer's protocol. Procedures for assay sensitivity, precision and accuracy were based on the overnight protocol. Briefly, minimum detectable concentration is calculated by the StatLIA® immunoassay analysis software from Brendan Technologies (Carlsbad, CA). Intra-assay precision is generated from the mean of the %CV's from 8 reportable results across two different concentration of cytokines in one experiment. Inter-assay precision is generated from the mean of the %CV's from 4 – 8 reportable results across two different concentrations of cytokine in 4 different experiments. For accuracy, the data represent mean percent recovery of six levels of spiked standards in serum matrices.

Identification of Candidate Genes

Evaluation of genes within the QTL region of chromosome 1 was conducted using a searching tool, PGMapper (http://www.genediscovery.org/pgmapper/index.jsp) [11]. The procedures used were to that previous published [12]. Briefly, query terms were the combination of the name of the gene with any of these key words: arthritis, inflammation, anti-inflammatory, inflammatory mediator, inflammatory cytokine, autoimmune, immune, joint damage, T-cell, macrophage, neutrophil, angiogenesis, synovial, synovial hyperplasia, synovial fibroblast, lymphocytic infiltrate, and cartilage degradation. For any potential candidates, at least the abstract of one reference was read by two authors to ascertain whether there was a reason to believe that there might be an association between the gene and arthritis. For a gene with more than one reference that indicated its candidacy, at least two references were reviewed to confirm the potential association between the gene and arthritis.

Bioinformatics Analysis of Candidate Genes

To obtain information on polymorphism and SNPs of every candidate, we searched every gene including at least 2 kb of DNA up and downstream using the Mouse SNP database at the Jackson Laboratory: http://informatics.jax.org/javawi2/servlet/WIFetch?page=snpQF. Input terms for SNP database are two strains, BALB/cj and DBA/1j and the name of each gene in every search. For the information on gene differential expression, we used the list of genes that are differentially expressed in the spleen between BALB/c and DBA/1 (for data in Table S2) in our previous publication [8]. Splenic gene expression patterns in the four arthritis susceptible- (BALB/c-based) and four resistant- (DBA/1-based) IL1ra knockout mice were analyzed using the Affymetrix microarray platform. Statistical analysis for gene expression (expressed as a P value) was determined using EDGE software [8]. To analyze correlation of expression between candidate genes and *Il1rn*, including *Il-1r*, *Il1a* and *Il1b*, we took full advantage of existing data of gene expression profiles at GeneNetwork at http://www.genenetwork.org/webqtl/main.py. We used gene expression data from spleens obtained from a set of recombinant inbred (RI) strains derived by crossing C57BL/6J (B6) and DBA/2J (D2) and inbreeding progeny for 20 or more generations [13]. Data was generated using the Affymetrix GeneChip Mouse Gene 1.0 ST array (http://www.genenetwork.org/webqtl/main.py?FormID=sharinginfo&GN_AccessionId=283). Correlation was examined between expression levels of probes of each gene and probes from *Il1rn*, including *Il-1r*, *Il1a*, and *Il1b*.

Results

Phenotyping of Congenic Strains

Shown in Figure 1 is the genetic mapping of the QTL on chromosome 1 that was most strongly associated with the development of arthritis in the F2 generation as we have previously described [8]. Strain BALB.D1-1 contains genetic DNA from DBA/1 between D1Mit55 and D1MitD1Mit209 inclusive. This region includes almost the entire DNA that was mapped within the QTL. Strain BALB.D1-2 contains genetic DNA from D1Mit359 through the end of the chromosome. Most of the DNA within this region is at the extreme distal end of the QTL. Thus, most of the transferred genomic region in BALB.D1-2 is not located within the QTL region identified in our initial mapping. The phenotype of BALB.D1-1$^{-/-}$ was determined by observation of 18 mice from this strain over a period of 100 days (Figure 2). This congenic strain has significantly delayed onset of disease and reduced severity. ($P<0.05$ for incidence and <0.01 for severity) as compared to the parental BALB/c$^{-/-}$ strain. The mean day of onset was delayed from day 60 to day 77. By 100 days of age, the congenic strain had an incidence of disease that was comparable to the parental strain i.e. 94% compared to 100%. However the severity was still reduced at 100 days in the congenic strain as compared to the parental. Shown in Figure 3 shows the histologic examination of a hind limb from a BALB/c$^{-/-}$ compared to a BALB.D1-1$^{-/-}$ mouse at 7 weeks of age. Both mice have arthritis but the severity of the BALB/c$^{-/-}$ is much greater than that of the congenic.

Although the etiology of the spontaneous arthritis has not yet been fully elucidated, it is known that disease is dependent on TNFα and IL-17 [14,15]. We therefore analyzed the response of splenocytes from each strain to determine if there was differential regulation of these cytokines (Figure 4). For each of the cytokines studied i.e. TNFα, IL-17, IL-6 and IFNγ there was up regulation in the susceptible BALB/c as compared to DBA/1. The congenic strain showed intermediate levels. These data suggest that the QTL controls a generalized regulator of the immune response and not a specific regulator of IL-17 or TNFα.

Congenic strain BALB.D1-2$^{-/-}$ contains a smaller piece of genomic region from DBA/1 than that in BALB.D1-1$^{-/-}$. The region covers the distal region of the chromosome extending from at least D1MitD1Mit359 to the end of the chromosome. However, the phenotype of BALB.D1-2$^{-/-}$ is similar to that of BALB.D1-1$^{-/-}$. The phenotype of BALB.D1-2$^{-/-}$ was established by observation of 15 mice. A control group consisted of 18 mice of the BALB/c$^{-/-}$ strain (Figure 2). This strain also had significantly reduced severity and delayed onset of arthritis (P value <0.01 and <0.05 1for severity and incidence, respectively). The mean day of onset was day 68 for the congenic compared to 50 for the parental strain. Similar to BALB.D1-1$^{-/-}$ by 100 days of age the congenic strain had an incidence disease of 87% compared to 100% for the parental controls. These data seem to confirm significant protection from spontaneous arthritis is conferred by the DNA from DBA/1 mice. The protection is only partial since the incidence of spontaneous arthritis ultimately approaches 100% in each of the strains tested. Interestingly, we did not find differential expression of *Il-1*, *Il-6* and *TNFα* between congenic strains and BALB/c$^{-/-}$ although spleen cells were hyper responsive in BALB/c as compared to DBA/1 and were intermediate for the congenic strains as noted above.

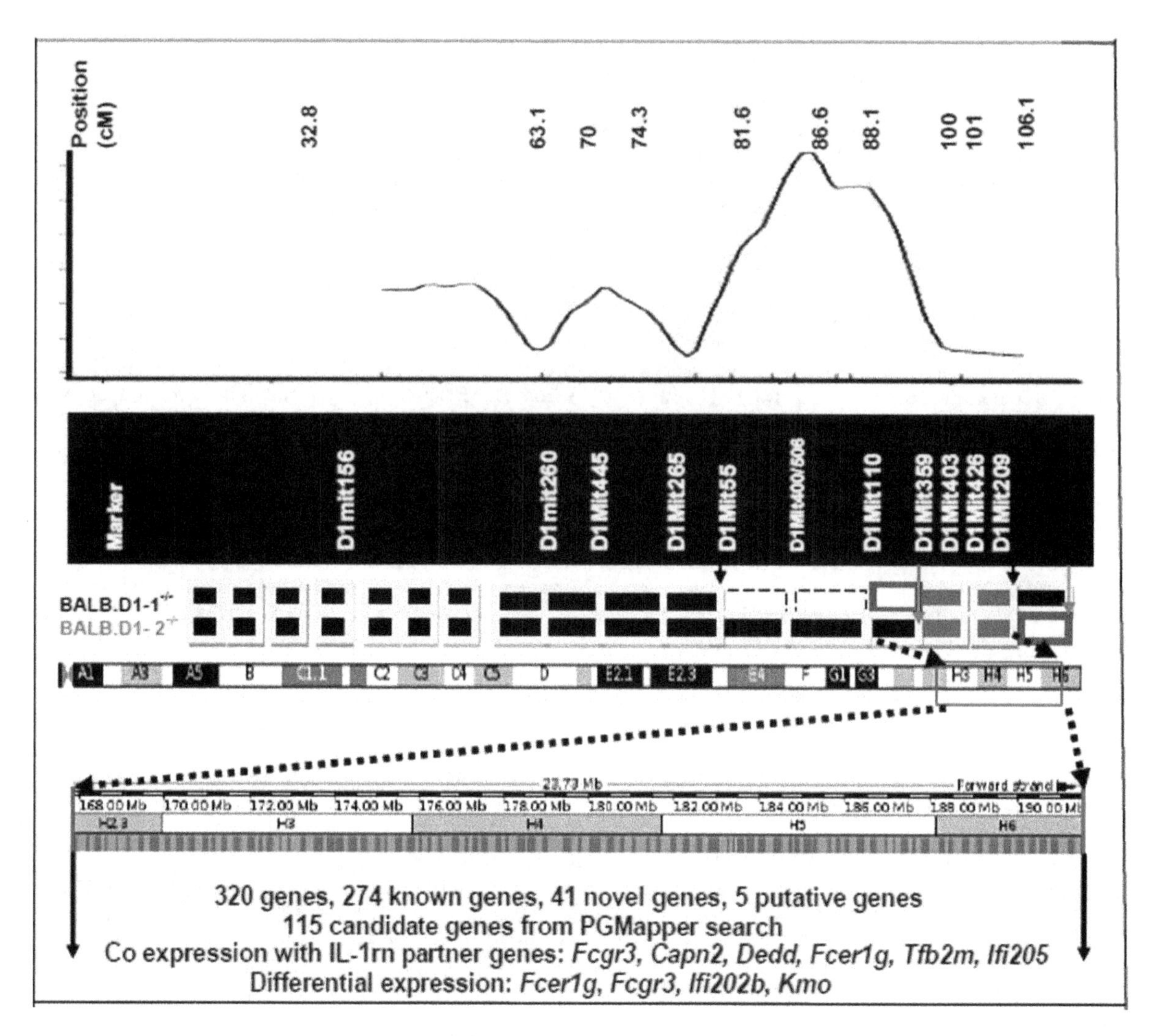

Figure 1. Genomic region of the QTL on chromosome 1 based on genetic markers in two congenic strains. Top is the location of initial mapping of the QTL for arthritis previously identified using the F2 generation. Below that is shown diagrammatically, the transferred genomic regions (white boxes) from DBA/1$^{-/-}$ into Balb/c$^{-/-}$ after the generation of 2 congenic strains. The genomic region from DBA/1 in BALB.D1-1$^{-/-}$ is flanked by two markers, D1Mit55 and D1Mit209. The transferred genomic regions (white boxes) from DBA/1$^{-/-}$ into Balb/c$^{-/-}$ in congenic strain BALB.D1-2$^{-/-}$ is flanked by marker D1Mit359 and the distal end of the chromosome. The minimum overlap region of the transferred genomic regions from DBA/1$^{-/-}$ into Babl/c$^{-/-}$ in congenic strains BALB.D1-1$^{-/-}$ and BALB.D1-2$^{-/-}$ is flanked by markers D1Mit359 and D1Mit209. The most likely QTL genomic region is between D1Mit110 and D1Mit209 with 23.73 Mb which contains 320 genetic elements. Among those genes, 115 are identified as genes relevant to arthritis and its potential pathways. Six genes *Fcgr3, Capn2, Dedd Fcer1g, Tfb2m and Ifi202* are co expressed with IL-1RN and related genes, four genes Fcer1g, Fcgr3, Ifi202b and Kmo are differentially expressed between Balb/c−/− and DBA/1$^{-/-}$.

Genomic Region of QTL

Because of the similarity of protection from spontaneous arthritis in both the BALB.D1-1$^{-/-}$ and BALB.D1-2$^{-/-}$strains we assume that the DBA/1 DNA that is common between the two congenic strains contains the genetic factor/s that are responsible for the prevention of disease. By comparing the genetic markers in these two congenic strains, the region of interest can be reduced to a final common transferred region (Figure 1). The minimum common transferred region is between D1Mit359 and D1Mit209. The maximum transferred region can be from D1Mit110 to the distal end of the chromosome. The other two possibilities are between D1Mit110 and D1Mit209 and between D1Mit359 and distal end of the chromosome. Genomic region between D1Mit110 and D1Mit209 coincides with the downstream tail of the peak region identified in the initial map [8]. Therefore, it is likely that the genomic region between D1Mit110 (or a position close to D1Mit110) and D1Mit359 contains the causal gene/s for the QTL of SAD., Although D1Mit110 has a BABL/c genotype in the second congenic strain this does not seem to affect the development of disease or disease severity. The genes on interest would appear to be located within the genomic region between 167758517 (D1Mit110) and 191493284 bp (D1Mit209). This region is sentenic to human chromosome 1 in two regions, 1:165171104 bp-158516918 bp and 1:1:240177648 bp-1:209788220 bp. An exception is the gene *Alyref2*, on mouse 1:171503478–171504750, which is located on human chromosome 17, between 17:79845713 bp-79849462 bp.

Candidate Genes within the QTL

Within the region of interest, there are 320 genes including 274 known genes, 41 novel genes, and 5 putative genes. Searching using key words (in our material and methods), PGMapper

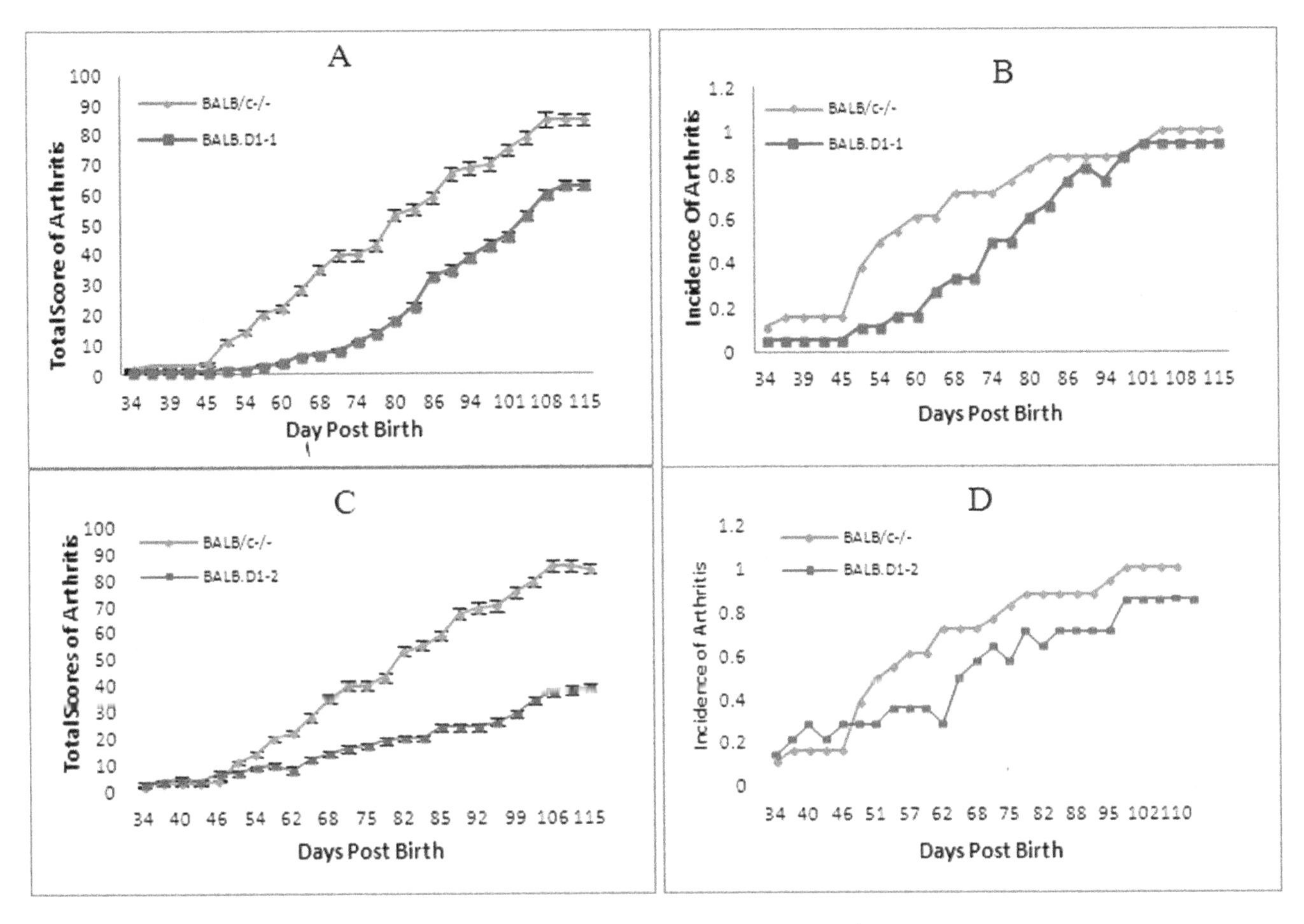

Figure 2. Comparison of arthritis in the parental and congenic strains. The BALB.D1-1$^{-/-}$congenic strain has much less severity (Figure 2A) and delayed onset (Figure 2B) of spontaneous arthritis as compared to the BALB/c$^{-/-}$ parental strain (P<=0.003 for severity and mean day of onset 74 vs 54 with a P<=0.05 for incidence). The Phenotype of BALB.D1-2$^{-/-}$ is similar to that of BALB.D1-1$^{-/-}$ even through this strain contains a smaller piece of the genomic region from DBA/1. The phenotype of BALB.D1-2$^{-/-}$ was investigated with 15 mice. In comparison to BALB/c$^{-/-}$, thise congenic strain also had both reduced severity (Figure 2C) and delayed onset (Figure 2D) of arthritis (P value <=0.007 for severity and mean day of onset 68 vs 54 with P<=0.01 for incidence). However, the overall incidence of disease in both congenic strains and BALB/c$^{-/-}$ reaches to almost 100% by 105 days of age. For measurement of severity the data are expressed as a percentage of the maximal total score.

revealed 115 candidate genes according PubMed reports based on Ensembl (NCBI m37) (Table S2). However, the candidate genes favored based on our initial mapping [6], (*Mr1, Pla2g4a, Fasl, Prg4,* and *Ptgs2*) are not included in those genes. Analysis of known single nucleotide polymorphisms (SNPs) with the region of interest established that 32 of the 115 candidate genes have at least 1 polymorphism between BALB/cj and DBA/1j (Table S2). By analyzing the correlation of expression of those genes with *Il1m* and its related genes, *Il1r, Il1a,* and *Il1b,* we fund none of the 115 genes has a high correlation with *Il1m* and its related genes (Table S3). However, we found that 6 genes have correlation R value higher than 0.40 to at least one of the *Il1m* related gene. Those 6 genes are *Fcgr3, Capn2, Dedd, Fcer1g, Tfb2m,* and *Ifi205.* Because we have previously established a list of 241 differentially expressed genes in comparison between BALB/c$^{-/-}$ and DBA/1$^{-/-}$ $-/-$ mice, we compared the list of 115 candidates and the 241 differentially expressed genes. Our comparison indicated that 4 candidate genes are among the differentially expressed genes. Those 4 genes are *Fcer1g, Fcgr3, Ifi202b,* and *Kmo.* Interestingly, *Fcer1g, Fcgr3* are both show a correlation with gene expression of *Il1m* as well as differential expression between spontaneous arthritis mice and healthy BALB/c mice. However, SNP data indicated that there are no known polymorphic SNPs between BALB/c and DBA/1 within those two genes and 2 kbp up and down stream.

Discussion and Conclusion

Our congenic breeding was successful in identifying a QTL associated with the development of spontaneous arthritis. When a fragment of DNA from the DBA/1 strain was introduced onto a BALB/c background, arthritis was delayed in onset and was less severe. The congenic strains provide a unique tool for evaluating specific genetic factor/s that regulates the spontaneous onset of spontaneous arthritis. The genetic mapping of QTL using a F2 population, especially with a relatively small population, usually identifies approximate locations of genetic loci for a complex trait. Using the congenic strains that we developed, the genomic region of the originally identified QTL has been redefined into a region that is downstream from the peak region of our original mapping. In our previous study, in our F2 mapping we used 137 microsatellite markers with initially 191 F2 and then 561 F2 mice. Within the 561 F2 population, there is sex ratio of 1:2 between male and female. This data emphasizes the importance of confirmation of QTL regions using additional breeding techniques including the development of congenic strains or other approaches.

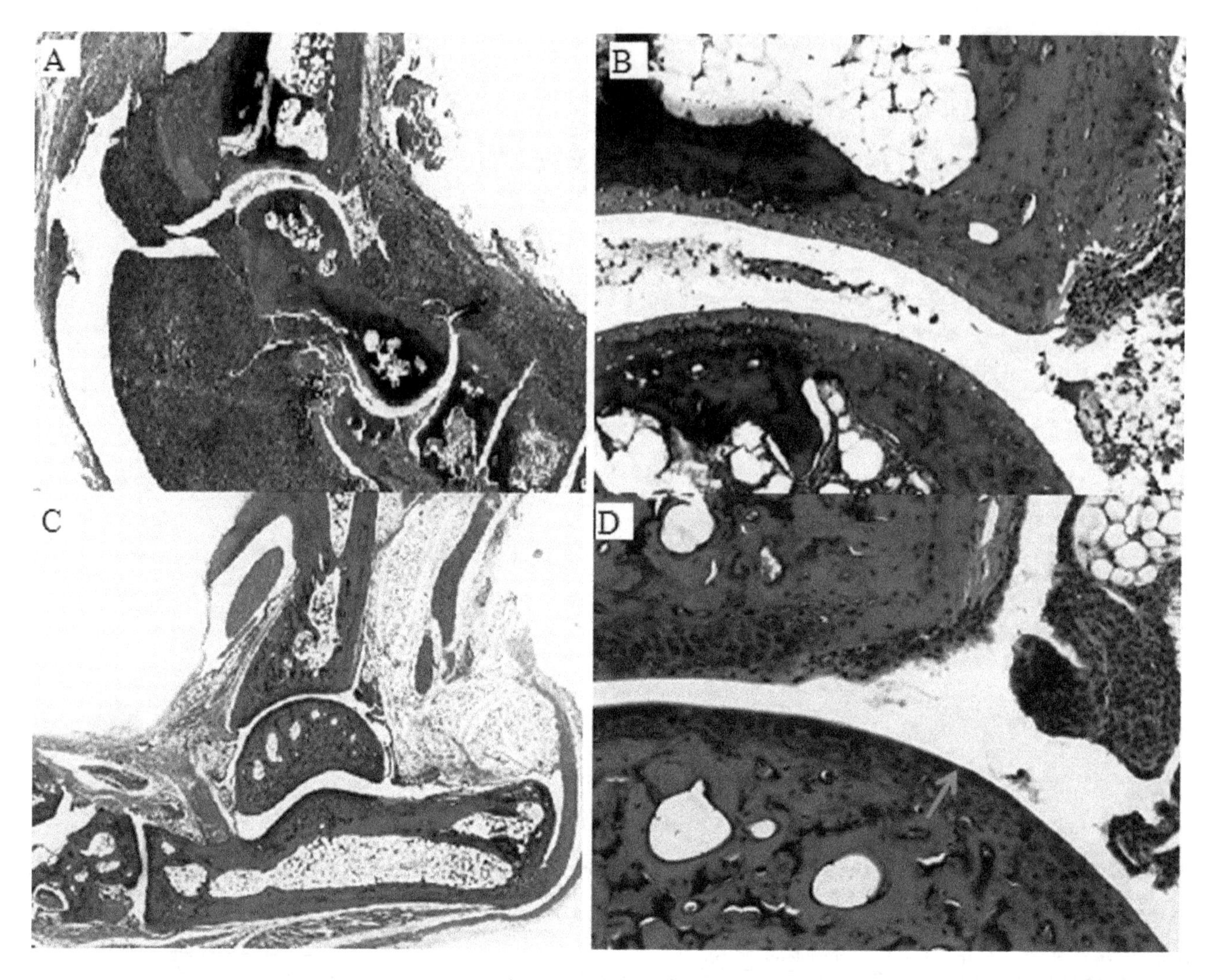

Figure 3. Comparison of arthritis severity in a BALB/c$^{-/-}$ (A, B) and Congenic BALB/c.D1-1$^{-/-}$ mouse (C, D) Panels A and C are cross section of the ankle joint. Panels B and D are higher power views to illustrate the development of an early erosion in the BALB/c$^{-/-}$ mouse whereas the comparable area of the congenic mouse shows only synovitis without erosive disease.

Understanding the molecular mechanisms underlying the phenotype of the congenic strains has two potentially profound consequences. First, it may enable us to identify novel pathways that contribute to the development of inflammatory arthritis. It has been recognized that IL-1 signaling is a key component of many forms of human inflammatory arthritis, in the development of joint erosions, and in development of osteoporosis. Our data support the possibility that erosions are at least delayed in the congenic strains. They have not as yet been analyzed for bone density and we do not have QTL mapping of bone density in BALB/c$^{-/-}$ X DBA/1$^{-/-}$ mice. Second, because spontaneous arthritis develops independently of TNF, identification of novel signaling pathways may help to explain why some patients fail to respond to TNF inhibitors, Our analysis indicated that the QTL did not result changes in *Il1, Il6* and *Tnfα* or *Th17* and *Il17* based on expression levels in the splenocytes used for analysis. When splenocytes were isolated, cultured, stimulated by anti-CD3CD28 beads and analyzed for cytokine expression by quantitation of protein levels, it was evident that there was potential for different levels of expression based on the genotype. This data suggests that these cytokines may be coordinately regulated by genes within the QTL but further work will be needed to determine the specific pathways involved. Study of the molecular basis of this QTL may identify a complementary approach to what is the most widely implemented biologic therapy for inflammatory arthritis in humans. Although we have not yet identified the causal gene/s within the QTL, those genes with polymorphisms and differential expression levels between BALB/c and DBA/1 deserve detailed examination in the future. The mouse model in this study has important implications for understanding rheumatoid arthritis in humans and potentially other human diseases. We anticipate that the candidate gene/s in the QTL that regulate the spontaneous arthritis in the IL-1RN$^{-/-}$ knockout mice may also be involved in human RA. 1) The effects of treatment of several inflammatory disorders with anakinra have been reported. Both positive and negative effects have been described [16–18]. Detailed study on the molecular function of *Il1m* and its interaction with other genes or genetic factors is essential for development therapeutic application using *Il1m.* 2) The genetic factors that interact with *Il1m* may be an ideal target for the development therapeutic application.

We analyzed the common transferred region 2.4 Mbp (between 167758517 and 191493284 bp) (Figure 1) based in the overlapping markers of two congenics. This should be considered as the

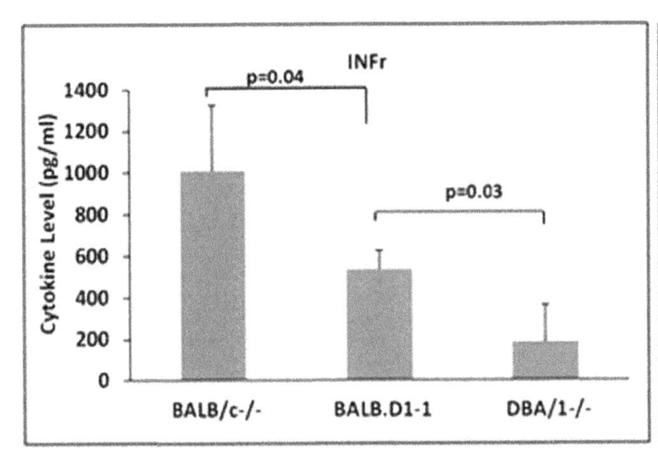

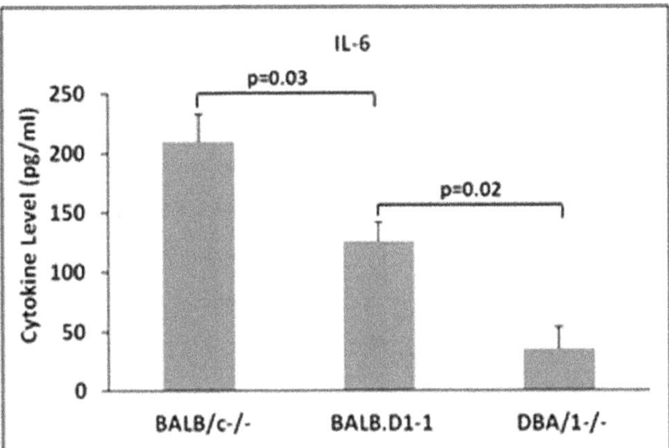

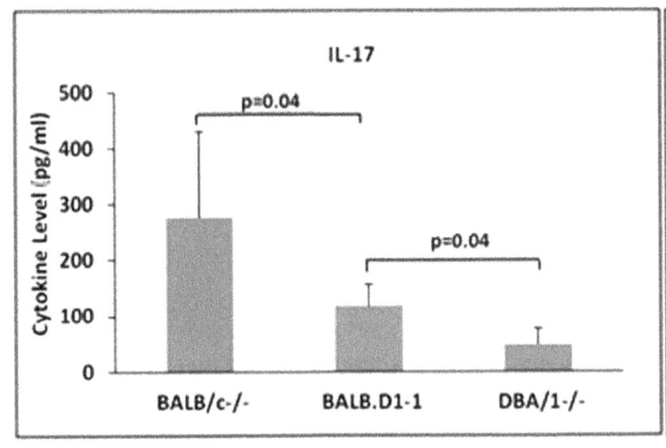

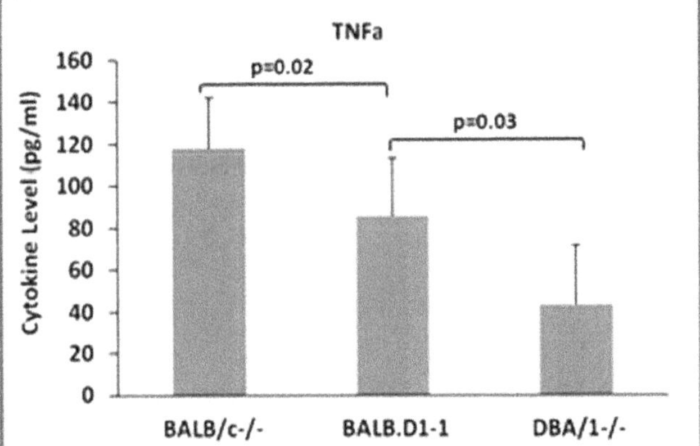

Figure 4. Production of cytokines by spleen cells from 4 month old mice. Cells were harvested from individual mice of each strain and cultured with CD3CD28 stimulation beads. Supernatant fluids were harvested after 48 hours of culture and assayed for the cytokines noted. Each bar represents the mean and standard deviation of 3 biologic replicates. Differences between goups was calculated by Student's T test.

maximum size of the QTL. The actual size of the QTL could be between D1Mit 359 (located between 177285202-177285317) and D1Mit209, or between D1Mit110 and D1Mit426 (180411709–180411792) with the size of approximately 1.5 Mbp and 1.1 Mbp, respectively. Further break down within the current transferred region in the congenic strains will greatly reduce the number of the candidate genes for this QTL.

Acknowledgments

The authors thank Ms Yue Huang for the technique assistance for this study.

Author Contributions

Conceived and designed the experiments: WG YC JMS KH. Performed the experiments: YC XL ND YJ YM. Analyzed the data: YC WG ND YJ JMS KH. Contributed reagents/materials/analysis tools: WG JMS YC. Wrote the paper: YC WG JMS.

References

1. Horai R, Saijo S, Tanioka H, Nakae S, Sudo K, et al. (2000) Development of chronic inflammatory arthropathy resembling rheumatoid arthritis in interleukin 1 receptor antagonist-deficient mice. J Exp Med 191: 313–320.
2. Zhou F, He X, Iwakura Y, Horai R, Stuart JM (2005) Arthritis in mice that are deficient in interleukin-1 receptor antagonist is dependent on genetic background. Arthritis Rheum 52: 3731–3738.
3. Aksentijevich I, Masters SL, Ferguson PJ, Dancey P, Frenkel J, et al. (2009) An autoinflammatory disease with deficiency of the interleukin-1-receptor antagonist. N Engl J Med 360: 2426–2437.
4. Langdahl BL, Løkke E, Carstens M, Stenkjaer LL, Eriksen EF (2000) Osteoporotic fractures are associated with an 86-base pair repeat polymorphism in the interleukin-1–receptor antagonist gene but not with polymorphisms in the interleukin-1beta gene. J Bone Miner Res 15: 402–414.
5. El-Omar EM, Carrington M, Chow WH, McColl KE, Bream JH, et al. (2000) Interleukin-1 polymorphisms associated with increased risk of gastric cancer. Nature 404: 398–402.
6. Ikonomidis I, Tzortzis S, Lekakis J, Paraskevaidis I, Dasou P, et al. (2011) Association of soluble apoptotic markers with impaired left ventricular

deformation in patients with rheumatoid arthritis. Effects of inhibition of interleukin-1 activity by anakinra. Thromb Haemost 106: 959–967. doi: http://dx.doi.org/10.1160/TH11-02-0117.

7. Merlin E, Berthomieu L, Dauphin C, Stéphan JL (2011) Cardiac tamponade in a child with systemic-onset juvenile idiopathic arthritis: dramatic improvement after interleukin-1 blockade by anakinra. Pediatr Cardiol 32: 862–863.
8. Jiao Y, Jiao F, Yan J, Xiong Q, Shriner D, et al. (2011) Identifying a major locus that regulates spontaneous arthritis in IL-1ra-deficient mice and analysis of potential candidates. Genet Res 18: 1–9.
9. Armstrong NJ, Brodnicki TC, Speed TP (2006) Mind the gap: analysis of marker-assisted breeding strategies for inbred mouse strains. Mamm Genome 17: 273–287.
10. Markel P, Shu P, Ebeling C, Carlson GA, Nagle DL, et al. (1997) Theoretical and empirical issues for marker-assisted breeding of congenic mouse strains. Nat Genet 17(3): 280–284.
11. Xiong Q, Qiu Y, Gu W (2008) PGMapper: a web-based tool linking phenotype to genes. Bioinformatics 24: 1011–1013. doi: http://dx.doi.org/10.1093/bioinformatics/btn002.
12. Xiong Q, Jiao Y, Hasty KA, Stuart JM, Postlethwaite A, et al. (2008) Genetic and molecular basis of quantitative trait loci of arthritis in rat: genes and polymorphisms. J Immunol 181: 859–864.
13. Peirce JL, Lu L, Gu J, Silver LM, Williams RW (2004) A new set of BXD recombinant inbred lines from advanced intercross populations in mice. BMC Genet 5:7. doi: http://dx.doi.org/10.1186/1471-2156-5-7.
14. Horai R, Nakajima A, Habiro K, Kotani M, Nakae S, et al. (2004) TNF-alpha is crucial for the development of autoimmune arthritis in IL-1 receptor antagonist-deficient mice. J Clin Invest 114: 1603–1611.
15. Nakae S, Saijo S, Horai R, Sudo K, Mori S, et al. (2003) IL-17 production from activated T cells is required for the spontaneous development of destructive arthritis in mice deficient in IL-1 receptor antagonist. Proc Natl Acad Sci U S A 100: 5986–5990.
16. Rigante D, Leone A, Marrocco R, Laino ME, Stabile A (2011) Long-term response after 6-year treatment with anakinra and onset of focal bone erosion in neonatal-onset multisystem inflammatory disease (NOMID/CINCA). Rheumatol Int 31(12): 1661–1664. doi: http://dx.doi.org/10.1007/s00296-010-1787-5.
17. Lin Z, Hegarty JP, Lin T, Ostrov B, Wang Y, et al. (2011) Failure of anakinra treatment of pyoderma gangrenosum in an IBD patient and relevance to the PSTPIP1 gene. Inflamm Bowel Dis 17: E41–42. doi: http://dx.doi.org/10.1002/ibd.21684.
18. Mahamid M, Paz K, Reuven M, Safadi R (2011) Hepatotoxicity due to tocilizumab and anakinra in rheumatoid arthritis: two case reports. Int J Gen Med 4: 657–660. doi: http://dx.doi.org/10.2147/IJGM.S23920.

A *COL7A1* Mutation Causes Dystrophic Epidermolysis Bullosa in Rotes Höhenvieh Cattle

Annie Menoud[1], Monika Welle[2,5], Jens Tetens[3], Peter Lichtner[4], Cord Drögemüller[1,5]*

1 Institute of Genetics, Vetsuisse Faculty, University of Bern, Bern, Switzerland, 2 Institute of Animal Pathology, Vetsuisse Faculty, University of Bern, Bern, Switzerland, 3 Institute for Animal Breeding and Husbandry, Christian-Albrechts-University Kiel, Kiel, Germany, 4 Institute of Human Genetics, Helmholtz Zentrum München - German Research Center for Environmental Health, Neuherberg, Germany, 5 DermFocus, Vetsuisse Faculty, University of Bern, Bern, Switzerland

Abstract

We identified a congenital mechanobullous skin disorder in six calves on a single farm of an endangered German cattle breed in 2010. The condition presented as a large loss of skin distal to the fetlocks and at the mucosa of the muzzle. All affected calves were euthanized on humane grounds due to the severity, extent and progression of the skin and oral lesions. Examination of skin samples under light microscopy revealed detachment of the epidermis from the dermis at the level of the dermo epidermal junction, leading to the diagnosis of a subepidermal bullous dermatosis such as epidermolysis bullosa. The pedigree was consistent with monogenic autosomal recessive inheritance. We localized the causative mutation to an 18 Mb interval on chromosome 22 by homozygosity mapping. The *COL7A1* gene encoding collagen type VII alpha 1 is located within this interval and *COL7A1* mutations have been shown to cause inherited dystrophic epidermolysis bullosa (DEB) in humans. A SNP in the bovine *COL7A1* exon 49 (c.4756C>T) was perfectly associated with the observed disease. The homozygous mutant T/T genotype was exclusively present in affected calves and their parents were heterozygous C/T confirming the assumed recessive mode of inheritance. All known cases and genotyped carriers were related to a single cow, which is supposed to be the founder animal. The mutant T allele was absent in 63 animals from 24 cattle breeds. The identified mutation causes a premature stop codon which leads to a truncated protein representing a complete loss of COL7A1 function (p.R1586*). We thus have identified a candidate causative mutation for this genetic disease using only three cases to unravel its molecular basis. Selection against this mutation can now be used to eliminate the mutant allele from the Rotes Höhenvieh breed.

Editor: Pal Bela Szecsi, Lund University Hospital, Sweden

Funding: The project and AM were supported by a grant of the H.W. Schaumann Stiftung, Hamburg, Germany. The funder had no role in study design, data collection and analysis, decision to publish, or preparation of the manuscript.

Competing Interests: The authors have declared that no competing interests exist.

* E-mail: cord.droegemueller@vetsuisse.unibe.ch

Introduction

Epidermolysis bullosa (EB) is a family of heritable mechanobullous disorders affecting the integrity of the skin and mucosa [1]. Abnormalities of macromolecules, which anchor the dermis to the epidermis lead to diminished cohesion of the skin layers, blister formation, and fragility. The condition is triggered by frictional movement as well as minor trauma. The severity of skin manifestations can be highly variable and is dependent on the mode of inheritance and the underlying mutation [1]. Mutations in 14 genes have been identified as causes of human EB [1,2]. EB affects the basement membrane zone of the skin and has traditionally been divided into three broad categories based on the cleavage levels of skin: (a) the simplex forms (EBS) demonstrate tissue separation within the basal keratinocytes; (b) the junctional forms (JEB) depict tissue separation within the lamina lucida of the dermal – epidermal basement membrane, and (c) the dystrophic forms (DEB) show tissue cleavage below the lamina densa within the upper papillary dermis [1,3]. Inherited forms of EB are well known in humans and in several domestic animal species. Various forms of EB have been described in sheep [4,5], cattle [6,7,8,9], horse [10], dog [11,12,13] and cat [14]. In only five EB cases in domestic animals the responsible gene mutation is known [5,7,10,12,13]. A dominant inherited EBS in Friesian x Jersey crossbred cattle was associated with a *KRT5* mutation [7]. Recessive inherited JEBs were reported in Belgian draft horses (*LAMC2* nonsense mutation [10]), in German pointer dogs (*LAMA3* missense mutation [12]) and in Black headed mutton sheep (*LAMC2* nonsense mutation [5]). Recessive inherited DEB in Golden retriever dogs is caused by *COL7A1* mutations [13] and in Swiss white alpine sheep an inherited *COL7A1* defect is probably causative but the mutation has not been described [4].

Currently, there is no drug therapy available to treat the EB disease. Therefore, studying suitable genetically defined animal models may be useful for the identification of therapeutic targets and approaches [2,12,15].

In 2010, an outbreak of EB was noticed in a local German cattle breed. We assessed the clinical presentation of the disease and carried out histopathological examination to define the disease as subepidermal bullous dermatosis. Considering the age of the affected animals dystrophic epidermolysis bullosa (DEB) was suspected. Subsequently, we employed a positional cloning approach to identify the causative mutation in the bovine *COL7A1* gene.

Results

Clinical presentation

We ascertained six affected calves in total. The clinical findings from a detailed examined affected bull calf were as follows: the calf was bright immediately post-partum. It was able to stand up, but once standing it was reluctant to walk and therefore could suck his dam only with assistance. It showed extensive epidermal loss at the four fetlocks (Figure 1A). As a consequence of frictional movements the skin defects rapidly extended subsequently involving the entire distal limbs, the whole muzzle and the oral cavity (Figure 1B). The calf showed appetite but was unable to suck, because of the pain it caused. It also showed skin defects around the eyes and at the base of the ears. The skin without epidermis was reddened, edematous, and hairless. The calf developed dysungulation affecting two claws during the first hours of life (Figure 1C). There were no apparent neurological defects or muscle abnormalities. After 8 hours, the calf was recumbent and couldn't stand up anymore because of pain. The calf was euthanized on humane grounds due to the severity, extent and progression of the lesions.

Clinical findings in the other five affected cases were similar and all calves had skin defects at birth at the four fetlocks and skin lesions rapidly extended to the muzzle, the oral cavity and the claws. They all showed extreme fragility of the normal appearing skin. Additional features have been observed in individual cases reported by the owner. In addition, two calves showed ear deformities (one with a closed ear), two other affected calves showed missing dewclaws and one calf had also skin lesions at the tail.

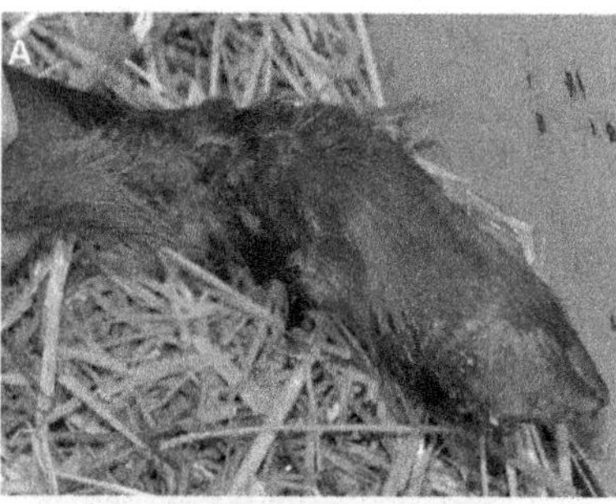

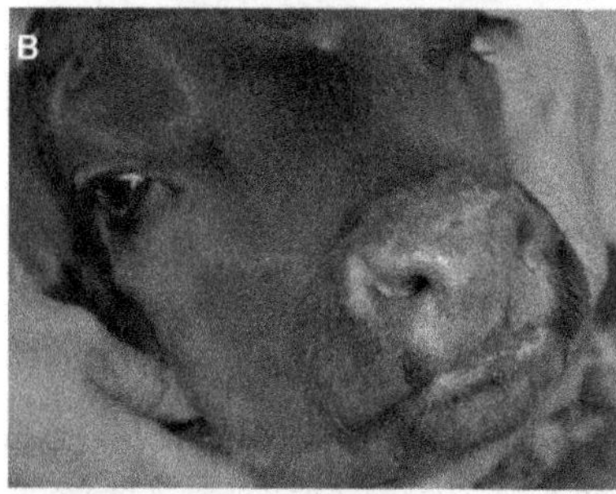

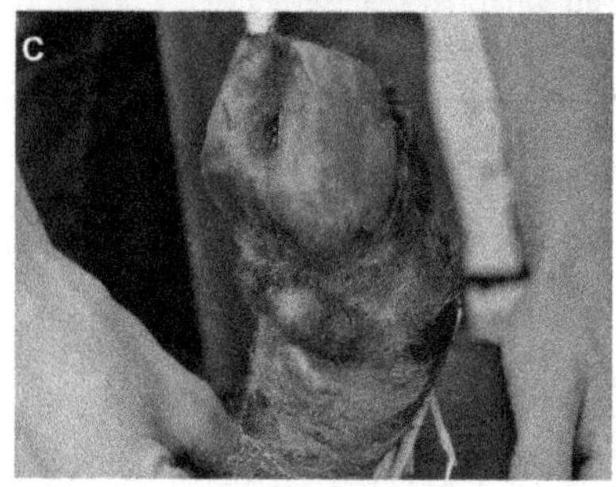

Figure 1. Clinical features of a DEB affected Rotes Höhenvieh calf. Typical signs are the extensive epidermal loss with ulcerations at the fetlocks (A) and muzzle (B) and dysungulation (C).

Histopathology

Histopathological examination revealed in all but one biopsies, a complete absence of the epidermis (Figure 2A–D). Only in one from the ear a long stretch of epidermis was completely detached from the dermis and the clean separation from the underlying dermis indicates that the epidermis is part of a blister roof (Figure 2A). In the same biopsy larger vacuoles and small vesicles along the basement membrane zone are visible (Figure 2B). In the biopsies where hair follicles were present the infundibular epithelium was missing in most cases (Figure 2B, 2C). The denuded dermal surface was covered by homogenous eosinophilic material and in some biopsies nuclear debris and cocci were present within this proteinaceous layer. In the biopsies of another affected calf a mild to severe necrosis of the superficial dermis was present. The necrotic dermis was characterized by fibrinous exudation, cellular debris, variable amounts of degenerate inflammatory cells, congested blood vessels, and hemorrhage (Figure 2D).

Pedigree analysis

Initially, we collected samples from three DEB affected calves, a single sire, and three dams of Rotes Höhenvieh cattle on a single farm. The owner reported three additional cases showing similar signs, which occurred during the last 12 months (Figure 3). The parents of the affected cattle showed no clinically visible skin anomalies. Analysis of the pedigree data revealed that two distantly related natural service sires (Hannibal, Oska; Figure 3) had affected offspring among their progeny. The pedigree of the affected calves shows many inbreeding loops (Figure 3). Analysis of the pedigree data revealed that all affected individuals trace back, on both the maternal and paternal path, to a single founder cow (Hanne) born in the year 1985 (Figure 3). The disease was recognized in the year 2010, 3 to 6 generations later. The breeding history was consistent with a monogenic autosomal recessive mode of inheritance.

Genome-wide homozygosity mapping of the DEB mutation

Assuming a monogenic recessive inheritance the epidermolysis bullosa affected calves were expected to be identical by descent (IBD) for the causative mutation and flanking chromosomal segments. Therefore, we decided to apply a homozygosity mapping approach to determine the position of the mutation in the bovine genome. We genotyped approximately 777,000 SNPs distributed across the entire genome in 3 cases. We analyzed the 3 genotyped DEB cases for extended regions of homozygosity with simultaneous allele sharing. A total of 17 genomic regions larger than 150 SNP and 0.3 Mb fulfilled our search criteria (Figure 4). The size of homozygous blocks ranged between 0.327 Mb and 18.600 Mb with a mean size of 1.860 Mb and a median of 0.595 Mb. The 3 genotyped cases showed a single large homozygous chromosome region on cattle chromosome (BTA) 22, containing 5,632 SNP markers corresponding to an 18.6 Mb interval from 38.8–57.4 Mb (Figure 4). Analyzing the data of 10 controls from the Rotes Höhenvieh breed genotyped at 54,000 evenly spaced SNPs revealed no shared homozygous regions greater than 300 kb.

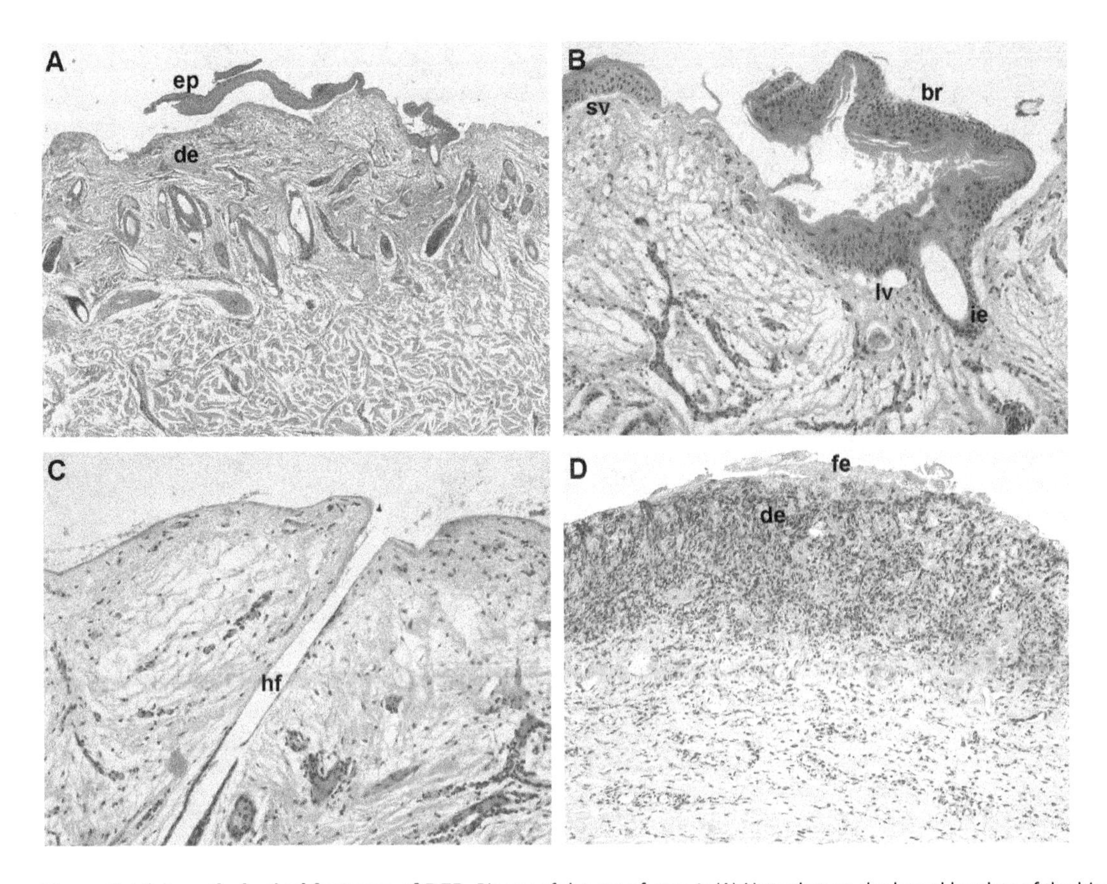

Figure 2. Histopathological features of DEB. Biopsy of the ear of case 1: (A) Note that on the lateral borders of the biopsy the epidermis (ep) is missing and, where present, the epidermis is detached from the underlying dermis (de). The epidermis is cleanly separated from the dermis and the basal cells are intact. H&E, 25×. (B) Higher magnification: Note the intact basal cells of the blister roof (br), larger vacuoles (lv) and a small vesicle (sv) along the basement membrane zone and separation of the infundibular epithelium (ie) from the surrounding connective tissue. H&E, 200× (C) Note that the epidermal and infundibular epithelium of the hair follicle (hf) is missing and the surface is covered by homogenous proteinaceous material. H&E, 200×. Biopsy of the right hind leg of case 3: (D) Necrotic superficial dermis (de) characterized by fibrinous exsudation (fe), cellular debris, dense amounts of degenerate inflammatory cells, and hemorrhage. H&E, 100×.

Identification of a functional candidate gene and mutation analysis

As the quality of the bovine genome annotation is not yet perfect, we inferred the gene annotation of the mapped interval from the corresponding human interval. The bovine DEB interval on BTA 22 corresponds to segments on human chromosome (HSA) 3. A careful inspection of HSA 3 genes and database searches of their presumed function revealed *COL7A1* encoding for the collagen type VII alpha 1 as a functional candidate gene within the critical interval at 48.6 Mb on HSA 3 and 51.8 Mb on BTA 22, respectively.

We performed a mutation analysis in three animals: a single DEB affected calf (case 3, Figure 3), its sire (Hannibal), and its mother (Daisy). We re-sequenced a contiguous genomic interval of more than 30 kb spanning the complete *COL7A1* gene including 118 exons of a total spliced coding length of 8802 bp encoding a protein of 2932 amino acids. Mutation analysis in the three re-sequenced animals revealed 14 single nucleotide polymorphisms (SNPs) in comparison to the cattle reference genome sequence (Table S1). Three SNPs were perfectly associated with the assumed recessive DEB inheritance. Two intronic SNPs were immediately excluded based on the fact that the affected calf was homozygous for the wildtype allele. The remaining SNP is within exon 49 (c.4756C>T; Figure 5). All three available affected calves were homozygous mutant T/T and all 4 supposed carriers were heterozygous C/T for this variant (Table 1). None of 143 healthy Rotes Höhenvieh cattle had the homozygous T/T genotype. All 42 detected C/T carriers of the DEB mutation were directly related to the assumed founder cow (Figure 3). The mutant T-allele was absent from 329 control cattle from 24 diverse cattle breeds (Table 1). The c.4756C>T mutation is predicted to result in a stop codon at amino acid residue 1586 in the bovine COL7A1 protein sequence (p.R1586*).

Discussion

In this study we applied a positional cloning approach to unravel the molecular basis of a recently recognized outbreak of dystrophic epidermolysis bullosa (DEB) in Rotes Höhenvieh cattle. Using just a few samples of well diagnosed cases showed again that the availability of genome sequences and high-density SNP genotyping microarrays enables the rapid mapping of causative mutations for monogenic diseases in domestic animal species. The long region of homozygosity surrounding the mutation suggests that the DEB mutation is quite young and that we have correctly identified a single putative founder cow. We note that the mapping of older mutations may require many more cases, because the associated IBD haplotype is usually much smaller, due to independent recombination events over several generations. Thus, we could quickly identify a genomic interval for the DEB mutation in Rotes Höhenvieh cattle containing the *COL7A1* gene, a good

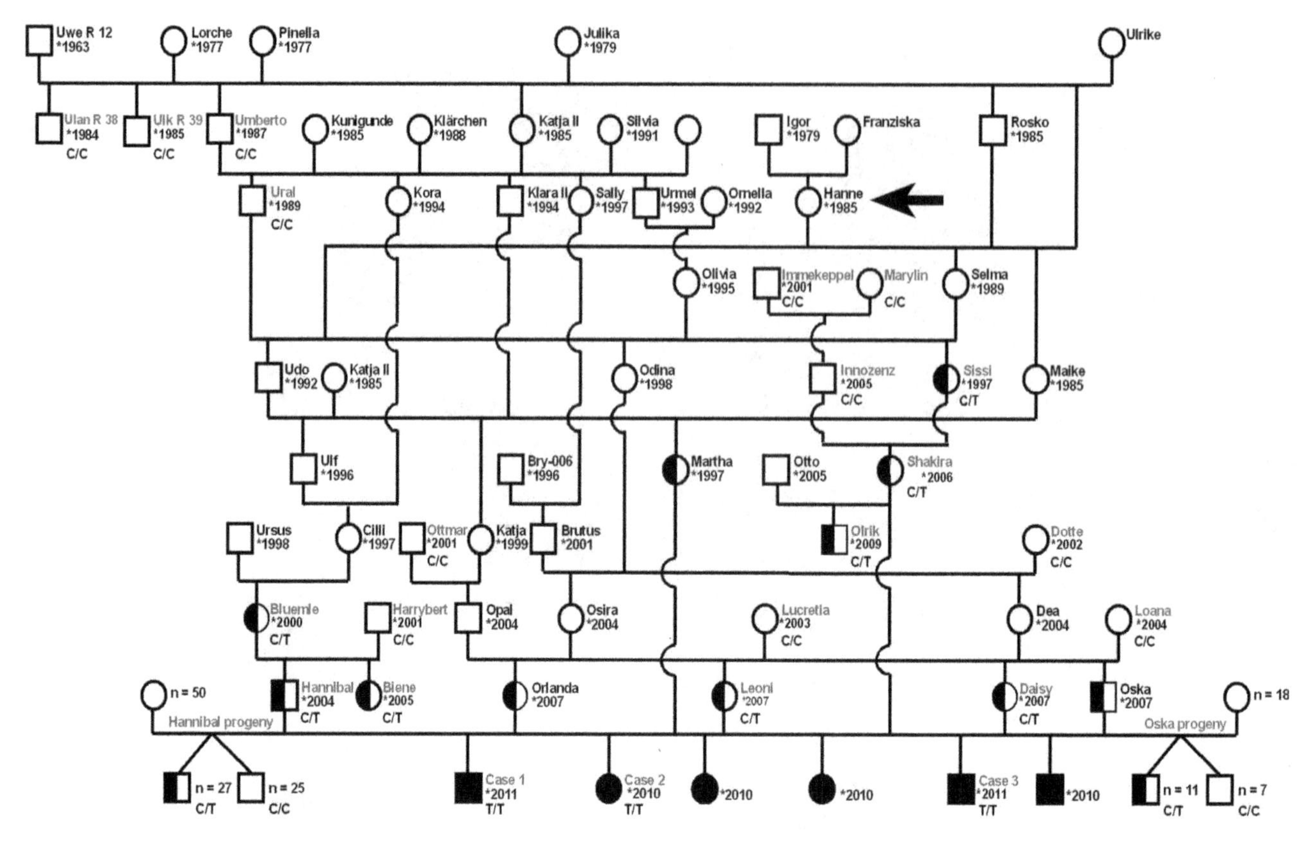

Figure 3. Pedigree of Rotes Höhenvieh cattle with DEB. DNA samples were available only from red labeled cattle. The genotypes for the *COL7A1* c.4756C>T exon 49 SNP are given below the symbols. The arrow indicates the cow, which is supposed to be the founder animal. We genotyped all available male offspring of the two carrier bulls Hannibal and Oska and detected 51% and 61% carriers, respectively.

functional and positional candidate gene. Based on the recessive mode of inheritance, we expected to find a loss-of-function mutation affecting the coding sequence of *COL7A1*. DNA sequencing revealed a nonsense mutation in the *COL7A1* gene, which is perfectly associated with the DEB phenotype in Rotes Höhenvieh cattle. Possibly nonsense-mediated decay selectively recognizes and degrades mRNAs whose open reading frame is truncated by a premature translation termination codon. We were not able to get fresh skin biopsies for RNA extraction. RT-PCR experiments using extractions from the formalin fixed paraffin embedded material failed. Furthermore, we were not able to detect the COL7A1 protein by immunofluorescence using a human COL7A1 antibody on paraffin sections from skin of affected and normal control animals. Unfortunately, all our samples showed strong background signals of the dermis, therefore we couldn't distinguish specific staining signals from background noise (data not shown). While it is unclear whether the mutant protein of 1585 residues is actually expressed, with more than 46% of the normal COL7A1 protein missing including several collagen triple helix repeats, it is very unlikely that the mutant protein fulfills any physiological function. The dramatically truncated polypeptides likely impair anchoring fibril formation in the homozygous patients. The main component of the anchoring fibrils is collagen VII, a homotrimeric collagen synthesized by keratinocytes and fibroblasts [2,13,16]. The type VII procollagen monomer consists of three alpha 1 (VII) polypeptide chains folded into a triple helix. Two monomers form an antiparallel dimer, from which the non-collagen pro-peptides are removed proto-lytically. Finally, the mature dimers laterally aggregate into anchoring fibrils [17].

Human DEB is caused by mutations in the *COL7A1* gene encoding collagen type VII alpha 1 [1,16,18]. Altered expression of collagen type VII results in blisters caused by intradermal separation occurring beneath the lamina densa, at the level of the anchoring fibrils [19,20]. Loss of collagen VII functions in DEB leads to absence or anomalies of the anchoring fibrils and to dermal-epidermal tissue separation. Numerous recessive inherited severe DEB forms caused by *COL7A1* gene mutations are reported in humans [16,17,19,20,21]. These severe phenotypes involve extracutaneous lesions in the oral cavity and the gastrointestinal tract. On the other hand, the dominantly inherited *COL7A1* associated DEB phenotypes in humans show milder symptoms [8]. The global severity of the EB diseases in humans can be explained most likely by the fact that the skin is not protected by hair and thus is more susceptible to epidermal detachment as a result of microtrauma. The hair follicles covering most of the body surface in animals may possibly contribute to a better coherence of the skin layers [4,22]. This may explain the more generalized severe skin lesions in human DEB patients. The clinical appearance of the calves and the histopathologic findings closely correlate with severe generalized recessive dystrophic epidermolysis bullosa in humans.

The observed phenotype resembled cases of epidermolysis bullosa and epitheliogenesis imperfecta reported in different breeds of cattle [23]. Epitheliogenesis imperfecta (EI) is known in veterinary medicine as differential diagnosis for EB but is still not genetically defined. In the past, several cases with similar clinical signs comparable to the recent cases were diagnosed as EI. Because they showed, like the presented DEB cases, only little or no characteristic blistering as seen in milder EB forms, the skin

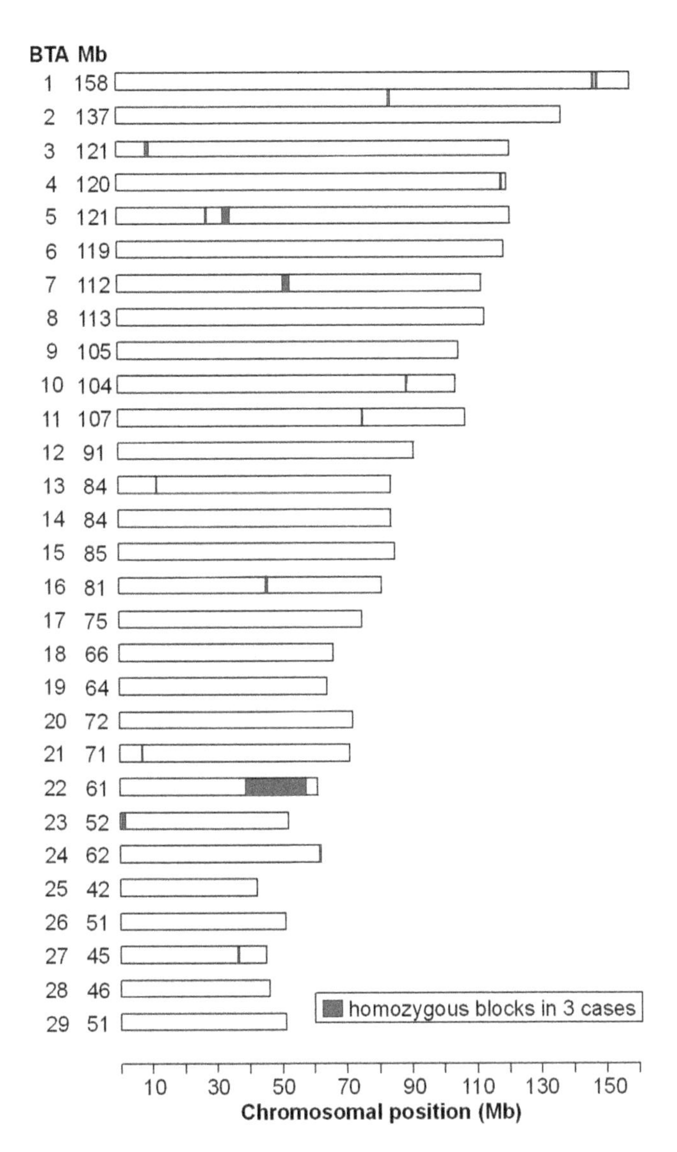

Figure 4. Genome-wide homozygosity mapping of the DEB mutation. After genotyping approximately 777,000 uniformly distributed SNP markers homozygous blocks >0.1 Mb were identified across 3 DEB affected cattle. Only on BTA 22 a very large homozygous block was identified.

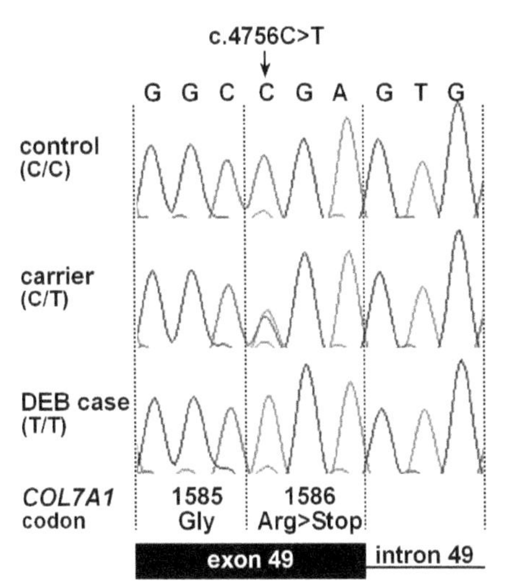

Figure 5. *COL7A1* mutation analysis. Electropherograms of the *COL7A1* c.4756C>T mutation. Representative sequence traces of PCR products amplified from genomic DNA of 3 cattle with the different genotypes are shown. The mutant T allele in homozygous form is present only in DEB affected calves and leads to a premature stop codon.

defects were already present at birth and sometimes the individuals showed no ear deformities [23]. Therefore, we suppose that in the past cases of DEB were misdiagnosed as EI.

The Rotes Höhenvieh cattle breed is a traditional German dual purpose cattle breed that was very widespread till the 1950s in Germany. This breed became nearly extinct in the 1980s. A dedicated breeding association (Bundesarbeitgemeinschaft Rotes Höhenvieh) was created to rescue the breed, first to preserve the genetic diversity and, second, because of their particular combination of characteristics like great vitality, longevity, high fertility and efficiency, which are valued traits in animal husbandry. Starting with just a few individuals, the population increased in the 1990s with subsequent breeding of consanguineous individuals. Today there are a total of approximately 100 Rotes Höhenvieh bulls and 1,000 cows. Due to the intense use of very few single artificial insemination (AI) sires the inbreeding rate of the current population is quite high [24]. The pedigree of the presented DEB cases shows inbreeding (Figure 3). Therefore, this study represents an example that this breeding practice can lead to high frequencies of deleterious recessive alleles within a few generations [25]. Homozygosity mapping identified a quite large IBD segment on BTA 22. The long region surrounding the mutation suggests that the identified *COL7A1* mutation is quite young. The founding mutation presumably happened in the cow Hanne (Figure 3). All affected calves and all genotyped carriers of the *COL7A1* mutation can be traced back to this cow born in 1985. The popular AI sire Uwe-R12, which had a significant impact during the consolidation phase of this breed in the 1990s is not involved in the dissemination of the DEB mutation.

In conclusion, we have identified a nonsense mutation of the bovine *COL7A1* gene as the likely causative mutation for DEB in Rotes Höhenvieh cattle. The knowledge on *COL7A1* mutations in human DEB patients suggests that this loss of function mutation is also responsible for the inherited DEB in Rotes Höhenvieh cattle, although we have no functional proof. Our study is consistent with the critical function of collagen type VII alpha 1 for formation of anchoring fibrils in the dermal-epidermal junction.

Our finding enables genetic testing and the eradication of this genetic disease from the breeding population. It is essential that domestic animal populations with small effective population sizes are continuously monitored for the appearance of recessive defects, so that selection against deleterious alleles can be implemented as early as possible.

Materials and Methods

Ethics Statement

All animal work has been conducted according to the national and international guidelines for animal welfare. There is no permit number as this study is not based on an invasive animal experiment. The cattle owner agreed that the samples can be used for our study. The data were obtained during diagnostic procedures that would have been carried out anyway. This is a very special situation in veterinary medicine. As the data are from client-owned cattle that underwent veterinary exams, there was no

Table 1. *COL7A1* genotype frequencies.

	Rotes Höhenvieh				Other breeds*
c.4756C>T Genotype	DEB affected (n = 3)	DEB carrier (n = 4)	Control, related (n = 86)	Control, unknown relationship (n = 57)	Controls (n = 329)
CC			44	57	
CT		4	42		
TT	3				329

*Angus (n = 18), Aubrac (n = 1), Ayshire (n = 1), Belgian blue (n = 3), Blonde d'Aquitaine (n = 2), Brown Swiss (n = 35), Charolais (n = 17), Chianina (n = 19), Dutch belted (n = 18), Eringer (n = 16), Evolenard (n = 10), Gelbvieh (n = 1), Galloway (n = 20), Hereford (n = 3), Scotish Highland (n = 4), Holstein (n = 35), Jersey (n = 3), Limousin (n = 17), Montbéliarde (n = 4), Nelore (n = 3), Ongola (n = 1), Piedmontese (n = 2), Pinzgauer (n = 10), Pustertaler Sprinzen (n = 10), Romagnola (n = 18), Salers (n = 1), Simmentaler (n = 38), Tyrolean Grey (n = 18), Domestic Yak (n = 1).

"animal experiment" according to the legal definitions in Germany.

Animals

We collected blood and skin samples from 3 dystrophic epidermolysis bullosa affected calves (2 male, 1 female) from a single farm. We did a full clinical exam of one affected calf. Additional information about the phenotype was obtained from owner records. Two cases were seen by a local veterinarian. Additionally, we collected 4 samples recorded as parents (1 sire, 3 dams) of affected offspring which were classified as obligate carriers and 143 healthy Rotes Höhenvieh cattle resulting in a total of 150 samples from this breed. Furthermore, we sampled 329 healthy control cattle from 24 genetically diverse cattle breeds for the re-sequencing of *COL7A1* exon 49 (Table 1). Genomic DNA was isolated using the Nucleon Bacc2 kit (GE Healthcare).

Histopathological examination of skin biopsies

Immediately after euthanasia 6 mm punch biopsies were taken from the skin of two affected calves. Biopsies from one DEB affected calf (case 1, Figure 3) were taken from the right and left foreleg, the right hind leg, the ear and the muzzle on day one after birth. Biopsies from another DEB calf (case 3, Figure 3) were taken one day after birth from lesional skin of the right hind and the right foreleg, the muzzle and the medial canthus of the eye. Skin biopsies were fixed in 10% buffered formalin, processed routinely and the resulting 4 µm sections were stained with haematoxylin and eosin (H&E).

Genome-wide homozygosity mapping of the DEB mutation

Genomic DNA from 3 cases was genotyped using Illumina's BovineHD BeadChip with 777,962 SNPs [26]. Genomic DNA from 10 controls was genotyped using Illumina's BovineSNP50 BeadChip with 54,001 SNPs [27]. The results were analyzed with PLINK [27]. After removing 12,098 SNPs with low genotyping success (failed calls>0.1) the average genotyping rate per individual was 99.9% for the three cases. After removing 6,287 SNPs with low genotyping success (failed calls>0.1) the average genotyping rate per individual was 98.2% for the ten controls. To identify extended homozygous regions with allele sharing across all affected animals the options –homozyg-group and –homozyg-match were applied. All given bovine genome positions correspond to the UMD3.1 cattle assembly.

Mutation analysis of the bovine *COL7A1* gene

Genomic DNA of a trio (a single case and both parents, which were assumed to be heterozygous carriers of the mutation) was used for mutation analysis (primer sequences available on request). We amplified PCR products using QIAGEN Multiplex PCR Kit (Qiagen). PCR products were directly sequenced on an ABI 3730 capillary sequencer (Applied Biosystems) after treatment with exonuclease I and shrimp alkaline phosphatase. We analyzed sequence data with Sequencher 4.9 (GeneCodes). All given positions correspond to the bovine *COL7A1* mRNA reference sequence XM_002697051. For the detection of the *COL7A1* exon 49 (c.4756C>T) mutation we amplified a PCR product of 605 bp using a forward primer located in intron 48 (5′-GGCTGATCGTCTTTGTCACC-3′) and reverse primer located in intron 49 (5′-TCAGTCCTGATCCCCAACTC-3′) which was subsequently sequenced as described above.

Supporting Information

Table S1 Polymorphisms and genotypes of 3 Rotes Höhenvieh cattle in the region of the *COL7A1* gene.

Acknowledgments

The authors would like to thank Leena Bruckner-Tuderman, Brigitta Colomb, and Tosso Leeb for expert technical assistance and helpful discussions. The authors would also like to thank all the Rotes Höhenvieh cattle breeders and the Bundesarbeitsgemeinschaft Rotes Höhenvieh for donating samples and sharing pedigree data.

Author Contributions

Conceived and designed the experiments: CD. Performed the experiments: AM MW JT PL CD. Analyzed the data: AM MW CD. Contributed reagents/materials/analysis tools: MW JT PL. Wrote the paper: AM CD.

References

1. Fine JD, Robin AJ, Bauer EA, Bauer JW, Bruckner-Tuderman L, et al. (2008) The classification of inherited epidermolysis bullosa (EB): report of the third international consensus meeting on diagnosis and classification of EB. J Am Acad Dermatol 58: 931–950.
2. Fine JD (2010) Inherited epidermolysis bullosa: recent basic and clinical advances. Curr Opin Pediatr 22:453–458.
3. Fine JD, Eady RA, Bauer EA, Briggaman RA, Bruckner-Tuderman L (2000) Revised classification system for inherited epidermolysis bullosa: report of the second international consensus meeting on diagnosis and classification of epidermolysis bullosa. J Am Acad Dermatol 42: 1051–1066.

4. Bruckner-Tuderman L, Guscetti F, Ehrensperger F (1991) Animal model for dermolytic mechanobullous disease: sheep with recessive dystrophic epidermolysis bullosa lack collagen VII. J Invest Dermatol 96: 452–458.
5. Mömke S, Kerkmann A, Wöhlke A, Ostmeier M, Hewicker-Trautwein M, et al. (2011) A frameshift mutation within LAMC2 is responsible for Herlitz type junctional epidermolysis bullosa (HJEB) in black headed mutton sheep. PloS ONE 6: e18943.
6. Foster AP, Skuse AM, Higgins RJ, Barrettx DC, Philbeyx AW, et al. (2010) Epidermolysis bullosa in calves in the United Kingdom. Science Direct 142: 336–340.
7. Ford CA, Stanfield AM, Spelman RJ, Smits B, Ankersmidt-Udy AE, et al. (2005) A mutation in bovine keratin 5 causing epidermolysis bullosa simplex, transmitted by a mosaic sire. J Invest Dermatol 124:1170–1176.
8. Thompson KG, Crandell RA, Rugeley WW, Sutherland RJ (1985) A mechanobullous disease with sub-basilar separation in Brangus calves. Vet Pathol 22: 283–285.
9. Medeiros GX, Riet-Correa F, Armién AG, Dantas AF, de Galiza GJ, et al. (2012) Junctional epidermolysis bullosa in a calf. J Vet Diagn Invest 24: 231–234.
10. Spirito F, Charlesworth A, Linder K, Ortonne JP, Baird J, et al. (2002) Animal models for skin blistering conditions: absence of laminin 5 causes hereditary junctional mechanobullous disease in the belgian horse. J Invest Dermatol 119: 684–691.
11. Palazzi X, Marchal T, Chabanne L, Spadafora A, Magnol JP, et al. (2000) Inherited dystrophic epidermolysis bullosa in inbred dogs: a spontaneous animal model for somatic gene therapy. J Invest Dermatol 115: 135–137.
12. Guaguere E, Capt A, Spirito F, Meneguzzi G (2003) Junctional epidermolysis bullosa in the german shorthaired pointer: a spontaneous model for junctional epidermolysis bullosa in man. Bull Acad Vet France 157: 47–51.
13. Magnol JP, Pin D, Palazzi X, Lacour JP, Gache Y, et al. (2005) Characterization of a canine model of dystrophic bullous epidermolysis (DBE). Development of a gene therapy protocol. Bull Acad Natle Med 189: 107–121.
14. Olivry T, Dunston SM, Marinkovich MP (1999) Reduced anchoring fibril formation and collagen VII immunoreactivity in feline dystrophic epidermolysis bullosa. Vet Pathol 36: 616–618.
15. Baldeschi C, Gache Y, Rattenholl A, Bouillé P, Danos O, et al. (2003) Genetic correction of canine dystrophic epidermolysis bullosa mediated by retroviral vectors. Human Molecular Genetics 15: 1897–1905.
16. Christiano AM, Greenspan DS, Hoffman G, Zhang X, Tamai Y, et al. (1993) A missense mutation in type VII collagen in two affected siblings with recessive dystrophic epidermolysis bullosa. Nat Genet 4: 62–66.
17. Kern JS, Kohlhase J, Bruckner-Tuderman L, Has C (2006) Expanding the COL7A1 mutation database: novel and recurrent mutations and unusual genotype-phenotype constellations in 41 patients with dystrophic epidermolysis bullosa. J Invest Dermatol 126: 1006–1012.
18. Uitto J (2004) Epidermolysis bullosa: the expanding mutation database. Soc Inv Derm 123, xii–xiii.
19. Christiano AM, Suga Y, Greenspan DS, Ogawa H, Uitto J (1995) Premature termination codons on both alleles of the type VII collagen gene (COL7A1) in three brothers with recessive dystrophic epidermolysis bullosa. J Clin Invest 95: 1328–1334.
20. Hovnanian A, Hilal L, Blanchet-Bardon C, De Prost Y, Christiano AM, et al. (1994) Recurrent nonsense mutations within the type VII collagen gene in patients with severe recessive dystrophic epidermolysis bullosa. Am J Hum Genet 55: 289–296.
21. Kern JS, Grüninger G, Imsak R, Müller ML, Schumann H, et al. (2009) Forty-two novel COL7A1 mutations and the role of a frequent single nucleotide polymorphism in the MMP1 promoter in modulation of disease severity in a large european dystrophic epidermolysis bullosa cohort. Br J Dermatol 161: 1089–1097.
22. Fritsch A, Loeckermann S, Kern JS, Braun A, Bösl MR, et al. (2008) A hypomorphic mouse model of dystrophic epidermolysis bullosa reveals mechanisms of disease and response to fibroblast therapy. J Clin Invest 118:1669–1679.
23. Bähr C, Drögemüller C, Distl O (2004) Epitheliogenesis imperfecta bei deutschen Holsteinkälbern (German). Tierärztl Prax 32 (G): 205–211.
24. Kehr C, Klunker M, Fischer R, Groeneveld E, Bergfeld U (2010) Untersuchungen zu einem Monitoring genetischer Diversität bei Nutztierrassen: Ergebnisse zum Roten Höhenvieh (German). Züchtungskunde 82:387–399.
25. Drögemüller C, Reichart U, Seuberlich T, Oevermann A, Baumgartner M, et al. (2011) An unusual splice defect in the mitofusin 2 gene (MFN2) is associated with degenerative axonopathy in Tyrolean Grey cattle. PLoS ONE 6: e18931.
26. Homepage Illumina (2011). http://www.illumina.com., Accessed 28 December 2011
27. Homepage PLINK (2011). Available: http://pngu.mgh.harvard.edu/̃purcell/plink/., Accessed 28 December 2011

4

High Prevalence of Bovine Tuberculosis in Dairy Cattle in Central Ethiopia: Implications for the Dairy Industry and Public Health

Rebuma Firdessa[1¶¤], Rea Tschopp[1,4,6¶], Alehegne Wubete[2¶], Melaku Sombo[2], Elena Hailu[1], Girume Erenso[1], Teklu Kiros[1], Lawrence Yamuah[1], Martin Vordermeier[5], R. Glyn Hewinson[5], Douglas Young[4], Stephen V. Gordon[3], Mesfin Sahile[2], Abraham Aseffa[1], Stefan Berg[5*]

1 Armauer Hansen Research Institute, Addis Ababa, Ethiopia, **2** National Animal Health Diagnostic and Investigation Center, Sebeta, Addis Ababa, Ethiopia, **3** School of Veterinary Medicine, University College Dublin, Dublin, Republic of Ireland, **4** Centre for Molecular Microbiology and Infection, Imperial College London, London, United Kingdom, **5** Department for Bovine Tuberculosis, Animal Health and Veterinary Laboratories Agency, Weybridge, Surrey, United Kingdom, **6** Swiss Tropical and Public Health, Basel, Switzerland

Abstract

Background: Ethiopia has the largest cattle population in Africa. The vast majority of the national herd is of indigenous zebu cattle maintained in rural areas under extensive husbandry systems. However, in response to the increasing demand for milk products and the Ethiopian government's efforts to improve productivity in the livestock sector, recent years have seen increased intensive husbandry settings holding exotic and cross breeds. This drive for increased productivity is however threatened by animal diseases that thrive under intensive settings, such as bovine tuberculosis (BTB), a disease that is already endemic in Ethiopia.

Methodology/Principal Findings: An extensive study was conducted to: estimate the prevalence of BTB in intensive dairy farms in central Ethiopia; identify associated risk factors; and characterize circulating strains of the causative agent, *Mycobacterium bovis*. The comparative intradermal tuberculin test (CIDT), questionnaire survey, post-mortem examination, bacteriology, and molecular typing were used to get a better understanding of the BTB prevalence among dairy farms in the study area. Based on the CIDT, our findings showed that around 30% of 2956 tested dairy cattle from 88 herds were positive for BTB while the herd prevalence was over 50%. Post-mortem examination revealed gross tuberculous lesions in 34/36 CIDT positive cattle and acid-fast bacilli were recovered from 31 animals. Molecular typing identified all isolates as *M. bovis* and further characterization by spoligotyping and MIRU-VNTR typing indicated low strain diversity within the study area.

Conclusions/Significance: This study showed an overall BTB herd prevalence of 50% in intensive dairy farms in Addis Ababa and surroundings, signalling an urgent need for intervention to control the disease and prevent zoonotic transmission of *M. bovis* to human populations consuming dairy products coming from these farms. It is suggested that government and policy makers should work together with stakeholders to design methods for the control of BTB in intensive farms in Ethiopia.

Editor: Gordon Langsley, Institut national de la santé et de la recherche médicale - Institut Cochin, France

Funding: This study was sponsored by the Wellcome Trust (UK) under their "Animal health in the developing world initiative" (grant number 075833/A/04/Z). The authors acknowledge the Schweizerische Stiftung für Medizinisch-Biologische Stipendien (SSMBS) and the Swiss National Science Foundation (SNSF) for their financial support of Rea Tschopp during the analysis and writing phase. The funders had no role in study design, data collection and analysis, decision to publish, or preparation of the manuscript.

Competing Interests: The authors have declared that no competing interests exist.

* E-mail: Stefan.berg@ahvla.gsi.gov.uk

¶ These authors contributed equally to this work.

¤ Current address: University of Würzburg, Würzburg, Germany

Introduction

The population of Ethiopia has increased dramatically in the last two decades, growing from approximately 55 million people in 1992 to a current estimate of around 85 million [1]. Increased population size has led to an inexorable increase in demand for food, putting pressure on the agricultural sector in which 85% of the work force is employed. Ethiopia has the largest livestock population in Africa, including an estimated ~52 million cattle [2] that contributes to the livelihoods of 60–70% of the population [3]. The vast majority of the cattle are indigenous zebu (*Bos indicus*) managed under traditional husbandry systems (grazing in the field) in rural areas. However, in recent years the number of dairy cattle of highly productive exotic (*Bos taurus*, mainly Holstein-Friesian) and cross breeds has been on the rise, particularly in urban and peri-urban areas in response to the increasing demand for milk products and the Ethiopian government's effort to improve productivity in the livestock sector. The population of dairy cows accounts for 6.3 million animals (around 12% of the total cattle population) and the estimated total national milk production per year is 2.6 billion litres [4] of which the urban and peri-urban dairy farmers produce 2%.

In a country such as Ethiopia, where livestock are extremely important for people's livelihood, animal diseases can be a real threat to animal productivity and thus negatively impact on the agricultural sector and economic development. Bovine tuberculosis (BTB), caused by *Mycobacterium bovis*, is a chronic and contagious disease of cattle and other domestic and wild animals [5,6]. BTB is prevalent worldwide but prevalence data is scarce in most developing countries due to lack of active control programmes. Several studies conducted since 2006 have confirmed that BTB is endemic in Ethiopia with prevalence rates varying from 0.8% to around 10% in extensive rural farming systems [7–11], while higher prevalence rates have been reported from regions in Ethiopia where intensive husbandry systems are more common [12–14]. As well as causing a high morbidity, BTB can also be a financial burden to farmers owning infected cattle; it has been suggested that cattle with BTB have a reduced productivity affecting milk yield and carcass value [15] as well as through reduced pulling power in traditional farming system [16]. BTB has also zoonotic potential [17,18] - mainly through consumption of unpasteurised milk products - and its prevalence in Ethiopian cattle can therefore be a contributing factor to the human burden of TB in Ethiopia that currently is ranked as the 7th highest in the world [19,20].

This study was designed to get a better understanding of the prevalence of BTB among intensive dairy farms in the urban and peri-urban belt of Addis Ababa (central Ethiopia), areas from where the vast majority of commercial dairy products in the country are produced. Furthermore, we show molecular typing results of *M. bovis* strains circulating in this region of Ethiopia and discuss the importance of our findings for informing BTB control policies in Ethiopia.

Materials and Methods

Study sites and ethical considerations

Statistics from the Ethiopian Ministry of Agriculture and Rural Development (MoARD) as well as data from regional agricultural offices recognised a high density of intensive dairy farms in a 50 km radius region from Addis Ababa city centre. Most dairy farms in this region had access to one of the five main roads that lead into the capital. Therefore, the region was divided into six geographically defined study areas, including Addis Ababa city, Debre Zeit, Sebeta, Holeta, Sululta and Sendafa (Figure 1). The recruitment of dairy farms within each study area was based on the existence of high numbers of Holstein-Friesian and cross breed cattle, as well as variation in herd size, with the latter being categorized into three groups; small herds (1–10 cattle), medium herds (11–50 cattle), and large herds (>50 cattle). This study was approved by the MoARD and was based on voluntary participation, and consent to participate was given from all farm owners after being informed about the study by local agricultural office authorities as well as by study representatives. Ethical approval for the study was given by the institutional ethical committee at the Armauer Hansen Research Institute (AHRI) and the All Africa Leprosy, Tuberculosis and Rehabilitation Training centre (ALERT). As part of the commitment of this study, one week training was given to 63 of the farm owners and/or their staff by professionals from National Animal Health Diagnostic and Investigation Center (NAHDIC), AHRI, the National Artificial Insemination Center (NAIC), and the Ministry of MoARD Extension Directorate and Oromia Special Zone Surrounding Finfinne Livestock Development & Marketing Agency Office. The training was done in the Amharic language and included topics such as mastitis, abortion, dystocia, calcium deficiency, BTB, brucellosis, heat detection, artificial insemination, vaccination, and herd management (e.g. annual planning, dairy management).

Comparative intradermal test and questionnaire

All cattle older than six months within a herd, except clinically sick animals and cows one month pre-and post partum, were tested by CIDT: Two sites, 12 cm apart, horizontally of the mid-neck of the animal were shaved and the skin thickness was measured with a calliper. In calves with short neck, the sites were shaven vertically. Aliquots of 0.1 ml of 2500 IU/ml bovine purified protein derivative (PPD) and 0.1 ml of 2500 IU/ml avian PPD (Synbiotics Corporation, Lyon, France) were injected separately into the respective shaved site. The thickness of the skin at each injection site was measured again after 72 hours. The test results were interpreted in line with OIE recommendations (2004): if the difference between skin thickness at the bovine site of injection and the avian site of injection was <2 mm, between 2 mm and ≤4 mm, or >4 mm, the animal was classified as negative, doubtful (inconclusive), or positive for BTB, respectively. A herd with at least one positive reactor was considered as "PPD positive". *M. avium* positivity was defined as previously described [9] with the avian PPD injection site showing a reaction of >4 mm between the testing day and the reading day. In addition, we calculated how many animals showed an avian skin reaction of >4 mm as compared to a bovine skin reaction [(A2-A1) - (B2-B1)>4 mm]. Information on farm structure and management were collected from 80 out of the 88 investigated farms using a standardised questionnaire (Questionnaire S1) with closed questions. The owner or manager of each farm was interviewed in the local language at the time the CIDT was performed. Sanitation status was judged as poor, medium (satisfactory), or good, based on aspects such as odour, waste drainage, cleanness of floor and animals, barn ventilation and light source, and animal stocking.

Ante and post mortem examination

All cattle tested by CIDT were recorded for age, sex, breed, and given a body condition score (poor, medium, or good). Selected CIDT positive cattle from farms of all six study areas were purchased and subjected to post mortem examination performed by an experienced veterinarian. The selection criteria for these purchased animals were based on strong PPD response, unproductive animal, and willingness of farmer to sell the animal. The pathology scoring system was based on the semi-quantitative procedure developed by Vordermeier et al. [21]. Each lobe of the lung was inspected externally and then sliced into 2 cm-thick slices to facilitate detection of any typical TB lesion. Similarly, liver, spleen, and kidney as well as other lymph nodes (mandibular, medial retropharyngeal, bronchial, mediastinal, and mesenteric) were sliced into thin sections and inspected for the presence of typical TB lesions. The severity scores of the gross lesions and the pathology scores for lymph nodes, lungs, and other organs were added up to determine the total pathology score per animal. Furthermore, lesion type (caseous or calcified) and TB stage (localised or generalized) were documented.

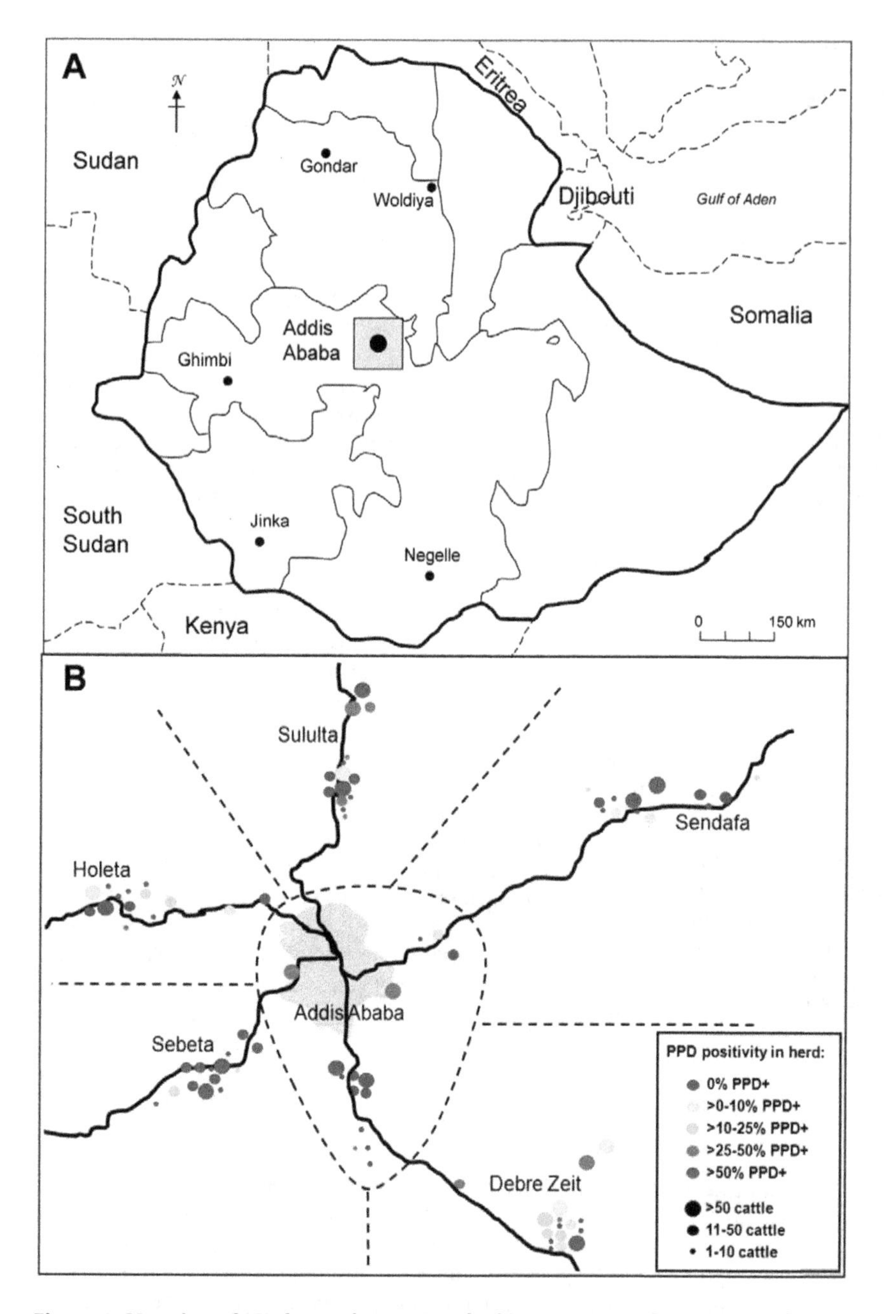

Figure 1. Mapping of (A) the study area (marked in grey) and other sites in Ethiopia where *M. bovis* isolates used in this study were collected, and (B) BTB prevalence in investigated farms in the five study areas. PPD positivity and farm size category are defined by colour and size of respective circle.

Sample collection, processing and culturing of mycobacteria

During post-mortem, up to seven suspected visible TB-like lesions were collected per animal in individual 50 ml sterile universal tubes (containing sterile PBS buffer) and transported at 4°C to the AHRI for further processing. At AHRI, samples were stored at 4°C when processed within 48 hours or at −20°C if processed later than two days after sample collection. All samples were processed and cultured for mycobacteria as previously described [7]. Samples were inoculated on three different media, including two Löwenstein-Jensen (LJ) media (supplemented with either glycerol or pyruvate) and a modified Middlebrook 7H11 medium optimised for the culture of *M. bovis* [22].

Identification and molecular typing of mycobacteria

Bacterial colonies from culture positive samples were Ziehl-Neelsen stained according to standardized protocol to identify Acid-Fast Bacilli (AFB). Slants with no growth at week 8 were considered negative. Heat-killed cells of each AFB positive isolate were prepared by mixing ~2 loopfuls of cells (~20 µl cell pellet) in 200 µl dH_2O followed by incubation at 80°C for 1 hour and then stored at 4°C until used for molecular typing. AFB positive isolates were also prepared as 20% glycerol stocks and stored at −80°C. Heat-killed AFB positive isolates were investigated by multiplex PCR for the presence or absence of RD4 [23], a chromosomal deletion characteristic of *M. bovis*. PCR amplification for RD4 typing was as follows: reactions were performed in a total volume of 20 µl consisting of 10 µl HotStarTaq Master Mix (Qiagen, United Kingdom), 7.1 µl distilled H_2O, 0.3 µl of each oligonucleotide primer (100 µM), and 2 µl DNA template (heat-killed cells, see above). Isolates genetically typed as *M. bovis* were spoligotyped as previously described [24]. Selected *M. bovis* isolates were typed for 24-loci MIRU-VNTR by Genoscreen, France [25,26].

Data analysis

All collected field and laboratory data were double entered into Microsoft Access database. Data validation was done with EpiInfo software ("Data compare" package). Data analysis was performed with STATA Version 10.1 (StataCorp, Texas, USA). We used logistic regression models for the analysis of BTB prevalence and the univariate analysis of parameters associated with positive farms. A GEE model (Generalized Estimating Equations) with binary outcome was used to analyse potential risk factors for individual reactors, in which farms were used as random effect and farm size as fixed effect with adjusted odds ratio (OR). In both cases, results were presented with 95% confidence interval for OR and p-values.

Results

Study areas and farm recruitment

As shown in Figure 1, six study areas were chosen in the region of Addis Ababa in central Ethiopia; these were Addis Ababa, Sendafa, Holeta, Sululta, Debre Zeit and Sebeta. A total of 88 farms were included in the study, with a distribution of 14–16 farms per study area (Table 1). At least five small farms (1–10 cattle) and five medium size farms (11–50 cattle) were included from each study area, while the number of large dairy farms (>50 cattle) varied between two and five due to limited large herds in the Holeta (3), Sebeta (2), and Sendafa (2) study areas.

Dairy farm management

The vast majority (95%) of the investigated farms were privately owned while the remaining 5% were owned by a cooperative or the government. Seventeen of the 88 farms were owned by women. In our survey 80 out of these 88 farms answered our questionnaire (Questionnaire S1). These 80 farms were intensive dairy farms with similar husbandry practice and nearly all cattle herds were kept in indoor barns; only three small herds shared the same house as the owners and one farm kept its animals outdoors. Herds were strictly kept separate from others at 78 out of 80 farms while owners of the remaining two farms allowed mixing with animals from other farms. Sanitation status was assessed for the barns on each farm. The sanitation was mostly poor (20%) to medium (65%), while only 12 farms (15%) showed good sanitation standards. All farms included in this study used mainly their own bull or artificial insemination for restocking the herd. However, about 1/3 of the farms purchased animals during 2008 and 2009 and they did so within the area where this study was conducted (Figure 1). Fifty of the farm owners sold cattle during these two years of which 13 found a buyer in other locations of Ethiopia (outside the described study area) such as Tigray in the north, Ambo and Jimma in the west, and Nazareth, Hawassa, Woleita, Hosana, and Shashemene southwards of Addis Ababa. More than half of the farms that sold animals to regions outside the study area were medium-sized farms, whereas only ~1/4 of both small and large farms had such behaviour. Animals were sold particularly when they showed low productivity (N = 50/80; 62%). Other motives for selling animals included being old (5%), weak and emaciated (5%), or being diseased (6.3%). Seven owners (8.9%) reported to have sold while still at high productivity.

Trading and consumption of milk

The vast majority of milk produced in the investigated farms was either sold (N = 36/80; 45%), or partly sold and partly used

Table 1. Individual and herd prevalence of BTB based on CIDT and individual PPD-A response, both stratified by farm size and study site.

Farm size		Addis Ababa	Debre Zeit	Sebeta	Holeta	Sululta	Sendafa	Total
Small#	Farm number	6	6	5	6	5	6	34
	Animal number	40	25	37	36	34	46	218
	Indiv. prev. (%)	0	16	16.2	13.9	11.8	2.2	9.2
	Herd prev. (%)	0	33	40	16.7	40	17	23.5
	PPD-A resp. (%)*	0 (0)	11 (0)	2.7 (0)	0 (0)	6.2 (0)	2.2 (0)	3.2 (0)
Medium#	Farm number	5	5	7	6	5	5	33
	Animal number	187	106	177	181	154	137	942
	Indiv. prev. (%)	0.5	20.7	38.4	14.9	47.4	1.5	20.5
	Herd prev. (%)	20	100	71	67	80	40	63.6
	PPD-A resp. (%)*	3.2 (0.5)	1.9 (0)	8.5 (1.1)	2.8 (0)	22.7 (0.6)	0.7 (0)	6.8 (0.4)
Large#	Farm number	5	5	2	2	4	3	21
	Animal number	531	518	235	144	210	158	1796
	Indiv. prev. (%)	36.2	23.5	72.8	1.4	53.8	89.9	41.3
	Herd prev. (%)	80	100	100	100	100	67	90.5
	PPD-A resp. (%)*	17.1 (0.2)	11 (1.3)	11.1 (0)	9 (0.7)	10 (0.5)	24.7 (0.6)	13.8 (0.6)

#Farm size definition: Small, 1–10 animals; Medium, 11–50 animals; Large, >50 animals.
*The percent of animals with response to PPD-A alone [(A2-A1)>4 mm] regardless of the PPD-B response. In bracket, percentage of animals with a skin reaction defined as [(A2-A1) – (B2-B1)>4 mm].

for own consumption (50%). The remaining 5% of the farms produced milk for their own consumption only. Thirty-five farms sold their milk to processing plants, 23 to cafés and restaurants, 19 to intermediate caterers, and 16 directly to private individuals.

The number of people living on a farm varied between one and 53. However, the majority (81%) of farms accommodated less than ten people while half had only between one and four people living on the farm. The majority (86%) of the farm owners did not drink raw milk and 71% knew that BTB is an animal and a zoonotic disease.

Tuberculin testing

Over a period of 13 months (January 2009 to January 2010), a total of 2956 cattle from 88 dairy farms were tuberculin tested by CIDT. The vast majority (94%) of these cattle were female and either of Holstein-Frisian breed (37%) or cross breeds thereof (62%) with local zebu cattle (Table 2). The remaining tested animals were of either zebu or Jersey breed, confined to a few farms in Holeta, Sendafa, Sebeta, and Debre Zeit.

The result of the tuberculin testing is presented in Table 1 as prevalence per study area and herd size. The CIDT result shows that BTB was prevalent in all six study areas. It also demonstrates that the prevalence increased with farm size category. The average individual prevalence of small, medium, and large farms was 9.2%, 20.5%, and 41.3%, respectively, and a similar trend was seen in herd prevalence with the extreme of over 90% infected herds among the large farms (Table 1). The overall number of herds with at least one reactor animal was 48 (55%); however, 38 farms (43%) had ≥10% of its animals infected (Figure 1B).

More detailed analysis of small and medium sized farms indicated that the areas of Addis Ababa and Sendafa (Figure 1) had the lowest individual animal prevalence; only one out of 86 tested animals from small farms in those two areas combined was a reactor, and in the medium-sized farms only three out of 324 animals were positive according to the CIDT. Small and medium farms tested in the remaining four study areas had significantly higher individual prevalence, ranging between 12–16% in small farms while it varied more in medium farms (between 15–47.4%); elevated prevalence was seen in medium farms in Sebeta and Sululta. The distribution of tuberculin reactors between farms is reflected in the herd prevalence (Table 1). More than 1/3 of small farms and over 2/3 of medium farms were infected in Debre Zeit, Sebeta, and Sululta. The latter was valid also for the medium farms in Holeta. We conclude that BTB was widely distributed among the tested small and medium farms in at least Debre Zeit, Sebeta, and Sululta.

The results obtained from the analysis of large dairy farms showed a somewhat different picture. The individual BTB prevalence in any tested large farm varied from 0% and up to 93%. By study area, Holeta showed the lowest prevalence with only two CIDT positive animals between the two large farms tested. The individual prevalence in all other study areas varied between 23.5–90%, (Table 1). Overall, 18 out of the 21 large farms were infected with BTB (overall herd prevalence: 85.7%) of which 11 had an animal prevalence of over 30%. The crude overall individual prevalence was 32.3% (30.6%–34.0%) (not considering stratification of farm size or study sites).

Sensitisation to environmental mycobacteria

Although the CIDT is used to detect BTB infection with high specificity, reactivity to the avian PPD indicates infection with or exposure to species of the *Mycobacterium avium* complex (MAC) or other environmental mycobacteria. In this study 318 out of the 2956 tested animals (~11%) reacted positive to PPD-A. The PPD-A response was lowest in Holeta (5%; 95%CI: 3.1; 7.8) and highest in Sululta (14.6%; 95%CI: 11.5; 18.5) (Table 1), regardless of farm

Table 2. Univariate analysis of risk factors for having positive reactors using a GEE model with binary outcome.

Variable	Category	Animal No.*	BTB reactors*	OR (95%CI)	p-value
Herd size	Small	216 (7.3)	20 (2.1)		
	Medium	942 (31.9)	193 (20.2)	2.5 (0.91; 6.85)	0.07
	Large	1796 (60.8)	742 (77.7)	5.9 (2.06; 16.84)	0.001
Breed	Crossbreed	1837 (62.8)	351 (37.2)		
	Holstein	1011 (34.6)	582 (61.8)	1.2 (0.69; 2.23)	0.47
	Zebu	45 (1.5)	3 (0.3)	1.1 (0.89; 1.33)	0.39
	Jersey	33 (1)	7 (0.7)	1.1 (0.73; 1.71)	0.59
Sex	Female	2783 (94)	919 (96.3)		
	Male	170 (6)	35 (3.7)	0.9 (0.76; 1.23)	0.8
Body condition score	Poor	212 (7.2)	62 (7)		
	Medium	1454 (49.5)	409 (43)	0.8 (0.57; 1.06)	0.1
	Good	1272 (43.3)	481 (50)	0.8 (0.58; 1.13)	0.2
Age class	<12 months	139 (5.2)	27 (3)		
	≥12 mo - 3 yrs	823 (31)	225 (24.3)	2.2 (0.91; 5.51)	0.07
	>3 yrs - 10 yrs	1642 (61.8)	656 (71)	3.6 (1.43; 9.30)	0.007
	>10 yrs	54 (2)	16 (1.7)	4.1 (1.59; 10.56)	0.003
PPD-A status	Negative	2636 (89.2)	755 (79)		
	Positive	318 (10.8)	200 (21)	1.6 (1.01; 2.60)	0.04

*A number in brackets shows percentage of animals recorded in a category of one variable. The total number of animals recorded for in one variable may vary as some data are missing.

size. However, 200 out of the 318 PPD-A positive animals reacted positive also to PPD-B (Table 2) and the PPD-A responses could be down to cross reactivity between *M. bovis* antigens and antigens from environmental mycobacteria. Interestingly, PPD-A responders with a stronger response than to PPD-B were few (between 0 and 1.3%; Table 1).

Factors associated with BTB

The univariate analysis (Table 2) showed that reactor animals differed statistically according to age categories. Adult animals between 3 and 10 years old were over threefold more at risk of being tuberculin positive (with adults older than 10 years fourfold more at risk) as compared to young animals (OR = 3.6, p-value = 0.007 and OR = 4.1, p-value = 0.003, respectively). Medium and large farms had a significant risk increase of having positive reactors (OR = 2.5 and OR = 5.9 respectively, p-value: 0.001). Positive PPD-A reactors had an OR of 1.6 for being BTB reactors (p-value: 0.04).

Table 3 shows the univariate analysis of parameters potentially affecting the BTB herd status: Herd size and education level of the owners were factors associated with having positive farms; medium herds had a statistically significant fivefold increase of tuberculin positivity (OR = 5.4; p-value: 0.002) compared to small herds. Higher education of owner was linked with BTB positive farms (secondary school with OR = 4.8, p-value: 0.04 and diploma/degree with OR = 4.5, p-value: 0.05). The majority of the owners (65.4%) had an education above junior secondary school (secondary school, diploma, degree). However, 87.5% of the owners having large farms had gone through secondary school and/or higher education (diploma, degree). On the other hand 50% of the owners with no education at all or primary school attendance governed over small farms. When stratified by farm size, there was no statistical difference between level of education and BTB farm status (data not shown).

There was no significant difference in BTB status amongst farms with different sanitation status and amongst the type of cattle housing (data not shown).

Ante- and post-mortem investigation

Over 90% of all investigated animals had medium or good body condition. However, no significant difference of the body condition score was observed between tuberculin positive and negative cattle (Table 2). To further confirm that the recorded high rates of BTB were due to *M. bovis* and to get a better understanding of the population structure of the causative agent, we purchased and slaughtered a minimum of three PPD positive animals per study area. In total 36 cattle from 19 farms (up to three animals per farm) were purchased and investigated by post-mortem and each animal was given a pathology score based on disease progression (Table 4). A trend of inverted correlation between pathology score and PPD response was observed; animals with strong response to the tuberculin test had in general a lower than average pathology score. Suspect TB lesions from up to seven infected sites per animal were collected from 34 animals. In addition, samples were taken from mediastinal and bronchial lymph nodes from two animals with non-visible lesions. Thirteen out of the 34 (32%) animals with lesions had generalised disease while the remaining 21 had lesions localised to one or a few infection sites. Figure 2 shows the distribution of lesions found in the 34 animals. Visible lesions were most frequent in mediastinal lymph nodes (N = 30; 88%), lungs (74%), bronchial lymph nodes (62%), and head lymph nodes (50%), but prevalent also in mesenteric and hepatic lymph nodes, and in liver. Calcification was the most commonly observed stage of lesion (Table 4).

Isolation and molecular typing of mycobacteria

Mycobacterial culturing on selected media yielded in total 107 AFB isolates from samples corresponding to 31 animals with visible lesion. No AFB were cultivated from the remaining three

Table 3. Univariate analysis for potential risk factors being a positive farm (logistic regression with likelihood ratio test).

Variable	Category	Farm No.	BTB+ farms	OR (95% CI)	p-value
Herd size	Small	32 (40.5)	9 (19.6)		
	Medium	31 (39.2)	21 (45.6)	5.4 (1.82; 15.76)	0.002
	Large	16 (20.3)	16 (34.8)	-	-
Owner education level	No school	11 (14)	3 (6.7)		
	Primary school	9 (11.6)	5 (11)	3.3 (0.51; 21.58)	0.2
	Secondary school	31 (39.8)	20 (44.5)	4.8 (1.06; 22.10)	0.04
	Diploma and degrees	27 (34.6)	17 (37.8)	4.5 (0.97; 21.14)	0.05
BTB knowledge	Yes	57 (71.2)	36 (76.6)		
	No	23 (28.8)	11 (23.4)	0.5 (0.20; 1.42)	0.21
Owner gender	Male	62 (78.5)	39 (84.8)		
	Female	17 (21.5)	7 (15.2)	0.4 (0.13; 1.23)	0.11
Source of cattle purchase	Dairy farms	55 (71.4)	31 (67.4)		
	Markets, unknown origin	12 (15.6)	8 (17.4)	1.5 (0.41; 5.75)	0.51
	Gift	2 (2.6)	1 (2.2)	0.8 (0.04; 13.02)	0.85
	Research stations	5 (6.5)	3 (6.5)	1.2 (0.17; 7.51)	0.87
	Dairy farms and markets	3 (3.9)	3 (6.5)	-	-

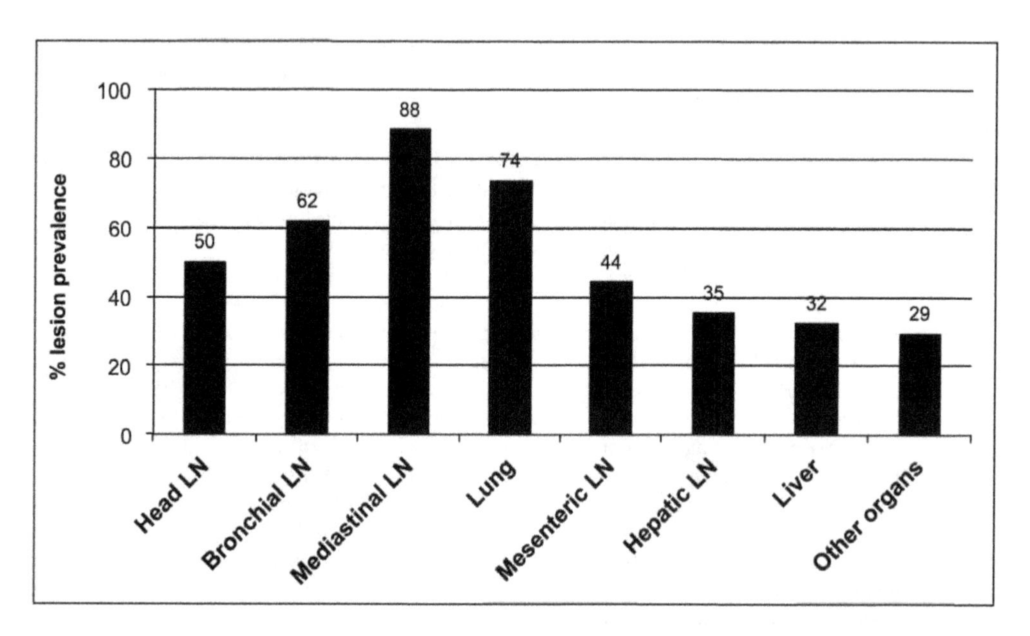

Figure 2. Organ association of suspect TB lesions in 34 cattle with visible lesions (LN, lymph node).

animals with suspect TB lesions or from samples of the two animals with non-visible lesions (Table 4).

Heat-killed cells of 67 AFB isolates originating from 31 animals were enrolled in a typing scheme initiated by RD4 deletion typing. The result showed that all isolates could be identified as *M. bovis* as they had the RD4 region deleted [23]. Spoligotyping was then employed to further differentiate these isolates and the outcome is shown in Table 4. Overall, four different spoligotype patterns were recognized among these *M. bovis* isolates and defined as of type SB0134, SB1176, SB0133, and SB1477 at the international spoligotyping database www.mbovis.org. None of the 31 animals were found to be infected with more than one *M. bovis* spoligotype suggesting no dual infection among these animals. On the other hand, *M. bovis* isolates of two different spoligotype patterns were collected from three farms (out of 19). The geographical distribution of these four spoligotypes suggested three main clusters; SB0134 was the most dispersed and found in all six study areas, SB1176 was represented in all areas but Sendafa, while SB0133 was only found in Sululta, Sendafa and Debre Zeit. The one isolate of type SB1477 was collected in a farm from Addis Ababa.

In attempts to find possible transmission links between dairy farms included in this study we performed 24-loci MIRU-VNTR typing to further discriminate the collected *M. bovis* isolates. MIRU-VNTR results were generated for 27 isolates of the four different spoligotypes (Table 4). All 11 isolates of type SB0134 had the same genotype (spoligotype plus MIRU-VNTR type) and that was also the case for four isolates of SB0133. Only the collected isolates of type SB1176 showed diversity; two different genotypes were identified among 11 isolates, however the two types differed in only one out of the 24 loci (Locus 2531; Table 4). Overall, this genotyping result suggested very low diversity within each spoligotype cluster among the collected strains.

In a previous study on BTB in Ethiopia [7] we isolated and identified *M. bovis* from several abattoirs in the country. Here we performed 24-loci MIRU-VNTR typing of 15 *M. bovis* isolates from that study with spoligotypes SB0133, SB1176, SB0134, the three most frequent types isolated in the present study (Table 4) with the aim to investigate genotype diversity among strains collected from different parts of Ethiopia. Table 5 summarises the different *M. bovis* genotypes and their frequency from each collection site. Two distinct MIRU-VNTR features were seen among strains within spoligotype SB0133 as well as within SB1176, while strains compared within SB0134 showed no diversity.

Discussion

With the recent shift towards larger intensive farming systems, mainly in urban areas, it is clear that the farming practices are becoming more favourable for BTB transmission. In this study we explored the prevalence of BTB in cattle in intensively managed dairy farms located in and around Addis Ababa and that supply milk to the capital. Based on the CIDT test, all six surveyed areas (Figure 1) recorded a high rate of the disease and we concluded that, overall, approximately one third of the tested animals and more than half of the investigated herds were infected with BTB (if stratification or random effect is not taken into account). However, the prevalence fluctuated significantly between the three investigated farm size categories. Both the overall individual and overall herd prevalence of medium and large farms were double and fourfold, respectively, as compared to the corresponding numbers of small farms (Table 1). Since as many as 20 out of 21 large herds (>50 animals/herd) were infected, we concluded that BTB had a strong-hold in dairy farms in this investigated region of central Ethiopia.

Risk factors

Several past and recent studies have shown that susceptibility to BTB can vary between cattle breeds [12,27] with suggestions that cattle of *Bos indicus* (Zebu) breeds are more resistant to BTB than *Bos taurus* (European breeds). The vast majority of the dairy cattle included in this study were cross breeds (62%) and Holstein-Frisian (34%), however the BTB prevalence did not differ significantly between these two categories (Table 2). The number of cattle recorded as Jersey and Zebu breeds in this study were too few to make relevant comparisons on susceptibility. On the other hand, Ameni et al. [28] have demonstrated that animal husbandry

Table 4. Post-mortem results of 36 animals from 19 farms.

Study area	Farm ID: Animal ID	CIDT (Δmm)	PM score	Type of TB	Type of lesion	AFB	Strain ID	*M. bovis* Spoligotype	24-loci MIRU-VNTR type*
Addis Ababa	1:A	24	4	localized	caseous	Yes	BTB-2269	SB0134	35223 40554 21415 24222 2XXX
Addis Ababa	1:B	26	15	localized	calcified	No	-	-	-
Addis Ababa	2:A	22	23	generalized	calcified/caseous	Yes	BTB-2850	SB1477	36223 32454 21445 24222 3213
Addis Ababa	2:B	18	23	generalized	calcified	Yes	BTB-2840	SB1176	35223 32554 21435 24222 3323
Holeta	3:A	30	10	generalized	caseous	Yes	BTB-2352	SB1176	35223 32554 21435 24222 3323
Holeta	3:B	26	0	-	-	No	-	-	-
Holeta	3:C	12	29	generalized	calcified	Yes	BTB-2400	SB1176	35223 32554 21435 24222 3323
Holeta	4:A	29	5	generalized	calcified	Yes	BTB-2349	SB0134	-
Holeta	5:A	11	20	localized	calcified	Yes	BTB-2350	SB0134	35223 40554 21415 24222 2332
Sululta	6:A	16	14	localized	calcified/caseous	Yes	BTB-2762	SB0133	23223 32454 21425 24222 3323
Sululta	6:B	18	21	generalized	calcified	Yes	BTB-2822	SB0133	23X23 32454 21425 24222 3323
Sululta	7:A	15	46	generalized	calcified	Yes	BTB-2466	SB1176	35223 32554 21435 24222 3323
Sululta	7:B	28	12	localized	calcified	Yes	BTB-2785	SB1176	35223 32554 21435 24222 3323
Sululta	7:C	14	43	generalized	calcified	Yes	BTB-2469	SB1176	35223 32554 21435 24222 3323
Sululta	8:A	38	5	localized	calcified	Yes	BTB-2465	SB1176	35223 32554 21435 24222 3323
Sululta	8:B	24	1	localized	calcified	Yes	BTB-2461	SB0134	35223 40554 21415 24222 2332
Sendafa	9:A	21	14	generalized	calcified/caseous	Yes	BTB-2847	SB0134	35223 40554 21415 24222 2332
Sendafa	9:B	13	8	localized	caseous	Yes	BTB-2849	SB0134	35223 40554 21415 24222 2332
Sendafa	10:A	17	2	localized	caseous	Yes	BTB-2862	SB0133	-
Sendafa	10:B	41	4	localized	calcified/caseous	Yes	BTB-2861	SB0133	23223 32454 2X425 24222 3X23
Debre Zeit	11:A	17	20	generalized	calcified	Yes	BTB-2266	SB1176	-
Debre Zeit	11:B	16	19	generalized	calcified	Yes	BTB-2261	SB0134	35223 40554 21415 24222 2332
Debre Zeit	12:A	4	9	localized	calcified	Yes	BTB-2260	SB1176	35223 32552 21435 24222 3323
Debre Zeit	13:A	23	19	generalized	calcified	No	-	-	-
Debre Zeit	13:B	11	8	localized	calcified	Yes	BTB-2257	SB0133	23223 32454 21425 24222 3323
Sebeta	14:A	6	2	localized	calcified	No	-	-	-
Sebeta	14:B	20	12	localized	caseous	Yes	BTB-2813	SB1176	3522X 32554 2X435 24X22 3323
Sebeta	15:A	18	37	generalized	calcified	Yes	BTB-2803	SB1176	35223 32552 21435 24222 3323
Sebeta	15:B	18	6	localized	caseous	Yes	BTB-2805	SB1176	35223 32552 21435 24222 3323
Sebeta	16:A	20	0	-	-	No	-	-	-
Sebeta	16:B	22	4	localized	calcified	Yes	BTB-2768	SB1176	-
Sebeta	17:A	21	8	localized	calcified	Yes	BTB-2838	SB0134	35223 40554 21415 24222 2332
Sebeta	17:B	16	8	localized	calcified/caseous	Yes	BTB-2765	SB0134	35223 40554 21415 24222 2332
Sebeta	18:A	13	6	localized	caseous	Yes	BTB-2757	SB0134	35223 40554 21415 24222 2332
Sebeta	19:A	22	9	localized	calcified	Yes	BTB-2773	SB0134	35223 40554 21415 24222 2332
Sebeta	19:B	20	13	localized	calcified/caseous	Yes	BTB-2766	SB0134	35223 40554 21415 24222 2332

The average CIDT value and post-mortem score [21] were 20 mm and 13, respectively.
*Consecutive order of the 24-loci MIRU-VNTR: 2165; 2461; 577; 580; 3192; 154; 960; 1644; 2059; 2531; 2687; 2996; 3007; 4348; 802; 424; 1955; 2163b; 2347; 2401;3171; 3690; 4052; 4156; X = Typing failed.

conditions are a major influence on the prevalence of BTB. Cattle kept under high-intensity conditions showed significantly higher skin-test prevalence as compared to cattle kept under extensive conditions. Intensification, stressed animals, and overcrowding are all possible explanations for such relationship. The main routes of BTB transmission are through aerosol as gross lesions usually involve the lungs and thoracic lymph nodes [29], and therefore BTB transmission benefits from overcrowded herds. The post-mortem data collected in this study support this as most animals had TB lesions in lungs and/or lung associated lymph nodes

Table 5. Geographical diversity of *Mycobacterium bovis* genotypes in Ethiopia.

Spoligotype	Region	Collection site	24 loci MIRU-VNTR type*	Frequency
SB0133	North Ethiopia	Woldiya	23223 32454 21425 24222 3323	1
	Central Ethiopia	This study	23223 32454 21425 24222 3323	3
	Central Ethiopia	Addis Ababa	23223 32454 214**3**5 24222 3323	1
	South-East Ethiopia	Negelle	23223 32454 21425 24222 3**2**23	1
	South-East Ethiopia	Negelle	23223 32454 21425 24222 3323	2
	South Ethiopia	Jinka	**36**223 **2**2**53**4 214**36** 24222 **4**323	1
	South Ethiopia	Jinka	**36**223 **2**2**54**4 214**36** 24222 **4**323	4
	West Ethiopia	Ghimbi	**36**223 **2**2**54**4 214**36** 24222 **4**323	1
SB1176	North Ethiopia	Woldiya	35223 32554 2143**4** 24222 3323	2
	Central Ethiopia	This study	35223 32554 21435 24222 3323	8
	Central Ethiopia	This study	35223 3255**2** 21435 24222 3323	3
	South Ethiopia	Jinka	35223 3**3**5**7**4 **3**1**223** 24222 33**7**3	1
SB0134	North Ethiopia	Gondar	35223 40554 21415 24222 2332	1
	Central Ethiopia	This study	35223 40554 21415 24222 2332	11

Repeats of locus highlighted in bold diverge between isolates of the same spoligotype. All collection sites are shown in Figure 1.
*Consecutive order of the 24-loci MIRU-VNTR: 2165; 2461; 577; 580; 3192; 154; 960; 1644; 2059; 2531; 2687; 2996; 3007; 4348; 802; 424; 1955; 2163b; 2347; 2401; 3171; 3690; 4052; 4156.

(Figure 2). Once an animal is infected with BTB, it usually takes several months or longer for this chronic disease to develop clinical signs. It is also likely that infected cattle, similarly to humans, enter an asymptomatic phase after infection with *M. bovis*, in which no clinical signs are developed for years [30]. Our ante- and post-mortem investigation also confirmed this common attribute of BTB. The body condition scoring of the nearly 3,000 animals suggested no significant differences between tuberculin reactors and non-reactors (Table 2) and a poor correlation was also seen among the 36 slaughtered animals; despite that the majority of these animals were highly diseased with severe pathology, little or no clinical signs typical for BTB were observed. Thus, if BTB gets into a herd where no diagnostic measure is in place (such as CIDT), then it can be difficult to detect infected animals on a visual basis solely and to prevent further transmission.

Most owners of medium and large farms had higher educational level while illiterate farmers tended to have smaller farms. Although this leads to an association in our dataset between higher education and higher prevalence of BTB in the herds (Table 3), this is more likely to be an expression of higher educated people engaging in dairy farm business as an investment, hence larger farms being associated with higher education.

Farmers participating in the current study said that the main means of restocking their herd was through artificial insemination. However, frequently animals were purchased or sold between farms within the study areas. Trading of animals beyond the study area occurred too (mainly by medium sized farms) but seemed unidirectional; dairy cattle were not purchased but only sold to other regions across Ethiopia. The response from farmers in this survey also suggested that some animals were sold because they were weak or diseased. As no official test-and-movement policy is in place, selling of infected animals from this highly affected region of Ethiopia may consequently contribute to spreading BTB to other parts of the country.

The causative agent

The mycobacterial strains isolated from cattle in this study were solely *M. bovis* and the CIDT test was confirmed by culture in 86% of the slaughtered cattle. This culturing yield can be compared with a 90% yield normally seen in slaughterhouse cases in the UK from animals with visible lesions and of which the vast majority were caused by *M. bovis* [31]. No strain of *M. tuberculosis* was isolated in our study in contrast to several other Ethiopian studies [7,13,32]. It can be argued though that only ~3% of all cattle that were tuberculin positive in this study were investigated by post-mortem, leaving out a large group of animals that could have been infected with *M. tuberculosis* as CIDT testing do not distinguish between *M. tuberculosis* and *M. bovis*. However, the dairy cattle of this study were managed under intensive husbandry systems with animals staying in separate barns, as compared to management systems common in the rural areas where domestic animals often sleep in the same house as people. It has previously been suggested that human-to-cattle transmission of *M. tuberculosis* is possible in settings where farm members in high-burden TB areas are in close contact with their animals [32].

Molecular typing of the 31 *M. bovis* strains isolated from the slaughtered animals identified four different spoligotype patterns which previously have been shown as common in several regions of Ethiopia [7]. Three of these four spoligotypes (SB1176, SB1477, and SB0133) carry a specific spoligotype feature (spacers 4–7 missing) typical for members of a clonal complex identified as *M. bovis* African 2 (Af2) so far only found in East Africa [33]. The fourth spoligotype (SB0134) do not belong to the Af2 clonal complex and is therefore likely to have a different epidemiological history. *M. bovis* isolates with spoligotype SB0134 has been found also in Europe. However, the phylogenetic relationship between the African and European strains of SB0134 is not known. Future chromosomal sequencing may shed light over their relationship.

Further discrimination of the *M. bovis* isolates performed by MIRU-VNTR typing suggested low diversity within each isolated spoligotype. This is in agreement with other studies as in high prevalence settings the genetic diversity of sampled strains is usually low, reflecting on-going local transmission events (due to high transmission rate) [34]. However, in comparison with genotypes of *M. bovis* strains that were collected from other parts of Ethiopia about 2–3 years earlier [7], some variations were observed and possible epidemiological links between specific regions of the country can be suggested (Table 5). Two distinct MIRU-VNTR features were seen within spoligotype SB0133 of which the one identified in central Ethiopia (this study) was also seen in sites along the Rift Valley, from Woldiya in the north to Negelle in the south. The other genotype of SB0133 was isolated from Jinka (south-west Ethiopia) and Ghimbi (west Ethiopia). Similarly, the single strain of type SB1176 collected in Jinka deviated in the MIRU-VNTR pattern from those seen in Woldiya and in central Ethiopia. However, no genotype diversity was seen between the one isolate of SB0134 from Gondar (North-west) when compared to the 11 *M. bovis* strains of SB0134 isolated in central Ethiopia (this study). Overall, based on these molecular typing results of our *M. bovis* collection, epidemiological links can be suggested for strains collected within the North/Central Ethiopian highlands and through the rift valley down to Negelle, while strains isolated from Jinka and Ghimbi in the south/west of Ethiopia diverged. This may reflect on past and ongoing cattle movements and trading patterns in the country. The low strain frequency from some collection sites means that these results should be interpreted with care. Nevertheless, this genotype comparison gives an indication of the diversity and the epidemiology of *M. bovis* in Ethiopia. Future studies may give better clarification on this subject. The 24-loci MIRU-VNTR typing employed for this study showed no diversity in at least 14 of the 24 loci (variation seen between spoligotypes; Table 5). This is in agreement with other studies [35], which has shown that VNTR typing of only a selective set of loci is informative for generating a better resolution between *M. bovis* isolates. This suggests that any future MIRU-VNTR typing of *M. bovis* strains from Ethiopia could consider including only loci that show diversity in its local strain population, however, a more comprehensive evaluation of what loci to select is recommended.

Reasons for controlling BTB

At least three important reasons for controlling BTB should be considered: animal welfare, the financial burden to farmers with diseased animals, and the risk of zoonotic transmission. This chronic disease can take years to develop. Animals infected with BTB can therefore slowly develop symptoms and may, if not carefully observed by animal keepers, suffer unnecessarily. Depending on which organ or associated lymph node is affected, animals can display a wide range of symptoms e.g. coughing, dyspnoea, gastrointestinal problems, bone deformation, and emaciation [36].

Few thorough studies have been conducted to investigate loss of productivity due to BTB, but it is probable that there is a productivity impact in animals with BTB due to the nature of the disease and thus this is likely linked to economical losses [37]. It has previously been suggested that reduced milk production, the food value of the carcass, and reproduction (or fertility) are factors that affects the animal productivity [38].

The zoonotic risk of BTB is often associated with consumption (ingestion) of dairy products based on unpasteurised milk infected with *M. bovis*. However, aerosol transmission (inhalation) from cattle-to-human should also be considered as a potential risk factor. Ethiopian milk consumers generally prefer raw milk (as compared to treated milk) because of its taste, availability and lower price. Therefore the demand for fresh whole milk in rural areas is and will probably remain high, and in general, the demand from these consumers is likely to be satisfied by home production or by purchasing from neighbouring producers. On the other hand, the markets for surplus milk are mainly found in the urban population, of which currently ~65% is accounted for by the capital Addis Ababa and its neighbouring districts [39].

The annual total milk supply to Addis Ababa has been estimated at 65 million litres and the major sources are from small private farms and smallholder urban dairies in and around the city that own upgraded breeds [40]. These over 5,000 dairy farms produce around 35 million litres of milk per year. Grossly 10% is used to feed calves, 10% is used for home consumption, 8% is processed into butter and cheese, while the remaining ~2/3 is sold to the market as un-boiled or unpasteurised milk [41]. Other important sources of milk to the capital city are medium to large dairy plants run privately or as enterprises/cooperatives. Today there are around eight major dairy plants in central Ethiopia with a total production capacity of around 100,000 litres per day (~35 million litres per year) (survey by Alehegne Wubete, unpublished), but the actual yearly production is likely to be less as animal product consumption can vary through the year due to cultural and religious habits such as fasting. These plants collect milk from farmers within a 150 km radius around Addis Ababa and provide standardized pasteurized milk in plastic sachets to the markets [39]. Our survey suggests that many dairy farms from the investigated area sell their milk to restaurants and catering companies, but it is difficult to estimate how much of that milk is pasteurised or boiled before consumption. Overall, the available information suggests that a significant volume of milk may be distributed without being pasteurised.

In a recent Ethiopian study of the rate of *M. bovis* causing TB in humans [42] we reported that the rate of TB in humans due to *M. bovis* is currently below 1%. However, we concluded that areas where BTB prevalence in cattle is high (such as the "hot-spot" area of central Ethiopia shown in this study) could generate an increase of BTB in the human population in such areas. It is important to note that existing human TB control methods relying on the directly observed treatment short course (DOTS) strategy will not directly impact on BTB transmission. The high rate of BTB in the dairy farms of this study is therefore likely to pose a serious risk to public health and deserves a targeted intervention as early as possible.

Conclusion and Control strategies

This study suggests that the overall prevalence of BTB in intensive dairy farms in Addis Ababa and its surroundings is very high and requires urgent intervention to control the disease. To comprehensively control BTB regardless of the method chosen, a robust system of individual animal and herd identification is required. The powers to restrict movements to prevent disease dissemination and the ability to trace movements to catch up with translocation are essential components of an eradication strategy. Most developed countries where BTB is a problem have introduced a test-and-slaughter policy to control/eradicate the disease. The CIDT, although not 100% sensitive, is the standard test widely used in developed countries for this purpose. Most importantly, infected animals have to be detected and removed at an early stage to minimise further transmission. As such policy is considered costly it has rarely been implemented in developing countries. No compensation scheme for elimination of infected animals is currently in place in Ethiopia and due to financial constraints changes toward such policy might not be feasible in the near future. However, a test-and-segregation policy of tuberculin positive animals can be suggested and pursued and the milk of infected animals strictly and consequently pasteurized before selling and consumption. Additional interventions by the government authorities could be to encourage farmers to regularly test their animals for BTB and create incentives to keep their herds free from BTB. It could be compulsory for dairy farms that are supplying milk to the public to test their cattle for BTB, and create incentives to TB free herds – "good milk certificate" – that could lead to better economic values for farmers. Also dairy plants should be engaged and could e.g. pay a higher rate for milk from herds certified as being BTB free. Pasteurization at a bigger scale has to be promoted, either directly at farm level before selling the milk or at household level through boiling the milk before consumption (e.g. awareness campaigns).

As the dairy industry in Ethiopia has expanded in recent years and is expected to continue doing so, significant number of high productive exotic and cross breed animals are likely to be traded from the urban areas around the capital to the rural areas where dairy cattle numbers are still relatively low. Such movement is linked with increasing financial well-being of farmers and amelioration of artificial insemination technique/delivery. However, without any control strategy the risk of spreading BTB by such movements is high and may create new hot-spots of BTB in other parts of the country. Several recommendations can be made to minimize the risk of spreading the disease between farms and regions. First and most important is to spread knowledge about BTB and make farmers and people involved in trading of cattle aware of the risks involved in trading animals. On a smaller scale, our field work and specific training to the farmers aimed on this, informing them in how to handle cattle infected with BTB and other diseases, and encouraging them not to trade with diseased animals. Another way to decrease risk of transmission is to encourage farmers who are restocking by purchasing animals to request that the animals should be tested for BTB. The CIDT test is relatively inexpensive and would reduce the risk to the farmer of purchasing an infected animal. A possibility for Ethiopia to in part meet the demand for cattle of upgraded breeds could be to establish farms with dairy cattle free of BTB and from which farmers could restock their farms without the risk of introducing the disease into their herds. This might imply scaling up of existing artificial insemination centres to improve productivity in the national dairy herd and possibly purchase of new stocks of exotic breeds from abroad.

Acknowledgments

We thank all farm owners and staff that participated in this study, and we appreciate laboratory and database support from AHRI staff.

Author Contributions

Conceived and designed the experiments: RF SB AW M. Sombo GE RT AA SVG DY MV M. Sahile RGH. Performed the experiments: RF AW M. Sombo EH GE TK RT. Analyzed the data: RT RF SB AW MV. Contributed reagents/materials/analysis tools: AW RT LY AA M.Sahile. Wrote the paper: SB RT RF AW. Revised the manuscript: M. Sombo AA SVG DY MV.

References

1. Bureau PR (2010) World Population Data Sheet. Available: http://www.prb.org/Publications/Datasheets/2010/2010wpds.aspx.
2. CSA (2011) Report on Livestock and Livestock Characteristics. Agricultural Sample Survey 2011/2012. Addis Ababa: Central Statistical Agency of Ethiopia.
3. Halderman M (2004) The Political Economy of Pro-Poor Livestock Policy-Making in Ethiopia. In: Initiative P-PLP, editor. Rome, Italy. pp. 1–59.
4. CSA (2007) Report on Livestock and Livestock Characteristics. Agricultural Sample Survey 2006/2007. Addis Ababa: Central Statistical Agency of Ethiopia.
5. Cosivi O, Grange JM, Daborn CJ, Raviglione MC, Fujikura T, et al. (1998) Zoonotic tuberculosis due to *Mycobacterium bovis* in developing countries. Emerg Infect Dis 4: 59–70.
6. Ayele WY, Neill SD, Zinsstag J, Weiss MG, Pavlik I (2004) Bovine tuberculosis: an old disease but a new threat to Africa. Int J Tuberc Lung Dis 8: 924–937.
7. Berg S, Firdessa R, Habtamu M, Gadisa E, Mengistu A, et al. (2009) The burden of mycobacterial disease in ethiopian cattle: implications for public health. PLoS One 4: e5068.
8. Tschopp R, Schelling E, Hattendorf J, Aseffa A, Zinsstag J (2009) Risk factors of bovine tuberculosis in cattle in rural livestock production systems of Ethiopia. Prev Vet Med 89: 205–211.
9. Tschopp R, Schelling E, Hattendorf J, Young D, Aseffa A, et al. (2010) Repeated cross-sectional skin testing for bovine tuberculosis in cattle kept in a traditional husbandry system in Ethiopia. Vet Rec 167: 250–256.
10. Gumi B, Schelling E, Firdessa R, Aseffa A, Tschopp R, et al. (2011) Prevalence of bovine tuberculosis in pastoral cattle herds in the Oromia region, southern Ethiopia. Trop Anim Health Prod 43: 1081–1087.
11. Demelash B, Inangolet F, Oloya J, Asseged B, Badaso M, et al. (2009) Prevalence of bovine tuberculosis in Ethiopian slaughter cattle based on post-mortem examination. Trop Anim Health Prod 41: 755–765.
12. Ameni G, Aseffa A, Engers H, Young D, Gordon S, et al. (2007) High prevalence and increased severity of pathology of bovine tuberculosis in Holsteins compared to zebu breeds under field cattle husbandry in central Ethiopia. Clin Vaccine Immunol 14: 1356–1361.
13. Tsegaye A, Aseffa A, Mache A, Mengistu Y, Berg S, et al. (2010) Conventional and Molecular Epidemiology of bovine tuberculosis in dairy farms in Addis Ababa City, the capital of Ethiopia. J Appl Res Vet Med 8: 143–151.
14. Elias K, Hussein D, Asseged B, Wondwossen T, Gebeyehu M (2008) Status of bovine tuberculosis in Addis Ababa dairy farms. Rev Sci Tech 27: 915–923.
15. Meisinger G (1970) Economic effects of the elimination of bovine tuberculosis on the productivity of cattle herds. 2. Effect on meat production. Monatsh Veterinarmed 25: 7–13.
16. Tschopp R, Aseffa A, Schelling E, Zinsstag J (2010) Farmers' perceptions of livestock, agriculture, and natural resourses in the rural Ethiopian highlands. Mountain Research and Development (MRD) 30: 381–390.
17. Grange JM (2001) *Mycobacterium bovis* infection in human beings. Tuberculosis (Edinb) 81: 71–77.
18. de la Rua-Domenech R (2006) Human *Mycobacterium bovis* infection in the United Kingdom: Incidence, risks, control measures and review of the zoonotic aspects of bovine tuberculosis. Tuberculosis (Edinb) 86: 77–109.
19. Gumi B, Schelling E, Berg S, Firdessa R, Erenso G, et al. (2012) Zoonotic Transmission of Tuberculosis Between Pastoralists and Their Livestock in South-East Ethiopia. Ecohealth 9: 139–149.
20. WHO (2011) Global tuberculosis control: World Health Organisation report 2011. pp. 1–258.
21. Vordermeier HM, Chambers MA, Cockle PJ, Whelan AO, Simmons J, et al. (2002) Correlation of ESAT-6-specific gamma interferon production with pathology in cattle following *Mycobacterium bovis* BCG vaccination against experimental bovine tuberculosis. Infect Immun 70: 3026–3032.
22. Gallagher J, Horwill DM (1977) A selective oleic acid albumin agar medium for the cultivation of *Mycobacterium bovis*. J Hyg (Lond) 79: 155–160.
23. Brosch R, Gordon SV, Marmiesse M, Brodin P, Buchrieser C, et al. (2002) A new evolutionary scenario for the *Mycobacterium tuberculosis* complex. Proc Natl Acad Sci U S A 99: 3684–3689.
24. Kamerbeek J, Schouls L, Kolk A, van Agterveld M, van Soolingen D, et al. (1997) Simultaneous detection and strain differentiation of *Mycobacterium tuberculosis* for diagnosis and epidemiology. J Clin Microbiol 35: 907–914.
25. Allix-Beguec C, Supply P, Wanlin M, Bifani P, Fauville-Dufaux M (2008) Standardised PCR-based molecular epidemiology of tuberculosis. Eur Respir J 31: 1077–1084.
26. Allix-Beguec C, Fauville-Dufaux M, Supply P (2008) Three-year population-based evaluation of standardized mycobacterial interspersed repetitive-unit-variable-number tandem-repeat typing of *Mycobacterium tuberculosis*. J Clin Microbiol 46: 1398–1406.
27. Carmichael J (1939) Bovine tuberculosis in the tropics with special reference to Uganda, part 1. J Comp Pathol Therap 52: 322–335.
28. Ameni G, Aseffa A, Engers H, Young D, Hewinson G, et al. (2006) Cattle husbandry in Ethiopia is a predominant factor affecting the pathology of bovine tuberculosis and gamma interferon responses to mycobacterial antigens. Clin Vaccine Immunol 13: 1030–1036.
29. Thoen CO, Bloom BR (1995) Pathogenesis of *Mycobacterium bovis* In: Thoen CO, Steele JH, editors. *Mycobacterium bovis* Infection in Animals and Humans. First ed. Ames, Iowa: Iowa State University Press. pp. 3–14.

30. Phillips CJ, Foster CR, Morris PA, Teverson R (2003) The transmission of *Mycobacterium bovis* infection to cattle. Res Vet Sci 74: 1–15.
31. Liebana E, Johnson L, Gough J, Durr P, Jahans K, et al. (2008) Pathology of naturally occurring bovine tuberculosis in England and Wales. Vet J 176: 354–360.
32. Ameni G, Vordermeier M, Firdessa R, Aseffa A, Hewinson G, et al. (2011) *Mycobacterium tuberculosis* infection in grazing cattle in central Ethiopia. Vet J 188: 359–361.
33. Berg S, Garcia-Pelayo MC, Muller B, Hailu E, Asiimwe B, et al. (2011) African 2, a clonal complex of *Mycobacterium bovis* epidemiologically important in East Africa. J Bacteriol 193: 670–678.
34. Supply P, Warren RM, Banuls AL, Lesjean S, Van Der Spuy GD, et al. (2003) Linkage disequilibrium between minisatellite loci supports clonal evolution of *Mycobacterium tuberculosis* in a high tuberculosis incidence area. Mol Microbiol 47: 529–538.
35. Le Fleche P, Fabre M, Denoeud F, Koeck JL, Vergnaud G (2002) High resolution, on-line identification of strains from the *Mycobacterium tuberculosis* complex based on tandem repeat typing. BMC Microbiol 2: 37.
36. OIE (2009) Bovine tuberculosis. Terrestrial manual: Office International des Epizootie.
37. Tschopp R, Hattendorf J, Roth F, Choudhoury A, Shaw A, et al. (2012) Cost Estimate of Bovine Tuberculosis to Ethiopia. Curr Top Microbiol Immunol. DOI: 10.1007/82_2012_245.
38. Thoen CO, Steele JH, editors (1995) *Mycobacterium bovis* infection in animals and humans. 1 ed. Ames, Iowa: Iowa State University Press. 355 p.
39. SNV (2008) Dairy Investment Opportunities in Ethiopia. Addis Ababa: The Netherlands Development Organisation. pp. 1–48.
40. Abreha T (2006) Dairy production in Addis Ababa. Paper presented at a regional workshop organized by ASARECA, Kampala, Uganda.
41. Tegegne A, Sileshi Z, Tadesse M, Alemayhu M (2002) Scoping Study of Urban and Peri-urban Livestock Keepers in Addis Ababa: Part I. Literature Review on Dairy Production, Management and Marketing. ILRI and EARO: Addis Ababa.
42. Firdessa R, Berg S, Hailu E, Schelling E, Gumi B, et al. (2013) Mycobacterial Lineages Causing Pulmonary and Extrapulmonary TB in Ethiopia. Emerging Infectious Diseases In press.

5

Genetic Diversity in the Modern Horse Illustrated from Genome-Wide SNP Data

Jessica L. Petersen[1]*, James R. Mickelson[1], E. Gus Cothran[2], Lisa S. Andersson[3], Jeanette Axelsson[3], Ernie Bailey[4], Danika Bannasch[5], Matthew M. Binns[6], Alexandre S. Borges[7], Pieter Brama[8], Artur da Câmara Machado[9], Ottmar Distl[10], Michela Felicetti[11], Laura Fox-Clipsham[12], Kathryn T. Graves[4], Gérard Guérin[13], Bianca Haase[14], Telhisa Hasegawa[15], Karin Hemmann[16], Emmeline W. Hill[17], Tosso Leeb[18], Gabriella Lindgren[3], Hannes Lohi[16], Maria Susana Lopes[9], Beatrice A. McGivney[17], Sofia Mikko[3], Nicholas Orr[19], M. Cecilia T Penedo[5], Richard J. Piercy[20], Marja Raekallio[16], Stefan Rieder[21], Knut H. Røed[22], Maurizio Silvestrelli[11], June Swinburne[12,23], Teruaki Tozaki[24], Mark Vaudin[12], Claire M. Wade[14], Molly E. McCue[1]

1 University of Minnesota, College of Veterinary Medicine, St Paul, Minnesota, United States of America, **2** Texas A&M University, College of Veterinary Medicine and Biomedical Science, College Station, Texas, United States of America, **3** Swedish University of Agricultural Sciences, Department of Animal Breeding and Genetics, Uppsala, Sweden, **4** University of Kentucky, Department of Veterinary Science, Lexington, Kentucky, United States of America, **5** University of California Davis, School of Veterinary Medicine, Davis, California, United States of America, **6** Equine Analysis, Midway, Kentucky, United States of America, **7** University Estadual Paulista, Department of Veterinary Clinical Science, Botucatu-SP, Brazil, **8** University College Dublin, School of Veterinary Medicine, Dublin, Ireland, **9** University of Azores, Institute for Biotechnology and Bioengineering, Biotechnology Centre of Azores, Angra do Heroísmo, Portugal, **10** University of Veterinary Medicine Hannover, Institute for Animal Breeding and Genetics, Hannover, Germany, **11** University of Perugia, Faculty of Veterinary Medicine, Perugia, Italy, **12** Animal Health Trust, Lanwades Park, Newmarket, Suffolk, United Kingdom, **13** French National Institute for Agricultural Research-Animal Genetics and Integrative Biology Unit, Jouy en Josas, France, **14** University of Sydney, Veterinary Science, New South Wales, Australia, **15** Nihon Bioresource College, Koga, Ibaraki, Japan, **16** University of Helsinki, Faculty of Veterinary Medicine, Helsinki, Finland, **17** University College Dublin, College of Agriculture, Food Science and Veterinary Medicine, Belfield, Dublin, Ireland, **18** University of Bern, Institute of Genetics, Bern, Switzerland, **19** Institute of Cancer Research, Breakthrough Breast Cancer Research Centre, London, United Kingdom, **20** Royal Veterinary College, Comparative Neuromuscular Diseases Laboratory, London, United Kingdom, **21** Swiss National Stud Farm, Agroscope Liebefeld-Posieux Research Station, Avenches, Switzerland, **22** Norwegian School of Veterinary Science, Department of Basic Sciences and Aquatic Medicine, Oslo, Norway, **23** Animal DNA Diagnostics Ltd, Cambridge, United Kingdom, **24** Laboratory of Racing Chemistry, Department of Molecular Genetics, Utsunomiya, Tochigi, Japan

Abstract

Horses were domesticated from the Eurasian steppes 5,000–6,000 years ago. Since then, the use of horses for transportation, warfare, and agriculture, as well as selection for desired traits and fitness, has resulted in diverse populations distributed across the world, many of which have become or are in the process of becoming formally organized into closed, breeding populations (breeds). This report describes the use of a genome-wide set of autosomal SNPs and 814 horses from 36 breeds to provide the first detailed description of equine breed diversity. F_{ST} calculations, parsimony, and distance analysis demonstrated relationships among the breeds that largely reflect geographic origins and known breed histories. Low levels of population divergence were observed between breeds that are relatively early on in the process of breed development, and between those with high levels of within-breed diversity, whether due to large population size, ongoing outcrossing, or large within-breed phenotypic diversity. Populations with low within-breed diversity included those which have experienced population bottlenecks, have been under intense selective pressure, or are closed populations with long breed histories. These results provide new insights into the relationships among and the diversity within breeds of horses. In addition these results will facilitate future genome-wide association studies and investigations into genomic targets of selection.

Editor: Hans Ellegren, University of Uppsala, Sweden

Funding: This work was supported by National Research Initiative Competitive Grants 2008-35205-18766, 2009-55205-05254, and 2012-67015-19432 from the United States Department of Agriculture-National Institute of Food and Agriculture (USDA-NIFA); Foundation for the Advancement of the Tennessee Walking Show Horse and Tennessee Walking Horse Foundation; National Institutes of Health-National Institute of Arthritis and Musculoskeletal and Skin Diseases (NIH-NAIMS) grant 1K08AR055713-01A2 (MEM salary support), and 2T32AR007612 (JLP salary support), American Quarter Horse Foundation grant "Selective Breeding Practices in the American Quarter Horse: Impact on Health and Disease 2011–2012"; Morris Animal Foundation Grant D07EQ-500; The Swedish Research Council FORMAS (Contract 221-2009-1631 and 2008-617); The Swedish-Norwegian Foundation for Equine Research (Contract H0847211 and H0947256); The Carl Tryggers Stiftelse (Contract CTS 08:29); IBB-CBA-UAç was supported by FCT and DRCT and MSL by FRCT/2011/317/005; Science Foundation Ireland Award (04/Y11/B539) to EWH; Volkswagen Stiftung und Niedersächsisches Ministerium für Wissenschaft und Kultur, Germany (VWZN2012). The funders had no role in study design, data collection and analysis, decision to publish, or preparation of the manuscript.

Competing Interests: Matthew M. Binns is employed by Equine Analysis, June Swinburne by Animal DNA Diagnostics Ltd., and Emmeline W. Hill, Nickolas Orr, and Beatrice A. McGivney are associated with Equinome Ltd. There are no competing interests including patents, products in development, or marketed products to declare in relationship to this work.

* E-mail: jlpeters@umn.edu

Introduction

With a world-wide population greater than 58 million [1], and as many as 500 different breeds, horses are economically important and popular animals for agriculture, transportation, and recreation. The diversity of the modern horse has its roots in the process of domestication which began 5,000–6,000 years ago in the Eurasian Steppe [2–4]. Unlike other agricultural species such as sheep [5] and pigs [6,7], archaeological and genetic evidence suggests that multiple horse domestication events occurred across Eurasia [2,8–12]. During the domestication process, it is believed that gene flow continued between domesticated and wild horses [13] as is likely to also have been the case during domestication of cattle [14,15]. Concurrent gene flow between domestic and wild horses would be expected to allow newly domestic stock to maintain a larger extent of genetic diversity than if domestication occurred in one or few events with limited individuals.

Prior genetic work aimed at understanding horse domestication has shown that a significant proportion of the diversity observed in modern maternal lineages was present at the time of domestication [2,8,16]. The question of mitochondrial DNA (mtDNA) diversity was further addressed by recent sequencing of the entire mtDNA genome. These studies estimate that, minimally, 17 to 46 maternal lineages were used in the founding of the modern horse [2,17]; however, those data were unable to support prior studies suggesting geographic structure among maternal lineages [9,18]. Recent nuclear DNA analyses have utilized "non-breed" horses sampled across Eurasia to attempt to understand the population history of the horse. These microsatellite-based studies suggest a weak pattern of isolation by distance with higher levels of diversity in, and population expansion originating from Eastern Asia [13,19]. High diversity as observed by both mtDNA and microsatellites and the absence of strong geographical patterns is likely a result of continued gene flow during domestication, the high mobility of the horse, and its prevalent use for transportation during and after the time of domestication. Interestingly, while significant diversity is observed in maternal lineages, paternal input into modern horse breeds appears to have been extremely limited as shown by a lack of variation at the Y-chromosome [20,21].

Diversity in the founding population of the domestic horse has since been exploited to develop a wealth of specialized populations or breeds. While some breeds have been experiencing artificial selection for hundreds of years (e.g. Thoroughbred, Arabian), in general, most modern horse breeds have been developed recently (e.g. Quarter Horse, Paint, Tennessee Walking Horse) and continue to evolve based upon selective pressures for performance and phenotype (Table 1). Horse breeds resulting from these evolutionary processes are generally closed populations consisting of individual animals demonstrating specific phenotypes and/or bloodlines. Each breed is governed by an independent set of regulations dictated by the respective breed association. Not all breeds are closed populations. Some breed registries allow admixture from outside breeds (e.g. Swiss Warmblood, Quarter Horse), and others are defined by phenotype (e.g. Miniature). Finally, some populations that are often referred to as breeds are classified simply by their geographic region of origin and may not be actively maintained by a formal registry (e.g. Mongolian, Tuva) (Table 1). Those breeds that may be free ranging and experience lesser degrees of management may more appropriately be termed "landrace populations." Therefore, genetic characteristics within horse breeds are expected to differ based upon differences in the definition of the breed, the diversity of founding stock, the time since breed establishment, and the selective pressures invoked by breeders. The extent of gene flow not only varies within breed, but among horse breeds, the direction and level of gene flow is influenced by breed restrictions/requirements, and potentially by geographic distance.

Considering modern breeds, unlike mtDNA, nuclear markers can discern breed membership [12]. However, studies of nuclear genetic diversity of modern breeds to date have most commonly focused on a single population of interest, sets of historically related breeds, or breeds within a specific geographic region [22–36]. Additionally, these analyses of nuclear genetic diversity in horse breeds are largely based upon microsatellite loci, which do not often permit consolidation of data across studies. Thus, large, across-breed investigations of nuclear diversity in the modern, domestic horse are lacking.

The Equine Genetic Diversity Consortium (EGDC), an international collaboration of the equine scientific community, was established in an effort to quantify nuclear diversity and the relationships within and among horse populations on a genome-wide scale. The development of this consortium has facilitated the collection of samples from 36 breeds for genotyping on the Illumina 50K SNP Beadchip. The breeds included in this report represent many of the most popular breeds in the world as well as divergent phenotypic classes, different geographic regions of derivation, and varying histories of breed origin (Table 1). The standardized SNP genotyping platform permits the compilation of data across breeds at a level never before achieved. Results of this collaboration now allow for the detailed description of diversity and assessment of the effects of genetic isolation, inbreeding, and selection within breeds, and the description of relationships among breeds. These data will also facilitate future across breed genome-wide association studies as well as investigations into genomic targets of selection.

Results

Samples

Of the 38 populations sampled, two breeds were represented by geographically distinct populations: the Thoroughbred was sampled in the both the United States (US) and the United Kingdom and Ireland (UK/Ire), and the Standardbred was sampled in the US as well as in Norway. Eight Standardbred horses sampled from the US were noted to be pacing horses as opposed to the Norwegian and remaining US individuals that were classified as trotters. In addition, the International Andalusian and Lusitano Horse Association Registry (IALHA) in the US maintains one stud book but designates whether the individual was derived from Spanish (Pura Raza Española) or Portuguese (Lusitano) bloodlines, or a combination of both. Of the Andalusian horses collected in the US, five were noted to have Portuguese bloodlines.

Phenotypic classifications of the horse breeds include those characterized by small stature (Miniature Horse, pony breeds), breeds characterized by large stature and/or large muscle mass in proportion to size (draft breeds), light horse or riding breeds, gaited breeds, rare breeds, breeds founded in the past 80 years, and populations that are relatively unmanaged ("landrace"). The number of samples, sampling location, region of breed origin, and a list of primary breed characteristics are found in Table 1.

After pruning of individuals for genotyping quality and relationships (see methods), and keeping a similar number of individuals per breed, 814 of the 1,060 horses remained in the analysis. Of the horses removed, 12 had known pedigree relationships at or more recent to the grandsire/dam level, 44

Table 1. Populations (breeds) included in the study, region of breed origin and sampling location, notes on population history relevant to diversity statistics, and breed classification based upon use and phenotype.

Breed	Geographic Origin	Region Sampled	Population size (approx)	Population Notes	Classification(s)
Akhal Teke	Turkmenistan	US & Russia	3,500	Pedigree records began-1885, Stud book-1941	Riding horse, endurance
Andalusian	Spain	United States	185,000	US registry formed in 1995 including Pura Raza Española & Lusitano bloodlines	Riding horse, sport
Arabian	Middle East	United States	1 million	Arabian type bred for over 3,500 years; US stud book-1908	Riding horse, endurance
Belgian	Belgium	United States	common	US Association began-1887	Draft
Caspian	Persia	United States	rare	Rediscovered in 1965 with N~50, no breeding records prior; Stud book-1966	Riding and driving pony
Clydesdale	Scotland	US & UK	5,000	Registry formed-1877 in Scotland; Stud book-1879	Draft
Exmoor	Great Britain	United Kingdom	2,000	Exmoor Pony Society-1921	Riding and driving pony
Fell Pony	England	United Kingdom	6,000	Fell Pony Society began in 1922; outcrossed with Dale's pony until 1970s	Light draft pony
Finnhorse	Finland	Finland	19,800	Stud book-1907	Light draft; riding horse; trotting
Florida Cracker	United States	United States	rare	Introduced to US in 1500s; association began-1989 with 31 horses	Riding horse, gaited
Franches-Montagnes	Switzerland	Switzerland	21,000	Official stud book-1921; Current breeding association established-1997	Light draft, riding horse
French Trotter	France	France	common	Population closed-1937 although allows some Standardbred influence	Riding horse, trotting
Hanoverian	Germany	Germany	20,000 (Germany)	Outcrossing allowed	Riding horse
Icelandic	Iceland	Sweden	180,000	Isolated >1,000 years; Federation of Icelandic Horse Association began-1969	Riding horse, gaited
Lusitano	Portugal	Portugal	12,000	Stud book-1967 after split from Spanish Andalusian breed	Riding horse, sport
Mangalarga Paulista	Brazil	Brazil	common	Registry began-1934	Riding horse
Maremmano	Italy	Italy	7,000	Breed identification based upon conformation and inspection	Riding horse
Miniature	United States	United States	185,000	Two US registries founded in 1970s; Maximum height restrictions for registration	Driving pony, extreme small size
Mongolian	Mongolia	Mongolia	2 million	Many types based upon purpose and geography	Riding horse, landrace
Morgan	United States	United States	100,000	Founding sire born in 1789; Registry-1894	Riding and driving horse
New Forest Pony	England	United Kingdom	15,000	Stud book-1910 with a variety of sires; No outcrossing since 1930s	Light draft, riding pony, landrace
North Swedish Horse	Sweden	Sweden	10,000	Breed association-1894; Stud book-1915	Draft
Norwegian Fjord	Norway	Norway	common	Stud book-1909	Riding and light draft
Paint	United States	United States	1 million	Registry-1965; One parent can be Quarter Horse or Thoroughbred	Riding horse, stock horse
Percheron	France	United States	20,000	Stud book-1893	Draft
Peruvian Paso	Peru	United States	25,000	Breed type over 400 years old; Closed population	Riding horse, gaited
Puerto Rican Paso Fino	Puerto Rico	Puerto Rico	250,000	Breed type ~500 years old; Association founded-1972	Riding horse, gaited
Quarter Horse	United States	United States	4 million	Association formed-1940; One parent may be Paint or Thoroughbred	Riding horse, stock horse, racing
Saddlebred	United States	United States	75,000	Breed type founded in late 1700s; Association began-1891	Riding and driving horse, some gaited
Shetland	Scotland	Sweden	common	Stud book-1891	Riding pony
Shire	England	United States	7,000	1st Shire organization-1877 (UK); stud book-1880; US assoc-1885	Draft
Standardbred	United States	Norway	common	Stud book-1871; Some outside trotting bloodlines (French Trotter) allowed	Riding horse, harness racing (trot)

Table 1. Cont.

Breed	Geographic Origin	Region Sampled	Population size (approx)	Population Notes	Classification(s)
Standardbred	United States	United States		Stud book-1871; Harness racing in early 1800s included pacing horses	Riding horse, harness racing (trot or pace)
Swiss Warmblood	Switzerland	Switzerland	15,000	Stud book-1921; Crossed with European Warmbloods, Thoroughbreds, Arabians	Riding horse, sport
Tenn Walking Horse	United States	United States	500,000	Registry-1935; Blood typing and parentage verification mandated in 1993	Riding horse, gaited
Thoroughbred	England	UK & Ireland	common	Stud book-1791; Closed population	Race horse, riding horse, sport
Thoroughbred	England	United States			Race horse, riding horse, sport
Tuva	Siberia	Russia	30,000	Different types depending on region	Light draft, landrace

individuals were removed at random from overrepresented breeds to equalize sample size across breeds, 4 failed to genotype at a rate greater than 0.90, and 186 were removed due to pi hat values (pairwise estimates of identity by descent) above the allowed threshold. Of those last 186 horses that were removed, 122 were from disease studies where relationships were common due to sampling bias.

Within Breed Diversity

Diversity indices were calculated using 10,536 autosomal SNPs that remained after pruning for minor allele frequency (MAF), genotyping rate, and linkage disequilibrium (LD) across breeds (referred to as the primary SNP set). Diversity indices were also calculated using three other SNP sets, resulting from different levels of LD-based pruning (see methods). Individuals noted as outliers in parsimony and cluster analyses (see below) were excluded from within-breed diversity calculations.

Using the primary SNP set, diversity, as measured by expected heterozygosity (H_e), ranged from 0.232 in the Clydesdale, to 0.311 in the Tuva (Table 2). Considering the SNP sets pruned less stringently for LD, the diversity within the Thoroughbred increased in relationship to the other breeds, as did that of nine other breeds. Mean and total heterozygosity increased with increased number of loci and less stringent LD pruning (Table 2). Inbreeding coefficients (F_{IS}) calculated on the primary SNP set showed significant excess homozygosity in 17 populations, which was greatest in the Andalusian (0.065). Three of the four lowest F_{IS} values were found in the Thoroughbred samples (UK/Ire, US, and when considered together) (Table 2).

Inbreeding coefficients (f) calculated for each individual based upon observed and expected heterozygosity showed several individuals with significant loss of heterozygosity. The highest individual value of f (0.56) was found in an Exmoor pony. Within breeds, average individual estimates of f were greatest in the Clydesdale, Mangalarga Paulista, and Exmoor while the lowest breed means were found in the landrace populations (Table 2).

Effective population size (N_e), as estimated by LD [37] using an autosomal SNP set pruned within each breed for quality, was lowest (143) in the UK/Ire sample of the Thoroughbred (UK/Ire) but also low in the other racing breeds as well as the Clydesdale (Table 2). Highest values of N_e were observed in the Eurasian landrace populations, the Mongolian (743) and Tuva (533), and also in the Icelandic (555), Finnhorse (575), and Miniature (521). Breed-specific decay of LD essentially mirrors the results of the N_e calculation given the relationship between the statistics. A plot of LD across 2 Mb in a subset of the breeds that represent the range of N_e estimates is found in Figure S1.

Parsimony and Principal Component Analyses

With a domestic ass designated as the outgroup, parsimony analysis of 10,066 loci pruned for LD of $R^2 = 0.2$ (see methods) resulted in generally tight clustering and monophyly of samples within breeds, supported by high bootstrap values (Figure 1). Major clades of the tree show grouping of the Iberian breeds (Lusitano and Andalusian), ponies (Icelandic, Shetland, Miniature), Scandinavian breeds (Finnhorse, North Swedish Horse, Norwegian Fjord), heavy draft horses (Clydesdale, Shire, Belgian, Percheron), breeds recently admixed with and/or partly derived from the Thoroughbred (Paint, Quarter Horse, Maremmano, Swiss Warmblood, Hanoverian), modern US breeds (American Saddlebred (hereafter "Saddlebred" and Tennessee Walking Horse), trotting breeds (Standardbred and French Trotter), and Middle Eastern breeds (Akhal Teke and Arabian). Exceptions to monophyly include the Paint and Quarter Horse as well as the Hanoverian and Swiss Warmblood, which are mixed in clades surrounding the Thoroughbred and Maremmano. In addition, the Clydesdale was placed as a clade within the Shire breed and the Shetlands as a clade within the Miniatures. Strong bootstrap support for monophyly is present within a subset each of Lusitanos (83%), and Andalusians (87%); however the remainder of individuals from these breeds were intermixed. No structure was found within the US sample regarding individual Andalusians noted to have Portuguese bloodlines opposed to those with Spanish bloodlines (Figure S2). The Mongolian and most Tuva horses were grouped together while a subset of the Tuvas fell out as a sister clade to the Caspians. Several individuals were not positioned in the clades that represented the majority of the other individuals in the breed (Figure 1). These include three Shires, two Mongolians, a Caspian, and a Norwegian Fjord. In each instance, the outlier status of these individuals was also supported by cluster analysis (see below).

Principal component analysis (PCA) also serves to visualize individual relationships within and among breeds. The plot of PC1 vs. PC2 shown in Figure S3 illustrates relationships similar to those shown by parsimony, including the placement of outliers outside of their respective breeds. All Thoroughbred samples, regardless of origin, are separated from the others by PC1 and form a cluster at the top of the figure. Intermediate between the Thoroughbred and central cluster of breeds are the Hanoverian, Swiss Warmblood, Paint, and Quarter Horse. The Shetland, Icelandic, and Miniature split from the remainder of samples in PC2, falling out in the lower

Table 2. Number of samples (N), effective population size (N_e), individual inbreeding estimates (f), inbreeding coefficient (F_{IS}), and expected heterozygosity (H_e) from four SNP sets pruned based upon varying levels of LD.

				Individual inbreeding (f)			Expected Heterozygosity (He)			
							r2 0.1	R2 0.1	r2 0.2	r2 0.4
Breed	**N**	**Ne**	**FIS**	**Min**	**Max**	**Mean**	**10,536**	**6,028**	**18,539**	**26,171**
Akhal Teke	19	302	0.015*	0.015	0.297	0.101	0.287	0.281	0.303	0.311
Andalusian	18[a]	329	0.065*	0.028	0.274	0.114	0.296	0.293	0.308	0.312
Arabian	24[a]	346	0.033*	0.060	0.060	0.060	0.287	0.280	0.302	0.310
Belgian	30[b]	431	−0.002	0.039	0.166	0.111	0.278	0.276	0.284	0.284
Caspian	18	351	−0.022	−0.033	0.136	0.041	0.294	0.292	0.305	0.308
Clydesdale	24	194	0.004	0.128	0.323	0.261	0.232	0.225	0.238	0.236
Exmoor	24	216	0.034*	0.055	0.556	0.239	0.247	0.242	0.253	0.252
Fell Pony	21	289	0.002	0.069	0.178	0.114	0.278	0.272	0.285	0.285
Finnhorse	27	575	−0.004	0.011	0.100	0.052	0.296	0.296	0.302	0.301
Florida Cracker	7	171	0.026*	0.004	0.359	0.159	0.270	0.263	0.284	0.291
Franches-Montagnes	19[a]	316	0.003	0.018	0.203	0.095	0.284	0.279	0.297	0.301
French Trotter	17[a]	233	−0.018	0.064	0.173	0.105	0.275	0.262	0.295	0.307
Hanoverian	15[a]	269	−0.010	0.002	0.087	0.052	0.294	0.280	0.320	0.335
Icelandic	25[c]	555	0.006*	0.043	0.234	0.083	0.289	0.288	0.290	0.288
Lusitano	24	391	0.039*	0.008	0.220	0.090	0.296	0.292	0.309	0.315
Maremmano	24	341	−0.012	−0.015	0.109	0.038	0.298	0.287	0.318	0.329
Miniature	21	521	0.005	0.043	0.161	0.075	0.291	0.292	0.296	0.295
Mangalarga Paulista	15	155	−0.011	0.176	0.320	0.242	0.235	0.228	0.246	0.250
Mongolian	19[a]	751	0.001	−0.034	0.055	0.015	0.309	0.308	0.314	0.314
Morgan	40	448	0.040*	0.003	0.307	0.090	0.296	0.287	0.310	0.317
New Forest Pony	15	474	0.000	−0.022	0.066	0.025	0.304	0.300	0.316	0.319
Norwegian Fjord	21[a]	335	−0.003	0.053	0.168	0.122	0.274	0.274	0.278	0.277
North Swedish Horse	19	369	0.011*	0.069	0.210	0.133	0.275	0.276	0.279	0.278
Percheron	23	451	0.003	0.043	0.143	0.086	0.287	0.284	0.292	0.293
Peruvian Paso	21	433	0.002	0.008	0.134	0.055	0.298	0.293	0.306	0.310
Puerto Rican Paso Fino	20	321	−0.003	0.004	0.298	0.103	0.280	0.278	0.287	0.290
Paint	25	399	0.006*	−0.013	0.101	0.040	0.302	0.289	0.324	0.337
Quarter Horse	40[a]	426	0.011*	−0.012	0.144	0.047	0.302	0.290	0.323	0.336
Saddlebred	25[d]	297	−0.008	0.051	0.145	0.103	0.279	0.268	0.297	0.306
Shetland	27	365	0.032*	0.108	0.370	0.182	0.264	0.268	0.268	0.266
Shire	23	357	0.024*	0.130	0.258	0.187	0.261	0.252	0.268	0.267
Standardbred - Norway	25[e]	232	−0.004	0.063	0.202	0.130	0.272	0.255	0.289	0.298
Standardbred - US	15	179	0.039*	0.097	0.222	0.153	0.276	0.262	0.293	0.303
Standardbred - all	40	290	0.022*	−0.028	0.323	0.130	0.276	0.260	0.293	0.303
Swiss Warmblood	15[a]	271	0.005	0.023	0.117	0.059	0.296	0.281	0.322	0.337
Thoroughbred - UK/Ire	19[a]	143	−0.028	0.089	0.171	0.133	0.264	0.245	0.292	0.309
Thoroughbred - US	17[a]	163	−0.015	0.093	0.182	0.134	0.267	0.250	0.295	0.313
Thoroughbred - all	36	190	−0.019	0.089	0.182	0.134	0.266	0.248	0.294	0.312
Tuva	15	533	0.016*	−0.028	0.116	0.022	0.311	0.309	0.320	0.322
Tennessee Walking Horse	19	230	0.008*	0.065	0.276	0.148	0.269	0.256	0.284	0.291
Mean	22.3	341	0.007	0.039	0.204	0.107	0.282	0.275	0.295	0.300
Total	814						0.313	0.303	0.329	0.336
Min			−0.028	−0.034	0.055	0.015	0.232	0.225	0.238	0.236
Max			0.005	0.176	0.556	0.261	0.311	0.309	0.324	0.337

[a]Individuals from this breed also included in [41];
[b]20 of these individuals were also reported in [41];
[c]17 of these individuals were also reported in [41];
[d]21 of these individuals were also reported in [41];
[e]19 of these individuals were also reported in [41].
F_{IS} and f were calculated based upon the primary SNP set (10,536 loci). Samples also used in [41] are indicated in the footnotes.
*indicates significance at $\alpha<0.05$ determined by 10,000 permutations.

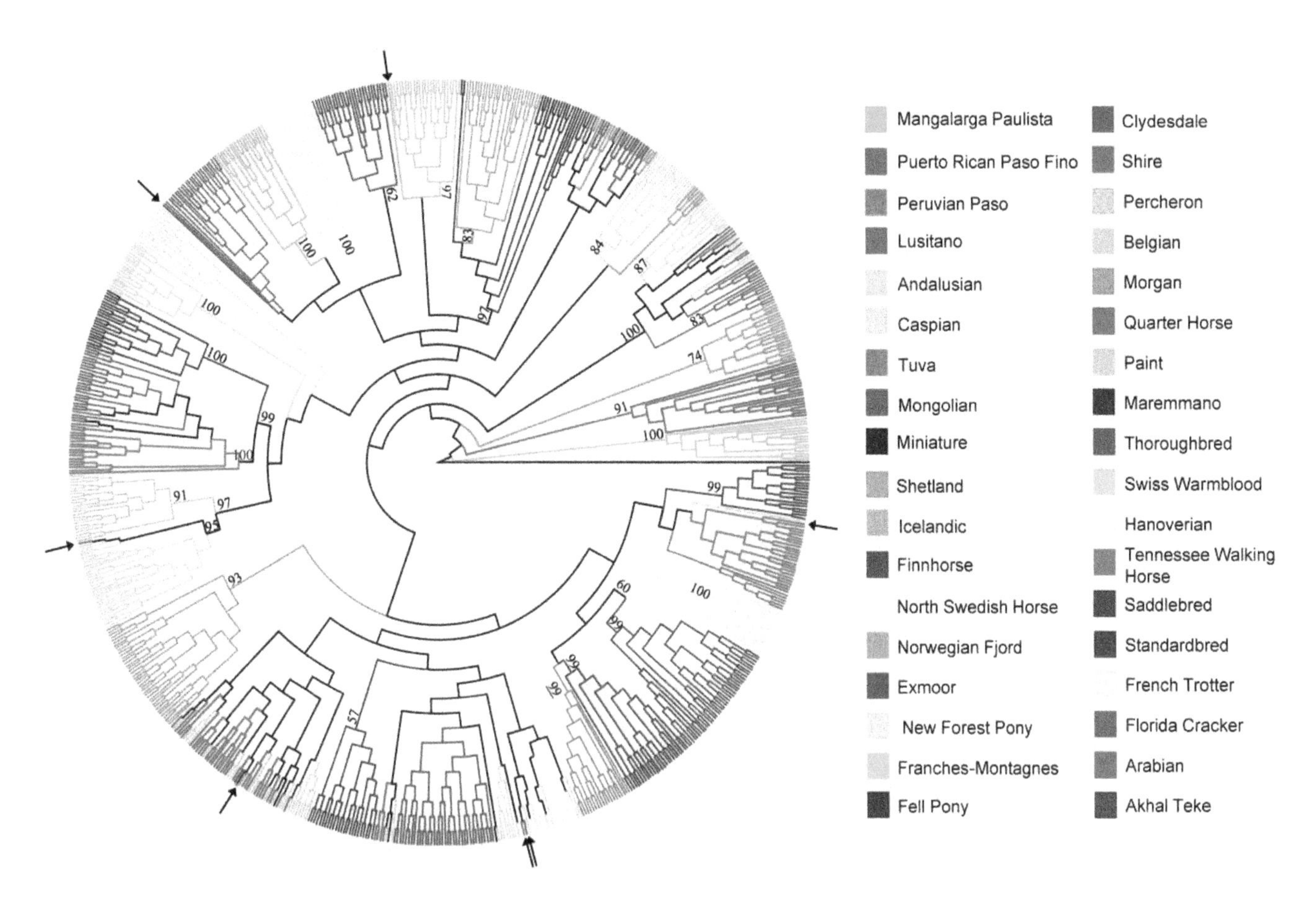

Figure 1. Individual and breed relationships among 814 horses illustrated by parsimony. Parsimony tree created from 10,066 SNPs and rooted by the domestic ass. Breeds are listed in the legend in order starting from the root and working counterclockwise. Individual outliers with respect to their breeds are noted with arrows. Bootstrap support calculated from 1,000 replicates is shown for major branches when greater than 50%.

left corner, and the British drafts anchor the figure at the lower right. While most breeds cluster tightly, several are dispersed across one or both PCs. The Hanoverian, Swiss Warmblood, Paint, and Quarter Horse, as noted above, are extended along PC1, while the Arabian and Franches-Montagnes show similar spreading, also along PC1. The Tuva, Clydesdale, and Shire individuals also are not as tightly clustered as other populations despite the low within breed diversity of the latter two.

Distance Analysis

An unrooted neighbor joining (NJ) tree of Nei's distance [38] was constructed using SNP frequencies within breeds from the 10,536 SNP data set (Figure 2). The relative placement of breeds reflects that seen in the parsimony tree with several exceptions. The Paint, Quarter Horse, Swiss Warmblood, Hanoverian, Maremmano, and Thoroughbred, are found in one large branch of the tree, although the Maremmano is placed outside of the clade containing the aforementioned breeds. The position of the Morgan with the Saddlebred and Tennessee Walking Horse also deviates from parsimony analysis but reflects historic records of relationships among these breeds. The Scandinavian breeds remain in one branch of the clade, which also includes the Shetland and Miniature. Unlike the parsimony cladogram, the Caspian falls in a clade with the other Middle Eastern breeds, the Arabian and Akhal Teke. Finally, the Exmoor, a British breed, is placed with another British breed, the New Forest Pony, rather than with the Scandinavian breeds as in the parsimony analysis. Each branch shows support of over 50%, with many clades being supported by over 99% of the 1,000 bootstrap replicates.

Cluster analysis

Likelihood scores for runs of various K in Structure showed an increase in overall mean ln P(X|K) until K = 35 (Figure S4). A clear "true" value of K is not obvious examining the likelihood scores or using the Evanno method [39] (data not shown); however, variance among runs begins to increase with a diminishing increase in likelihood scores after K = 29, which is near the peak of the curve. The value of the highest proportion (breed average q-value) of assignment of each breed for each value of K, as well as the cluster to which it assigns is shown in Table S1. Additionally, the proportion assignment at K = 29 for each of the breeds is found in Table S2.

The first breeds to have all individuals assign strongly to one cluster are the Thoroughbred and Clydesdale (with Shire) at K = 2, followed by the Shetland at K = 3; these four breeds do not show signs of admixture at any K value analyzed. Evidence of weak geographic grouping is observed at K = 4, which consists of: 1, the Middle Eastern and Iberian breeds (pink); 2, the Thoroughbred and breeds to which it continues to be or was historically crossed (yellow); 3, breeds developed in Scandinavia and Northern Europe (orange); and 4, the British Isles draft breeds (blue) (Figure 3).

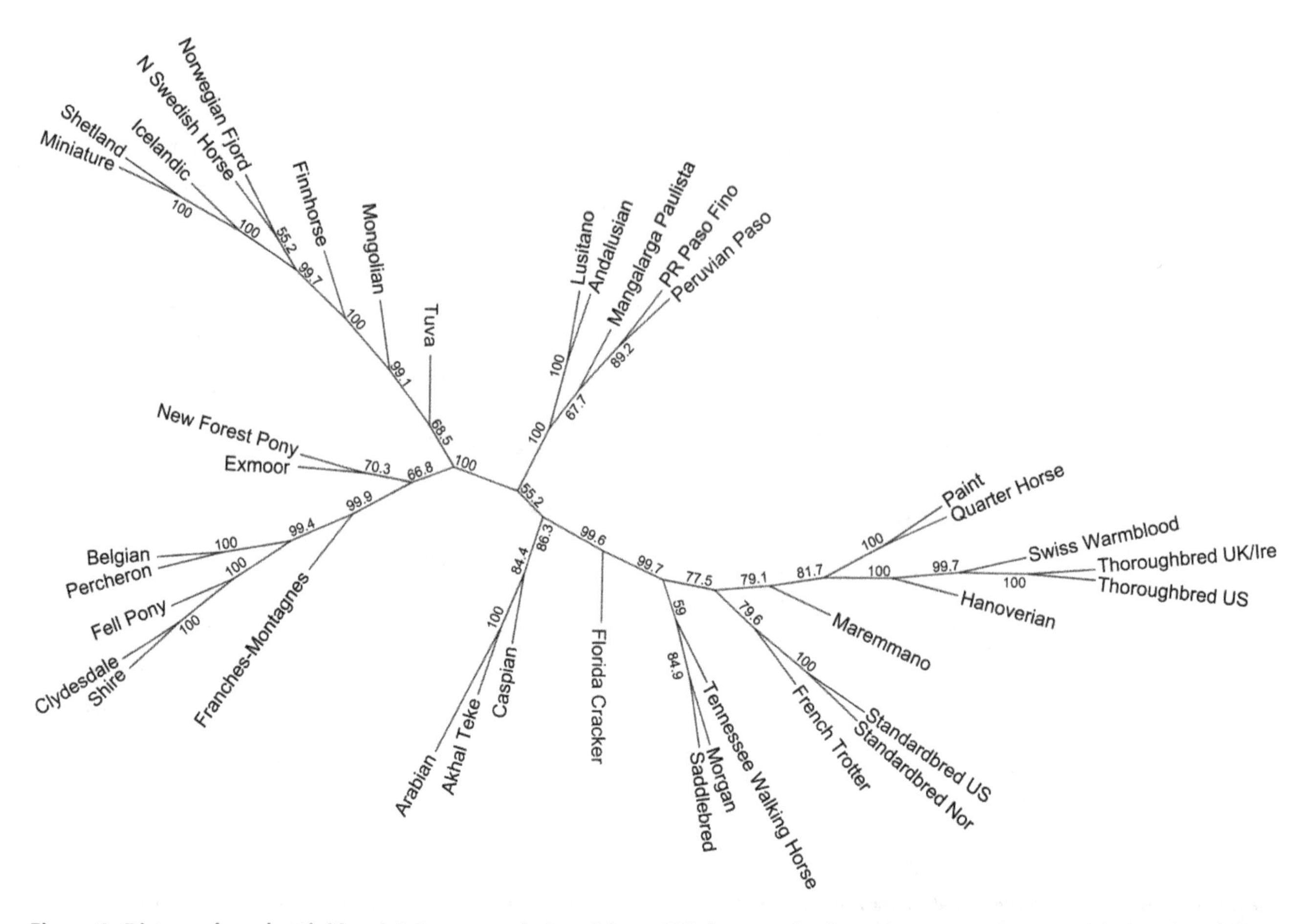

Figure 2. Distance based, neighbor joining tree calculated from SNP frequencies in 38 horse populations. Majority rule, neighbor joining tree created from 10,536 SNP makers using Nei's genetic distance and allele frequencies within each population. Percent bootstrap support for all branches calculated from 1,000 replicates is shown.

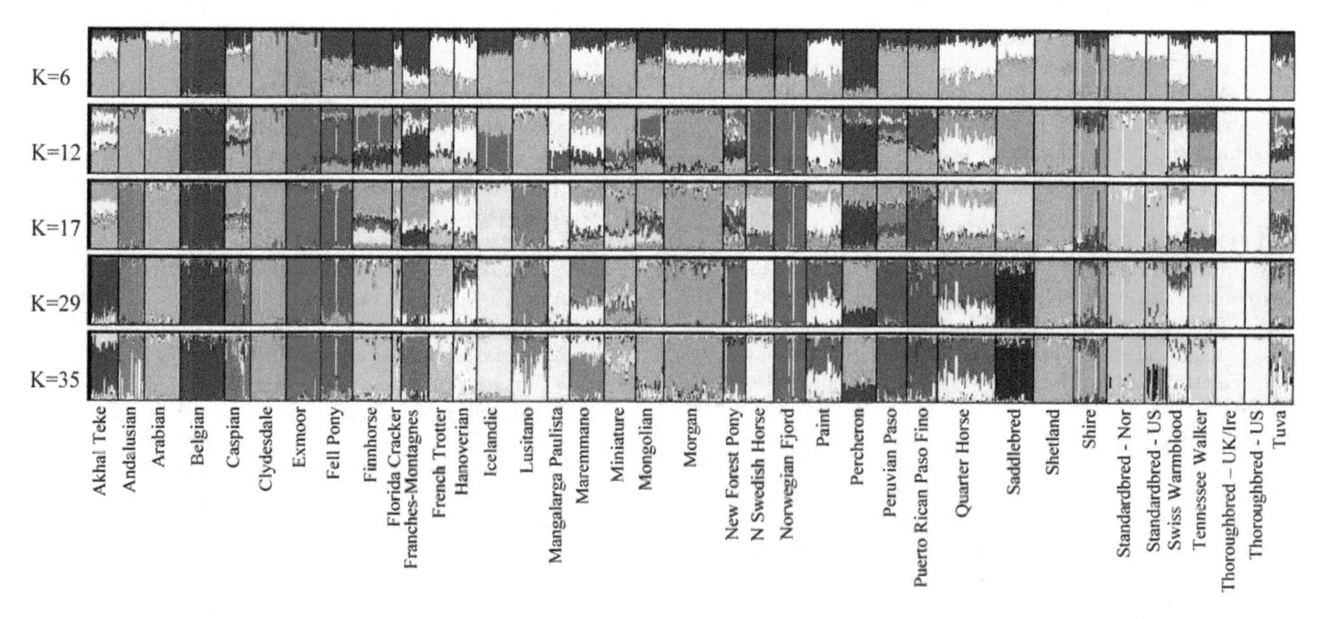

Figure 3. Bayesian clustering output for five values of K in 814 horses of 38 populations. Structure output for five values of K investigated. Each individual is represented by one vertical line with the proportion of assignment to each cluster shown on the y axis and colored by cluster. Other values of K are shown in Figure S1 and a summary of assignment of each breed in Tables S1 and S2.

Middle Eastern and Iberian Breeds

As also observed in the NJ tree, clustering of the Iberian and Middle Eastern breeds with the Mangalarga Paulista, Peruvian Paso, and Puerto Rican Paso Fino ($q > 0.5$) is observed until $K = 8$, after which point the Mangalarga Paulista assigns with $q = 0.93$ to another cluster. The remaining breeds cluster together until $K = 12$, at which time the Middle Eastern breeds (Arabian, Akhal Teke, and Caspian) are assigned to their own cluster, leaving the Iberian breeds clustered with the Peruvian Paso and Puerto Rican Paso Fino. At low values of K (*i.e.* $K < 6$) the Florida Cracker, Saddlebred, Standardbreds, Morgan, and Tennessee Walking Horse fall into the cluster with the Iberian and Middle Eastern breeds with breed mean $q > 0.5$. At $K = 29$, each of these breeds is assigned with $q > 0.72$ to an individual cluster with the exception of the Lusitano and Andalusian, which remaining clustered together.

Thoroughbreds and Thoroughbred Crossed Breeds

Relationships described by the NJ tree among the Thoroughbred, Hanoverian, Swiss Warmblood, Paint, Quarter Horse, and Maremmano are also seen in cluster analysis. Clustering of those breeds with the Thoroughbred is observed throughout the values of K examined although at moderate frequencies (Figure 3, Table S1, Figure S5). At $K = 29$, the Hanoverian and Swiss Warmblood remain assigned to the cluster defined by the Thoroughbred but with assignment probabilities of 0.51 each. The Quarter Horse and Paint also assign to this cluster with q-values of 0.30 and 0.34, respectively. Neither the Quarter Horse, Paint, Hanoverian, or Swiss Warmblood populations assign to any cluster with $q > 0.62$ at $K = 29$. No evidence of population substructure is observed between the US and UK/Ire Thoroughbreds as also shown by PCA and parsimony analyses (Figure S6).

Scandinavian and Northern European Breeds

As in the NJ and parsimony trees, the Finnhorse, Icelandic, Miniature, North Swedish Horse, Norwegian Fjord, and Shetland are parsed into the same cluster (q-value >0.5) through $K = 5$. However, unlike the NJ tree, at $K = 4$, the highest value of assignment places the Belgian and Percheron into this cluster although with $q < 0.5$ (0.42 and 0.38, respectively). The relationship remains until $K = 6$, at which time the Miniature, Icelandic, and Shetland fall into a different cluster. At $K = 10$, the Icelandic clusters again with the North Swedish Horse and Norwegian Fjord. The Norwegian and United States Standardbred populations, which at $K = 4$ assign with $q > 0.5$ to the cluster containing the Scandinavian breeds, separate from the Scandinavian breeds at $K = 5$. At $K = 31$, substructure appears in the Standardbred samples, which correlates to those individuals identified as pacers and that fall into an individual clade in the parsimony tree (Figure S7). At $K = 29$, the Miniature and Shetland continue to be assigned to the same cluster (q-values $= 0.55$ and 0.95, respectively). The next highest proportions of assignment of the Miniature horse are to the clusters described by the New Forest Pony ($q = 0.20$) and Icelandic ($q = 0.11$). No value of K evaluated eliminated signals of admixture from all populations in the dataset at $K = 38$ (the actual number of populations sampled) or any value of K through 45 (data not shown).

British Isles Draft

The Clydesdale and Shire cluster together, and apart from the other breeds beginning at $K = 3$. In addition, the Fell Pony, which is placed within the same clade in the NJ and parsimony trees, and proximal to the Clydesdale and Shire in PCA, shows moderate assignment to this cluster ($0.29 < q < 0.41$) for several values of K from 4 to 14. At $K = 29$, the Shire assigns to the same cluster as the Clydesdale with $q = 0.69$. The individual outliers from the Shire breed also noted in parsimony analysis are evident beginning at $K = 3$. Excluding these outliers, at $K = 29$, the proportion of assignment for the Shires to the cluster with the Clydesdale increases to 0.74.

F_{ST}

All pairwise F_{ST} values calculated between the 37 populations (excluding the Florida Cracker) were significant as tested by 20,000 permutations (Figure 4). The lowest level of differentiation was found between the Paint and Quarter Horse populations ($F_{ST} = 0.002$), while the greatest divergence was observed between the Clydesdale and Mangalarga Paulista ($F_{ST} = 0.254$). The two Thoroughbred populations had an F_{ST} value of 0.004, while the two Standardbred populations had 10-fold greater divergence ($F_{ST} = 0.020$) than the minimum observed value in this dataset; this value is similar to that observed between the Lusitano and Andalusian (0.021). An F_{ST} value of 0.006 was identified between the Tuva and Mongolian populations. The global F_{ST} value was 0.100. AMOVA computed on the set of 37 samples (excluding the outliers identified in Structure and the Florida Cracker) showed that 10.03% of the variance was accounted for among populations ($p = < 0.001$), 0.53% of the variance was among individuals within populations ($p = 0.19$), and 89.44% of the variation was within individuals ($p < 0.001$).

Discussion

These data are gathered from populations that represent tremendous diversity in phenotype and breed specialization. With breeds sampled across four continents, the resulting relationships observed largely reflect similarities of geographic origin, documented breed histories, and shared phenotypes. In general the highest within breed diversity was observed in breeds that are recently derived, continue to allow introgression of other populations, those that have a large census population size, and landrace populations that experience a lesser degree of controlled breeding. Not surprisingly, low diversity is observed in breeds with small census size, relatively old breeds with closed populations, and those with documented founder effects, whether due to population bottlenecks or selective breeding.

A total of seven individuals were identified by parsimony and cluster analysis as outliers with respect to the breed to which they were assigned. The pedigrees of these individuals were unknown. Because it is possible these horses were unknowingly crossbred or subject to mishandling in the field or laboratory, they were excluded from the within-breed analyses to avoid potential bias in indices of diversity. In addition, the potential impact of SNP ascertainment bias on diversity calculations must be acknowledged. The reference genome is from a Thoroughbred mare [40] and SNP identification was based upon the reference genome and data from seven other horses representing six breeds. Therefore, SNPs are generally derived to identify modern variation within the Thoroughbred as well as between the Thoroughbred and these other breeds. Thus, the SNPs identified may reflect an upward bias in diversity indices in the Thoroughbred and closely related breeds [41]. It seems that ascertainment bias may have particularly influenced the results when considering the data sets that have an increased number of loci resulting from relaxed LD pruning. These results show an increase in the relative diversity of the Thoroughbred, breeds with which the Thoroughbred contin-

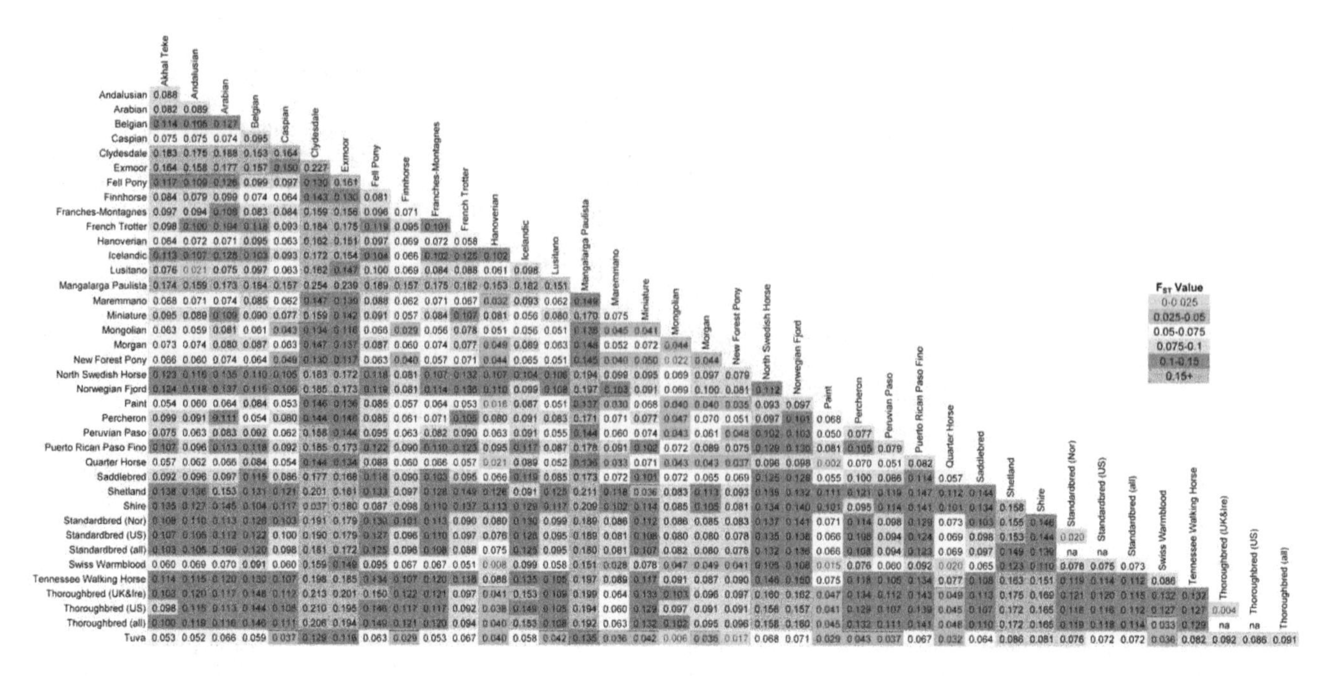

Figure 4. Pairwise F_{ST} values based upon 10,536 SNPs in 37 horse populations. Pairwise F_{ST} values as calculated in Arlequin using 10,536 autosomal SNPs and significance tested using 20,000 permutations. All pairwise values are significantly different from zero. (individual outliers were removed from this analysis).

ues to actively interbreed, and the other SNP discovery breeds, with respect to other breeds in the study. This is opposite of what may be expected given the high levels of genome-wide LD in the Thoroughbred. Without considering SNP ascertainment bias, it is expected that measured diversity would increase in breeds with low LD more quickly than in those with high LD, due to greater independence of markers in the former breeds. These SNPs, derived largely from the Thoroughbred, are apparently detecting a higher proportion of Thoroughbred-specific, rare variants and it appears that as more loci are included, more of these Thoroughbred-based variants are assayed, resulting in the observed increase in variation in the Thoroughbred, Thoroughbred-influenced breeds, and breeds used in SNP ascertainment.

The majority of the analyses were performed using 10,536 SNP markers pruned across breeds for LD of $r^2>0.2$ as well as MAF of 0.05 or above. Even though additional markers could have been used for analysis, many population-level statistics assume independence of loci. The stringent pruning for LD was therefore undertaken to eliminate bias in the test statistics that may result from substantial breed-specific differences in LD [40,41]. A truncated data set also helped to make calculations, especially cluster analysis, computationally feasible. On the other hand, diversity indices were calculated after pruning the full data set to $r^2 = 0.2$ and $r^2 = 0.4$ (using pairwise correlation), and one replicate setting the threshold to $R^2<0.1$ (using the variance inflation factor), to examine the effect of allowing for varying levels of LD and therefore varying numbers of loci (see methods).

Within-breed Diversity

Even considering SNP ascertainment, low diversity as measured by H_e was observed in the Thoroughbred as well as the Standardbred, which both experience high selective pressures and are closed populations. Low diversity was also observed in breeds that have undergone a severe population bottleneck, such as the Exmoor and Clydesdale, and breeds that have small census population sizes, such as the Florida Cracker. Although the Thoroughbred is a large population that is widely distributed on a geographic scale, historic records suggest that one sire is responsible for 95% of the paternal lineages in the breed and as few as 30 females make up 94% of maternal lineages [28]. In addition, the population has been largely closed to outside gene flow since the formation of the first stud book in 1791 [42] and individuals within the breed are subject to selective pressure for racing success; therefore low, within-breed diversity is not at all surprising.

Using LD-based calculations, the estimated N_e for the Thoroughbred was similar to that found in a UK sample [43] and among the lowest of the study set despite the large census population size and geographic distribution of this breed. Individual inbreeding values based upon observed vs. expected homozygosity indicate that individual Thoroughbred horses show signs of inbreeding, with a mean loss of heterozygosity of 16.3%. This value is slightly larger than that found in [28] (13.9%). Using the same SNP array, [44] also showed inbreeding in the Thoroughbred, and specifically an increase in inbreeding over time. The only breeds with higher f values were the Exmoor, Clydesdale, Mangalarga Paulista, and Shire. Despite low individual diversity, F_{IS} values do not show significant inbreeding in either of the Thoroughbred populations as a whole, or in the Norwegian Standardbred although F_{IS} is significant in the US Standardbred population (discussed below).

The Clydesdale and Exmoor, in addition to having high individual estimated coefficients of inbreeding, also show the lowest within-breed diversity observed in the dataset. A lack of diversity in the Clydesdale and another British draft breed, the Shire, is likely a result of a severe population bottleneck observed in most draft breeds with the onset of industrialization and after the conclusion of World War II (WWII) as well as selection for size and color [45,46]. The Exmoor pony, considered to be one of the purest native breeds of Britain, has been naturally selected for survival in harsh winter conditions on the moors in southwest England [45,47]. Similar to the draft breeds, the Exmoor population decreased significantly after WWII to approximately 50 individuals, undoubtedly influencing the diversity observed in this study. The effect of low population size and selection is also reflected in extremely high individual estimates of f within some individuals. Finally, the Mangalarga Paulista shows low levels of heterozygosity, and as discussed below, the greatest divergence as measured by pairwise F_{ST} of all breeds in the study. While these results could be due to geographic distance between this and other breeds, and/or genetic drift, unfortunately these horses were all sampled from only two farms and likely do not represent the entirety of the diversity present in the breed; therefore we cannot rule out sampling error which would inflate the estimated level of divergence between these individuals and the other breeds and result in a decrease in H_e. However, a lack of diversity in sampling of the breed would not have an effect on estimates of individual inbreeding coefficients, which were among the highest of the entire data set.

Converse to the above examples, high levels of diversity as measured by both H_e and N_e, accompanied by low estimates of inbreeding (f and F_{IS}), are observed in the Mongolian, Tuva, and New Forest Pony. The Mongolian and Tuva are unique in that they represent landrace populations that are less managed than the popular breeds of Western Europe and North America; they occupy a diverse range of habitat, have been selected for meat and milk in addition to use in transportation, and originate in the region where domestication was likely to have occurred. The population of Mongolian horses is large and individuals are phenotypically diverse [48]. In 1985, approximately two million Mongolian horses of four different types were estimated to live within the country [45]. The Tuva is not as numerous as the Mongolian but is similar in its purpose and also has high within-breed phenotypic diversity. In addition, it is suggested that the Tuva has experienced outcrossing in order to increase its size and stamina [45] as may also be the case in the Mongolian [49]. Similarly, the New Forest Pony was historically a free-ranging population in Great Britain, but was crossbred until the 1930's. These traits: old populations, large population size, outcrossing, high phenotypic diversity, and lesser artificial selection/management, result in the high levels of genetic diversity observed. This extent of diversity appears to diminish as populations are restricted by selective pressures into formal breeds.

Other population characteristics are likely the cause of the diversity observed in the Finnhorse, Icelandic, and Miniature. In the case of the Icelandic, the high level of diversity was possibly maintained by a large census population size despite isolation for almost a thousand years and several population bottlenecks due to natural disasters [45]. In the case of the Finnhorse, diversity may be due to within-breed substructure into four sections of the studbook established in 1970: the work horse (draft), trotters, riding horse, and pony [50]. Finally, high diversity in breeds such as the Miniature is likely a result of a diverse founding stock [45,51,52]; horses of small size from a variety of geographic regions and bloodlines were utilized in founding the breed, which is defined by phenotype.

All of these three factors, large population size, phenotypic diversity within the breed, and a diversity of founding stock, also lead to the relatively high levels of diversity observed in the Paint and Quarter Horse; in addition, these breeds both allow continued outcrossing between themselves and with the Thoroughbred and have experienced a tremendous population expansion since the formal foundation of the breeds within the past 45–75 years. Due to the relative infancy of these populations, it could be argued that the Paint, Quarter Horse, and other, newly-derived breeds, have not yet had time to undergo the evolutionary processes necessary to be genetically distinct populations as is observed in breeds with longer histories and closed studbooks. However, even with high within-breed diversity and large census population sizes (over 1 million and 4 million worldwide for the Paint and Quarter Horse, respectively), N_e for these breeds account for only a fraction of the census size, demonstrating non-random mating and selection. Outcrossing is also continued in the Swiss Warmblood and Hanoverian breeds, which show similar trends in diversity measures as the Quarter Horse and Paint. The relatively low N_e in these breeds, accompanied by moderate H_e may partially be due to significant crossing with the Thoroughbred, which would contribute long blocks of LD [40,41], resulting in decreased estimates of N_e.

Of note in breeds such as the Quarter Horse, Lusitano, and Andalusian, is that despite moderate to high relative levels of H_e, and low to moderate estimates of f, F_{IS} values in each breed are significantly positive. Significant F_{IS} was also previously observed in the Iberian breeds using microsatellite markers [53]. While selection and inbreeding may be responsible for significant values of F_{IS} in some of these breeds, another instance in which F_{IS} may be significantly positive is in the presence of subpopulation structure within the sample. Evidence of this in the Lusitano and Andalusian is present in parsimony analysis where individuals of the two breeds fall into one clade, but within that clade are two highly supported branches represented by a subset of each breed. In addition, when forcing high values of K in Structure, such as observed at K = 35, Andalusian and Lusitano individuals fall into one of two clusters with q-value >0.5 in a nonbreed-specific manner (data not shown). These results support [54], which showed potential subpopulation structure in the Lusitano via microsatellite analysis. In the Quarter Horse, subpopulation structure is evident through the evaluation of bloodlines and the selection of popular sires for diverse performance classes. This population substructure in the Quarter Horse has also been demonstrated by marked differences in allele frequencies among performance types (cutting, western pleasure, halter, racing, etc.) [55]. A similar instance is found in the US population of the Standardbred, which also has significant excess homozygosity (F_{IS}). Unlike Standardbreds in Europe, which are raced at a trot, those in the US are divergently selected for racing at either the pace or the trot, creating structure within the breed [56].

Finally, several rare populations are included in this dataset. The Caspian is one of the oldest breeds in the Middle East and was thought to be extinct until its recent rediscovery in 1965. The Florida Cracker, a now rare breed, was developed in the United States from feral stock of Iberian descent [57]. The sample size of the Florida Cracker limits the conclusions that can be drawn regarding within-breed diversity. However, the Caspian shows high N_e, H_e, and estimates of f, given its rarity. After rediscovery of the breed, which historically was believed to represent a type of landrace population, [58] describes a three-year survey, which found approximately 50 individuals remaining, noting that many could not be considered "pure." In addition, [33] were unable to show evidence of a recent bottleneck in the Caspian breed. The diversity observed in what are now considered Caspian horses likely stems from high levels of diversity within those individuals that founded the modern population.

Among-breed Diversity

The expectation of homogeneity within breeds due to closed populations and selection is supported by the results of AMOVA, which show significant variation present among populations, but a non-significant proportion of variance within. However, the variation among samples lends information about current and historic relationships. Observed trends include patterning based upon geographic origin and/or phenotypic similarities, and relatively low divergence observed in comparisons that include breeds with high within-breed diversity. In Structure analysis, K = 29 was chosen as the most likely value of K; however no single value stood out as the "best" number of clusters to describe these data. Regardless, patterns observed in clustering were also supported by pairwise F_{ST} values, parsimony, PCA, and NJ dendograms.

High Diversity and Low Divergence - Landrace Breeds

The Mongolian and Tuva populations are believed to have been influential in the spread of horses across Asia and Europe [45,59]; these landrace populations, harboring high levels of within-breed diversity, were found to be similar to one another, with a pairwise F_{ST} value of 0.006. In addition, with the exception of six Tuva individuals that fell into a clade with the Caspians, both parsimony and NJ analyses place the Tuva and Mongolian into the same clade of each tree. Examining all comparisons, low F_{ST} values were observed between the Mongolian and Tuva compared to the other breeds in this study, supporting the potential role of Eurasian horses of similar type in founding modern stocks. This also aligns with high microsatellite diversity observed in Eastern Eurasian "non-breed" (landrace) populations in [19]. On the other hand, breeds with high diversity in general show lower levels of divergence as measured by F_{ST}, while those with low diversity show higher values of F_{ST}. Low divergence in breeds with high diversity is expected as variation within a breed may indicate outcrossing with other populations, and high variation also makes these breeds more likely to share variation with others by chance. In contrast, if a breed has little within-breed variation, it is less likely to share genetic variation with another breed by chance, especially with another breed that is relatively homogeneous itself. As demonstrated in human literature, source populations are expected to contain greater diversity than those populations which they found [60,61]; this is also suggested in the horse by [11], which showed greater mtDNA diversity in Iberian breeds than the recently founded American breeds. If the argument is made that the low F_{ST} values of Tuvas and Mongolians supports their role in founding modern breeds, the same argument could be made for the Quarter Horse, which also shows low levels of pairwise divergence; however that argument would be unreasonable as the Quarter Horse was developed in only the past century. Therefore, the relative values of F_{ST} are informative, but these F_{ST} values and data, which represent modern breeds generally derived from limited founding stock, and subjected to intense artificial selective pressures, cannot be used independently to elucidate the evolution of the modern horse. Also, as is the case for other analyses, while the relationships observed can shed light on the history of breeds, they cannot distinguish between recent admixture and shared ancestry.

Thoroughbred-influenced Breeds

The Thoroughbred is believed to have founding sires of Arabian, Turk, and Barb ancestry [42], and [28] found that two sires, noted as being Arabian (Godolphin Arabian and Darley Arabian) together contributed to over 20% of the modern population. However, it is likely that the "Arabian" foundation stallions were not Arabians as the breed is known today. It is noted in [62] that the Godolphin Arabian was a Turkoman stallion with partial Arabian blood, while in other work it is suggested he was a Barb [45]. Regardless of the true ancestry of these stallions, restrictions placed upon the export of purebred Arabians during the 16th and 17th centuries, as well as the general use of the term "Arab" for horses of Middle Eastern descent, it is likely other "Arabian" horses with influence on the Thoroughbred breed also had Turkoman, and Barb bloodlines [45,62]. The pairwise F_{ST} values between the Thoroughbred and the Arabian do not suggest any less divergence than observed between the Thoroughbred and a majority of the other breeds. In addition, at K = 29 the Arabian assigns to the Thoroughbred cluster at only 2.3%. If the Arabian did have significant influence on the Thoroughbred breed, there are several possible explanations for why the supposed Arabian influence is not more apparent. The first is related to SNP ascertainment and the bias of SNPs toward modern variation in Thoroughbred. It is possible that the genes derived from the Arabian are at or near fixation in the Thoroughbred, which would reduce the chance that these SNPs, and the variation described within them are present in the dataset. Another possibility is that the current Arabian sample, taken from the United States, may not reflect the Arabian lineage(s) that were influential in the founding of the Thoroughbred. Finally, as noted above and also suggested elsewhere [63], it may simply be that Arabian bloodlines were not as instrumental in the Thoroughbred breed as once thought or that the initial Arabian influence (and genes) have been selected against or lost to drift during the development of the modern Thoroughbred racehorse.

Within the Thoroughbred itself, divergence between the US and European samples had a significant F_{ST} of 0.004, similar to that observed between the Hanoverian and Swiss Warmblood (0.008) and Mongolian and Tuva (0.006), but larger than the minimally observed value seen between the Paint and the Quarter Horse (0.002). Although artificial insemination is prohibited in the Thoroughbred and would be anticipated to limit gene flow to some extent, the founder effect in the original European Thoroughbred by few high-impact sires and dams, accompanied by shared selective pressures, relatively recent importation of the breed to the United States, and ongoing shipment of horses between continents are likely contributing to the lack of geographic population structure identified by parsimony and cluster analyses.

While within-breed diversity of the Thoroughbred was relatively low, and a notable relationship with the Arabian was not observed, among-breed analysis shows a clear influence of the Thoroughbred on many other breeds. Placed with the Thoroughbred in parsimony analysis are the Hanoverian, Maremmano, and Swiss Warmblood. The Maremmano, an Italian breed, shows a q-value of assignment to the Thoroughbred clade of 0.26 at K = 29. This is not surprising given reports that Thoroughbreds contributed over 13% of the maternal lineages to the stallion lines within the stud book [64]. Low differentiation of the Maremmano compared to the Hanoverian was also reported in [29], which is logical given similar influence of the Thoroughbred on the Hanoverian. This and the allowed crossing of Thoroughbreds into the Swiss Warmblood population is reflected in minimal measures of divergence between these samples. The continued influence of the Thoroughbred on the Paint and Quarter Horse is also reflected both by low F_{ST} values as well as greater than 30% assignment of the Paint and Quarter Horse to the Thoroughbred cluster. As each of these breeds experience continued gene flow from the Thoroughbred, and had Thoroughbred founding stock, these results are not unexpected. While outcrossing of these breeds is allowed (with restrictions), it is likely that even in cases of breeds with closed stud books, some outcrossing, intentional, unintentional, and/or undocumented, has occurred; it has been demonstrated that historical pedigrees, while helpful, are not always accurate [65]. Issues of outcrossing and individual identification can now be more easily addressed using genetic testing and have the potential to assist managers in decisions regarding breeding and registration.

Finally, the Standardbred samples, which also represent two continents, had a significant pairwise F_{ST} value of 0.020, five-fold greater than that observed between the two Thoroughbred samples. This comparison may reflect geographic structure, the influence of French Trotter bloodlines in the European sample, and selection for pacing horses in the US population. Within the horses included in this study, eight were noted to be pacers. These pacers all fall within one clade of the parsimony tree, supported by a bootstrap value of 98%. The limited sample size does not allow a thorough comparison of the pacing vs. trotting Standardbreds, however significant genetic differentiation between horses of the two racing groups has been reported [56].

Middle Eastern and Iberian Breeds

The Middle Eastern breeds, Arabian, Akhal Teke, and Caspian, were placed into a single clade of the NJ tree, with the Arabian and Akhal Teke in their own, highly supported clade. This relationship was supported by low values of K in Structure, which placed the Iberian and Middle Eastern breeds into the same cluster until K = 12. However, parsimony analysis did not support this relationship between the Middle Eastern and Iberian breeds as the Arabian and Akhal Teke individuals were placed into one clade, apart from the Iberian samples and from the Caspians.

In Europe, the Iberian breeds (Lusitano and Andalusian) have only recently been distinguished from one another depending upon the region in which they are bred and divergent selective pressures. In the US, horses of each breed are occasionally interbred [51]. The close relationship between the Andalusian and Lusitano samples in this study is reflected in the minimum F_{ST} value observed in either breed, 0.021. The parsimony tree and PCA also shows that individuals cannot necessarily be distinguished from one another regardless of whether they were sampled in the US or Portugal. Of the two clades that appear to suggest population substructure, one includes only Portuguese Lusitanos while the other includes only US samples, although two are of Portuguese ancestry.

Iberian and Gaited Breeds

A horse that is considered to be "gaited" naturally moves in a means other than the traditional walk, trot, canter, and gallop. Alternative gaits in horses are distinguished from traditional gaits by their unique footfall pattern and/or rhythm. The genetic basis of gait has recently been investigated and suggests that all modern gaited breeds share a common ancestor as supported by a shared, extended haplotype spanning a variant significantly associated with the ability to pace [66;67]. There is a great deal of historical evidence that the shared ancestry of gaited breeds traces back to Iberian bloodlines, in particular to the Spanish Jenette [45,51]. Influence of the Iberian breeds on modern gaited breeds is seen in early clustering in Structure analysis as well as in the NJ tree where the Puerto Rican Paso Fino and Peruvian Paso are placed on the same branch as the Andalusian and Lusitanos. In addition to Iberian lines, the Narragansett Pacer is often named as instrumental in the founding of American breeds that may gait including the Saddlebred, Standardbred, and Tennessee Walking Horse [45]. Within those breeds, the Tennessee Walking Horse was documented to be greatly influenced by the Saddlebred, Standardbred, and Morgan [68,69]; and the Standardbred itself had influence from the Thoroughbred and Morgan (among others) [70]. While we do not have samples of the now extinct Narragansett Pacer, our data set does support a close relationship between the Tennessee Walking Horse and Saddlebred, as observed in the NJ and parsimony analysis as well as with the Morgan and Standardbred. At low values of K in cluster analysis, the Tennessee Walking Horse and Saddlebred cluster strongly. In NJ analysis, the Florida Cracker, which has many individuals that demonstrate the ability to gait, is found intermediate to the Iberian and the modern US gaited breeds. Interestingly, the Icelandic, a four- or five-gaited breed, does not show any significant affinity to the other gaited breeds although they share the recently identified major locus that appears to be essential to the ability to perform alternate gaits [66;67]. It thus seems that the genetic variant associated with the gait phenotype arose well before the separation of breeds. Instead of clustering with the other gaited breeds, the Icelandic clusters with the Shetland through K = 16 and also is within a highly supported branch of the NJ tree with the Shetland and Miniature. Finally, the influence of the Shetland on the development of the Miniature is observed at all values of K as well as in the parsimony tree where both breeds occupy the same clade.

Drafts

The Shire and Clydesdale populations share assignment to the same cluster throughout Structure runs. The similarity between these breeds is also seen in a lack of monophyly in the parsimony tree, the sharing of a branch of the NJ tree, positioning in PCA, and a pairwise F_{ST} value of 0.037. The Fell Pony, a British breed, falls out as sister taxa to the British draft horses, the Shire and Clydesdales. However F_{ST} values show that divergence between the Fell Pony and either the Clydesdale or Shire is not significantly less than seen with most other populations. The other branch of the "draft" clade of the NJ tree contains the breeds from the European mainland, the Belgian, Percheron, and Franches-Montagnes; each of these breeds shows monophyly in parsimony analysis. In addition, similarities among draft and light draft breeds are reflected in cluster analyses at K = 6, which show the populations from the European mainland and Scandinavian Peninsula (Belgian, Finnhorse, Franches-Montagnes, North Swedish Horse, Norwegian Fjord, and Percheron) assign to one cluster with $q > 0.5$. The grouping of the Scandinavian breeds is similar to that previously reported [31]. Geographic relationships are also suggested by the two British breeds, the New Forest Pony and Exmoor that fall just basal to the draft clade in the NJ tree.

Summary

This data set resulting from a large international collaboration represents the first study in the horse to provide an extensive overview of nuclear genetic diversity within, and relationships among a diverse sample of breeds and landrace populations. These data are now available for use in subsequent studies of population-level relationships and provide a baseline for monitoring changes in breed diversity. With high mtDNA diversity but limited paternal input during domestication, this increased understanding of nuclear diversity within the horse will allow for the identification of genomic regions of importance to breed derivation and will be instrumental in guiding across-breed gene discovery projects.

Methods

Ethics Statement

DNA sampling was limited to the collection of blood by jugular venipuncture performed by a licensed veterinarian or from hairs pulled from the mane or tail by the horse owner or researcher. All animal work was conducted in accordance with and approval from the international and national governing bodies at the institutions in which samples were collected (the University of Minnesota

Institutional Animal Care and Use Committee (IACUC); the University of Kentucky IACUC; the University College Dublin, Animal Research Ethics Committee; Swiss Law on Animal Protection and Welfare; the Ethical Board of the University of Helsinki; the Animal Health Trust Clinical Research Ethics Committee; Norwegian Animal Research Authority; UK Home Office License; and the Lower Saxon state veterinary office).

Samples and Genotyping

Tissue samples and previously collected genotypes from 1,060 horses were obtained from members of the EGDC or were obtained by our laboratory. 814 samples representing the 38 populations included in this study were selected from the EDGC sample collection with the goal of obtaining as random of a sample as possible and to minimize close relationships among individuals. In some cases, genotypes were available from breeds collected for genome-wide association studies (GWAS). In all cases, when pedigree information was available, no relationships were allowed at or more recent to the grandsire/dam level. If no pedigrees were available, once genotyping was performed, individuals were removed from the analyses to reduce genome sharing as measured by autosomal estimates of identity by descent (pi hat) values in PLINK [71] greater than 0.3 (after pruning for MAF>0.05). In samples that were obtained as a result of GWAS, "control" individuals were preferentially chosen for inclusion in these analyses. When necessary, DNA isolation from hair roots took place using a modification of the Puregene (Qiagen) protocol for DNA purification from tissue. Modifications include the addition of 750 µl of isopropanol rather than 300, increasing the precipitation spin time to 15 m at 4°C, and washing the pellet twice. Approximately 1 µg of DNA was used for SNP genotyping using the Illumina SNP50 Beadchip according to the manufacturer's protocol. All genotype calls were extracted from the raw intensity data using GenomeStudio (Illumina) with the minimum gencall score threshold of 0.15. The raw intensity scores were available for all populations with the exception of the Lusitano and Maremmano.

Data Pruning

SNP discovery was conducted using horses from seven breeds (Akhal Teke, Andalusian, Arabian, Icelandic, Quarter Horse, Standardbred, Thoroughbred) as well as the reference genome of a Thoroughbred mare [41]. To eliminate ascertainment bias as much as possible, horses from the discovery breeds were removed from the dataset, which was then pruned to exclude SNPs with MAF less than 0.05. All horses were then replaced and those SNPs removed from all analyses. In this new, complete data set, SNP markers that failed to genotype in at least 99% of the individuals and SNPs that had a MAF of 0.05 or less across all samples were removed as well as SNPs on ECAX. SNPs that were in LD across breeds were also removed; files used for basic diversity indices were pruned for $r^2<0.1$ in PLINK [71] considering 100 SNP windows and moving 25 SNPs per set (–indep-pairwise 100 25 1.11). Allowing for additional LD, data sets were also created for r2<0.2 and 0.4. An additional data set, used for Structure analysis was pruned for $R^2<0.1$ in Plink (–indep; R = multiple correlation coefficient), which is similar to the above method but instead of analyzing pairwise relationships of SNPs as in the former method, uses a multiple regression approach upon the SNPs in the analysis window. Files were converted for usage between analyses programs using PLINK, perl script, CONVERT [72] and/or PGDSpider 2.0.1.4 [73].

Within-breed Diversity

Expected heterozygosity (H_e) and AMOVA were calculated in Arlequin3.5 [74] on all four data sets. AMOVA was conducted on the primary data set with breeds designated as populations and excluding the Florida Cracker due to small sample size. Analyses were also conducted grouping the two Thoroughbred and Standardbred samples together by breed. F_{IS} was calculated and significance tested on the primary data set, with 10,000 permutations of the data in Genetix [75]. Individual inbreeding coefficients (f) were calculated in PLINK based upon loss of heterozygosity (–het).

Among Breed Relationships

Pairwise F_{ST} values were calculated on the primary, 10,536 SNP dataset in Arlequin3.5 [74] using Reynolds' distance [76] with significance tested using 20,000 permutations.

A neighbor joining (NJ) cladogram was built using breed allele frequencies calculated from the primary SNP set using the packages seqboot, gendist, neighbor, and consense and Nei's genetic distance [38] in PHYLIP ver3.69 [77]. Bootstrap support from 1,000 iterations of the data was used to assess support for the resulting majority rule consensus cladogram.

A parsimony cladogram was constructed using 10,066 SNP markers pruned from the original data set using the MAF and genotyping rate criteria as above and allowing for $R^2<0.2$ (–indep-pairwise). The domestic ass was included as an outgroup for traditional and new-technology searches in TNT [78].

Principal component analysis (PCA) was conducted in snpStats in R (http://cran.r-project.org) on the full SNP set consisting of all 814 individuals and 38,755 autosomal SNPs (pruned only for MAF and genotyping rate).

Cluster Analysis

Clustering of breeds into genetic groups was examined using the program Structure 2.3.3 [79,80] assuming K = 1 to 45. The Structure algorithm included the admixture model and correlated allele frequencies. Three iterations of each K value were conducted with 35,000 MCMC repetitions (15,000 burn-in). The convergence of Structure runs was evaluated by equilibrium of alpha and likelihood scores. The value of K most suitable to explain the diversity in these data was predicted by the highest mean estimated ln P(X|K) while minimizing variance and also making biological sense [80,81]. The replicates from each run of K were input into CLUMPP [82] and the average cluster membership calculated using the LargeK Greedy algorithm. Output from CLUMPP was visualized in Distruct [83].

Effective Population Size

To estimate effective population size (N_e), the full set of 54,602 SNP markers was pruned within each population to remove those with MAF <0.01 and genotyping rate of <0.05. Pairwise r^2 values between remaining SNPs were calculated in Haploview [84], for each population considering intermarker distances from 0 to 4 Mb in 50 kb increments. Values of N_e were calculated using the method of [37], which includes a correction for small sample size and the assumption that 1 Mb = 1 cM.

Data Access

All SNP genotype data are available at the NAGPR Community Data Repository (animalgenome.org) for the purpose of reconstructing the analyses. The only exception is the data collected from the Tennessee Walking Horse, which, under agreement from the granting agency (to the University of

Minnesota from the Foundation for the Advancement of the Tennessee Walking Show Horse (FAST) and the Tennessee Walking Horse Foundation (TWHF)), is only available under a Material Transfer Agreement (MTA) between interested individuals and the University of Minnesota.

Supporting Information

Figure S1 Example of LD decay over 2 Mb in 9 breeds of horse. Decay of linkage disequilibrium over 2Mb for 9 of the 36 breeds. Landrace populations such as the Mongolian, and large and/or diverse breeds such as the Finnhorse and Quarter Horse, show more rapid decay than those with small population sizes and less diversity (e.g. Clydesdale, Tennessee Walking Horse).

Figure S2 Parsimony relationship among Lusitano and Andalusian individuals. Portion of the parsimony clade shown in Figure 1 consisting of the Lusitano (dark blue) and US Andalusian (light blue) individuals. Bootstrap values greater than 50% are shown. Asterisks indicate horses sampled in the US which were noted to be of Portuguese ancestry.

Figure S3 Principal component 1 vs. 2 as determined from 814 horses from 38 populations. Principal components 1 and 2 as determined from 38,755 SNPs (pruned for MAF and genotyping rate). All 814 individuals are included in the plot.

Figure S4 Mean of estimated ln P(X|K) for each run in Structure. Mean of estimated ln P(X|K) for each of the three runs for K = 1−45 in Structure.

Figure S5 Bayesian clustering output for additional values of K in 814 horses of 38 populations. Structure output for additional values of K. Each individual is represented by one vertical line with the proportion of assignment to each cluster shown on the y axis and colored by cluster.

Figure S6 Parsimony relationship among Thoroughbred horses from the US and UK/Ire. The branch of the parsimony clade shown in Figure 1 containing the US and UK/Ire Thoroughbreds. Horses sampled in the UK/Ire are noted with an asterisk. Bootstrap values >50% are shown.

Figure S7 Parsimony relationship among Standardbreds from the US and Norway. The branch of the parsimony clade shown in Figure 1 containing the US (yellow) and Norwegian (green) Standardbreds. Bootstrap values >50% are shown. The asterisks indicate individuals that are pacing horses.

Table S1 Cluster to which each population maximally assigns and corresponding q-value for K = 2 to 45. Highest breed q-value of assignment and cluster identity (#) for each value of K examined in Structure. The cluster ID # is not carried through across values of K.

Table S2 Proportion of assignment for 38 horse populations to each of K = 29 clusters. Proportion of assignment to each of K = 29 clusters as determined in Structure. The largest proportion of assignment for each population is outlined and shown in bold; those with 30–50% assignment are shown in italic. The top row notes the breed(s) with >50% of assignment to each of the 29 clusters. This analysis was performed without removal of outlier individuals.

Acknowledgments

The computational support provided by Rob Schaefer and Aaron Rendahl is gratefully acknowledged. Finally, the EGDC thanks the Havemeyer Foundation for their continued support of equine genomics research.

Author Contributions

Conceived and designed the experiments: JLP JRM MEM. Performed the experiments: JLP MEM. Analyzed the data: JLP. Contributed reagents/materials/analysis tools: JLP JRM EGC LSA JA EB DB MMB ASB PB ACM OD MF LFC KTG GG BH TH KH EWH TL GL HL MSL BAM SM NO MCTP RJP MR SR KHR MS JS TT MV CMW MEM. Wrote the paper: JLP JRM EGC MEM.

References

1. FAOSTAT (2010). Available: http://www.faostat.fao.org.Accessed 2012 Jun 5.
2. Lippold S, Matzke NJ, Reissmann M, Hofreiter M (2011) Whole mitochondrial genome sequencing of domestic horses reveals incorporation of extensive wild horse diversity during domestication. BMC Evol Biol 11: 328.
3. Ludwig A, Pruvost M, Reissmann M, Benecke N, Brockmann GA, et al. (2009) Coat color variation at the beginning of horse domestication. Science 324: 485.
4. Outram AK, Stear NA, Bendrey R, Olsen S, Kasparov A, et al. (2009) The earliest horse harnessing and milking. Science 323: 1332–1335.
5. Pedrosa S, Uzun M, Arranz JJ, Gutierrez-Gill B, Primitivo FS, et al. (2005) Evidence of three maternal lineages in near eastern sheep supporting multiple domestication events. P R Soc B 272: 2211–2217.
6. Larson G, Dobney K, Albarella U, Fang M, Matisoo-Smith E, et al. (2005) Worldwide phylogeography of wild boar reveals multiple centers of pig domestication. Science 307: 1618–1621.
7. Wu GS, Yao YG, Qu KX, Ding ZL, Li H, et al. (2007) Population phylogenomic analysis of mitochondrial DNA in wild boars and domestic pigs revealed multiple domestication events in East Asia. Genome Biol 8: R245.
8. Cieslak M, Pruvost M, Benecke N, Hofreiter M, Morales A, et al. (2010) Origin and history of mitochondrial DNA lineages in domestic horses. PLOS One 5: e15311.
9. Jansen T, Forster P, Levine MA, Oelke H, Hurles M, et al. (2002) Mitochondrial DNA and the origins of the domestic horse. PNAS 99: 10905–10910.
10. Lei CZ, Su R, Bower MA, Edwards CJ, Wang XB, et al. (2009) Multiple maternal origins of native modern and ancient horse populations in China. Anim Genet 40: 933–944.
11. Lira J, Linderholm A, Olaria C, Brandstrom Durling M, Gilbert MT, et al. (2010) Ancient DNA reveals traces of Iberian Neolithic and Bronze Age lineages in modern Iberian horses. Mol Ecol 19: 64–78.
12. Vila C, Leonard JA, Gotherstrom A, Marklund S, Sandberg K, et al. (2001) Widespread origins of domestic horse lineages. Science 291: 474–477.
13. Warmuth V, Eriksson A, Bower MA, Barker G, Barrett E, et al. (2012) Reconstructing the origin and spread of horse domestication in the Eurasian steppe. PNAS 109: 8202–8206.
14. Beja-Pereira A, Caramelli D, Lalueza-Fox C, Vernesi C, Ferrand N, et al. (2006) The origin of European cattle: evidence from modern and ancient DNA. PNAS 103: 8113–8118.
15. Gotherstrom A, Anderung C, Hellborg L, Elburg R, Smith C, et al. (2005) Cattle domestication in the Near East was followed by hybridization with aurochs bulls in Europe. P R Soc B 272: 2345–2350.
16. Keyser-Tracqui C, Blandin-Frappin P, Francfort HP, Ricaut FX, Lepetz S, et al. (2005) Mitochondrial DNA analysis of horses recovered from a frozen tomb (Berel site, Kazakhstan, 3rd Century BC). Anim Genet 36: 203–209.
17. Achilli A, Olivieri A, Soares P, Lancioni H, Kashani BH, et al. (2012) Mitochondrial genomes from modern horses reveal the major haplogroups that underwent domestication. PNAS 109: 2449–2454.
18. McGahern A, Bower MA, Edwards CJ, Brophy PO, Sulimova G, et al. (2006) Evidence for biogeographic patterning of mitochondrial DNA sequences in Eastern horse populations. Anim Genet 37: 494–497.
19. Warmuth V, Manica A, Eriksson A, Barker G, Bower M (2012) Autosomal genetic diversity in non-breed horses from eastern Eurasia provides insights into historical population movements. Anim Genet. In press. doi:10.1111/j.1365-2052.2012.02371.x.

20. Lindgren G, Backstrom N, Swinburne J, Hellborg L, Einarsson A, et al. (2004) Limited number of patrilines in horse domestication. Nat Genet 36: 335–336.
21. Ling Y, Ma Y, Guan W, Cheng Y, Wang Y, et al. (2010) Identification of y chromosome genetic variations in Chinese indigenous horse breeds. J Hered 101: 639–643.
22. Aberle KS, Hamann H, Drogemuller C, Distl O (2004) Genetic diversity in German draught horse breeds compared with a group of primitive, riding and wild horses by means of microsatellite DNA markers. Anim Genet 35: 270–277.
23. Achmann R, Curik I, Dovc P, Kavar T, Bodo I, et al. (2004) Microsatellite diversity, population subdivision and gene flow in the Lipizzan horse. Anim Genet 35: 285–292.
24. Bjornstad G, Gunby E, Roed KH (2000) Genetic structure of Norwegian horse breeds. J Anim Breed Genet 117: 307–317.
25. Bömcke E, Gengler N, Cothran EG (2010) Genetic variability in the Skyros pony and its relationship with other Greek and foreign horse breeds. Genet and Mol Biol 34: 68–76.
26. Canon J, Checa ML, Carleos C, Vega-Pla JL, Vallejo M, et al. (2000) The genetic structure of Spanish Celtic horse breeds inferred from microsatellite data. Anim Genet 31: 39–48.
27. Cothran EG, Canelon JL, Luis C, Conant E, Juras R (2011) Genetic analysis of the Venezuelan Criollo horse. Genet Mol Res 10: 2394–2403.
28. Cunningham EP, Dooley JJ, Splan RK, Bradley DG (2001) Microsatellite diversity, pedigree relatedness and the contributions of founder lineages to thoroughbred horses. Anim Genet 32: 360.
29. Felicetti M, Lopes MS, Verini-Supplizi A, Machado Ada C, Silvestrelli M, et al. (2010) Genetic diversity in the Maremmano horse and its relationship with other European horse breeds. Anim Genet 41 Suppl 2: 53–55.
30. Glowatzki-Mullis ML, Muntwyler J, Pfister W, Marti E, Rieder S, et al. (2005) Genetic diversity among horse populations with a special focus on the Franches-Montagnes breed. Anim Genet 37: 33–39.
31. Leroy G, Callede L, Verrier E, Meriaux JC, Ricard A, et al. (2009) Genetic diversity of a large set of horse breeds raised in France assessed by microsatellite polymorphism. Genet Sel Evol 41: 5.
32. Marletta D, Tupac-Yupanqui I, Bordonaro S, Garcia D, Guastella AM, et al. (2006) Analysis of genetic diversity and the determination of relationships among western Mediterranean horse breeds using microsatellite markers. J Anim Breed Genet 123: 315–325.
33. Shasavarani H, Rahimi-Mianji G (2010) Analysis of genetic diversity and estimation of inbreeding coefficient within Caspian horse population using microsatellite markers. Afr J Biotechnol 9: 293–299.
34. Solis A, Jugo BM, Meriaux J-C, Iriondo M, Mazon LI, et al. (2005) Genetic diversity within and among four South European native horse breeds based on microsatellite DNA analysis: Implications for conservation. J Hered 96: 670–678.
35. Tozaki T, Takezaki N, Hasegawa T, Ishida N, Kurosawa M, et al. (2003) Microsatellite variation in Japanese and Asian horses and their phylogenetic relationships using a European horse outgroup. J Hered 94: 374–380.
36. Luis C, Juras R, Oom MM, Cothran EG (2007) Genetic diversity and relationships of Protuguese and other horse breeds based on protein and microsatellite loci variation. Anim Genet 38: 20–27.
37. Weir BS, Hill WG (1980) Effect of mating structure on variation in linkage disequilibrium. Genetics 95: 477–488.
38. Nei M (1972) Genetic distance between populations. Amer Nat 106: 283–292.
39. Evanno G, Regnaut S, Goudet J (2005) Detecting the number of clusters of individuals using the software STRUCTURE: a simulation study. Mol Ecol 14: 2611–2620.
40. Wade CM, Giulotto E, Sigurdsson S, Zoli M, Gnerre S, et al. (2009) Genome sequence, comparative analysis, and population genetics of the domestic horse. Science 326: 865–867.
41. McCue ME, Bannasch DL, Petersen JL, Gurr J, Bailey E, et al. (2012) A high density SNP array for the domestic horse and extant perissodactyla: utility for association mapping, genetic diversity, and phylogeny studies. PLOS Genet 8: e1002451.
42. Weatherby and Sons (1791) An Introduction to a General Stud Book. London, UK.
43. Corbin LJ, Blott SC, Swinburne JE, Vaudin M, Bishop SC, et al. (2010) Linkage disequilibrium and historical effective population size in the Thoroughbred horse. Anim Genet 41 Suppl 2: 8–15.
44. Binns MM, Boehler DA, Bailey E, Lear TL, Cardwell JM, et al. (2012) Inbreeding in the Thoroughbred horse. Anim Genet 43: 340–342.
45. Hendricks BL (2007) International Encyclopedia of Horse Breeds. Norman, OK: University of Oklahoma Pres. 486 p.
46. Weatherley L (1978) Great horses of Britain. Hindhead: Spur Publications. viii, 269 p.
47. Gates S (1979) A study of the home ranges of free-ranging Exmoor ponies. Mammal Rev 9: 3–18.
48. Peilieu C (1984) Livestock Breeds of China. FAO: Animal Production and Health paper 46: 1–217.
49. Hund A (2008) The Stallion's Mane: The Next Generation of Horses in Mongolia. Available: http://digitalcollections.sit.edu/cgi/viewcontent.cgi?article=1543&context=isp_collection. Accessed 2012 Dec 31.
50. Ticklen M, editor (2006) Get to know the Finnhorse. Available: http://www.suomenhevonen.info/hippos/sh2007/pdf/SHjulkaisu_englanti_nettiin.pdf. Accessed 2012 Dec 31.
51. Bowling AT, Ruvinsky A (2000) The genetics of the horse. Wallingford: CABI Pub. viii, 527 p.
52. Lynghaug F (2009) The Official Horse Breeds Standards Guide: The Complete Guide to the Standards of All North American Equine Breed Associations. Stillwater, MN: Voyageur Press.
53. Conant EK, Juras R, Cothran EG (2012) A microsatellite analysis of five Colonial Spanish horse populations of the southeastern United States. Anim Genet 43: 53–62.
54. Lopes MS (2011) Molecular tools for the characterisation of the Lusitano horse. Angra do Heroísmo: University of Azores. 219 p.
55. Tryon RC, Penedo MC, McCue ME, Valberg SJ, Mickelson JR, et al. (2009) Evaluation of allele frequencies of inherited disease genes in subgroups of American Quarter Horses. J Am Vet Med Assoc 234: 120–125.
56. Cothran EG, MacCluer JW, Weitkamp LR, Bailey E (1987) Genetic differentiation associated with gait within American standardbred horses. Anim Genet 18: 285–296.
57. Florida Cracker Horse Association. Available: http://www.floridacrackerhorses.com/index.htm. Accessed 2012 Jun 5.
58. Firouz L (1969) Conservation of a domestic breed. Biol Conserv 1: 1–2.
59. Bjornstad G, Nilsen NO, Roed KH (2003) Genetic relationship between Mongolian and Norwegian horses? Anim Genet 34: 55–58.
60. Conrad DF, Jakobsson M, Coop G, Wen X, Wall JD, et al. (2006) A worldwide survey of haplotype variation and linkage disequilibrium in the human genome. Nat Genet 38: 1251–1260.
61. Li JZ, Absher DM, Tang H, Southwick AM, Casto AM, et al. (2008) Worldwide human relationships inferred from genome-wide patterns of variation. Science 319: 1100–1104.
62. Mackay-Smith A (2000) Speed and the thoroughbred : the complete history. Lanham, MD: Derrydale Press. xxvii, 193 p.
63. Bower MA, Campana MG, Whitten M, Edwards CJ, Jones H, et al. (2011) The cosmopolitan maternal heritage of the Thoroughbred racehorse breed shows a significant contribution from British and Irish native mares. Biol Letters 7: 316–320.
64. Silvestrelli M (1991) The Maremmano horse. Animal Genetic Resources Information (FAO/UNEP) 8: 74–83.
65. Bower MA, Campana MG, Nisbet RER, Weller R, Whitten M, et al. (2012) Truth in the bones: Resolving the identify of the founding elite Thoroughbred racehorses. Archaeometry 54: 916–925.
66. Andersson LS, Larhammar M, Memic F, Wootz H, Schwochow D, et al. (2012) Mutations in DMRT3 affect locomotion in horses and spinal circuit function in mice. Nature 488: 642–646.
67. Petersen JL, Mickelson JR, Rendahl AK, Valberg SJ, Andersson LS, et al. (2012) Genome-wide analysis reveals selection for important traits in domestic horse breeds. PLOS Genet. In press.
68. Fletcher JL (1946) A study of the first fifty years of Tennessee walking horse breeding. J Hered 37: 369–373.
69. Tennessee Walking Horse Breeders and Exhibitors Association. Available: http://www.twhbea.com/breed/history.php. Accessed 2012 Jun 5.
70. MacCluer JW, Boyce AJ, Dyke B, Weitkamp LR, Pfennig DW, et al. (1983) Inbreeding and pedigree structure in Standardbred horses. J Hered 74: 394–399.
71. Purcell S, Neale B, Tood-Brown K, Thomas L, Ferreira MAR, et al. (2007) PLINK: a toolset for whole-genome association and population-based linkage analysis. Amer J Hum Genet 81: 559–575.
72. Glaubitz JC (2004) CONVERT: A user-friendly program to reformat diploid genotypic data for commonly used population genetic software packages. Mol Ecol Notes 4: 309–310.
73. Lischer H, Excoffier L (2012) An automated data conversion tool for connecting population genetics and genomics programs. Bioinformatics 28: 298–299.
74. Excoffier L, Laval G, Schneider S (2005) Arlequin (version 3.0): An integrated software package for population genetics data analysis. Evol Bioinform 1: 47–50.
75. Belkhir KP, Borsa P, Chikhi L, Raufaste N, Bonhomme F (2001) GENTETIX, logiciel sous Windows TM pour la genetique des population. Universite de Montpellier II, Montpellier, France.
76. Reynolds J, Weir BS, Cockerham CC (1983) Estimation of the coancestry coefficient: basis for a short-term genetic distance. Genetics 105: 767–779.
77. Felsenstein J (1989) PHYLIP-Phylogeny Inference Package (version 3.2) Cladistics 5: 164–166.
78. Goloboff PA, Farris JS, Nixon K (2003) TNT: tree analysis using new technology. Syst Biol 54: 176–178.
79. Falush D, Stephens M, Pritchard JK (2003) Inference of population structure using multilocus genotype data: linked loci and correlated allele frequencies. Genetics 164: 1567–1587.
80. Pritchard JK, Stephens M, Donnelly P (2000) Inference of population structure using multilocus genotype data. Genetics 155: 945–959.
81. Pritchard JK, Wen X, Falush D (2010) Documentation for Structure software: version 2.3.38.
82. Jakobsson M, Rosenberg NA (2007) CLUMPP: a cluster matching and permutation program for dealing with label switching and multimodality in analysis of population structure. Bioinformatics 23: 1801–1806.
83. Rosenberg NA (2004) DISTRUCT: a program for the graphical display of population structure. Mol Ecol Notes 4: 137–138.
84. Barrett JC, Fry B, Maller J, Daly MJ (2005) Haploview: analysis and visualization of LD and haplotype maps. Bioinformatics 21: 263–265.

6

A Dig into the Past Mitochondrial Diversity of Corsican Goats Reveals the Influence of Secular Herding Practices

Sandrine Hughes[1]*, Helena Fernández[2], Thomas Cucchi[3,4], Marilyne Duffraisse[1], François Casabianca[5], Daniel Istria[6], François Pompanon[2], Jean-Denis Vigne[3], Catherine Hänni[1], Pierre Taberlet[2]

1 Paléogénomique et Evolution Moléculaire, Institut de Génomique Fonctionnelle de Lyon, Université de Lyon, Université Lyon 1, CNRS UMR 5242, INRA, Ecole Normale Supérieure de Lyon, 46 allée d'Italie, 69364 Lyon Cedex 07, France, **2** Laboratoire d'Ecologie Alpine, CNRS UMR 5553, Université Joseph Fourier, B.P. 53, 38041 Grenoble Cedex 9, France, **3** Centre National de la Recherche Scientifique, UMR 7209, Muséum National d'Histoire Naturelle, «Archéozoologie, Archéobotanique: Sociétés, Pratiques et Environnements», Département "Ecologie et Gestion de la Biodiversité" CP 56, 75005 Paris, France, **4** Department of Archaeology, University of Aberdeen, Aberdeen, United Kingdom, **5** Institut National de la Recherche Agronomique, UR 045 Laboratoire de Recherches sur le Développement de l'Elevage, Quartier Grossetti, 20250 Corte, France, **6** Laboratoire d'Archéologie Médiévale Méditerranéenne, CNRS UMR 6572, 5 rue du château de l'Horloge, BP 647, 13094 Aix-en-Provence, France

Abstract

The goat (*Capra hircus*) is one of the earliest domesticated species ca. 10,500 years ago in the Middle-East where its wild ancestor, the bezoar (*Capra aegagrus*), still occurs. During the Neolithic dispersal, the domestic goat was then introduced in Europe, including the main Mediterranean islands. Islands are interesting models as they maintain traces of ancient colonization, historical exchanges or of peculiar systems of husbandry. Here, we compare the mitochondrial genetic diversity of both medieval and extant goats in the Island of Corsica that presents an original and ancient model of breeding with free-ranging animals. We amplified a fragment of the Control Region for 21 medieval and 28 current goats. Most of them belonged to the A haplogroup, the most worldwide spread and frequent today, but the C haplogroup is also detected at low frequency in the current population. Present Corsican goats appeared more similar to medieval goats than to other European goat populations. Moreover, 16 out of the 26 haplotypes observed were endemic to Corsica and the inferred demographic history suggests that the population has remained constant since the Middle Ages. Implications of these results on management and conservation of endangered Corsican goats currently decimated by a disease are addressed.

Editor: Marco Salemi, University of Florida, United States of America

Funding: This work was supported by the Centre National de la Recherche Scientifique and by the CHRONOBOS project from the French "Agence Nationale de la Recherche" (ANR-05-GANI-004, coordinator SH). The funders had no role in study design, data collection and analysis, decision to publish, or preparation of the manuscript. No additional external funding received for this study.

Competing Interests: The authors have declared that no competing interests exist.

* E-mail: Sandrine.Hughes@ens-lyon.fr

Introduction

In the Near East, cradle of the domestication process, goats were among the first to be domesticated ca. 10,500 years ago [1–4]. Several thousands years later, domestic goats (*Capra hircus*) were dispersed beyond the natural distribution of its wild ancestor (*Capra aegagrus*). They spread in Anatolia and Europe (starting from 8,800 calBP) throughout the Neolithic dispersal, along with pigs, cattle and sheep [5,6]. Today, goats are present all over the world with more than 867 million of individuals [7]. In order to better assess the historical processes of domestication, goat mitochondrial genetic diversity has been largely studied across the old world (Europe, Asia, and more recently Africa). It is structured in six different haplogroups A, B, C, D, F and G [3,8–16], with more than 90% of goats solely from the A haplogroup [14]. Moreover, a very weak phylogeographic structure is observed at the continent scale [8] contrary to other domestic species, such as sheep. Thus, A and C haplogroups have a worldwide distribution although B is mostly present in Asia. Some genetic structure is suspected however at a more restricted geographical scale [16] and some haplogroups, such as G and F, are now restricted to small regions (Middle-East [14] and Sicily [12] respectively).

If many studies have been dedicated to the characterization of the mainland diversity few were pursued on islands [12,17]. However, large Mediterranean islands are of particular importance in the description of the genetic diversity of domestic species since they are considered as biodiversity hot spots, have a high degree of endemism and present a reservoir of cultural practices that have disappeared from the Mainland. In the case of goats, the study of genetic diversity on large Mediterranean islands is highly relevant for several reasons. First, goats are found on most of these islands from the beginning of the Neolithic diffusion and can serve as testimony of this spread [18]. Second, imported domestic goats, are physically isolated from their Mainland relatives. Third, breeding and husbandry practices on the Mediterranean islands are usually different to those on the continent because islands present large but restricted areas and preserve traditional practices mentioned previously. From these observations, we expect that mitochondrial diversity observed on islands can present evidence of historical events or ancient diffusion that would be lost elsewhere, such as the presence of the F haplogroup in Sicily that is unique outside the wild ancestor's area of distribution [3,12,14].

Here we characterize goat mitochondrial DNA (mtDNA) diversity through time by studying modern, but above all historical, goat populations from Corsica. We compared them with current continental or island breeds of the Mediterranean Basin to document the microevolution and the influence of insularity and husbandry practices on their genetic diversity. Corsica is an 183 km long and 83 km wide island located in the Northwestern part of the Mediterranean Basin. It formed a unique block with another island, Sardinia, during most of the Pleistocene until the land masses split approximately around 11,000 years ago. The presence of domestic Caprinae is attested in Corsica from the beginning of the Neolithic, ca. 7,600 calBP [19]. The herding system in Corsica, although close to other free-ranging and seasonal transhumance characteristic of other Mediterranean islands, displays a very interesting peculiarity namely the "wandering" of flocks in the mountains for weeks, between the end of the lactation period and the beginning of the births, under very loose surveillance from the herder [20]. These free-ranging practices in Corsica are not recent as Polybius had already mentioned them in his book XII during the IInd century Before Christ. Nonetheless, being able to ascertain that transhumance and free ranging were common herding practices in Corsica since the Neolithic is a difficult task for zooarchaeologists [19]. Goats are known to be hardy, tough and able to adapt to very difficult habitats compared to other livestock. More than 200,000 goats were still present in Corsica less than 80 years ago but this number has decreased to 30,000 individuals during the last decades. Nevertheless, goats have retained a particular status in Corsica where pastoralism is still strongly established [21]. Besides, a Corsican breed has been recently recognized by the French CNAG (Commission Nationale d'Amélioration Génétique depending on Ministry of Agriculture) in 2003 and by decree of the French Ministry of Agriculture in 2007. It is a dairy breed, with relatively long hairs, either of uniform colour or multicoloured, characterized by its rustic character and ability to adapt to difficult grounds. This was possible because efforts were made to protect and promote local breeds by avoiding mixings with commercial breeds [22]. However, the traditional Corsican breed is now endangered because the traditional husbandry system seems difficult to maintain. The strains on this traditional system are not only economic but also social. Traditional goat husbandry practices implicate daily mobility from the herder, which most of the new generation does not wish to pursue [23]. More recently, the occurrence of Johne's disease (i.e., paratuberculosis) presents another very serious threat and has decimated flocks.

This study aims to (i) better characterize the current and past (medieval) mitochondrial genetic diversity of Corsican goats using a Control Region (CR) fragment; (ii) bring information about goat dispersal in the Neolithic by testing the congruence between scenario proposed and data observed on this island; (iii) explain the maintenance of endemic variability in Corsica; (iv) discuss implications for conservation of the Corsican breed. For these purposes, we gathered samples of 28 present-day individuals and 29 bones dated from the Middle Ages (XIIth and XIVth centuries).

Methods

The archeological site of Rostino

The castrum (strong hold) of Rostino [24] is situated in the North East of Corsica (Figure 1). Occupied between the XIIth and XIVth centuries AD, this late medieval site has yielded the largest assemblage of Caprinae in Corsica in a good state of preservation [25], which is rare in the acidic soil of Corsica [19]. Among the domestic species of the XIVth century deposit, Caprinae represent more than 70% of the identified mammal remains [25]. The economy of the castrum relies on specialized caprine exploitation where the production of sheep and goat complement each other: sheep for meat production and goat for milk and hair production [25]. During the XIVth century, the caprine exploitation specialized in goats with *Capra hircus* representing more than 70% of the total caprine remains [25]. This large amount of late medieval Corsican *Capra hircus* represents a great opportunity to investigate the consequence of the secular herding practices and the selective choices made by herders to renew their flock given the genetic diversity of goats in large Mediterranean islands.

Archeological samples

We analysed 29 bone fragments from Rostino: 17 were excavated from a XIIth century deposit and 12 from a deposit dated to the XIVth century. These bones have been identified as *Capra hircus* using morphoscopic criteria on both dental [26,27] and appendicular [28–30] characters (Table S1). There is no ambiguity about the origin of bones from domestic animals as the wild ancestor (*Capra aegagrus*) has never been present in Europe [31]. According to the type of bone (mandible, radius, humerus ...), the laterality (left, right) and detailed information from the excavation, we were able to clearly identify 25 different individuals from the 29 fragments. The four other fragments very probably came from previously identified individuals (Table S1). Two samples were subsequently identified as sheep by the molecular analyses. This is not surprising as the inter-specific distinction between sheep and goat on fragmented bones cannot be ascertained with a 100% reliability [29]. Molecular analyses have precisely proven to be useful in this case [32] as has, more recently, the analysis of collagen by mass spectrometry [33].

Ancient DNA extracts and PCR amplifications

Retrieval of the DNA preserved in bones was performed in ultra clean rooms dedicated to ancient DNA experiments (French National Platform of Paleogenetics PALGENE, CNRS, ENS Lyon). No more than 4 *capra* samples were treated in the same session of DNA extraction along with a bone from another species (cervids, ursids) and a blank control [34,35]. 100 to 500 milligrams of each bone fragments were reduced in powder and suspended in 5 or 10 ml of EDTA buffer as described in [36]. We extracted the DNA using one, or both, of the two following protocols: a classical phenol-chloroform approach [36,37] or direct purification using Qiaquick column (Qiagen kit) [34]. 25 out of the 29 samples were independently re-extracted in a second laboratory dedicated to ancient DNA in another city, Grenoble, using the Qiaquick protocol.

A 130 bp fragment of the CR (HVI) was amplified with the CapFII and CapRII primers with conditions identical to those described in [38]. At least 2 or 3 independent positive amplifications per sample were obtained, cloned and sequenced following protocols described in [34,35]. The final sequence of one individual was obtained by making the majority-rule consensus of all consensus of all different clones obtained from each of the independent amplification. More than 80% of the differences observed between clones were G to A or C to T punctual substitutions. This result is consistent with ancient DNA degradation profiles where deamination of cytosines is known as the major factor of artifactual substitutions [39].

Present-day Corsican goat sequences

We sampled 28 goats in 5 different localities in Corsica (7 from Moltifao, 6 from Tralonca, 11 from Corte, 3 from Altiani, 1 from Quenza; Figure 1, Table S1). 14 sequences were already published [14] and 14 were produced for this study. To amplify the CR fragment we used the same primers (CapFII and CapRII) than for

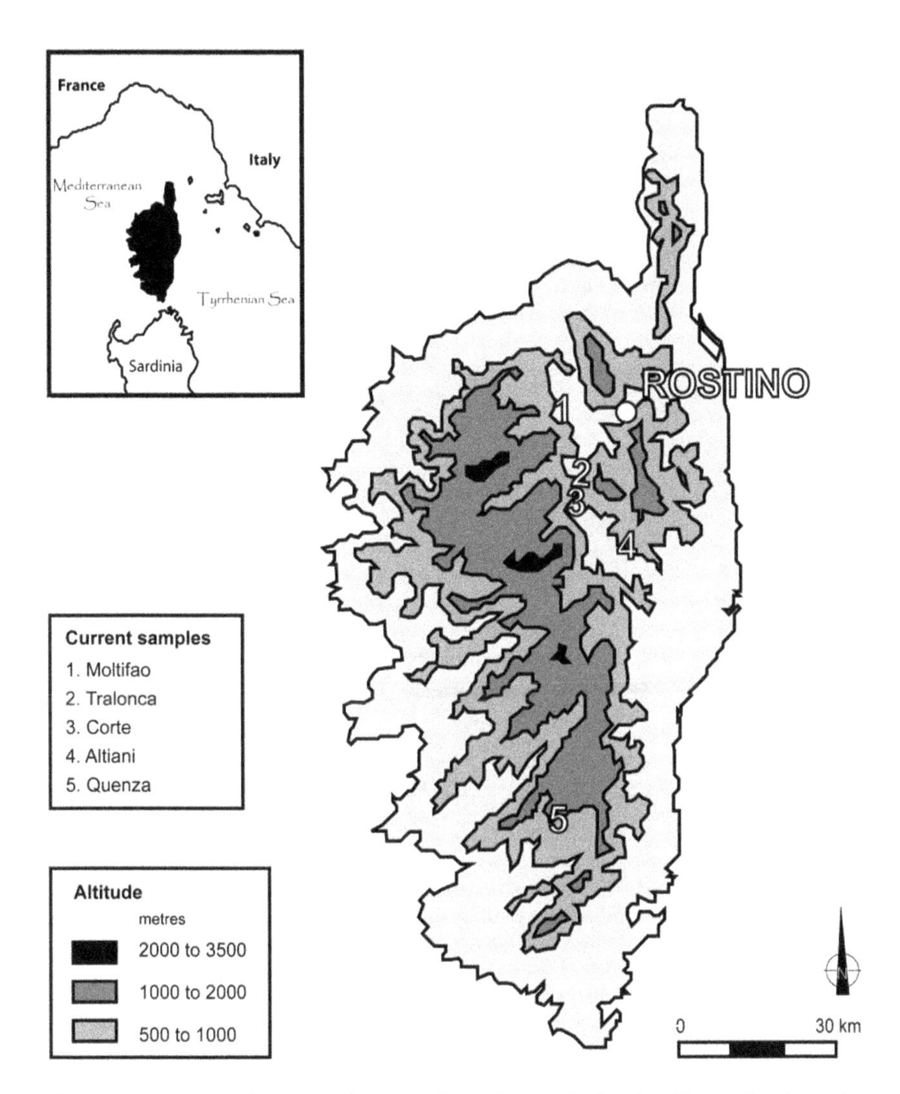

Figure 1. Ancient and present-day sampling of goats in Corsica. The medieval samples are all located in the archeological site of Rostino, in the northern part of the island. The present-day samples come from 5 different localities identified by numbers on the map.

medieval sequences with slightly modified conditions: 40 cycles were performed instead of 50–60 for ancient DNA and products were directly sequenced on both DNA strands.

Sequence analyses

All the sequences obtained were aligned (Seaview v4 [40]) and the primers removed leading to a fragment of 130 base pairs. Four data sets were constituted: (i) all medieval sequences, (ii) the XIIth century sequences, (iii) the XIVth century sequences and (iv) the present-day sequences.

Firstly, the mitochondrial haplogroup of each Corsican sequence found was determined by performing phylogenetic analyses. The different Corsican haplotypes were analyzed together with 20 haplotypes of known haplogroups. These 20 haplotypes corresponded to the 20 reference sequences that were different for the 130 bp fragment under study, in the dataset selected by Naderi et al. [14] to represent the worldwide variability of the whole HVI-control region (558 bp). Identical sequences, or haplotypes, were identified in the Corsican dataset by using Fabox [41]. After estimating the better model of evolution using jModeltest program [42] and the Akaike Information Criterion (AIC), we performed Bayesian analyses (BA) with MrBayes v3.1.2 [43,44] (independently confirmed by Maximum Likelihood analyses, not shown). The parameters used for BA were the following: GTR+I+G (nst = 6 and rates = invgamma), 5,000,000 generations sampled every 1000th generation, 4 chains, a burn-in period of 500 trees (i.e. 10% of generations) visually confirmed using Tracer v1.4.1 (developed by Rambaut A and Drummond A; available from http://beast.bio.ed.ac.uk/Tracer), allcompat option. Two independent runs were performed with an average standard deviation of split frequencies at completion of 0.006598. The average values obtained for alpha and proportion of invariable sites parameters were 0.287 and 0.323 respectively.

Secondly, we assessed the relationships between only medieval or all Corsican sequences, by performing median-joining networks, using the Network software ([45], available at fluxus-engineering.com) with default parameters. Network Publisher was used to manipulate the networks. To compare the Corsican mitochondrial genetic diversity with the Mediterranean or worldwide diversity, we defined supplementary datasets, one by haplogroup observed, gathering all the sequences published and covering the 130 bp fragment. These datasets combined with the Corsican sequences were used to draw median-joining networks (references are given in the legends of Figures S1 and S2).

Thirdly, to assess the past demographic history of the Corsican goats we performed a Bayesian Skyline Plot (BSP) using BEAST v.1.5.4 software [46]. All the Corsican sequences were used and average tip dates were given for all medieval sequences: 850 years BP for the XIIth century and 650 years BP for those of the XIVth century. BEAUti v1.5.3. was used to build the xml file by using the following parameters: HKY+I+G4 (best model for this dataset assessed by jModeltest and AIC criteria [42]); uncorrelated lognormal relaxed clock; 5 groups and 100 millions of iterations with parameters saved every 10 000 iterations; Burn-in: the first 10% were discarded. The results of 4 independent runs were analysed and the Bayesian Skyline Plot reconstructed with Tracer v.1.4.1 (Figure S3).

Finally, to describe the genetic diversity observed for the Corsican goats, we computed different classical parameters using either DnaSP v5 [47] or Arlequin v3.5.1.2 [48] including haplotype diversity and frequencies, sequence diversity, pairwise comparisons, and population comparisons (F_{ST}, Fu tests). We compared the different Corsican datasets with each other but also with other datasets corresponding to mainland or island populations (see [16] for accession numbers and geographical origin). We focused in particular on Sardinia's island (75 sequences, accession numbers FJ571522 to FJ571596, [17]; Corsica's closest island) and Portugal (the biggest dataset generated for mainland goats, 288 sequences, accession numbers AY961629 to AY961916, [11]). We also considered 4 different datasets (see Supporting Information for details and references) corresponding to the southern or the northern area of the Mediterranean Sea, the Mediterranean islands (Sicily and Sardinia), and finally 8 Neolithic goats of Baume d'Oullens [38]. To reduce possible biases due to large differences in the size of the datasets (low number of Corsican sequences), we randomly sampled 49 sequences of the non-Corsican sequences and repeated this operation at least three times. The genetic parameters estimated by DnaSP v5 [47] or Arlequin v3.5.1.2 [48] were then computed on these resampled datasets of equal size.

Results

Medieval goat samples

We analyzed 29 bones and obtained reproducible and congruent sequences of *Capra hircus* for 25 of them (Table S1). We are confident that these results are authentic as we obtained the same results in both laboratories. We also took the ancient DNA precautions recommended by the community as we are used to do (e.g. [34,35,38,49,50]). When different bones were supposed to be from a single individual (Ro-5, Ro-10 and Ro-22), systematically we obtained the same sequence confirming the first assessments. Finally, we determined 21 sequences coming from different individuals: 10 dated to the XIIth century and 11 dated to the first half of the XIVth century. All medieval sequences were from the A haplogroup as shown by the phylogenetic analysis (Figure 2). The six haplogroups appeared monophyletic and were supported by posterior probabilities (pp) higher than 0.9 for 3 of them B, C, G (F is not concerned as only one sequence is used). The A haplogroup that had the highest number of sequences and that was the more diverse received the lowest support (pp<0.5). Nevertheless, the clustering of the newly determined sequences inside the A haplogroup raises no doubt as confirmed by subsequent network analyses (see also mismatch distributions, Figure S4). Substantial diversity is observed among medieval haplotypes as seen on the network (Figure 3A). Among the 21 sequences, we detected 14 unique haplotypes (Table S1) with haplotype Ha 04 being the most frequent (5 individuals). Six out of the 14 haplotypes have never been described before. On average, the mean number of pairwise differences observed between sequences reaches 4.65±2.37 (Table 1, Figure S4). The diversity appeared not significantly different between the XIIth century with 9 haplotypes for 10 sequences (13 polymorphic sites) and the XIVth century with 8 haplotypes for 11 sequences (15 polymorphic sites). According to the network performed on all medieval sequences (14 haplotypes, 19 polymorphic sites), only 3 haplotypes were shared between both periods (Figure 3A). Two of them had a central position in the network (Ha 04 and 09; Figure 3A).

Present-day goat samples

28 individuals from 5 different localities were studied (Table S1 for details). Two different mitochondrial haplogroups were observed (A and C; Figure 2) with a higher proportion of A sequences (92.8% i.e. 26 out of 28 sequences). Considering all A and C sequences, the mean number of pairwise differences reaches 7.01±3.39 (29 polymorphic sites; Figure S4). However, when only sequences of the A haplogroup were considered, this value drops to 4.90±2.47 (19 polymorphic sites, Table 1) which is close to the one observed for A medieval goats. The two sequences from the C haplogroup came from the same locality, Tralonca (Figure 1), and shared the same haplotype Ha 26 that has not been described elsewhere (Figure 3B). 15 haplotypes were obtained for the 26 sequences of the A haplogroup among which, 10 were only observed in Corsica. The most frequent haplotypes were, like for the late medieval goats, Ha 04 and Ha 09, with 4 individuals from 3 different localities in both cases (Figure 3B).

Comparison of Corsican goat diversity through time

No significant difference was observed between the goats of the XIIth and the XIVth centuries (non significant F_{ST}-value; Table S2). The difference became significant when medieval and present goats were compared (0.036, p-value 0.027). This was due to the presence of the C haplogroup in the present-day sequences since the test was no more significant when only A sequences were taken into account (0.027, p-value 0.099). Finally, there were no significant differences between the medieval and present-day sequences, with or without considering the C sequences. Medieval and present-day goats shared 4 haplotypes (Ha 04, 06, 07 and 09, Figure 3B) but only one was specific to Corsica (Ha 06). The 3 remaining haplotypes were generally frequent in other populations.

No significant changes were observed in the demographic history of the Corsican population using the Bayesian Skyline Plot (Figure S3). Neither a sign of expansion nor of a crash were observed from the medieval times to date, as rather a constant population size pattern was obtained. This is congruent with the Tajima's D values obtained (Table 1) but not with all the Fu's Fs values computed. However, this latter parameter can reflect other factors than population growth (selection, bottleneck, … [16]).

Comparison of Corsican goat diversity with other geographical places

We compared the diversity observed in Corsica with other Mediterranean island populations and mainland breeds. The 75 sequences of Sardinia were clustered in 46 haplotypes that were all from the A haplogroup [17]. Similarly, the 288 Portuguese sequences were represented by 104 haplotypes, from which only one was from the C haplogroup and all the others from the A haplogroup [11]. By expanding the comparison to larger or other areas, we observed that the C haplogroup is described only in Europe (in present-day Northern Mediterranean area and already

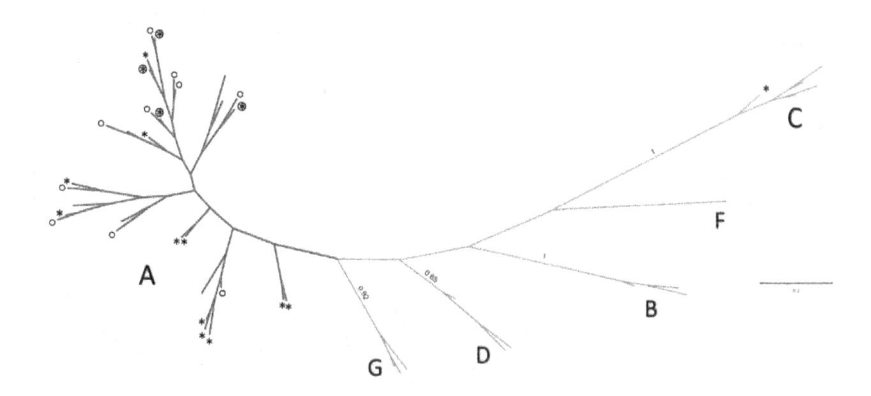

Figure 2. Mitochondrial diversity of Corsican goats compared to world-wide goats diversity. Ancient (circles) and current (stars) Corsican sequences have been used with 20 sequences of reference corresponding to all haplogroups described up to now [14] to reconstruct a Bayesian phylogenetic tree using a GTR+I+G model of evolution (see text). Only a single sequence by haplotype has been used, so haplotypes in common between medieval and today periods are indicated by both a circle and star. Only posterior probabilities higher than 0.6 are indicated.

at the Neolithic time, Table S3 and not shown). Both the medieval and present-day Corsican groups appeared significantly different from all other groups whatever the resampled datasets taken into account (Table 1, Tables S2 and S3). Similarly, all the non-Corsican groups of goats also appeared significantly different from each other (data not shown). However, four different haplotypes (Ha 05, 08, 09 and 10) were shared between Corsica and Sardinia islands and six (Ha 07, 08, 09, 10, 12 and 21) between Corsican and Portuguese goats.

The median-joining network performed on 39 worldwide sequences of the C haplogroup (Figure S1) revealed a classical expansion structure with European sequences on one side and Asian sequences on the other (with the single exception of one Swiss haplotype; [8]). As expected, the sequences obtained from Neolithic goats of the archeological site of Baume d'Oullens in France [38], among the first goats to have been diffused in Europe, appeared in the central part of the European cluster. The Corsican haplotype showed the highest number of substitutions with this

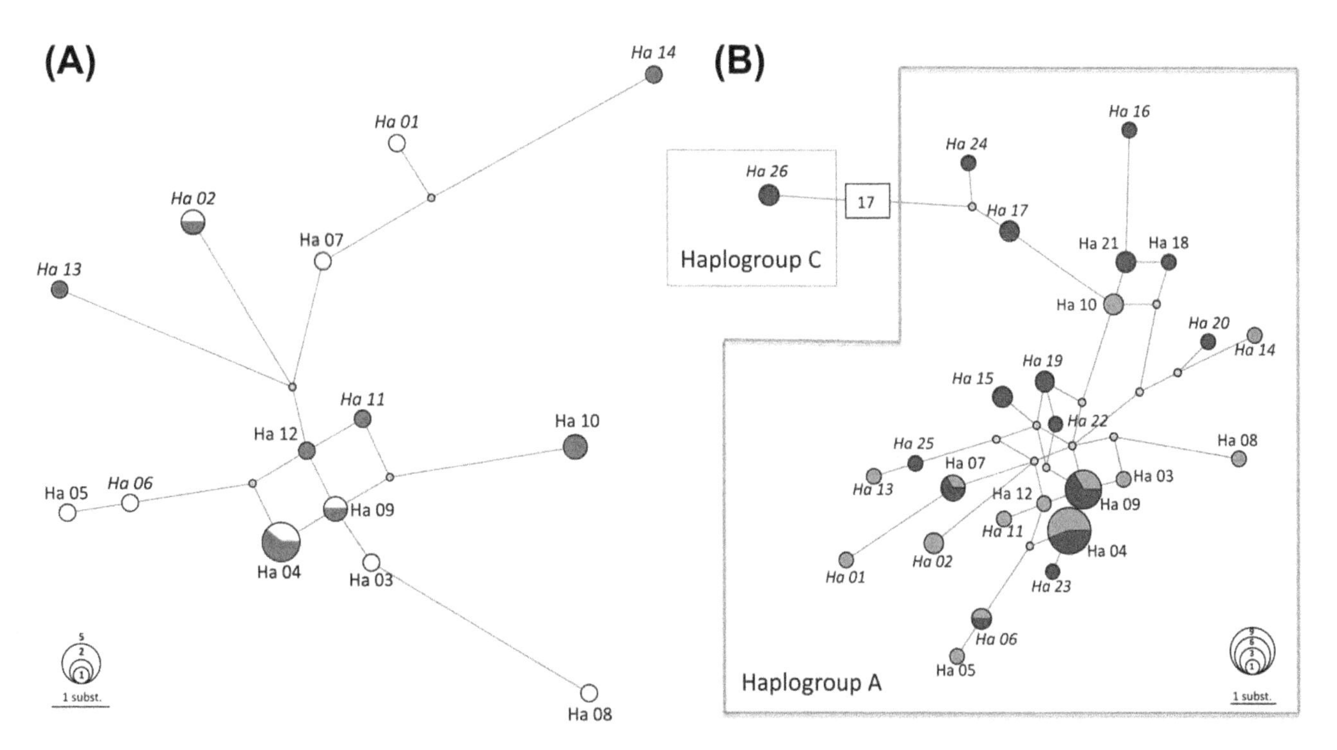

Figure 3. Networks generated with CR sequences of Corsican goats from A) medieval samples only or B) combined with present-day samples. The circles are proportional to the number of sequences obtained and colours indicated the period: yellow corresponds to XIIth century, red to XIVth century, orange to all medieval and blue to present-day sequences. See Table 1 for details about haplotypes and archeological samples. Haplotypes indicated in italics are those only observed in Corsica up to now. Haplotype Ha 26 is the only one from the haplogroup C and differs from the closer haplotype by 17 positions that is materialized by a square with this number.

Table 1. Genetic diversity parameters.

	Number of sequences	Number of haplotypes	Haplotype diversity	Mean number of pairwise differences	Nucleotide diversity	Fs (Fu)	Tajima D
Corsican goats							
XIIth century	10	9	0.9778±0.0540	4.82±2.57	0.0371±0.0224	−3.7113 (p: 0.0200)	0.2247 (p: 0.6240)
XIVth century	11	8	0.9273±0.0665	4.62±2.45	0.0355±0.0213	−1.6798 (p: 0.1530)	−0.4355 (p: 0.3440)
All Medieval	21	14	0.9381±0.0388	4.65±2.37	0.0358±0.0204	−5.0723 (p: 0.0130)	−0.4484 (p: 0.3590)
Present Corsican	28	16	0.9524±0.0208	7.01±3.40	0.0540±0.0291	−2.8511 (p: 0.1280)	−0.2150 (p: 0.4590)
Present Corsican haplogroup A only	26	15	0.9477±0.0238	4.90±2.47	0.0377±0.0211	−4.3581 (p: 0.0380)	−0.0555 (p: 0.5510)
All Corsican (Med+Pre.)	49	26	0.9473±0.0189	6.12±2.96	0.0471±0.0253	−9.8130 (p: 0.0010)	−0.3022 (p: 0.4350)
All Corsican (Med+Pre.) haplogroup A only	47	25	0.9436±0.0204	4.85±2.41	0.0373±0.0206	−11.9233 (p: 0.0000)	−0.0826 (p: 0.5300)
Other goats							
Present Sardinian	75	46	0.9827±0.0052	5.68±2.75	0.0437±0.0235	−25.3300 (p: 0.0000)	−0.7342 (p: 0.2470)
Present Portuguese	288	104	0.9790±0.0030	5.30±2.60	0.0414±0.0221	−24.8789 (p: 0.0000)	−0.7741 (p: 0.2560)

The different parameters for all datasets (see text) have been computed on a 130 bp fragment of the HVI region of the mitochondrial control region (CR). The values are given with their confidence interval (±) and tests with their p-value (p).

central node (3 substitutions). The same analysis performed on the sequences from the A haplogroup restricted to the Mediterranean area, showed a more complex history with no clear emerging pattern (Figure S2).

Discussion

DNA preservation in medieval samples

From the 29 archeological bones dated back to the XIIth and XIVth centuries, we obtained 21 sequences from different ancient goat individuals: 10 from the XIIth century deposit and 11 from the XIVth century deposit, which represents a surprisingly high DNA preservation for remains in Corsica. Such a good preservation of the DNA, which is here correlated to the good preservation of the bones themselves like in most late Medieval Corsican sites [19,51], is probably due to the recent age of the site.

Characterization through time of mitochondrial genetic diversity in Corsican goats

The comparison of Middle Ages and present mitochondrial diversity was carried out using 28 present-day goats from five different localities. This comparison may be slightly biased by differences existing in the time span and geographic distribution for either ancient or modern samples. Indeed, the sampled geographic area is larger for the modern goats (Figure 1) whereas the time span is longer for the ancient samples.

Our results tend to prove that the mitochondrial diversity of Corsican goats has remained relatively constant since the Middle Ages. Moreover, we detect no significant demographic changes (F_{ST}, BSP and Table 1) or decrease of genetic diversity. The only difference between both periods is the occurrence of two C haplotypes in the present-day samples, all other goats belonging to the A haplogroup. However, given a dataset of 21 medieval individuals and assuming a constant frequency of the C haplogroup in Corsica ($2/28 = 0.0714$), we have a 21% of chance of having missed this haplogroup in the Middle Ages sampling.

The occurrence of the C haplogroup in Corsican goats in the context of the Neolithic diffusion

The presence of the A and C haplogroups in Corsica is in agreement with the goat mitochondrial variability observed in European countries. Most of European goats are from the A haplogroup, with C haplotypes found at a rare frequency in Switzerland, Portugal, Spain and Slovenia [3,8,11,14]. This is also consistent with previous paleogenetic studies that already detected both haplogroups in goats from Southern France in the early Neolithic period [38]. The median-joining network performed on all the C haplotypes found worldwide (Figure S1) gave results congruent with the diffusion of goats in the Neolithic. Indeed we observed: i) a clear separation of the European haplotypes from the Asian ones with only one exception, a shared haplotype between China and Switzerland; ii) a star-like pattern for European haplotypes suggesting a population expansion with a central position for the Neolithic haplotypes [38]; iii) a divergent haplotype for Corsica compatible with a subsequent isolation.

Without a doubt, the origin of the C haplogroup in Europe can be traced back to the Neolithic spread where its frequency was probably higher than the one observed now [38]. Today, the C haplotypes are relatively scarce in goats and appear more like reminiscent testimonies of this first diffusion. Interestingly, not one has been detected yet in the Southern Mediterranean area or in other Mediterranean islands than Corsica (see Supporting Information). According to the few data we have, it is difficult to conclude on the ancient origin of the Corsican C haplotype or a more recent origin linked to subsequent exchanges with the

Northern Mediterranean mainland. However, its position in the network is compatible with the first explanation. Further ancient DNA studies, for a larger area and older period, would be very interesting to highlight this question.

Variability in Corsica vs other Mediterranean areas

The relative stability through time of goat diversity in Corsica could be explained by regular importation of goats from other continental areas or islands, e.g. for commercial trade, as many contacts by sea have been reported during the last centuries in the Mediterranean area [52]. However, a striking point is that the diversity observed in Corsica is substantial for both periods (Middle Ages and present-day, Figure 3A and Figure 3B respectively) and differs from that of other places (Table 1, data not shown); only about half of the Corsican haplotypes (14/26) are shared with goats from other geographical regions despite our study focused on a short CR fragment (130 bp). This result is not unexpected however. Previous studies - usually targeting the 480 bp fragment first described in [8] and covering our shorter fragment - have already shown that goats were more polymorphic than other livestock (cattle, sheep, pig) on the CR [11,15,53]. Moreover, the specific analysis of the A haplogroup (more than 90% of the modern goats, [14]) confirmed high haplotype diversities for 20 populations/countries with values close to one [16]. Pereira et al. [16] showed a strong correspondence between mitochondrial genetic and geographic distances suggesting that after the initial expansion, differentiation among regions has been established and maintained [16]. A similar conclusion was obtained with large-scale nuclear SNP analyses obtained for 16 breeds of goats [54] and from microsatellites analyses including Corsican and other European goat populations [55]. This seems in agreement with what we observed in Corsica when compared with other Mediterranean populations (F_{ST} comparisons, Table S2, data not shown).

Implication of the traditional Corsican husbandry in the maintenance of the variability

Except for some haplotypes that are common in many countries (Figure S2) and probably constitute traces of the initial diffusion, 46% of the Corsican haplotypes found have not been previously described elsewhere. The preservation through time of this endemic genetic diversity and its constant level since medieval time could suggest that relatively large effective population sizes have been maintained in domestic goats through exchanges of animals. But ethnographic insights into the herding practices carried out in the Niolu [20] provide another possible or complementary explanation for the preserved genetic diversity of the Corsican goat breed.

Typically in the past for the Mediterranean area, goats were usually moved according to the seasonal changes (transhumance) to gain access to more reliable food. Because fodder resources fluctuate in the wild according to different factors (e.g. annual climatic conditions), the system developed in Corsica has been extensive with goats left free-ranging most of the times although under the careful control of the herders. This particular system led to the characteristics observed in the Corsican goats [21,56]: i) the high diversity of coat colours is encouraged as individuals can be more easily identified by sight; ii) only the strongest and toughest goats can generally survive in this relatively hostile environment, explaining why the introduction of goats from industrial breeds usually failed; iii) large herds are usually managed to maximize the herd's productivity overall instead of individual productivity. Indeed, selection is not performed to optimize for instance, either milk productivity or the fecundity, but instead to obtain a constant productivity of the herd by year whatever the weather or difficulties encountered [57]. All these points lead to a strategy where the phenotypic diversity, and the underlying genetic diversity, is maximized in order to obtain herds that are highly adaptable, rustic and robust.

In order to do that, herders exert strong selection pressure while forming their herds to gain in productivity but also to favour behavioural traits. Indeed, along with the search for constant productivity, one of the main objectives of herders is to design flocks that will maintain "families" from the same maternal lineage [20]. A young descendant female is usually chosen according to its mother's and grandmother's family, taking into account its desirable productive traits as well as its abilities to endure the tough conditions. Animals coming from the same family/pool will live together and move together more easily and naturally. Such "familial" behaviour is extremely beneficial to the cohesion of herds under free-ranging exploitation and especially during the movements on the pasturelands (transhumance). On the contrary, introduction of new animals from other herds can induce significant disturbance in the herd movements by breaking the cohesion of the group. This practice of herders of course has the effect of reducing the genetic diversity. Hence, to insure the persistence of the genetic diversity and to "change the blood", "cambià u sangui" in the language of traditional Corsican herders, selected males of these different maternal lineages or "families" will be exchanged between flocks to avoid inbreeding and, without control, renew the pool of the mating males within a seasonal time span. As mitochondria are inherited maternally, this system will naturally lead to maintaining high mitochondrial genetic diversity between herds or "families". So, the diversity we observed would be not promoted by large herds with many exchanges of females between them but on the contrary by traditional practices. These ones rely on very few introductions of females coming from other herds to keep the kinship within the herd as tight as possible to reinforce cohesive behaviour during the ranging but also on a mixing of the genetic diversity through the exchanges of males only.

Such entangled practices in the Corsican husbandry system can explain why we do not observe change of the mitochondrial diversity in Corsican goats since the late Medieval period, and probably earlier if we had been able to investigate the genetic diversity of goats from earlier periods. It's thus highly probable that sustainable husbandry practices in today's Corsica that are so well adapted to their environment are, at least partly, the result of practices over millenaries contributing to the maintenance of a relatively high level of genetic diversity. Finally, analyses of additional samples and genetic markers coupled with population simulations using varying population genetic parameters should help to test between both hypotheses, regular importation of goats or millenaries husbandry practices.

Towards a protection of the Corsican goat breed

In this study, we observed that the mitochondrial diversity of goats in Corsica Island has been maintained since the Middle Ages to date. In a time where rustic breeds are endangered and industrial breeds tend to reduce the genetic resources [22], Corsican goats, by the model of husbandry used and the high diversity observed constitute an interesting pool to preserve for the future management of domestic genetic resources. All the actors concerned by the Corsican goat (regional association in charge of its management, public authorities, researchers and extension services) should so pay attention to make their preservation successful all the more because paratuberculosis has started to decimate flocks.

Ethic statement

The medieval bones of goats were excavated from the Rostino archeological site. Daniel Istria, in charge of the excavation, authorized their analyses. Tissue samples were collected in Corsica in the framework of the ECONOGEN project (European Union contract QLK5-CT-2001-02461), following the European ethical rules implemented in all European projects.

Supporting Information

Figure S1 Network generated with CR sequences of the C haplogroup (130 bp). Only domestic goats have been considered (35 sequences; Luikart et al. 2001 [8], Sultana et al. 2003, Joshi et al. 2004, Pereira et al. 2005 [11], Chen et al. 2005, Liu et al. unpublished, Naderi et al. 2007 [14]). Neolithic sequences coming from the archeological site of Baume d'Oullen (Fernández et al. 2006 [38]) were also taking into account (4 sequences).

Figure S2 Network generated with CR sequences of the A haplogroup (130 bp) for goats coming from the Mediterranean Sea around Corsica. 584 sequences were used for the median-joining network analysis coming from: North Mediterranean area (Italy, France, Spain), Mediterranean Islands (Malta, Sicily, Sardinia), South Mediterranean area (Morocco, Algeria, Tunisia). See Pereira et al. 2009 [16] for the accession numbers of the sequences used and their geographical origin. Positions were weighted inversely to the number of mutations observed by position on a first run-test.

Figure S3 Bayesian Skyline Plot. The analyses were performed using the 49 Corsican sequences with date-tips for the medieval sequences (see text for details). X-axis: Time in years; Y-axis: Population size (Neτ) in log-scale. Mean is plotted with the 95% HPD. 3 runs of 100 M of iterations were performed (ESS >200 for all parameters).

Figure S4 Mismatch distributions for medieval and extant Corsican goat populations. The numbers of pairwise differences are given on the x-axis and their frequency on the y-axis.

Table S1 Corsican goats sampling and mitochondrial genotyping results. A star indicates archeological samples for which the molecular identification was *Ovis aries* and a dash when no amplification was obtained.

Table S2 Population pairwise F_{ST}. Pairwise difference was used as distance method. F_{ST} values are given at upper right corner and corresponding p-values at bottom left. Significant p-values are highlighted in color (orange for intra-Corsica comparisons, pink for inter comparisons). NS: Non-significant, * p-value between 0.05 and 0.01, ** p-value between 0.01 and 0.001, *** p-value < to 0.001.

Table S3 Genetic diversity parameters. Comparison for A and C haplogroups between Corsican and other Mediterranean or Portuguese datasets (see Figures S1, S3 and text).

Acknowledgments

We thank Hamid Reza Rezaei and Saeid Naderi for having given us details about samples of Corsican goats they have published. We also thank members of the paleogenetics lab and the Palgene Platform (CNRS, ENS Lyon) for help during the experiments. We are also grateful to Javier Oliver and Benjamin Gillet for critical reading of the manuscript. Finally, we express gratitude to Joanne Burden who kindly proofread the English.

Author Contributions

Conceived and designed the experiments: PT HF TC SH CH. Performed the experiments: MD HF. Analyzed the data: SH TC. Contributed reagents/materials/analysis tools: DI J-DV PT CH. Wrote the paper: SH FP TC. Provided expertise for Corsican husbandry, ancient DNA and archeozoological data: FC CH J-DV.

References

1. Zeder MA, Hesse B (2000) The initial domestication of goats (Capra hircus) in the Zagros mountains 10,000 years ago. Science 287: 2254–2257.
2. Peters J, von den Driesch A, Helmer D (2005) The First Steps of Animal Domestication. New Archaeological Approaches Vigne JDPJ, Helmer D, eds. Oxford, UK: Oxbow Books.
3. Naderi S, Rezaei HR, Pompanon F, Blum MG, Negrini R, et al. (2008) The goat domestication process inferred from large-scale mitochondrial DNA analysis of wild and domestic individuals. Proc Natl Acad Sci U S A 105: 17659–17664.
4. Vigne JD (2011) The origins of animal domestication and husbandry: a major change in the history of humanity and the biosphere. C R Biol 334: 171–181.
5. Guilaine J (2003) De la vague à la tombe. La conquête néolithique de la Méditerranée. Paris.
6. Tresset A, Bollongino R, Edwards CJ, Hughes S, Vigne JD (2009) Early diffusion of domestic bovids in Europe: an indicator for human contacts, exchanges and migrations? In: Hombert J-MadE, F., eds. The Origin of Man, Language and Languages. Cambridge: McDonald Institute Monograph Series. pp 69–90.
7. FAOSTAT (2009) Preliminary data for production of Live Animals in 2009 in the World. FAOSTAT website. Available: http://faostat.fao.org/ Accessed 2011 May 10.
8. Luikart G, Gielly L, Excoffier L, Vigne JD, Bouvet J, et al. (2001) Multiple maternal origins and weak phylogeographic structure in domestic goats. Proc Natl Acad Sci U S A 98: 5927–5932.
9. Azor PJ, Monteagudo LV, Luque M, Tejedor MT, Rodero E, et al. (2005) Phylogenetic relationships among Spanish goats breeds. Anim Genet 36: 423–425.
10. Chen SY, Su YH, Wu SF, Sha T, Zhang YP (2005) Mitochondrial diversity and phylogeographic structure of Chinese domestic goats. Mol Phylogenet Evol 37: 804–814.
11. Pereira F, Pereira L, Van Asch B, Bradley DG, Amorim A (2005) The mtDNA catalogue of all Portuguese autochthonous goat (Capra hircus) breeds: high diversity of female lineages at the western fringe of European distribution. Mol Ecol 14: 2313–2318.
12. Sardina MT, Ballester M, Marmi J, Finocchiaro R, van Kaam JB, et al. (2006) Phylogenetic analysis of Sicilian goats reveals a new mtDNA lineage. Anim Genet 37: 376–378.
13. Fan B, Chen S-L, Kijas JH, Liu B, Yu M, et al. (2007) Phylogenetic relationships among Chinese indigenous goat breds inferred from mitochondrial control region sequence. Small Ruminant Research 73: 262–266.
14. Naderi S, Rezaei HR, Taberlet P, Zundel S, Rafat SA, et al. (2007) Large-scale mitochondrial DNA analysis of the domestic goat reveals six haplogroups with high diversity. PLoS One 2: e1012.
15. Wang J, Chen YL, Wang XL, Yang ZX (2008) The genetic diversity of seven indigenous Chinese goat breds. Small Ruminant Research 74: 231–237.
16. Pereira F, Queiros S, Gusmao L, Nijman IJ, Cuppen E, et al. (2009) Tracing the history of goat pastoralism: new clues from mitochondrial and Y chromosome DNA in North Africa. Mol Biol Evol 26: 2765–2773.
17. Vacca GM, Daga C, Pazzola M, Carcangiu V, Dettori ML, et al. (2010) D-loop sequence mitochondrial DNA variability of Sarda goat and other goat breeds and populations reared in the Mediterranean area. Journal of animal breeding and genetics = Zeitschrift fur Tierzuchtung und Zuchtungsbiologie 127: 352–360.
18. Vigne JD (1999) The large "true" Mediterranean islands as a model for the Holocene human impact on the European vertebrate fauna? Recent data and new reflections. In: Benecke N, ed. The Holocene history of the European vertebrate fauna Modern aspects of research Berlin Deutsches Archäologisches Institut, Eurasien-Abteilung (Archäologie in Eurasien, 6). pp 295–322.
19. Vigne JD Les Mammifères post-glaciaires de Corse, étude Archéozoologique (XXVIe suppl. à Gallia Préhistoire); CNRS, editor. Paris.

20. Ravis-Giordani G (1983) Bergers corses. Communautés villageoises du Niolu. Aix-en-Provence: Edisud.
21. Bouche R, Aragni CHJ, Bordeaux C (2009) Caprin extensif en Corse: Savoirs durables en quête de développement. Options Méditerranéennes, A 91: 209–213.
22. Taberlet P, Valentini A, Rezaei HR, Naderi S, Pompanon F, et al. (2008) Are cattle, sheep, and goats endangered species? Mol Ecol 17: 275–284.
23. Pernet F, Lenclud G (1977) Berger en Corse. Essai sur la question pastorale. Grenoble: Presses universitaires de Grenoble.
24. Istria D (2005) Pouvoirs et fortifications dans le Nord de la Corse, XIe-XIVe siècle. Ajaccio Editions Alain Piazzola. 517 p.
25. Cucchi T (2003) Production et distribution de biens d'origine animale en contexte seigneurial de Haute-Corse: la faune mammalienne de la salle 1 du castrum de Rostino (XIIIe-XIVe siècles). Bulletin d'Archéologie et d'Histoire de la Corse 1: 70–93.
26. Payne S (1985) Morphological distinction between the mandibular teeth of young sheep, Ovis and goats, Capra. Journal of Archaeological Science 12: 139–147.
27. Helmer D (2000) Discrimination des genres Ovis et Capra à l'aide des prémolaires inférieures 3 et 4 et interprétation des âges d'abattage: l'exemple de Dikili Tash (Grèce). Anthropozoologica 31.
28. Boessneck J (1969) Oesteological differences beetween sheep (Ovis aries Linne) and goat (Capra hircus Linne). In: D HEB, ed. Science in Archaeology. Londres: Thames and Huston. pp 331–358.
29. Clutton-Brock J, Dennis-Brian K, Armitage P-A, Jewel P (1990) Osteology of the Soay sheep. Bull Br Mus Nat Hist 56: 1–56.
30. Helmer D, Rocheteau M Atlas du squelette appendiculaire des principaux genres holocènes de petits ruminants du nord de la méditerranée et du Proche-Orient; APDCA, editor. Juan-les-Pins.
31. Poplin F (1979) Origine du Mouflon de Corse dans une nouvelle perspective paléontologique: par marronnage. Annales de Génétique et Sélection Animale 11: 133–143.
32. Loreille O, Vigne JD, Hardy C, Callou C, Treinen-Claustre F, et al. (1997) First distinction of sheep and goat archaeological bones by the means of their fossil mtDNA. Journal of Archaeological Science 24: 33–37.
33. Collins M, Buckley M, Grundy HH, Thomas-Oates J, Wilson J, et al. (2010) ZooMS: the collagen barcode and fingerprints. Spectroscopy Europe 22: 11–13.
34. Hughes S, Hayden TJ, Douady CJ, Tougard C, Germonpre M, et al. (2006) Molecular phylogeny of the extinct giant deer, Megaloceros giganteus. Mol Phylogenet Evol 40: 285–291.
35. Calvignac S, Hughes S, Tougard C, Michaux J, Thevenot M, et al. (2008) Ancient DNA evidence for the loss of a highly divergent brown bear clade during historical times. Mol Ecol 17: 1962–1970.
36. Loreille O, Orlando L, Patou-Mathis M, Philippe M, Taberlet P, et al. (2001) Ancient DNA analysis reveals divergence of the cave bear, Ursus spelaeus, and brown bear, Ursus arctos, lineages. Curr Biol 11: 200–203.
37. Orlando L, Bonjean D, Bocherens H, Thenot A, Argant A, et al. (2002) Ancient DNA and the population genetics of cave bears (Ursus spelaeus) through space and time. Mol Biol Evol 19: 1920–1933.
38. Fernández H, Hughes S, Vigne JD, Helmer D, Hodgins G, et al. (2006) Divergent mtDNA lineages of goats in an Early Neolithic site, far from the initial domestication areas. Proc Natl Acad Sci U S A 103: 15375–15379.
39. Hofreiter M, Jaenicke V, Serre D, Haeseler Av A, Paabo S (2001) DNA sequences from multiple amplifications reveal artifacts induced by cytosine deamination in ancient DNA. Nucleic Acids Res 29: 4793–4799.
40. Gouy M, Guindon S, Gascuel O (2010) SeaView version 4: A multiplatform graphical user interface for sequence alignment and phylogenetic tree building. Mol Biol Evol 27: 221–224.
41. Villesen P (2007) FaBox: an online toolbox for fasta sequences. Mol Ecol Notes 7: 965–968.
42. Posada D (2008) jModelTest: phylogenetic model averaging. Mol Biol Evol 25: 1253–1256.
43. Huelsenbeck JP, Ronquist F (2001) MRBAYES: Bayesian inference of phylogenetic trees. Bioinformatics 17: 754–755.
44. Ronquist F, Huelsenbeck JP (2003) MrBayes 3: Bayesian phylogenetic inference under mixed models. Bioinformatics 19: 1572–1574.
45. Bandelt HJ, Forster P, Rohl A (1999) Median-joining networks for inferring intraspecific phylogenies. Mol Biol Evol 16: 37–48.
46. Drummond AJ, Rambaut A, Shapiro B, Pybus OG (2005) Bayesian coalescent inference of past population dynamics from molecular sequences. Mol Biol Evol 22: 1185–1192.
47. Librado P, Rozas J (2009) DnaSP v5: a software for comprehensive analysis of DNA polymorphism data. Bioinformatics 25: 1451–1452.
48. Excoffier L, Lischer HEL (2010) Arlequin suite ver 3.5: A new series of programs to perform population genetics analyses under Linux and Windows. Mol Ecol Ressources 10: 564–567.
49. Orlando L, Calvignac S, Schnebelen C, Douady CJ, Godfrey LR, et al. (2008) DNA from extinct giant lemurs links archaeolemurids to extant indriids. BMC Evol Biol 8: 121.
50. Calvignac S, Hughes S, Hänni C (2009) Genetic diversity of endangered brown bear (Ursus arctos) populations at the crossroads of Europe, Asia and Africa. Diversity and Distributions 15: 742–750.
51. Ruas M-P, Vigne JD (1995) Perspectives archéobotaniques et archéozoologiques pour les périodes historiques en Corse. Patrimoine d'une île, 1: Recherches récentes d'archéologie médiévale en Corse: Ajaccio: Soc. Archéol. Corse-du-Sud. pp 113–126.
52. Charlet P, Le Jaouen JC (1977) Les populations caprines du bassin méditerranéen. Aptitudes et évolution. Options Méditerranéennes, A 35: 45–53.
53. Bruford MW, Bradley DG, Luikart G (2003) DNA markers reveal the complexity of livestock domestication. Nat Rev Genet 4: 900–910.
54. Pariset L, Cuteri A, Ligda C, Ajmone-Marsan P, Valentini A (2009) Geographical patterning of sixteen goat breeds from Italy, Albania and Greece assessed by Single Nucleotide Polymorphisms. BMC Ecol 9: 20.
55. Canon J, Garcia D, Garcia-Atance MA, Obexer-Ruff G, Lenstra JA, et al. (2006) Geographical partitioning of goat diversity in Europe and the Middle East. Anim Genet 37: 327–334.
56. Bouche R, Gambotti JY, Maestrini O (2004) L'avenir c'est de durer… Quand le changement en élevage extensif nécessite le maintien ou la réinvention de systèmes collectifs. Cas des caprins corses. L'évolution des systèmes de production ovine et caprine: Avenir des systèmes extensifs face au changement de la société, Alghero (ITA). Montpellier, France: CIHEAM. pp 319–327.
57. Hugot S, Bouche R (1999) Regards sur les pratiques de l'élevage extensif de la chèvre corse: Préalable à la mise en place d'un schéma de sélection. Options Méditerranéennes, A 38: 137–143.

Why Herd Size Matters – Mitigating the Effects of Livestock Crashes

Marius Warg Næss[1]*, Bård-Jørgen Bårdsen[2]

1 Center for International Climate and Environmental Research – Oslo (CICERO), Fram Centre, Tromsø, Norway, 2 Norwegian Institute for Nature Research (NINA), Arctic Ecology Department, Fram Centre, Tromsø, Norway

Abstract

Analysing the effect of pastoral risk management strategies provides insights into a system of subsistence that have persevered in marginal areas for hundreds to thousands of years and may shed light into the future of around 200 million households in the face of climate change. This study investigated the efficiency of herd accumulation as a buffer strategy by analysing changes in livestock holdings during an environmental crisis in the Saami reindeer husbandry in Norway. We found a positive relationship between: (1) pre- and post-collapse herd size; and (2) pre-collapse herd size and the number of animals lost during the collapse, indicating that herd accumulation is an effective but costly strategy. Policies that fail to incorporate the risk-beneficial aspect of herd accumulation will have a limited effect and may indeed fail entirely. In the context of climate change, official policies that incorporate pastoral risk management strategies may be the only solution for ensuring their continued existence.

Editor: Jon Moen, Umea University, Sweden

Funding: The present study is part of the ECOPAST project financed by the Research Council of Norway (the FRISAM program, project number 204174). Webpage: http://www.forskningsradet.no/en/Home_page/1177315753906. The funders had no role in study design, data collection and analysis, decision to publish, or preparation of the manuscript.

Competing Interests: The authors have declared that no competing interests exist.

* E-mail: m.w.nass@cicero.tromso.no

Introduction

More of the Earth's land surface is used for grazing than for any other purpose [1]. Extensive pastoral production occurs in 25% of the global land area from the drylands of Africa and the Arabian Peninsula, to the highlands of Asia and Latin America and the Arctic parts of Fennoscandia and Russia [2]. Specifically, grazing land covers 77% of Australia, 61% of Africa, 49% of Asia and 18% of Europe [1]. It has been estimated that pastoralists produce 10% of the world's meat, and supports some 200 million pastoral households who raise nearly 1 billion head of camel, cattle and smaller livestock [2]. The main livestock species kept by pastoralists are cattle, donkeys, goats and sheep, although they also keep, e.g., alpaca and llamas in the Andes, camels and horses in east-central Asia, the dromedary in Africa and West Asia, reindeer in northern Eurasia, and yak on the Tibetan Plateau [1]. Pastoralism is also economically important, especially in poor regions: compared to settled farmers in Africa, pastoralists produce 50–70% of all the milk, beef and mutton produced on the continent and while comprising only 1.5% of the total population of Iran, pastoralists keep 25% of the national herd [1]. From a global point of view, >1 billion people depend on livestock, and 70% of the 880 million rural poor living on less than USD 1.00 per day are at least partially dependent on livestock for their livelihoods [3]. Accordingly, Dong et al. [4] argue that pastoralism is important from a global point of view because of: (1) the human populations it supports; (2) the food and ecological services it provides; (3) the economic contributions it makes; and (4) the long-standing societies it helps to maintain.

Environmental Hazards and Pastoralism

Environmental hazards, such as drought, floods and icing significantly affect livestock survival and reproduction. For example, in Africa mortality rates for cattle during drought have been estimated to be between 35–75% [5] and 10–25% [6], and 25% on average [7]. Small stock losses have been found to range between 1–35% (mean = 24.2%) for sheep and between 5–30% (mean = 16.6%) for goats [6]. Drought has also been found to increase the number of stockless households from 7 to 12% [7]. As for Mongolia, it has been reported that icing in 1993 resulted in the deaths of three-quarter of a million head of livestock where 110 households lost all animals, and 2 090 households lost >70% of their herds [8]. Between 1999 and 2002, 12 million livestock died in winter disasters, and many thousands of households lost their livelihoods [9]. In Inner Mongolia, about 30% of households have lost nearly all their livestock since 2001 due to continuous drought conditions [10]. On the Tibetan Plateau, six harsh winters with heavy snowfall from 1955 to 1990 resulted in 20–30% livestock losses [11]. Specifically, during the winter of 1996–1997 nomads on the western part of the plateau experienced losses of up to 70% and 25% of juvenile and adult goats respectively, and 20% of their lambs [11]. For Tibet in general, some townships lost up to 70% of their total livestock population, and by April 1998 it was estimated that the region had lost over 3 million head of livestock, which represents an estimated loss of USD 125 million [11]. In northern Norway, the reindeer husbandry utilizes winter pastures characterized by a cold but stable continental climate [12]. Nevertheless, mass starvation due to severe winter conditions, i.e. icing events, have been reported to dramatically reduce reindeer populations:

In 1918 one reindeer population was, for example, reduced by a third [13], and adverse weather events, i.e. too much snow in late winter, also caused substantial reductions in 1958, 1962 and 1968 [14].

The effects of environmental hazards are especially important in the context of climate change because the frequency of extreme weather events are predicted to increase in the future (a trend that has already been observed empirically: e.g. [15,16]), and thus represents a significant challenge for pastoralists [17]. It is therefore important to increase our understanding of both impacts of environmental hazards and the effects of strategies aimed at dampening them to enhance the ability of pastoralists to deal with the negative impacts of climate change.

Predictions

Herd accumulation has been argued, as well as to some extent demonstrated empirically, to be an effective strategy for buffering environmental hazards for short periods of time because it seems that wealthier pastoral households weather calamities better than poorer ones [18]. Nevertheless, few studies have evaluated the long-term effect of herd accumulation (but see [19]) and a case has been made that the effectiveness of herd accumulation should be assessed by analysing changes in livestock holdings during crisis periods, i.e. when pastoralists experience a near collapse in livestock holdings [20].

Consequently, our study assess two predictions: First, while the relationship between pre-collapse herd size and losses due to the crisis should be positive, adding one animal to pre-collapse herd size should result in losing less than one animal during the crisis if herd accumulation is an effective strategy for countering the negative impacts of environmental hazards (this is in line with evidence from Africa, see [21]). Second, we also assess to what extent pre-collapse herd size predicts post-collapse herd size, i.e. the *per se* benefit that herders gain by adding animals to their herds [7,20]. In short, we expect pre-collapse herd size to be a positive predictor of post-collapse herd size, if herd accumulation is an effective strategy for countering the negative impacts of environmental hazards.

Materials and Methods

Ethics Statement

The data utilized in this study were provided by the Norwegian Institute of Nature Research as part of the participation in the project ECOPAST (http://pastoralism-climate-change-policy.com/projects/). The standard of ethics pertaining to the data has been approved by the Norwegian Social Science Data Services in connection with the project 'Beregning av produktivitet i reindrift' ('Calculation of productivity in the reindeer husbandry').

The Saami Reindeer Husbandry in Norway

Saami reindeer husbandry has been said to be the cornerstone of the Saami culture in northern Fennoscandia [22]. Although it is difficult to come up with accurate dating of the origin of reindeer husbanding as a pastoral economy, it developed at least 400 years ago [23] and probably evolved from a hunting culture based on wild reindeer. Traditionally, reindeer pastoralism was based on families, or households, which followed the herds year-round where the pastoral economy was based on reindeer products [24]. The reindeer husbandry has undergone major technological, economic and political changes; most notably the production system has changed from being subsistence based to a motorized and market-oriented industry [25]. During the late 1970s, the Norwegian Government became more directly engaged in the reindeer husbandry through subsidies and regulations. Reforms during the end of the 1970s and early 1980s aimed at increasing both production and co-management [26]. In 1976, negotiation between the Saami Reindeer Herders' Association of Norway and the Norwegian Government resulted in the General Agreement for the Reindeer Industry (GARI). Importantly, this laid the foundation for annual negotiations pertaining to official subsidies and development: an arrangement that continues to this day [26]. The Reindeer Management Act (RMA) from 1978 focused on: (1) the establishment of formal institutions for access to the reindeer husbandry and pasture management; and (2) co-management. Berg [27] argued that the RMA of 1978 and the GARI of 1976 provided the foundation for a change into a corporative reindeer husbandry, i.e. not only production of meat for subsistence and sale but also for official subsidies. Accordingly, in many areas different support and compensatory arrangements have provided around half of the income [27]. The RMA from 2007 broadened the focus on co-management by giving the industry more self-determination, influence and responsibility for its actions [28].

At present, the Saami reindeer husbandry can be distinguished into three different levels of social organization. The *husbandry unit*, lately designated as 'siida shares', is the basic unit of the social organization and consists of a government license that entitles a person to manage a herd of reindeer within a delimited area. The *siida* is a cooperative unit composed of one or more reindeer management families organized on the basis of kinship joined together in social and labour communities for keeping control of herds of reindeer through herding. A *district* is a formal management unit with responsibility to provide the Norwegian Reindeer Husbandry Administration with information as well as ensuring that the reindeer husbandry is managed in accordance with governmental regulations [18,29,30,31].

Data Material – Collapse in the Reindeer Husbandry

To evaluate the efficiency of herd accumulation as a risk reducing strategy the present study utilized data from the reindeer husbandry in Finnmark, Norway (see Text S1 for details). From an historical point of view, the number of reindeer in Finnmark has been characterized by considerable temporal variation: from the early 1900s, there was a decreasing trend that reached a minimum around the Second World War. Afterwards, while fluctuating there has been an upward trend that peaked in the early 1990s (from 90 000 animals in 1976 to 210 000 in 1988, [32]) and decreased until 2000/2001. More recently, reindeer abundance increased by ~40% from 2002 to 2010 [33] and again reached a historical high-level [34]. After the peak in the early 1990s, governmental subsidies resulted in increased harvest rates and a subsequent decline in reindeer abundance [35]. This downward trend was further compounded by the "[...] catastrophic winters with heavy snowfalls in 1997 and 2000 [...]" (Hausner et al. [14], p. 6), which resulted in a low point in reindeer abundance in 2001 (Fig. 1).

By designating the low point in 2001 (Fig. 1) as a 'collapse' in reindeer abundance (but see Text S2) it was possible to shed additional light on the effectiveness of herd accumulation as a risk management strategy by looking at the relationship between: (1) pre-collapse herd size and loss; and (2) pre-collapse and post-collapse herd size.

Study Protocol and Statistical Analyses

As in previous studies, e.g. [36], this empirical study is based on governmental statistics compiled and published annually by the Norwegian Reindeer Husbandry Administration. This dataset contains data on herd size (total number of reindeer in the spring per husbandry unit), covering the period 1998–2008 with data from 20 reindeer husbandry summer districts. Data pertaining to herd size are based on counts made by herders that are regularly checked by the authorities (for more details pertaining to dataset and design, see [18,29,36]). The utilized dataset contains the following variables:

$N_{post-collapse}$ (response).– A continuous (husbandry unit level) variable denoting the total herd size at the end of the period (i.e. 2008).

N_{loss} (response).– A continuous (husbandry unit level) variable denoting the number of reindeer lost from the *pre-collapse* (1998) to the *collapse* (2001) year. The variable was created by subtracting herd size in 2001(collapse) from herd size in 1998 (pre-collapse).

$N_{pre-collapse}$(predictor).– A continuous (husbandry unit level) variable denoting the total herd size at the beginning of the period (i.e. 1998).

To evaluate the efficacy of herd accumulation as a risk reducing strategy, we looked at the relationship between: (1) $N_{pre-collapse}$ and N_{loss}; and (2) $N_{pre-collapse}$ and $N_{post-collapse}$. Statistical analyses and plotting of results were carried out in *R* [37]. All tests were two-tailed and the null-hypothesis was rejected at an α-level of 0.05, and we used Wald statistics to test if estimated parameters were significantly different from zero. Regular linear regression was used to investigate the overall relationships between pre-collapse and post-collapse herd size (grouping effects, e.g. possible differences between districts with respect to natural and/ or social factors, were considered to be negligible, see Text S3 for details). Visual inspection of the data indicated problems related to the homoscedastic assumption [38]. We therefore fitted models with different variance structures in order to assess if violations of the homoscedastic assumption altered our conclusions [38]. As the conclusions from homoscedastic and heteroscedastic models were similar, we present the results from regular linear models in the main text (see Text S4 for the results based on heteroscedastic models).

Results

Pre-collapse herd size had a positive effect on the number of animals lost [effect of$N_{pre-collapse}$: 0.44], which means that husbandry units with more animals in 1998 tended to lose more animals from 1998 to 2001 (Fig. 2A). More specifically, if a herder increased his/her herd by one animal in 1998 his/her losses were expected to increase by 0.44 animals. This indicates that while herd accumulation is effective, it is a costly form of insurance against environmental variability. The heteroscedastic models revealed the same relationship although the estimated effect size was reduced (see Text S4 for details). For a similar analysis pertaining to reported losses, see Text S5.

Pre-collapse herd size had a positive effect on post-collapse herd size [effect of$N_{pre-collapse}$: 0.62], i.e. husbandry units that had more animals in 1998 also tended to have more animals in 2008 (Fig. 2B). While the relationship is not perfect, it provides the underlying rationale for herd accumulation: having a large herd prior to a collapse ensures a large herd after the collapse (a 1% increase in pre-collapse herd size predicts a 0.62% increase in post-collapse herd size, as both variables were $\log_e$-transformed). Herd accumulation thus seems to be an effective strategy for countering the negative impacts of environmental hazards. The heteroscedastic models revealed the same relationship, but again the effect size was reduced (see Text S4 for details).

Discussion

The positive relationship between pre- and post-collapse herd size indicates that herd accumulation is a rational response to environment-induced catastrophes, and provides the rationale for why both reindeer herders [29,31] and pastoralists in general [39] invest labour to increase herd size. Nevertheless, herd accumulation may also stem from cultural values: e.g. prestige or status [40,41,42]; conspicuous display [43]; and provision of bridewealth [44]. However, the fact that pre- and post-collapse herd size was positively correlated clearly shows the economic rationale and whether herd accumulation also results from cultural values is somewhat irrelevant [20]. This point of view is also shared by the herders in Finnmark since 51% of the herders 'agree' or 'strongly agree' that herds size is an important risk reducing strategy, while only 26% 'agree' or 'strongly agree' that herd size is important for social status [45].

A previous study found that herd size one year was positively correlated with herd size the next year [18]. Although the previous study was based on a shorter time series, the present study complements the former in showing the underlying rationale for herd accumulation: (1) a large herd one year result in a larger herd the next year [18]; and (2) a large pre-collapse herd also result in a larger post-collapse herd (this study). In short, since herders with large herds also have comparable larger herds from one year to the next and during crisis periods, herd accumulation maximizes long-term survival for pastoralists (see also [46]). Nevertheless, the positive relationship between pre-collapse herd size and loss indicates that herd accumulation is a costly form of insurance against environmental variability (see [21] for a similar result for Africa; and see [47] p. 36-8 for a discussion pertaining to 'loss' and the effectiveness of herd accumulation).

Herd accumulation may, however, not be the best available risk reducing strategy because of negative density dependence. For the reindeer husbandry it has been demonstrated that reindeer density and climatic conditions have negative effects on individual body mass [12,31,48,49,50] and, consequently, also survival [49,51]. Increased reindeer abundance also increases herders' vulnerability to unfavourable climate as the negative impacts of adverse climatic events increases with increasing reindeer density [12,49,52]. From a risk perspective, the best long-term strategy may thus be to invest in livestock body mass and not herd size. Næss et al. [29] hypothesised that one reason for why this strategy is not utilised by herders is because of competition for access to common winter pastures since access is to a large degree determined by herd size. Herd accumulation, through e.g. cooperative labour investment [29], may thus be a viable strategy for gaining access to winter pastures. More to the point, competition for access to winter pastures may explain why herd accumulation is the dominant and only viable risk reducing strategy in Finnmark. Herd accumulation may thus be taken to support a Tragedy of the Commons (ToC) [53] situation in Finnmark. ToC is a prisoners' dilemma where everybody is better off by coordinating strategies, e.g. by restricting the number of animals on the pastures, but the dominant solution is to always maximize herd size because the cost of high reindeer density is shared between all herders while the benefit of adding additional animals is attached to individual units [18]. Nevertheless, evidence supports both a *presence*, e.g. the aforementioned negative density dependence, and an *absence*, e.g. a positive relationship between slaughter undertaken by neighbouring herders and own slaughter [30], of ToC in Finnmark.

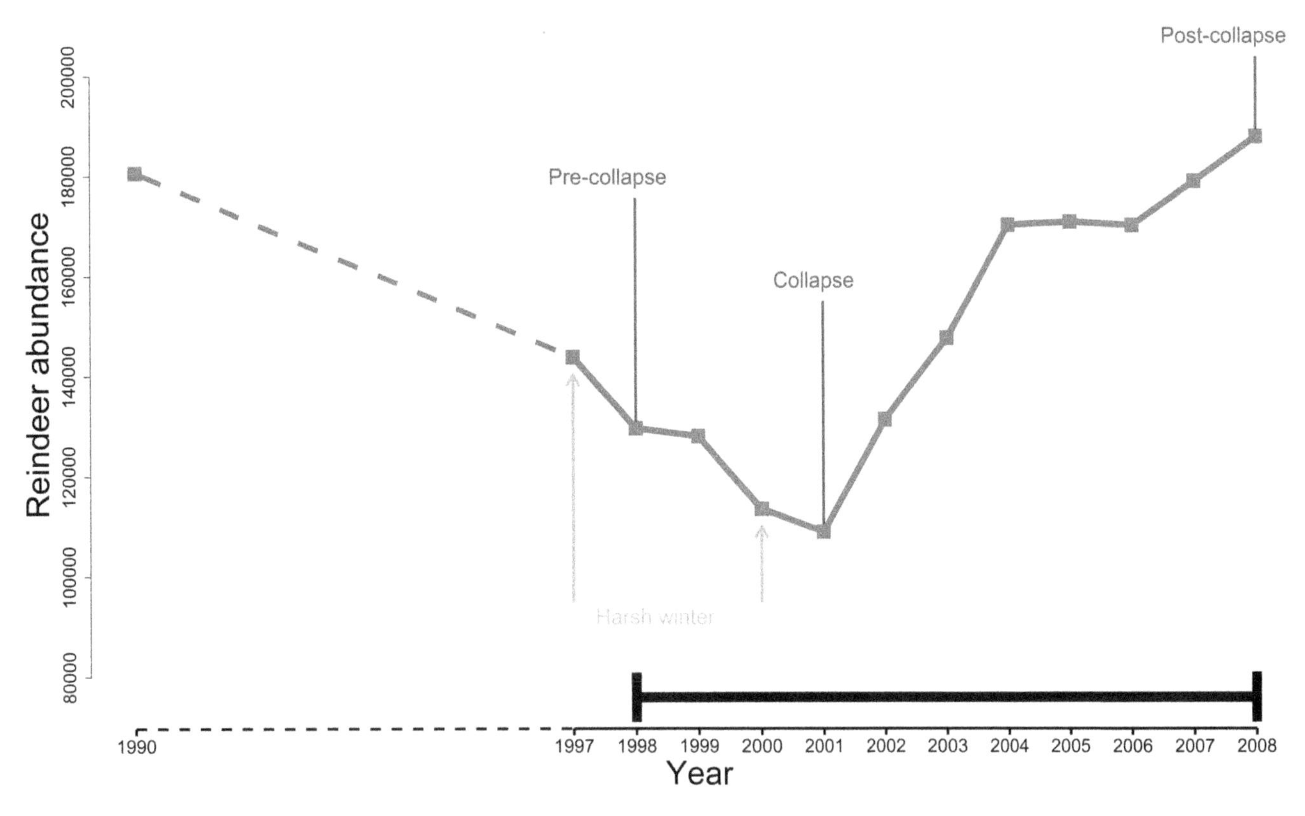

Figure 1. Temporal trend in reindeer abundance in Finnmark, Norway. Thick arrow indicates the period for which there exists official statistics pertaining to herd size for *individual husbandry units* (1998–2008), and thus restricts the period designations 'pre-collapse', 'collapse' and 'post-collapse' used in the analyses. Hatched line indicates missing data (i.e. abundance between 1990 and 1997). Abundance for 1990 from Tømmervik and Riseth [34], while abundance from 1997–2008 are per 31st March for each year from Anonymous [65,66,67].

Future Prospects and Management Implications

Scenarios for future climate change generally predict an increasing average, variance and even a changed distribution of important climatic variables like precipitation and temperature [15,54]. Climate change will most likely result in more frequent extreme weather events and there are indications that extreme events, such as heat waves and precipitation extremes, will increase and already have done so [15,16]. Nevertheless, there is limited information available pertaining to how these changes are going to affect the many pastoral cultures of the world. In Africa, climate change is predicted to increase the variability and frequency of rainfall at the same time as the proportion of arid and semiarid lands is likely to increase by 5–8% by 2080 [55]. Furthermore, in the Greater Horn of Africa droughts have now become the norm rather than the exception [56]. While previously pastoralists experienced one major long-term drought every decade coupled with minor occurrences every 3–4 years, droughts now occur annually [56]. As for the Arctic and Sub-Arctic, scenarios generated by most global climate models predict that the climate is likely to become increasingly unstable during the next half century with concomitant increases in the frequency of extreme weather conditions [57]. For Mongolia, regional climate predictions anticipate an increase in areas affected by droughts and in the frequency of extreme events [58]. Importantly, the frequency of droughts has almost doubled during the last 60 years and the worst droughts on record (covering over 50–70% of the country) have occurred during the last decade [58].

Considering the negative impact that environmental hazards have on livestock survival and reproduction, climate change thus represents a significant threat for the future of pastoral societies on a global scale. Nevertheless, it has been argued that by reinforcing the traditional strategies pastoralists have developed to deal with climate variability, in addition to introducing newer techniques, the economic, social, and cultural well-being of pastoral societies can be supported in the face of climate change [59]. Moreover, a case has been made that pastoralists are in a unique position to tackle climate change due to extensive experience managing environmental variability in marginal areas [2] and it has been argued that the ability to withstand environmental shocks is a *defining* feature of pastoralism [60].

Nevertheless, traditional pastoral risk management strategies, such as herd accumulation, may be insufficient for dealing with climate change [56]. While herd accumulation seems to be an efficient strategy, it is predicated on periods of recuperation when herd growth is possible. In fact, a delay in restocking after environment-induced losses is one of the main problems of pastoral production [61]. Herd accumulation can thus be expected to work less efficiently, if at all, when the frequency of extreme events increases.

As for Finnmark, it could be argued that the number of reindeer in Finnmark is unsustainable high, witnessed by the presence of negative density dependence (see above). Consequently, the Norwegian Government is aiming to reduce the number of reindeer so as to achieve a sustainable balance between pasture resources and number of reindeer [28]. The primary tool utilized are several subsidies that aims to increase slaughter and thus reduce herd size [62,63]. Considering the risk-beneficial aspects of having a large herd, this may, however be viewed as a short-term solution that, if followed, decreases long term viability by reducing the insurance potential of large herds. More to the point, official

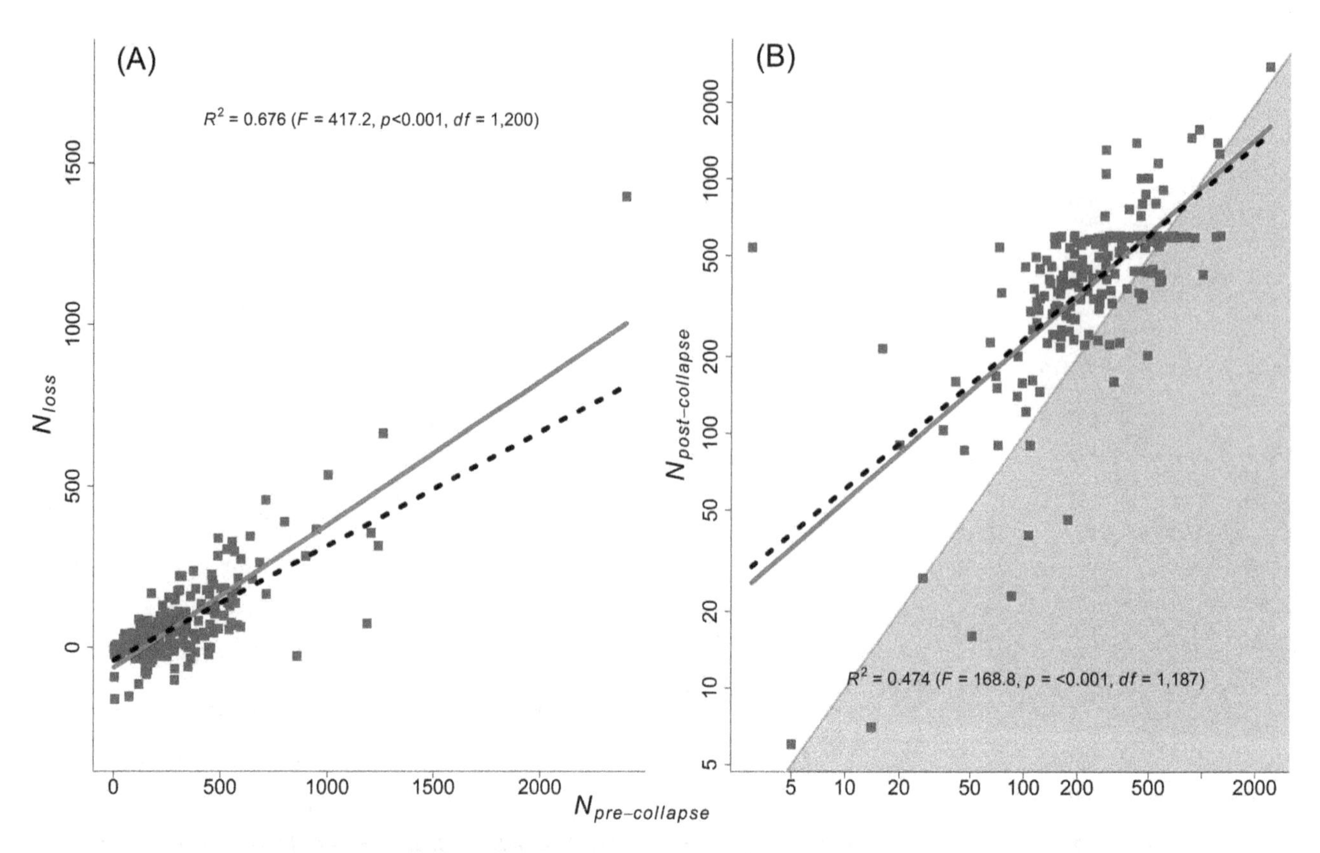

Figure 2. Showing the linear relationship between pre-collapse herd size ($N_{pre-collapse}$) and number of animals lost from pre-collapse to collapse (N_{loss}) (A). Model parameters: Intercept = 75.24 [95% confidence intervals (CI): 63.42, 87.07] and slope ($N_{pre-collapse}$) = 0.44 (95% CI: 0.40, 0.49). The positive relationship indicates that as herd size increases losses also increases: increasing herd size by one animal in 1998 increases the expected losses by 0.44 reindeer. Note that the model parameters are from fitting a model when centring $N_{pre-collapse}$ while the plot shows the relationship on the original scale. Hatched line show the relationship from a Generalized Least Square (GLS) model accounting for potential residual heterogeneity (see Text S4 for details). Showing the linear relationship (on $\log_e$-scale) between pre-collapse ($N_{pre-collapse}$) and post-collapse ($N_{post-collapse}$) herd size (**B**). Points above the shaded area indicate herd increase over the period, while a point on the 45-degree line means that pre- and post-collapse herd size was equal, and points in the shaded region indicate a decrease. The cloud of points above the 45-degree shaded area reflects the overall increase in reindeer abundance for the study area (Fig. 1). Model parameters: Intercept = 2.58 (95% CI: 2.06, 3.09) and slope ($N_{pre-collapse}$) = 0.62 (95% CI: 0.52, 0.71). The positive relationship indicates that as pre-collapse herd size increases so does post-collapse herd size: a 1% increase in $N_{pre-collapse}$ herd size predicts a 0.62% increase in$N_{post-collapse}$ herd size. Hatched line show the relationship from a GLS model accounting for potential residual heterogeneity (see Text S4 for details).

management strategies that fails to incorporate the risk-beneficial aspect of herd accumulation will have a limited effect and may indeed fail entirely [18,36]. Furthermore, production subsidies alone may not properly account for the decision problem facing herders: how to secure a reliable income while at the same time maximizing long-term survival. This decision problem is even more dire in the face of climate change: if herd accumulation works less efficiently when the frequency of extreme events increases, governmental support that incorporates (or preferably increase the effect of) pastoral risk management strategies may be the only solution for ensuring the continued existence of pastoralism. In the end, it may be official policies that disregard the inherent risk reducing logic of pastoral strategies rather than climate change *per se* that represents the greatest challenge for pastoral adaptability [17,64].

Supporting Information

Text S1 Study design and the reindeer husbandry. This text provides a more detailed description of the study design as well as the reindeer husbandry in Norway.

Text S2 Is there really a 'collapse' in reindeer abundance? This text investigates temporal trends in low points of reindeer abundance from 1845–2000.

Text S3 Possible grouping effects. This text investigates possible grouping effects, e.g. differences between summer districts with respect to natural and/or social factors.

Text S4 Finding the correct variance structure. This text explores models with different variance structures in order to assess if violations of the homoscedastic assumption altered the conclusions presented in the main text.

Text S5 Reported loss. This text replicates the analysis pertaining to loss presented in the main text but where 'reported loss' is used as a response.

Acknowledgments

We thank the Reindeer Husbandry Administration for access to data and Torkild Tveraa for preparing the data prior to the statistical analyses. We would also like to thank two anonymous reviewers for improving the manuscript and Matthew Thomas for correcting and improving the language.

Author Contributions

Analyzed the data: MWN BJB. Wrote the paper: MWN BJB.

References

1. Reid RS, Galvin KA, Kruska RS (2008) Global Significance of Extensive Grazing Lands and Pastoral Societies: An Introduction. In: Galvin KA, Reid RS, Behnke JRH, Hobbs NT, editors. Fragmentation in semi-arid and arid landscapes: consequences for human and natural systems. Dordrecht: Springer. 1–24.
2. Nori M, Taylor M, Sensi A (2008) Browsing on fences: Pastoral land rights, livelihoods and adaptation to climate change. Nottingham, UK: International Institute for Environment and Development 29 p.
3. Neely C, Bunning S, Wilkes A (2009) Review of evidence on drylands pastoral systems and climate change - Implications and opportunities for mitigation and adaptation. Rome: Food and Agriculture Organization of the United Nations.
4. Dong SK, Wen L, Liu SL, Zhang XF, Lassoie JP, et al. (2011) Vulnerability of Worldwide Pastoralism to Global Changes and Interdisciplinary Strategies for Sustainable Pastoralism. Ecology and Society 16(2): 10. Available: http://www.ecologyandsociety.org/vol16/iss2/art10/. Accessed 2013 Jun 26.
5. Nkedianye D, de Leeuw J, Ogutu JO, Said MY, Saidimu TL, et al. (2011) Mobility and livestock mortality in communally used pastoral areas: the impact of the 2005–2006 drought on livestock mortality in Maasailand. Pastoralism: Research, Policy and Practice 1: 17 Available: http://www.pastoralismjournal.com/content/1/1/17. Accessed 2013 Jun 24.
6. Little PD, Mahmoud H, Coppock DL (2001) When deserts flood: risk management and climatic processes among East African pastoralists. Climate Research 19: 149–159.
7. McPeak J, Little PD (2005) Cursed If You Do, Cursed If You Don't: The Contradictory Processes of Pastoral Sedentarization in Northern Kenya. In: Fratkin E, Roth EA, editors. As Pastoralists Settle Social, Health, and Economic Consequences of Pastoral Sedentarization in Marsabit District, Kenya. New York and London New York, Boston, Dordrecht, London, Moscow: Kluwer Academic Publishers. 87–104.
8. Templer G, Swift J, Payne P (1993) The changing significance of risk in the Mongolian pastoral economy. Nomadic Peoples 33: 105–122.
9. Janes CR (2010) Failed Development and Vulnerability to Climate Change in Central Asia: Implications for Food Security and Health. Asia-Pacific Journal of Public Health 22: 236S–245S.
10. Xie Y, Li W (2008) Why do herders insist on *otor*? Maintaining mobility in Inner Mongolia. Nomadic Peoples 12: 35–52.
11. Miller DJ (2000) Tough times for Tibetan nomads in Western China: Snowstorms, settling down, fences, and the demise of traditional nomadic pastoralism. Nomadic Peoples 4: 83–109.
12. Bårdsen B-J, Tveraa T (2012) Density dependence vs. density independence – linking reproductive allocation to population abundance and vegetation greenness. Journal of Animal Ecology 81: 364–376.
13. Bjørklund I (1990) Sami Reindeer Pastoralism as an Indigenous Resource Management System in Northern Norway: A Contribution to the Common Property Debate. Development and Change 21: 75–86.
14. Hausner VH, Fauchald P, Tveraa T, Pedersen E, Jernsletten J-LL, et al. (2011) The Ghost of Development Past: the Impact of Economic Security Policies on Saami Pastoral Ecosystems. Ecology and Society 16: 4. Available: http://dx.doi.org/10.5751/ES-04193-160304. Accessed 2013 Jun 24.
15. Sun Y, Solomon S, Dai AG, Portmann RW (2007) How often will it rain? Journal of Climate 20: 4801–4818.
16. Coumou D, Rahmstorf S (2012) A decade of weather extremes Nature Climate Change 2: 491–496.
17. Næss MW (2012) Tibetan Nomads Facing an Uncertain Future: Impacts of Climate Change on the Qinghai-Tibetan Plateau. In: Lamadrid A, Kelman I, editors. Climate Change Modeling for Local Adaptation in the Hindu Kush-Himalayan Region: Emerald Group Publishing Limited. 99–122.
18. Næss MW, Bårdsen B-J (2010) Environmental Stochasticity and Long-Term Livestock Viability-Herd-Accumulation as a Risk Reducing Strategy. Human Ecology 38: 3–17.
19. Fratkin E, Roth EA (1990) Drought And Economic Differentiation Among Ariaal Pastoralists Of Kenya. Human Ecology 18: 385–402.
20. McPeak J (2005) Individual and collective rationality in pastoral production: Evidence from Northern Kenya. Human Ecology 33: 171–197.
21. Lybbert TJ, Barrett CB, Desta S, Coppock DL (2004) Stochastic wealth dynamics and risk management among a poor population. Economic Journal 114: 750–777.
22. Bostedt G (2001) Reindeer husbandry, the Swedish market for reindeer meat, and the Chernobyl effects. Agricultural Economics 26: 217–226.
23. Paine R. (1994). *Herds of the Tundra: a portrait of Saami reindeer pastoralism*, Washington and London: Smithsonian Institution Press.
24. Vorren Ø. (1978). *Bosetning og ressursutnytting under veidekulturen og dens differensiering*, in NOU 18A, Finnmarksvidda-natur-kultur. Oslo: Universitetsforlaget. 145–62.
25. Riseth JÅ (2003) Sami Reindeer Management in Norway: Modernization Challenges and Conflicting Strategies. Reflections Upon the Co-management Alternative. In: Jentoft S, Minde H, Nilsen R, editors. Indigenous Peoples: Resource Management and Global Rights. Delft, Netherlands: Eburon Academic Publishers. 229–247.
26. Riseth JÅ, Vatn A (2009) Modernization and Pasture Degradation: A Comparative Study of Two Sàmi Reindeer Pasture Regions in Norway. Land Economics 85: 87–106.
27. Berg BÅ (2008) Utviklingen av reindriften i nordre Nordland 1750–2000. In: Evjen B, Hansen LI, editors. Nordlands kulturelle mangfold: etniske relasjoner i historisk perspektiv. Oslo: Pax. 151–191.
28. Ulvevadet B (2008) Management of reindeer husbandry in Norway - power-sharing and participation. Rangifer 28: 53–78.
29. Næss MW, Bårdsen B-J, Fauchald P, Tveraa T (2010) Cooperative pastoral production - the importance of kinship. Evolution and Human Behavior 31: 246–258.
30. Næss MW, Bårdsen B-J, Tveraa T (2012) Wealth-dependent and interdependent strategies in the Saami reindeer husbandry, Norway. Evolution and Human Behavior 33: 696–707.
31. Næss MW, Fauchald P, Tveraa T (2009) Scale Dependency and the "Marginal" Value of Labor. Human Ecology 37: 193–211.
32. Riseth JÅ, Johansen B, Vatn A (2004) Aspects of a two-pasture – herbivore model. Rangifer Special Issue 15: 65–81.
33. Anonymous (2012) Riksrevisjonens undersøkelse av bærekraftig bruk av reinbeiteressursene i Finnmark - Dokument 3: 14 (2011–2012). Norway: Riksrevisjonen. 89.
34. Tømmervik H, Riseth JÅ (2011) Historiske tamreintall i Norge fra 1800-tallet fram til i dag. NINA report 672. Tromsø: NINA.
35. Ulvevadet B, Hausner VH (2011) Incentives and regulations to reconcile conservation and development: Thirty years of governance of the Sami pastoral ecosystem in Finnmark, Norway. Journal of Environmental Management 92: 2794–2802.
36. Næss MW, Bårdsen B-J, Pedersen E, Tveraa T (2011) Pastoral Herding Strategies and Governmental Management Objectives: Predation Compensation as a Risk Buffering Strategy in the Saami Reindeer Husbandry. Human Ecology 39: 489–508.
37. R Development Core Team (2011) R: A language and environment for statistical computing. Vienna, Austria: R Foundation for Statistical Computing.
38. Zuur AF, Ieno EN, Walker N, Saveliev AA, Smith GM. (2009). *Mixed effects models and extensions in ecology with R*, New York: Springer.
39. Næss MW (2012) Cooperative Pastoral Production: Reconceptualizing the Relationship between Pastoral Labor and Production. American Anthropologist 114: 309–321.
40. Anonymous (2009) Sluttrapport for prosjektet: "Økosystem Finnmark", Tromsø: University of Tromsø.
41. Herskovits MJ (1926) The Cattle Complex in East Africa. American Anthropologist 28: 633–664.
42. Nilsen R, Mosli JH. (1994). *Inn fra vidda: hushold og økonomisk tilpasning i reindrifta i Guovdageaidnu 1960–1993*, Guovdageaidnu: BAJOS Utviklingsselskap AS/NORUT Samfunnsforskning AS, Tromsø.
43. Paine R. (2009). *Camps of the tundra: politics through reindeer among Saami pastoralists*, Oslo: Instituttet for sammenlignende kulturforskning.
44. McCabe JT. (2004). Cattle bring us to our enemies: Turkana ecology, politics, and raiding in a disequilibrium system, Ann Arbor: University of Michigan Press.
45. Johannesen AB, Skonhoft A (2011) Livestock as Insurance and Social Status: Evidence from Reindeer Herding in Norway. Environmental & Resource Economics 48: 679–694.
46. Mace R, Houston A (1989) Pastoralist strategies for survival in unpredictable environments: A model of herd composition that maximises household viability. Agricultural Systems 31: 185–204.
47. Næss MW (2009) Pastoral Risk Management - The Importance of Cooperative Production [PhD]. Tromsø: University of Tromsø.
48. Bårdsen B-J (2009) Risk sensitive reproductive strategies: the effect of environmental unpredictability [PhD]. Tromsø: University of Tromsø.
49. Bårdsen B-J, Henden J-A, Fauchald P, Tveraa T, Stien A (2011) Plastic reproductive allocation as a buffer against environmental stochasticity – linking life history and population dynamics to climate. Oikos 20: 245–257.
50. Tveraa T, Fauchald P, Yoccoz NG, Ims RA, Aanes R, et al. (2007) What regulate and limit reindeer populations in Norway? Oikos 116: 706–715.
51. Tveraa T, Fauchald P, Henaug C, Yoccoz NG (2003) An examination of a compensatory relationship between food limitation and predation in semi-domestic reindeer. Oecologia 137: 370–376.

52. Bårdsen B-J, Tveraa T, Fauchald P, Langeland K (2010) Observational evidence of a risk sensitive reproductive allocation in a long-lived mammal. Oecologia 162: 627–639.
53. Hardin GJ (1968) The Tragedy of the Commons. Science 162: 1243–1248.
54. Benestad RE (2011) A New Global Set of Downscaled Temperature Scenarios. Journal of Climate 24: 2080–2098.
55. Galvin KA (2009) Transitions: Pastoralists Living with Change. Annual Review of Anthropology 38: 185–198.
56. Blackwell PJ (2010) East Africa's Pastoralist Emergency: is climate change the straw that breaks the camel's back? Third World Quarterly 31: 1321–1338.
57. Brannlund I, Axelsson P (2011) Reindeer management during the colonization of Sami lands: A long-term perspective of vulnerability and adaptation strategies. Global Environmental Change-Human and Policy Dimensions 21: 1095–1105.
58. Marin A (2010) Riders under storms: Contributions of nomadic herders' observations to analysing climate change in Mongolia. Global Environmental Change-Human and Policy Dimensions 20: 162–176.
59. Secretariat of the Convention on Biological Diversity (2009) Connecting Biodiversity and Climate Change Mitigation and Adaptation: Report of the Second Ad Hoc Technical Expert Group on Biodiversity and Climate Change. Montreal, Technical Series No. 41. 126 p. Available: http://www.cbd.int/doc/publications/cbd-ts-41-en.pdf. Accessed 2013 Jun 26.
60. Hatfield R, Davies J (2006) Global Review of the Economics of Pastoralism. Nairobi: IUCN. 44 p. Available: http://cmsdata.iucn.org/downloads/global_review_ofthe_economicsof_pastoralism_en.pdf. Accessed 2013 Jun 26.
61. Bollig M, Göbel B (1997) Risk, Uncertainty and Pastoralism: An Introduction. Nomadic Peoples 1: 5–21.
62. Anonymous (2008) Ressursregnskap for reindriftsnæringen. Alta: Reindriftsforvaltningen. 164 p.
63. Anonymous (2007) St.prp. nr. 74: Om reindriftsavtalen 2007/2008, om dekning av kostnader vedrørende radioaktivitet i reinkjøtt, og om endringer i statsbudsjettet for 2007 m.m. Oslo: Det Kongelige Landbruks- og Matdepartement. 36.
64. Næss MW (2013) Climate Change, Risk Management and the End of *Nomadic* Pastoralism. International Journal of Sustainable Development and World Ecology 20: 123–133.
65. Anonymous (2001) Totalregnskap for reindriftsnæringen. Alta: Reindriftsforvaltningen. 145 p.
66. Anonymous (2004) Totalregnskap for reindriftsnæringen. Alta: Reindriftsforvaltningen. 143 p.
67. Anonymous (2010) Totalregnskap for reindriftsnæringen. Alta: Reindriftsforvaltningen. 132 p.

8

Pattern of Social Interactions after Group Integration: A Possibility to Keep Stallions in Group

Sabrina Briefer Freymond[1*], Elodie F. Briefer[2¤], Rudolf Von Niederhäusern[1], Iris Bachmann[1]

1 Agroscope Liebefeld-Posieux Research Station ALP-Haras, Swiss National Stud Farm SNSTF, Les Longs Prés, Avenches, Switzerland, 2 Queen Mary University of London, Biological and Experimental Psychology Group, School of Biological and Chemical Sciences, London, United Kingdom

Abstract

Horses are often kept in individual stables, rather than in outdoor groups, despite such housing system fulfilling many of their welfare needs, such as the access to social partners. Keeping domestic stallions in outdoor groups would mimic bachelor bands that are found in the wild. Unfortunately, the high level of aggression that unfamiliar stallions display when they first *encounter each other* discourages owners from keeping them in groups. However, this level of aggression is likely to be particularly important only during group integration, when the dominance hierarchy is being established, whereas relatively low aggression rates have been observed among stable feral bachelor bands. We investigated the possibility of housing breeding stallions owned by the Swiss National Stud in groups on a large pasture (5 stallions in 2009 and 8 stallions in 2010). We studied the pattern of agonistic, ritual and affiliative interactions after group integration (17–23 days), and the factors influencing these interactions (time after group integration, dominance rank, age or experience of group housing). We found that stallions displayed generally more ritual than agonistic and than affiliative interactions. The frequency of agonistic and ritual interactions decreased quickly within the first three to four days. The frequency of affiliative interactions increased slowly with time before decreasing after 9–14 days. A stable hierarchy could be measured after 2–3 months. The highest-ranking males had less ritual interactions than the lowest-ranking. Males had also less agonistic, ritual and affiliative interactions if they had already been housed in a group the previous year. Therefore, we found that breeding stallions could be housed together on a large pasture, because the frequency of agonistic interactions decreased quickly and remained at a minimal level from the fourth day following group integration. This housing system could potentially increase horse welfare and reduce labour associated with horse management.

Editor: Katie Slocombe, University of York, United Kingdom

Funding: E. Briefer is funded by a Swiss National Science Foundation fellowship. The funders had no role in study design, data collection and analysis, decision to publish, or preparation of the manuscript.

Competing Interests: The authors have declared that no competing interests exist.

* E-mail: sabrina.briefer@haras.admin.ch

¤ Current address: Institute of Agricultural Sciences, ETH Zürich, Zürich, Switzerland

These authors contributed equally to this work.

Introduction

Despite being social animals, domestic horses (*Equus caballus*) are very often kept in individual housing systems. This is especially true for expensive horses used for racing and other competitions, because of the potential risks of aggressive interactions such as kicks or bites that could occur when horses are housed together [1]. Stallions used for breeding are also traditionally housed individually, because the high level of aggression that unfamiliar males display towards one another when they first *encounter each other* discourages owners to keep them in groups [1–3]. However, individual housing systems can have several disadvantages for horse welfare, and particularly for their mental health, when they are not designed properly (e.g. inducing confinement and preventing social contact [4–6]).

Horses housed in individual stables are partially or even totally deprived of physical contact and of activities that are seen under natural conditions, such as locomotion and social behaviours [1,7–9]. Consequently, they display more stress-related behaviours than horses stabled in pairs [10]. They are also likely to develop stereotypies like weaving and cribbing, particularly in stables with minimised contact between neighbouring horses [4–6]. Furthermore, a lack of social contact, especially during ontogeny, may predispose horses to impairments in social skills and to an inability to cope with social challenges [2,11,12]. Keeping horses in stable groups and in adequate densities could improve welfare, because it would give them access to social interactions, such as affiliative interactions (e.g. play and allogrooming), which have rewarding properties and are indispensable behaviours [2,9,13].

Feral stallions (*Equus ferus*) are harem breeders that defend a group of females instead of a particular territory [14]. When they do not have a harem, most stallions form associations known as bachelor bands. These bands contain two to 15 individuals, and are relatively stable over time, although less stable than harem bands. They are composed of yearling or young stallions that have not yet acquired a harem, and are in an intermediate state of development between sexual and social maturity. Bachelor bands can also include older stallions that have lost their harem [15–17]. Agonistic and ritualized behaviours like fights, threats, avoidance and submissive behaviours occur among bachelor bands [18,19]. These aggressive interactions could play an important role in improving skills and physical stamina necessary for stallions to acquire and maintain a harem [16,19]. However, as in many other species, when they interact, stallions typically display the minimum amount of aggression required by the situation [3]. Therefore, aggression rates are relatively low in natural conditions and encounters rarely escalate into serious fights leading to injuries [18–21].

Agonistic interactions, which result in increased distance between two opponents through spontaneous displacement, non-contact or physical aggression, can be prevented by ritualized interactions [3,22]. Indeed, combat is typically preceded by ritual, threat display and mutual assessment using information about fighting ability from visual, olfactory or acoustical signals [23]. For example, information about familiarity is present in auditory signals such as vocalisations and in olfactory cues, available through behaviour such as dung sniffing [24,25]. As in many other ungulates (e.g. fallow deer, *Dama dama* [26,27]; red deer, *Cervus elaphus* [28]), vocalisations also provide information about individuality, body size and dominance status [23,24,29]. Ritualized displays, which refer to interactions that do no longer keep their initial function, are common between stallions [17,19,22]. These displays typically show a decrease in intensity and duration with time, and seem to facilitate stallions being able to graze side-by-side [19]. They play an important role in establishing and maintaining the hierarchy without involving physical aggression [3,19].

Housing stallions in outdoor groups is likely to have two main benefits, if enough space is available. First, it could increase horse welfare by allowing them to fully express their natural behaviours including social interactions and locomotion [1,2,13]. Second, it could potentially reduce labour required for housing cleaning and exercising horses (H. Besier and I. Bachmann, unpublished data). According to recent reviews on group housing [1,3], the main reason that prevents owners to keep horses in groups is the potential risk of physical aggression. Several studies have shown that stallions can be kept in stable groups, with few injuries linked to aggressive interactions [18–21]. However, physical aggression rates are likely to be particularly high during group integration, when stallions are interacting for the first time and when the dominance hierarchy is being established [3,30,31]. Because agonistic encounters and rituals play a role in establishing dominance relationships within a group, we expect their rate and intensity to decrease with time, although not disappear completely, in a stable bachelor group [18–21]. More studies are needed to fully evaluate if stallions can be housed in groups, in order to determine aggression levels associated with group integration [1,3].

In this study, we investigated the possibility of housing breeding stallions owned by the Swiss National Stud in groups on a large pasture. For this purpose, we observed the changes in social interactions over a period of 17–23 days after group integration. We differentiated ritual and affiliative interactions, which do not involve physical aggression, and agonistic interactions, which can potentially involve physical aggression [3,19]. A rapid decrease in the frequency of agonistic interactions with time would indicate that stallions can be housed in group, because the risk of physical aggression is low after these interactions reach their minimum rate. We also investigated if the final dominance rank, the age or the experience of group housing of stallions affected the frequencies of agonistic, ritual and affiliative interactions during group integration. Finally, we assessed when the dominance hierarchy stabilises.

Materials and Methods

Subjects and Management Conditions

The study was carried out at the Swiss National Stud Farm, Avenches, on two groups of Swiss breed stallions (Franches-Montagnes): one group of 5 individuals in 2009 and one group of 8 individuals in 2010. Four individuals were included in both 2009 and 2010 groups ($n = 9$ stallions in total). These stallions were 8–19 years old and had been kept at the Swiss National Stud for 5–16 years. They were used for breeding and for driving. They had all, but one, been regularly hitched next to each other for driving. Before the study, they had been housed on several occasions in adjacent stables, but they had never been in a group. Therefore, all the stallions used in this study were familiar with each other, but had no experience of group housing.

Because prior exposure can reduce aggression between horses during physical encounters [30], the stallions were housed for 14 days next to each other in indoor individual stables (9 m^2) separated by partitions with a rail at the top half, allowing them to interact. They could therefore hear, see, smell and partially touch each other. When housed in individual stables, in 2009, the stallions were individually put in a pasture for two hours per day. In 2010, they were exercised four by four in a horse walker for one hour per day. They were given feed mix three times a day and were provided with hay two times a day and straw.

Stallions were then moved together to an outdoor pasture (4 hectares) for six months. Horseshoes were removed before group integration in order to minimize the risks of injuries. In pasture, hay was distributed during winter according to horses' needs. Pasture fences and horse health was checked daily. Dung was cleaned once a week. In case of high summer heats, an insecticide was applied daily or weekly as required. Six wood shelters (5 of 9 m^2 and one of 15 m^2) with wide stabilised entrances and whose ground was covered with straw were available for horses within the pasture. The pasture did not contain any closed spaces. Food was well distributed to ensure that every horse could feed easily without being threatened or kicked by other horses. Finally, the group was housed in a pasture away from mares and other horses. After the study, stallions were put back in their previous individual stables and used for breeding.

Group Integration Procedure

Following a preliminary experiment in 2008, in which four stallions were successfully integrated together, we repeated the same procedure. In July 2009 and 2010, the stallions were handled individually on a halter and brought to the pasture. The persons handling the stallions walked once around the pasture and then released all the stallions at the same time. Ten people holding driving whips were present and ready to intervene in case of serious fight. The vet team of the Swiss National Stud Farm was present during the integration and checked horse health on a daily basis throughout the experiment.

Observations

Social interactions were scored daily either at 09:00 h–11:00 h, 13:00 h–15:00 h and 17:00 h–18:00 h, or at 07:00 h–09:00 h, 11:00 h- 13:00 h and 15:00 h–17:00 h from the first hour to the 557th hour (23 days) after group integration in 2009 and to the 413th hour (17 days) after group integration in 2010. Because the frequency of interactions was considerably higher during the first two days after integration, these data were analysed later from videos filmed by two experimenters. Data for the rest of the study were scored by direct observation by two experimenters. All data were collected from an observatory post, from which the whole pasture (i.e. all horses at all time) was visible. In total, the behaviour of each stallion was scored during 109 hr in 2009 and 87 hr of observation in 2010.

We scored the frequency of the following social interactions (defined in Table 1) continuously using the behaviour sampling rule, i.e. by observing the whole group and scoring every interaction with details of which individuals were involved: agonistic interactions; ritual/investigative interactions and affiliative interactions. Agonistic interactions were defined as non-contact or contact interactions that resulted in increased distance between two stallions (e.g. chase, push and kick; Table 1). Ritual/investigative interactions (thereafter "ritual interactions") were defined as non-contact interactions between two stallions used to assess each other's social status without fighting (i.e. faecal pile display, sniff and sniff and squeal; Table 1). Affiliative interactions (i.e. non agonistic and non ritual) included allogrooming (or mutual grooming) and play (Table 1 [3,17,19,22,32,33]). Interactions were analysed as frequencies per hour per horse.

Dominance Relationships

We tested dominance relationships once a month, during three months after group integration using pair feeding tests [34,35]. These tests consisted in placing a bucket of carrots between each possible pair of stallions. Videos of the tests were analysed and the stallion that chased the other one away to eat in the bucket was considered as dominant and the other horse as subordinate. Because dominance hierarchies in horses are generally linear, particularly in the case of small groups such as in our study [3,36], the dominance index for a given male was then calculated according to [37] as follows: [(number of horses that this male dominates – number of horses that this male is dominated by + group size + 1) / 2]. The male with the lowest index value in each year was assigned the rank of 1 and all other males were ranked accordingly. Therefore, higher values of rank indicate higher-ranking males. We used the final dominance rank measured after three months to investigate the effect of the hierarchy on the frequency of interactions.

Statistical Analyses

Social interactions in group. We used generalized linear mixed model (GLMM) fit by the Laplace approximation (lmer function in R [38]) to investigate the effects of the time after group integration, the age and dominance rank of stallions, the number of matings they performed and their experience of group housing on the frequency of social interactions.

We first tested if, independently of the time after integration, stallions favoured one category of interactions over the others (antagonist, ritual, affiliative). To this aim, we carried out a GLMM including the frequency of interactions (frequency per hour per horse; 109 frequencies per hour per horse in 2009 and 87 frequencies per hour per horse in 2010; mean$\pm$SE = 99.6$\pm$3.0) as a dependant variable, the time after integration (1–557 hours) as a control factor, and the category of interaction (antagonist, ritual, affiliative) as a fixed effect. We also included as random effects the

Table 1. List and description of the interactions scored after group integration.

Behaviour	Description
Agonistic interactions	
Chase	Chasing another horse, ears laid back with the neck extended and exposing the teeth.
Push	Pushing with the head the neck, shoulder, chest, body or rump of another stallion.
Kick threat	Raising a hind leg in the direction of another stallion, but without touching him, ears laid back.
Kick	Kicking another horse with one or the two hindlegs.
Strike	A rapid motion of one or both forelegs in the anterior direction.
Bite threat	Neck stretched, teeth exposed and ears laid back, pretending to bite without touching the other horse.
Bite	Biting another horse, lips retracted, ears laid back with the muzzle muscles tensed.
Nip	Biting another horse, but without the ears laid back and with the mouth less widely open than during a real bite.
Mount	Mounting another stallion, similarly as during copulation.
Lunge	One stallion rears with the forelegs in the direction of another horse, ears laid back.
Circling	Two stallions circle each other head-to-tail, trying to nip or bite each other's body parts.
Kneeling	Two stallions circle each other and drop on one or both of their knees.
Fleeing	Avoiding, retreating from another horse by walking, trotting or galloping, usually with ears laid back.
Following	Walking behind another horse, head low without any attempt to attack or bite. This behaviour was scored only in 2010.
Ritual/investigative interactions	
Sniff and squeal	Two stallions sniff each other's muzzle, body parts or genitals, with the neck arched and produce a squeal.
Faecal pile display	Sequence of behaviours associated with defecation onto a faecal pile. Typically, two or more stallions defecate on a faecal pile, turn around, sniff the pile and scratch the ground with a foreleg.
Sniffing	Olfactory investigation of another horse's muzzle, body parts or genitals, with the neck arched, but without squealing like during *sniff and squeal.*
Affiliative interactions	
Play	Two stallions nip each others' body parts, without their ears laid back, while moving or not.
Mutual grooming	Two stallions groom each others' neck, back or rump by gentle nipping, nuzzling, or rubbing while standing head-to-tail.

The categories of interactions that were included in the analyses are shown in bold and the behaviours scored are in italic. A short description of the behaviours is included when needed (see also [3,17,19,22,32,33,36]).

year of observation (2009 or 2010), to account for between year differences, and the individual identity of horses, to account for repeated measurements of the same individual within and between years. This model was fit with residual maximum likelihood estimation (REML). We carried out more GLMMs including the same fixed and random factors as described above for two-by-two comparisons and we applied a Bonferroni correction at $\alpha = 0.017$ (0.05/3).

We then used a model selection procedure based on the Akaike's information criterion adjusted for small sample size (AIC_C) to identify the factors (time after group integration, dominance rank, age or experience) that best explained each of the three categories of interactions (antagonist, ritual and affiliative; frequency per hour per horse [39]). All models were fit with maximum likelihood estimation (ML). We formulated one set of candidate models for each of the three interaction categories (Table 2). Within each set of models, the first model consisted of the random effects only (null model; model 0), which were the year of observation and horse identity. In the next model, we included the time after group integration (1–557 hours) as a fixed effect (model 1). Because this factor was highly significant (Table 2), it was included as a control factor in all the other models. In addition, we included as a fixed effect the final dominance rank after three months (1–5 in 2009 and 1–8 in 2010; model 2), the age of the stallions (8–19 years old; model 3), or their experience of group housing (*i.e.* if they had been housed in group already the year before: coded as 1 for 2010 horses that were in group in 2009 and 0 for the others; model 4).

Within each set of model, when the difference between the AIC_C values of two models (ΔAIC_C) is less than 2 units, both models have support and can be considered competitive. Models with ΔAIC_C ranging from 3 to 7 have considerably less support by the data, models with $\Delta AIC_C>10$ are poorly supported, and $\Delta AIC_C>20$ have no empirical support [39,40]. Akaike weights (*wi*) indicate the probability that a particular model has more or less support from the data among those included in the set of candidate models [39]. For each model, we also calculated the evidence ratio, defined as the ratio between the Akaike weight of the best model and the Akaike weights of the competing model, to determine to what extent it was better than another. Additionally, we used the likelihood-ratio tests (LRT) to compare models within a given set and to assess statistical significance of the factors, by comparing the model with and without the factor included (Table 2).

We fit fixed effects as linear, quadratic or log terms based on the lowest AIC_C value (Table 2). All categories of interactions were log-transformed and fit with a Gaussian family distribution and identity link function. Q-Q plots and scatterplots of the residuals of the dependent variables were inspected visually to ensure their normal distribution.

Stability of the hierarchy. To measure the stability of the hierarchy over time, we calculated, for each year, Kendall rank correlations ("Kendall's tau") between the dominance ranks of the stallions measured after one month and their ranks after two months, and between their ranks measured after two month and their ranks after three months (Table 3).

We carried out statistical analyses using R v.2.9.0 [41]. All means are given with standard errors (SEs).

Table 2. Models fit to investigate the effects of the time after group integration ("Hours"), the age ("Age") and the dominance rank ("Rank") of stallions, and their experience of group housing ("Experience") on the frequency of interactions (agonistic, ritual or affiliative).

Response variable	Model	Fixed effect(s)	AICc	ΔAIC_C	*wi*	ER	Model comparison	χ^2 (df)	*P*
Agonistic	0	None	630.75	590.95					
	1	log(Hours)	42.82	3.02	0.15	4.53	1 vs 0	589.95(1)	**<0.0001**
	2	log(Hours) + Rank	44.51	4.70	0.06	10.51	2 vs 1	0.34(1)	0.56
	3	log(Hours) + log(Age)	43.46	3.66	0.11	6.22	3 vs 1	1.39(1)	0.24
	4	**log(Hours) + Experience**	**39.80**	**0.00**	**0.68**	**1.00**	**4 vs 1**	**5.04(1)**	**0.025**
Ritual	0	None	651.31	700.02					
	1	log(Hours)	−42.63	6.07	0.04	20.80	1 vs 0	695.96(1)	**<0.0001**
	2	**log(Hours) + Rank²**	**−48.70**	**0.00**	**0.81**	**1.00**	**2 vs 1**	**10.11(2)**	**0.006**
	3	log(Hours) + Age²	−42.53	6.18	0.04	21.93	3 vs 1	3.94(2)	0.14
	4	log(Hours) + Experience	−44.81	3.89	0.12	7.00	4 vs 1	4.20(1)	**0.040**
Affiliative	0	None	−3153.89	40.69					
	1	Hours²	−3191.06	3.51	0.10	5.80	1 vs 0	41.21(2)	**<0.0001**
	2	Hours² + Rank²	−3192.89	1.68	0.26	2.32	2 vs 1	5.88(2)	0.053
	3	Hours² + Age²	−3189.62	4.95	0.05	11.09	3 vs 1	2.61(2)	0.27
	4	**Hours² + Experience**	**−3194.58**	**0.00**	**0.59**	**1.00**	**4 vs 1**	**5.54(1)**	**0.019**

Note. The response variable (category of interaction) and fixed effect(s) included in the models are indicated. The fit of the models is assessed by Akaike's information criterion corrected for small sample sizes (AIC_C): the lowest value for a given response variable (i.e. set of models) indicates the best fit (in bold). ΔAIC_C gives the difference in AIC_C between each model and the best model. The Akaike's weights (*wi*) assess the relative support that a given model has from the data, compared to other candidate models in the set. The evidence ratio (ER) is the ratio between the Akaike's weight of the best model and that of a competing one. Results of the likelihood-ratio tests (χ^2 and *p*) used to compare the various models ("Model comparison") and to assess statistical significance of the factors are indicated (significant results are in bold). Fixed effects: "Hours" indicates a linear term, "log(Hours)" a log term and "Hours²" a quadratic term (indicating that both linear and quadratic terms were included in the model).

Table 3. Dominance hierarchy after one, two and three months (final rank) following group integration.

Year	Stallion	Dominance rank after		
		One month	Two months	Three months (final)
2009	Havane	3	4	5
	Lordon	3	4	4
	Naguar	3	3	3
	Nico	2	2	2
	Valentino	1	1	1
2010	Havane	7	6	8
	Naguar	6	5	7
	Nico	4	4	6
	Lordon	5	4	5
	Laura	4	4	4
	Nestor	3	3	3
	Van Gogh	2	2	2
	Commodore	1	1	1

The hierarchy appeared stable after two (2010) to three months (2009) following group integration. Higher dominance ranks indicate higher-ranking males.

Ethics

Keeping horses in outdoor groups is a housing system allowed by welfare regulations. All animal work was conducted in accordance with the relevant local guidelines (Swiss law on animal protection and welfare). No experiment with animals has been performed in our study. The health of stallions was checked on a regular basis by veterinarians of the Swiss National Stud Farm. None of the stallions had to be removed from the group because of injuries caused by social interactions.

Results

Social Interactions in Group

The time after group integration (1–557 hours; GLMM: log term, $z = -34.16$, $p<0.0001$) and the interaction category (antagonist, ritual, affiliative; GLMM: $z = -31.19$, $p<0.0001$) had an effect on the frequency of interactions. Further tests showed stallions displayed, independently of the time after integration, more ritual interactions (4.60±0.20 interactions per hour) than agonistic interactions (3.17±0.22 interactions per hour) and than affiliative interactions (0.30±0.02 interactions per hour; $n = 1241$ frequencies for each interaction category; GLMM: ritual versus agonistic, $z = 10.92$, $p<0.0001$; ritual versus affiliative, $z = 34.15$, $p<0.0001$; agonistic versus affiliative, $z = -21.48$, $p<0.0001$; Bonferroni correction: $\alpha = 0.017$).

The model selection procedure based on AIC_C showed that the time after group integrations explained the largest amount of variation in the frequency of all categories of interactions (Table 2). Agonistic and ritual interactions decreased quickly with time, whereas affiliative interactions increased during the first days and decreased later on (Fig. 1). The experience of group housing was also a good predictor of all categories of interactions, with males having fewer interactions when experienced. The dominance rank of stallions was a good predictor of the frequency of ritual interactions (Fig. 2a).

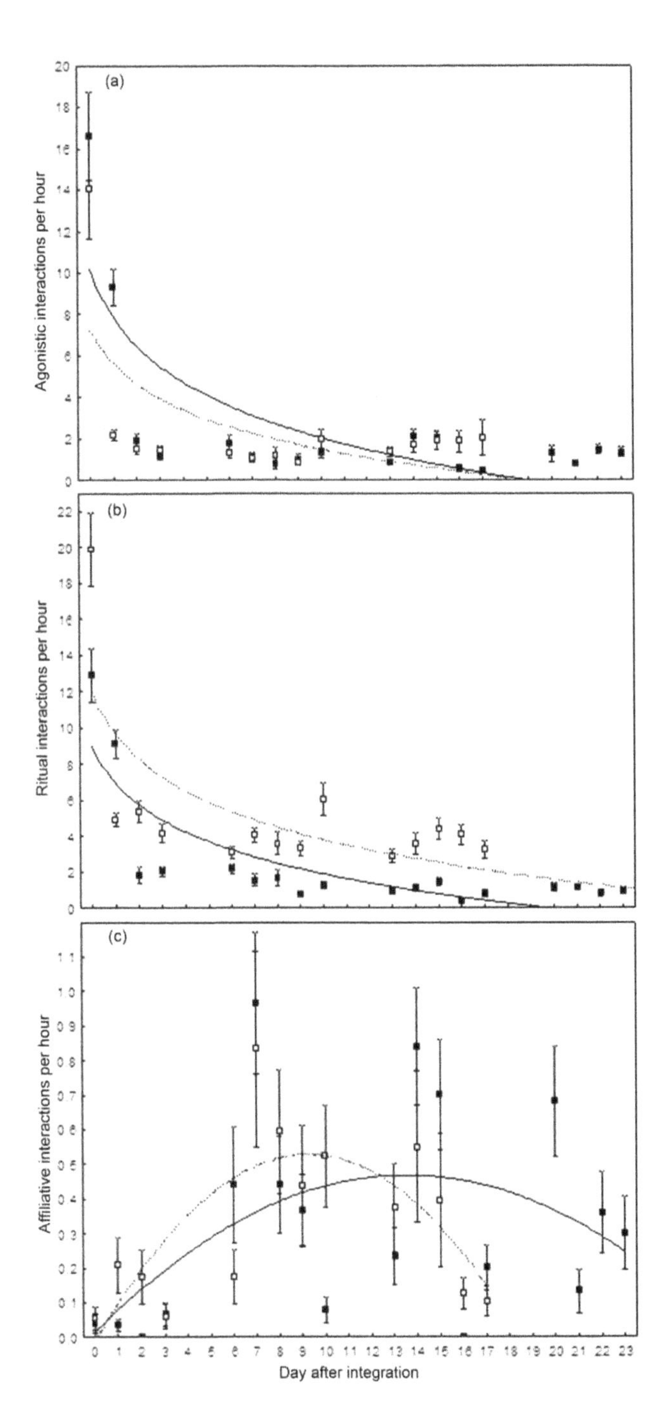

Figure 1. Changes with time in the frequency of social interactions after group integration. Frequency of interactions per hour (mean±SE per day; agonistic (a), ritual (b) and affiliative (b) interactions) as a function of time (days) in 2009 (black square) and in 2010 (empty squares). The best fit (log or quadratic) is indicated with a solid line for 2009 and dashed line for 2010 data.

Adding the time after integration to the null model significantly improved the models explaining all categories of interactions (model 1; Table 2). This parameter explained a large amount of variation in the data, particularly for agonistic and ritual interactions (ΔAIC_C between model 0 and model 1: agonistic interactions, 587.93; ritual interactions, 693.95; affiliative interactions, 37.17). The frequency of agonistic and ritual interactions decreased rapidly after group integration (3–4 first days) and was maintained at its lowest values for the rest of the study both in 2009 (mean interactions per hour before day 4: agonistic, 8.92±0.88; ritual, 7.84±0.64; $n = 160$ frequencies; after day 4: agonistic, 1.22±0.07; ritual, 1.17±0.06; $n = 385$ frequencies) and in 2010 (mean interactions per hour before day 4: agonistic, 5.28±0.77; ritual, 9.07±0.74; $n = 256$ frequencies; after day 4: agonistic, 1.56±0.15; ritual, 3.84±0.18; $n = 440$ frequencies; Fig. 1a and b). The frequency of affiliative interactions increased from day 0 to day 14 in 2009 and from day 0 to 9 in 2010 and decreased afterwards (mean±SE: 2009, 0.30±0.03; $n = 545$ frequencies; 2010, 0.30±0.04; $n = 969$ frequencies; Fig. 1c).

Adding the dominance rank (model 2) to model 1 significantly improved the model explaining ritual interactions, but not agonistic interactions (not significant) and affiliative interactions (trend only; Table 2). In 2010, the frequency of ritual interactions increased from ranks 1 to 4, and then decreased from ranks 5 to 8, whereas it mainly decreased with rank in 2009, with higher-ranking individuals having less ritual interactions (Fig. 2a). Adding the dominance rank tended to improved the model explaining the frequency of affiliative interactions, although this was only a trend (likelihood-ratio test: $X^2 = 5.88$, $p = 0.053$). In 2009, the frequency of affiliative interactions was higher in the top-ranking stallion (rank 5) compared to lower-ranking ones, whereas the opposite seemed to occur in 2010, with affiliative interactions being highest in males with rank 2 and 3 and decreasing in higher ranking stallions (Fig. 2b).

Adding the experience of group housing significantly improved model 1 for all categories of interactions (Table 2). In 2010, horses with no experience of group housing had more agonistic interactions (model residuals controlled for the effect of time after integration: −0.025±0.015), more ritual interactions (0.098±0.016) and more affiliative interactions (0.012±0.005; $n = 4$ horses and 1241 frequencies) than horses that were already in group in 2009 (agonistic interactions = −0.050±0.014; ritual interactions = 0.003±0.013; affiliative interactions = −0.015±0.003; $n = 4$ horses and 1241 frequencies).

As a result, the model that best explained the variation in the frequency of agonistic and affiliative interactions was the model including both the time after integration and the experience of group housing (model 4; Table 2). This model had 68% chance to be the best model within the set of models explaining agonistic interactions, and 59% chance to be the best model within the set of models explaining affiliative interactions. The model that best explained the variation in the frequency of ritual interactions was the model including both the time after integration and the dominance rank of stallions (model 2; Table 2). This model had 81% chance to be the best model within the set of models explaining ritual interactions. Within the set of models explaining affiliative interactions, the model including the time after integration and the dominance rank of stallions (model 2) was a close competitor of model 4 (ΔAICc<2; Table 2). This model had 26% chance to be the best model. All the other models had considerably less support by the data (ΔAICc>3). To summarize, the best model explaining the frequency of agonistic and affiliative interactions included the time after integration and the experience of group housing, and the best model explaining the frequency of ritual interactions included the time after integration and the dominance rank of stallions.

Stability of the Hierarchy

A stable hierarchy was established and could be measured after two (2010) to three months (2009; Table 3). In 2009, the hierarchy was stable after 3 months (correlation between dominance ranks measured after 2 and 3 months; Kendall's tau = 1.00, $n = 5$ horses, $p = 0.027$), but not after 2 months (correlation between dominance measured ranks after 1 and 2 months; Kendall's tau = 0.84, $n = 5$ horses, $p = 0.096$). In 2010, the hierarchy was already stable after 2 months (Kendall's tau = 0.96, $n = 8$ horses, $p = 0.002$), and was still stable after 3 months (Kendall's tau = 0.95, $n = 8$ horses, $p = 0.002$; Table 3).

Discussion

Unlike individual housing systems, group housing allows horses to fully express their natural behaviours [2,3,9,13]. The main reason that prevents owners to keep horses in groups is the potential risk of physical aggression, or a lack of suitable grazing land. The risk of physical aggression is likely to be particularly high during group integration, when the dominance hierarchy is being established. In this study, we investigated social interactions occurring after stallions had been integrated into a new group, in order to assess the potential risks of aggressive interactions such as kicks or bites between horses. We showed that stallions displayed generally more ritual than agonistic and than affiliative interactions. Agonistic and ritual interactions decreased within a few days following group integration (three to four days), while affiliative interactions increased slowly with time before decreasing later on. A stable hierarchy was established between group members after two to three months. The males at the top of this hierarchy after three months had less ritual interactions than the lower-ranking ones during the observation period (17 to 23 days after group integration). Males had also less agonistic, ritual and affiliative interactions if they had already been housed in group the previous year, suggesting an effect of social experience on interactions. Therefore, under the specific tested conditions, stallions can be kept in groups, because agonistic interactions are maintained at a minimum rate after the first few days following group integration, which corresponds to the rate observed among wild bachelor groups (1.4 interactions per hour in our study versus 1.5 in natural populations of Przewalski's horse, *Equus ferus przewalskii* [20]). We therefore encourage horse breeders with extensive pasture land to keep stallions in stable groups and in adequate densities [3,19–21], particularly for those that are not used for breeding the whole year around. This could potentially improve horse welfare and reduce labour associated with horse management (H. Besier and I. Bachmann, unpublished data).

Pattern of Social Interactions after Group Integration

We found that the time after group integration explained a large amount of the variance in the data. Agonistic and ritual interactions decreased quickly within the first three to four days after integration. These changes were very similar between the two groups studied in 2009 and 2010. After that, the frequency of agonistic interactions that we measured (1.40 h^{-1} per horse) was similar to the frequency measured by Christensen et al. [20] in a bachelor group of Przewalski's horses (1.46 h^{-1} per horse; $n = 13$ stallions), but higher than the frequency measured in Bourjade et al. [17] (0.2 h^{-1} per horse; $n = 9$ Przewalski's stallions) or in a smaller bachelor group (0.76 h^{-1} per horse; $n = 4$ Przewalski's stallions [42]). In contrast, affiliative interactions increased slowly with time and then decreased after 9–14 days.

Social interactions play an important role in the establishment and maintenance of hierarchies. Within a social group, a stable hierarchy functions to regulate aggression and thus reduce the number of serious fights [43]. When two males *encounter each other*, they perform a ritual that allows them to assess each other's fighting abilities using information contained in visual, olfactory or acoustical signals, without having to fight [23]. Accordingly, in our

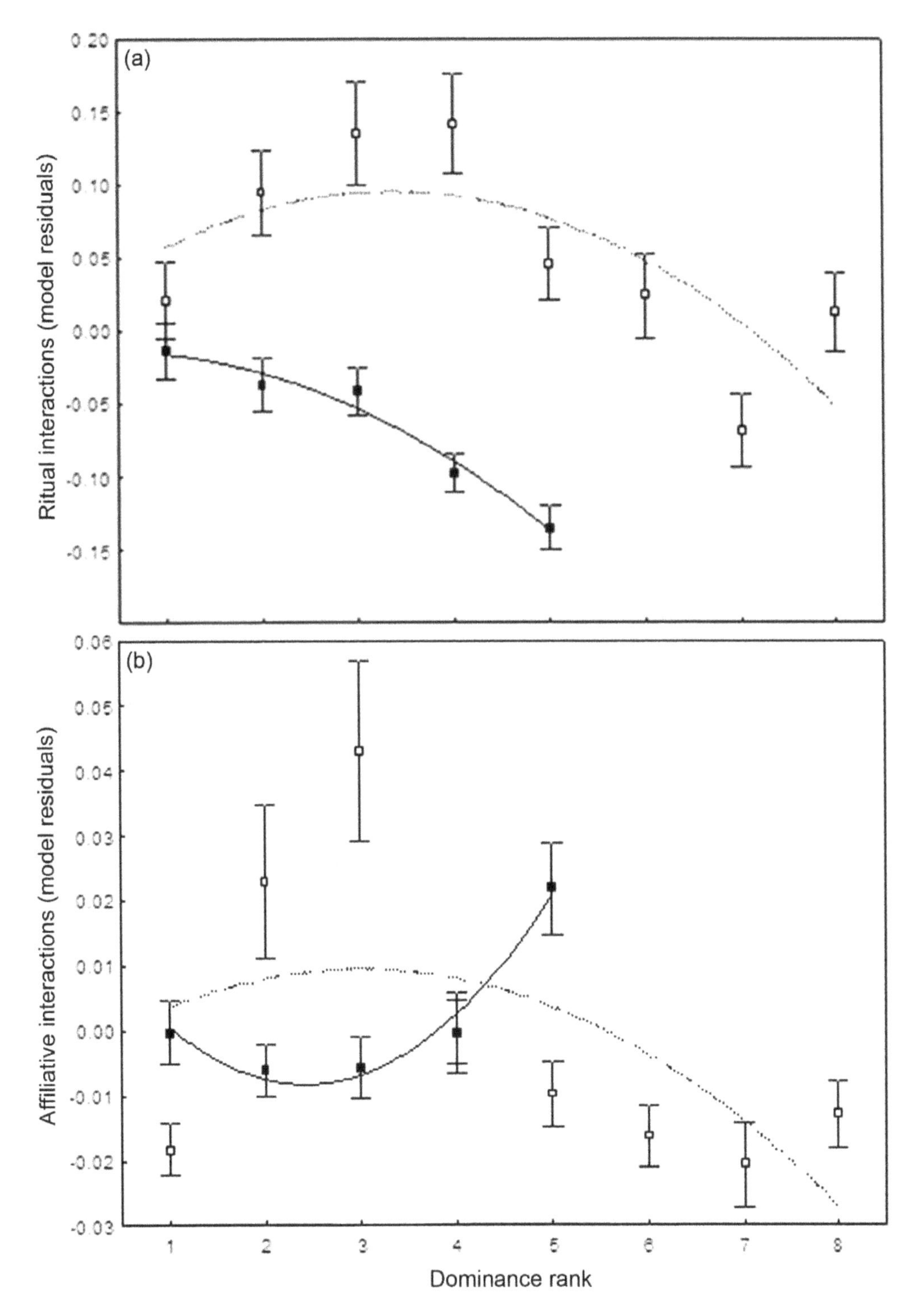

Figure 2. Relationship between the frequency of ritual (a) and affiliative (b) interactions per hour (model residuals controlled for the effect of the time after integration) and the dominance rank of stallions in 2009 (black square) and 2010 (empty squares; mean±SE per rank). The best fit (quadratic) is indicated with a solid line for 2009 and dashed line for 2010 data. Residuals represented stallions that had more interactions than predicted by the time after integration. Higher dominance ranks indicate higher-ranking individuals.

study, stallions had generally more ritual than agonistic interactions, thus preventing real fights [36]. These mutual assessments are effective alternatives to real aggression, but can escalate into serious fights over resources of any kind, when the degree of asymmetry in fighting abilities between the two individuals is low, or if there is an ambiguous hierarchy [3,21,43]. In contrast, the increase in the frequency of affiliative interactions at the beginning of the study indicated that social bonds were being established. In horses, typical affiliative behaviours are play, allogrooming and anti-parallel standing rest [9,32]. Play behaviour is particularly displayed in groups of males or mixed gender groups, compared to female groups [33]. The main function of affiliative relationships is to reduce social tension between group members and therefore, to increase group cohesion [2,13]. We suggest that the following decrease in affiliative interaction observed after 9–14 days in our study could be due to the fact that the frequency of affiliative interactions required to establish social bonds is higher than the frequency required to maintain these bonds. Therefore, once relationships have been established, the frequency of affiliative interactions could decrease [2].

Factors Affecting Social Interactions

The time after group integration was the main predictor of the frequency of interactions. However, other factors, such as the dominance rank of stallions and their experience of group housing also played a role. Ritual interactions were lower in higher-ranking stallions compared to lower-ranking ones. Stallions experienced in group housing had less agonistic, ritual and affiliative interactions than other stallions.

Our results showed that the frequency of ritual interactions, but not agonistic interactions, was influenced by the hierarchy. Similarly, in Przewalski's horse natural populations, lower-ranking stallions have been shown to engage more often in rituals than higher-ranking ones, which could indicate that they have a tendency for compromising rather than fighting [17]. High-ranking stallions win more fights, but do not to display higher rates of physical aggression than other males [3,17]. This suggests that the dominance rank of high-ranking males is rarely challenged. Threats, olfactory cues and vocal cues may suffice to maintain their dominance rank. Tilson et al. [18] found that conflict for rank was limited to the three lower-ranking Przewalski's stallions, within a group of eight bachelors. Because mutual assessments are more frequent when the degree of asymmetry in fighting abilities between two individuals is low [43], these results suggest that the degree of asymmetry in fighting abilities is generally lower at the bottom of the hierarchy, or in our 2010 group, within the stallions that were ranked 2–4.

Affiliative interactions tended to be affected by dominance rank, with higher ranking males displaying more affiliative interactions in 2009, and less affiliative interactions in 2010, than low-ranking ones (trend, $p = 0.053$). In other studies, affiliative interactions have been shown to be more often initiated by dominant individuals, as we found in 2009 (e.g. [44,45]). Low-ranking individuals might rarely initiate affiliative interactions with higher-ranking individuals, because of the elevated risks of provoking an agonistic interactions [44,46]. Therefore, dominant individuals are expected to contribute more than subordinate to affiliative relationships, because they can choose whom to bond with, whereas subordinates cannot [47]. In our 2010 group however, which was a larger group than in 2009, the relationship between rank and frequency of affiliative interactions was less clear. The frequency increased from ranks 1 to 3, followed by a decrease in high-ranking stallions. Our observations were collected during the first 17–23 days following group integration, while the hierarchy was being established. Indeed, our measures of the stability of the hierarchy revealed that it was stable after three months in 2009 and two months in 2010. We suggest that in large groups, while the hierarchy is being established, dominant individuals could have less affiliative interactions than their subordinates while trying to maintain their rank in the hierarchy.

Our results show that stallions had also less agonistic, ritual and affiliative interactions if they had already been housed in group the previous year. These results could be linked to an increase in familiarity between stallions. Indeed, Hartmann et al. [30] showed that pre-exposing unfamiliar horses by placing them in adjacent stables reduces both aggressive and non-aggressive interactions when they physically meet for the first time. However, the stallions used in our study had been regularly hitched next to each other for driving and had been housed at several occasions in adjacent stables before the first group integration. Therefore, all the stallions used in this study were already familiar when we first housed them in a group. An alternative explanation would be that these results are linked to stallions' experience of group living. Previously singly stabled stallions have been shown to display more aggressive interactions (e.g. bite threat), but also more affiliative interactions (allogrooming and play), than previously group housed ones [2]. These results could be due to a build-up of motivation during the period when horses are kept individually, suggesting that stallions are sensitive to social deprivation and that individual housing has long-term negative effects on social behaviour [2]. Furthermore, horses might need to acquire social competences in order to behave appropriately in group [3,11,48]. The proportion of "inappropriate" threats directed towards more dominant individuals decreases with age [44], indicating an important role of experience on social skills [3]. Horses that have been living in group have more refined social skills and are less aggressive towards other horses and even towards humans during training [2,12,49,50]. Therefore, these results suggest that the stronger the social experience of horses that are integrated in a group is, the lower the frequency of agonistic interactions would be. Further experiments, in which stallions are unfamiliar to each other before group integration, could help to disentangle the effects of familiarity and experience of group living on the frequency of interactions.

By definition, a natural behaviour is important for animal welfare if performing this behaviour improves the animal's physical or mental health [51]. A behaviour is considered as an "ethological need" if it is performed by all individuals, is self-rewarding, has a rebound effect and if chronic stress, which can lead to abnormal behaviour, is triggered when the performance of this behaviour is prevented [52]. In horses, allogroming, and to a lesser extend play, have been identified as ethological need because they meet all criteria [2,9]. A lack of social contacts triggers stress-related behaviours and stereotypies in horses. Social interactions should therefore be considered as crucial for welfare [4–6]. Many individual stables afford horses no opportunity to interact with neighbours. However, where possible, stables should be designed to allow adjacent neighbours to physically interact through, for example, partitions with vertical bars at the top half.

Conclusions

Housing horses in groups fulfils many of their welfare needs, including the access to social partners and the establishment of a social structure [1,9]. Such system could potentially increase horse welfare and reduce labour associated with horse management. In this study, we showed that stallions can be housed in groups under specific conditions, because agonistic interactions, which are potentially linked to physical aggression, decrease and are kept at a minimum rate after only three to four days following group integration.

Acknowledgments

We are grateful to K. Annen, M. Courtois, A. Fuchs, M. Linard, H. Zurkinden, to the vet team of the Swiss Institute of Equine Medicine "ISME", including F. Berruex, D. Burger, M. Federicci, A. Ramseyer, and to all the staff of the Swiss National Stud Farm for assistance. We are also grateful to B. Pitcher, H. Würbel, A. Zollinger for helpful comments on the manuscript.

Author Contributions

Revised the paper: IB. Conceived and designed the experiments: SBF EFB IB RVN. Performed the experiments: SBF. Analyzed the data: EFB. Wrote the paper: SBF EFB.

References

1. Hartmann E, Søndergaard E, Keeling LJ (2012) Keeping horses in groups. A review. Appl Anim Behav Sci 136: 77–87.
2. Christensen JW, Ladewig J, Søndergaard E, Malmkvist J (2002) Effects of individual versus group stabling on social behaviour in domestic stallions. Appl Anim Behav Sci 75: 233–248.
3. Fureix C, Bourjade M, Henry S, Sankey C, Hausberger M (2012) Exploring aggression regulation in managed groups of horses *Equus caballus*. Appl Anim Behav Sci 138: 216–228.
4. McGreevy PD, Cripps PJ, French NP, Green LE, Nicol CJ (1995) Management factors associated with stereotypic and redirected behaviour in the Thoroughbred horse. Equine Vet J 27: 86–91.
5. Cooper JJ, Mason GJ (1998) The identification of abnormal behaviour and behavioural problems in stabled horses and their relationship to horse welfare: a comparative review. Equine Vet J 30: 5–9.
6. Bachmann I, Audigé L, Stauffacher M (2003) Risk factors associated with behavioural disorders of crib-biting, weaving and box-walking in Swiss horses. Equine Vet J 35: 158–163.
7. Heleski C, Shelle A, Nielsen B, Zanella A (2002) Influence of housing on weanling horse behavior and subsequent welfare. Appl Anim Behav Sci 78: 291–302.
8. Rose-Meierhöfer S, Klaer S, Ammon C, Brunsch R, Hoffmann G (2010) Activity behavior of horses housed in different open barn systems. J Equine Vet Sci 30: 624–634.
9. van Dierendonck MC, Spruijt BM (2012) Coping in groups of domestic horses – Review from a social and neurobiological perspective. Appl Anim Behav Sci 138: 194–202.
10. Visser EK, Ellis AD, Van Reenen CG (2008) The effect of two different housing conditions on the welfare of young horses stabled for the first time. Appl Anim Behav Sci 114: 521–533.
11. Bourjade M, Moulinot M, Henry S, Richard-Yris MA, Hausberger M (2008) Could adults be used to improve social skills of young horses, Equus caballus? Dev Psychobiol 50: 408–417.
12. Ladewig J, Søndergaard E, Christensen JW (2005) Ontogeny: preparing the young horse for its adult life. In: Mills DS, McDonnell SM, editors. The domestic horse: the evolution, development and management of its behaviour. Cambridge: Cambridge University Press. 139–149.
13. van Dierendonck MC (2006) The importance of social relationships in horses Utrecht University. PhD thesis. Utrecht University.
14. Klingel H (1975) Social organization and reproduction in equids. J Reprod Fertil Suppl 23: 7–11.
15. McCort WD (1984) Behavior of feral horses and ponies. J Anim Sci 58: 493–499.
16. Berger J (1986) Wild Horses of the Great Basin: Social Competition and Population Size. Chicago, IL: University of Chicago Press.
17. Bourjade M, Tatin L, King SRB, Feh C (2009) Early reproductive success, preceding bachelor ranks and their behavioural correlates in young Przewalski's stallions. Ethol Ecol Evol 21: 1–14.
18. Tilson RL, Sweeny KA, Binczik GA, Reindl NJ (1988) Buddies and bullies: Social structure of a bachelor group of Przewalski horses. Appl Anim Behav Sci 21: 169–185.
19. McDonnell SM, Haviland JCS (1995) Agonistic ethogram of the equid bachelor band. Appl Anim Behav Sci 43: 147–188.
20. Christensen JW, Zharkikh T, Ladewig J, Yasinetskaya N (2002) Social behaviour in stallion groups (*Equus przewalskii* and *Equus caballus*) kept under natural and domestic conditions. Appl Anim Behav Sci 76: 11–20.
21. Heitor F, Vicente L (2010) Dominance relationships and patterns of aggression in a bachelor group of Sorraia horses (*Equus caballus*). J Ethol 28: 35–44.
22. Feist JD, McCullough DR (1976) Behavior patterns and communication in feral horses. Z Tierpsychol 41: 337–371.
23. Rubenstein DI, Hack MA (1992) Horse signals: The sounds and scents of fury. Evol Ecol 6: 254–260.
24. Lemasson A, Boutin A, Boivin S, Nlois-Heulin C, Hausberger M (2009) Horse (Equus caballus) whinnies: a source of social information. Anim Cogn 12: 693 – 704.
25. Krueger K, Flauger B (2010) Olfactory recognition of individual competitors by means of faeces in horse (*Equus caballus*). Anim Cogn 14: 245–257.
26. Vannoni E, McElligott AG (2008) Low frequency groans indicate larger and more dominant fallow deer (*Dama dama*) males. PLoS ONE 3: e3113.
27. Briefer E, Vannoni E, McElligott AG (2010) Quality prevails over identity in the sexually selected vocalisations of an ageing mammal. BMC Biol 8: 35.
28. Reby D, McComb K (2003) Anatomical constraints generate honesty: acoustic cues to age and weight in the roars of red deer stags. Anim Behav 65: 519–530.
29. Proops L, McComb K, Reby D (2009) Cross-modal individual recognition in domestic horses (*Equus caballus*). Proc Natl Acad Sci USA 106: 947–951.
30. Hartmann E, Christensen JW, Keeling LJ (2009) Social interactions of unfamiliar horses during paired encounters: Effect of pre-exposure on aggression level and so risk of injury. Appl Anim Behav Sci 121: 214–221.
31. Christensen JW, Søndergaard E, Thodberg K, Halekoh U (2011) Effects of repeated regrouping on horse behaviour and injuries. Appl Anim Behav Sci 133: 199–206.
32. McDonnell MD (2003) The equid ethogram, a practical field guide to horse behaviour. Lexington: The Blood-Horse.
33. Jørgensen GHM, Borsheim L, Mejdell CM, Søndergaard E, Bøe KE (2009) Grouping horses according to gender–Effects on aggression, spacing and injuries. Appl Anim Behav Sci 120: 94–99.
34. Houpt KA, Law K, Martinisi V (1978) Dominance hierarchies in domestic horses. Appl Anim Ethol 4: 273–283.
35. Lehmann K, Kallweit E, Ellendorff F (2006) Social hierarchy in exercised and untrained group-housed horses–A brief report. Appl Anim Behav Sci 96: 343–347.
36. Zharkikh TL, Andersen L (2009) Behaviour of bachelor males of the Przewalski horse (*Equus ferus przewalskii*) at the reserve Askania Nova. Zool Gart 78: 282–299.
37. Craig JV (1986) Measuring social behavior: social dominance. J Anim Sci 62: 1120–1129.
38. Bates D, Maechler M, Bolker B (2011) lme4: Linear mixed-effects models using S4 classes. Available:http://CRAN.R-project.org/package = lme4.
39. Burnham KP, Anderson DR (2002) Model selection and multimodel inference: a practical information-theoretic approach. New York: Springer.
40. Burnham KP, Anderson DR, Huyvaert KP (2011) AIC model selection and multimodel inference in behavioral ecology: some background, observations, and comparisons. Behav Ecol Sociobiol 65: 23–35.
41. R Development Core Team (2012) R Foundation for Statistical Computing. Vienna, Austria. Available: http://www.R-project.org.
42. Feh C (1988) Social behaviour and relationships of Prezewalski horses in Dutch semi-reserves. Appl Anim Behav Sci 21: 71–87.
43. McElligott AG, Mattiangeli V, Mattiello S, Verga M, Reynolds CA, et al. (1998) Fighting tactics of fallow bucks (*Dama dama*, Cervidae): reducing the risks of serious conflict. Ethology 104: 789 – 803.
44. Wells SM, von Goldschmidt-Rothschild B (1979) Social behaviour and relationships in a herd of Camargue horses. Z Tierpsychol 49: 363–380.
45. Heitor F, Vicente L (2010) Affiliative relationships among Sorraia mares: influence of age, dominance, kinship and reproductive state. J Ethol 28: 133–140.
46. Heitor F, do Mar Oom M, Vicente L (2006) Social relationships in a herd of Sorraia horses: Part II. Factors affecting affiliative relationships and sexual behaviours. Behav Proc 73: 231–239.
47. Sigurjónsdóttir H, van Dierendonck MC, Snorrason S, Thórhallsdóttir AG (2003) Social relationships in a group of horses without a mature stallion. Behaviour 140: 783–804.
48. Bourjade M, de Boyer des Roches A, Hausberger M (2009) Adult-young ratio, a major factor regulating social behaviour of young: a horse study. PLoS ONE 4: e4888.
49. Rivera E, Benjamin S, Nielsen B, Shelle J, Zanella AJ (2002) Behavioral and physiological responses of horses to initial training: the comparison between pastured versus stalled horses. Appl Anim Behav Sci 78: 235–252.
50. Søndergaard E, Ladewig J (2004) Group housing exerts a positive effect on the behaviour of young horses during training. Appl Anim Behav Sci 87: 105–118.
51. Dawkins MS (2003) Behaviour as a tool in the assessment of animal welfare. Zoology 106: 383–387.
52. Jensen P, Toates FM (1993) Who needs "behavioural needs"? Motivational aspects of the needs of animals. Appl Anim Behav Sci 37: 161–181.

9

Methods for Generating Year-Round Access to Amphioxus in the Laboratory

Èlia Benito-Gutiérrez*, Hermann Weber, Diana Virginia Bryant, Detlev Arendt*

Developmental Biology Unit, European Molecular Biology Laboratory (EMBL), Heidelberg, Germany

Abstract

Cephalochordates, commonly known as amphioxus, are key to understanding vertebrate origins. However, laboratory work suffers from limited access to adults and embryonic material. Here we report the design and experimental validation of an inland marine facility that allows establishing stable amphioxus colonies in the laboratory and obtaining embryos at any time of day and over almost the entire year, far exceeding natural conditions. This is achieved by mimicking the natural benthic environment, natural day- and moon- light, natural substrate and by providing a strictly controlled and seasonally fluctuating temperature regimen. Moreover, supplemented algae diets allow animals to refill their gonads in consecutive years. Spontaneous spawning, a major problem in previous setups, no longer occurs in our facility; instead, all breeding is induced and fertilization occurs fully *in vitro*. Our system makes amphioxus a standard laboratory animal model.

Editor: Hector Escriva, Laboratoire Arago, France

Funding: The authors are grateful for financial support to E.B.G. in the course of this investigation from La Generalitat de Catalunya, Beatriu de Pinós Program (2006BP-A10108), from EMBL, Interdisciplinary Postdocs Program (EIPOD) and from the European Community, ASSEMBLE grant agreement number 227799. The funders had no role in study design, data collection and analysis, decision to publish, or preparation of the manuscript.

Competing Interests: The authors have declared that no competing interests exist.

* E-mail: elia.begu@embl.de (EBG); arendt@embl.de (DA)

Introduction

The past decades have seen renewed interest in amphioxus as a model system that sheds light on the invertebrate-vertebrate transition and the molecular basis for the vertebrate radiation [1–3]. However, the use of amphioxus in molecular genetics and developmental biology has been hampered by technical limitations such as limited access to embryonic material, precluding experimentation on a daily basis. Consequently, current studies have remained largely descriptive (with a few exceptions [4–7].

To overcome these limitations, various approaches have been developed to facilitate access to amphioxus in the lab [8–14]. However, these studies disagree on the husbandry conditions in an artificial marine environment. Moreover, all systems previously described suffer from recurrent spontaneous spawning in the parental tanks, thus limiting the breeding period in captivity to only a couple of months of each year or just sporadically [9–12,14–16]. Since spontaneous spawning in these systems are semi-synchronous and in line with the natural breeding season (e.g. [14]), animals appear to still follow the natural cycle.

Here, we describe a new facility and robust protocols for long-term husbandry and efficient and controlled breeding of different amphioxus (species, genus or populations) of interest, including: a powerful refrigeration system that reproduces a broad range of temperatures from 7°C to 27°C, which covers virtually all naturally occurring temperature ranges in world-wide habitats; the ability to mimic any natural photoperiod, moonlight intensity and twilight (dusk and dawn); a self-purifying flow-through system, which allows efficient washout of organic matter and other contaminants.

We report the establishment of stable colonies for various populations of the European species, *Branchiostoma lanceolatum*. All populations show substantially improved long-term survival rates, with values above 95% survival in more than 2-years-long cultures, in contrast to the 83.3% survival previously achieved in 1 year-long cultures (with no survival reported for subsequent years) [9–14]. We obtain strongly expanded annual breeding seasons of up to nine months, in contrast to the two-month-breeding season previously achieved in *B.lanceolatum* [9–19]. This represents a 4.5-fold expansion respect from any other system previously reported and, in addition, allows controlling the beginning and the end of the expanded breeding season thus decoupling the animals from their natural cycle. Providing special dietary supplements, we observe gonad refill in line with the seasonal fluctuations programmed in the facility and animals spawning in consecutive years post collection. Furthermore, and in contrast to previous systems, the animals in our facility spawn after induction only. By manipulating the temperature and light conditions, spawning can be programmed at desired times and days according to the experimental needs.

Our facility dramatically improves the conditions of working with amphioxus and thus facilitates efforts to make it a model species fully accessible to *in vivo* experimental manipulation, imaging, stock keeping and any approach that requires vast amounts of animal material.

Materials and Methods

Ethics Statement

This research did not involve the use of any vertebrate animal. This research was carried on non-protected and non-endangered invertebrates only. No animals were sacrificed for the work reported here.

Branchiostoma lanceolatum specimens were collected in Helgoland (Germany), Roscoff or Argelès-sur-mer (France) by dredging. All animal collections were performed according to the recommendations and regulations in the respective Marine Stations: AWI, Station Biologique Roscoff and Observatoire Oceanologique de Banyuls-sur-mer. Research was done under the supervision of the responsible bodies in the respective Marine Stations and under the framework of the Association of European Marine Biological Laboratories (ASSEMBLE) approved by the European Union (FP7).

Facility design and husbandry conditions

In order to simulate the benthic marine ecosystem where wild amphioxus live, we engineered an automated marine facility with a closed circuit of constantly flowing-through natural sea water (Figure S2), with natural daylight, dusk and dawn, natural moon phases and sand-bedding (Fig. 1 and Figure S1). For a technical description of the basic framing, the encapsulated tank system, the closed water circuitry, light instalments, oxygen supply and water quality control see Materials and Methods S1.

Animals were cultured at a density of 25–30 animals per box. Annual temperature curves followed records of the SOLA station (SOMLIT Observatory Laboratoire Arago) close to the Argelès sampling site (http://somlit-db.epoc.u-bordeaux1.fr) (see Table S1). In all cases, pH and salinity values stabilized around pH 7.9–8.1 and 54–56 ms. Animals are fed on seven algal species grown in a custom-made algae facility (Figure S4).

Survival and Spawning Data

Routine inventories were carried out every six months and survivor numbers were determined by subtracting the number of dead animals from the number of inventoried animals. Adaptation was calculated on the number of dead animals in the first month subtracted from the number of animals collected. Spawning efficiency is defined as the percentage of spawning animals among the total subjected to heat shock. The raw data for survival and spawning records are displayed in Table S2 and Figure S5. Tables and graphs were done in 'Numbers' (iWork, Apple) and statistical analyses in R (The R foundation for Statistical Computing).

Results and Discussion

Stable amphioxus colonies in the laboratory

The major cause of mortality for cultured amphioxus is bacterial infection, a recurring problem in all previous attempts to culture amphioxus for longer periods [13]. To overcome this limitation, our approach mimics natural conditions as closely as possible (see Materials and Methods and Materials and Methods S1). Whilst most quantified parameters (e.g., salinity, pH, day and night ratios and twilight times) were kept constant throughout the

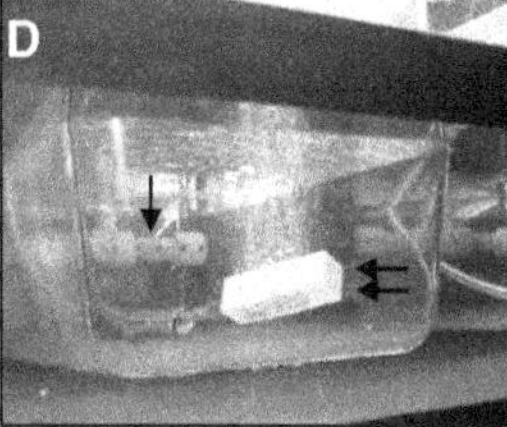

Figure 1. The amphioxus facility. (**A**) Lightproof cabinet of the facility, with open shutters, showing three shelves with two main tanks each. The upper distribution tank and UV sterilisation device are visible at the top outside the cabinet. The two-compartment lower reservoir tank and neighbouring filter unit are at the base of the facility inside the cabinet. (**B**) View of the six main tanks encapsulating eight amphioxus boxes each, distributed in two rows of four. All tanks in use are sand-bedded. (**C**) Encapsulated tank system: the amphioxus boxes are encapsulated into the main tank through a fitted PVC grid (arrow), leaving two thirds of their height immersed in the main tank. The unique unsealed part is the netted outlet (double arrow) of the amphioxus boxes, used for the flow-through circulation of the seawater. (**D**) All amphioxus boxes are equipped with inverted T-shape water jets, multi-perforated at the base (arrow). This was conceived to generate a unidirectional seawater wave washing out the entire width of the boxes. In addition, all amphioxus boxes are equipped with 5 cm air stones (double arrow) to constantly oxygenate the water by dispersion.

different conditions assayed, we varied culturing conditions ("prototype" and "optimized" facility; see materials and methods S1 and table S1) in two important points that proved to significantly improve our survival rates from an average of 20% to an average of 97.8% of survival in our cultures (see Fig. 2B): (1) Commercial sterile sand bedding was replaced with natural sand collected from the original habitats in the optimised facility; and (2) Natural annual temperature curves were lowered by three degrees in average in the optimised facility. Data on the performance of both facilities was collected (see raw data in Table S2) and further compared with regard to adaptation of newly collected animals, long term maintenance and resistance to fluctuating temperatures (important for controlled breeding, see below).

Improved Adaptation. Upon arrival all animals are closely monitored (see Protocol S1). As a measure of successful adaptation to the facility we calculated the survival rates during this period. While some populations consistently adapted better than others, in the optimized facility the population differences were reduced, resulting in overall improved survival rates (around 92%), independently of the site of origin ($p = 0.005$; Fig. 2A). In order to exploit a good-sized starting population this first month of farming is critical. In this regard, care should be emphasized during the collection and transportation of the animals, which might affect the health status of the animals and their capabilities to adapt to the new environment (See Protocol S1).

Long-term maintenance. While satisfactory short-term survival rates had been previously observed using other culturing systems, these usually decline dramatically over time [11,14]. In this regard, the most distinguishing feature of the optimized facility is that long-term survival rates stabilise at an even higher value than during adaptation, with an observed average survival of ~95% during the first year in the optimised facility (Fig. 2B). By contrast, in the prototype facility, first year-survival depended on population origin. Moreover, the survival rates in the optimised facility further improved in subsequent years, stabilising at a value close to 100% (Fig. 2C). Thus, the optimised facility enables the establishment of stable colonies of amphioxus with overall survival rates above 95% (Fig. 2D).

Resistance to fluctuating temperatures. Importantly, survival rates in the optimised facility were robust to fluctuations in the temperature at which the animals were cultured. This is important since such fluctuations typically arise in nature and are

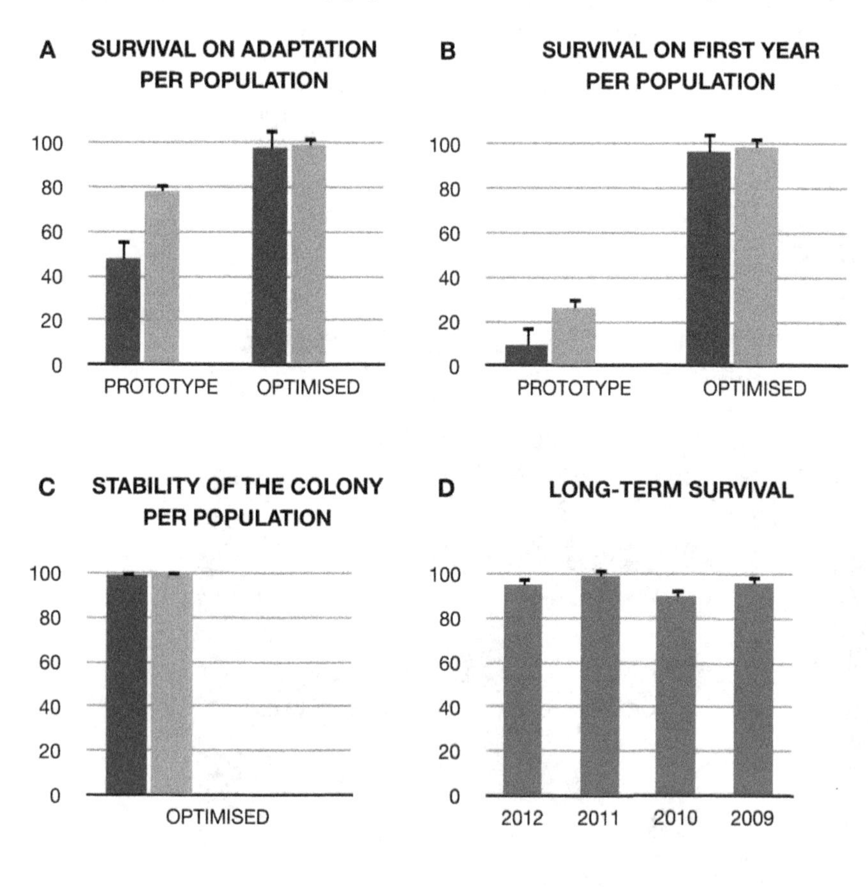

Figure 2. Long-term survival and stability of amphioxus colonies. (**A**) Survival rates during the first month of farming in the laboratory per population in "prototype" and "optimized" setups. (**B**) Survival rates during the first year of farming in the laboratory per population in the two culturing setups. (**C**) Survival of the different amphioxus populations in our optimised facility over 4 years of farming. (**D**) Average survival rates in 2012 of animals that were farmed for several years, meaning they were collected in 2012, 2011, 2010 or 2009. Confidence intervals on the binomial probabilities were generated using the R package "binom" with the exact method.

required for controlled maturation of the gonads and subsequent breeding (see below).

Gonad refill in captivity

Unsurprisingly, previous studies have identified proper feeding as playing a critical role in the refill of gonads of cultured amphioxus, (albeit, such refilling was restricted to the natural spawning season [9–10]. Beyond this, little is known about external cues and hormonal signals that induce gonad development and maturation in amphioxus and all efforts to boost gametogenesis have failed thus far [11]. As we initially observed stasis in the gonadal sacs of long-term surviving animals that had been fed on microalgae only, we tested different kinds of diets with increased variety of both micro- and macro-algae, cultured in situ (See Figure S4), alongside various supplements containing vitamins, essential amino acids, protein, fatty acids and iodine. Three different food mixes were tested; A) fresh algae mix, as in previous culturing systems; B) fresh algae supplemented with vitamins and essential amino acids; and C) as in B but also enriched in lipids, protein and iodine. This was fed into different tanks containing animals without gonads, with incipient gonads or with no gonads at all, either after collection or after spawning in the facility (independently of year), so we could observe the general status of gonad maturation per tank (see Fig. 3 for more information about the animal groups tested).

While, as previously observed [9–10] food mix A allowed completion of gonad maturation in animals freshly collected during the natural maturation phase (Fig. 3A1–A3), it was insufficient to trigger gametogenesis in animals that, at the time of collection, had not started gonad maturation (Fig. 3B1). This means that food mix A is only effective in about a 70% of the collected animals, which is the average percentage of animals with gonads in our annual collections (for dates see Table S2). By contrast, food Mix B enabled sustained gonad maturation in animals without visible gonads at the time of collection, meaning that mix B is sufficient to prepare an entire batch of freshly collected animals for spawning, independently of their gonad status at the time of collection. However, it did not allow gonad maturation in subsequent years (Fig. 3B1–B3). Only with food Mix C could we observe gonad maturation in captivity in the breeding seasons of subsequent years. (Fig. 3B and 3C). Food Mix C sustained gonad refill of both animals that had spawned in the natural breeding season in successive years and animals that had spawned outside of the natural breeding season. Gonad refill in captivity appears to occur as in nature [17] (Fig. 3A–D), but is obtained in full accordance with the artificial seasonal light and

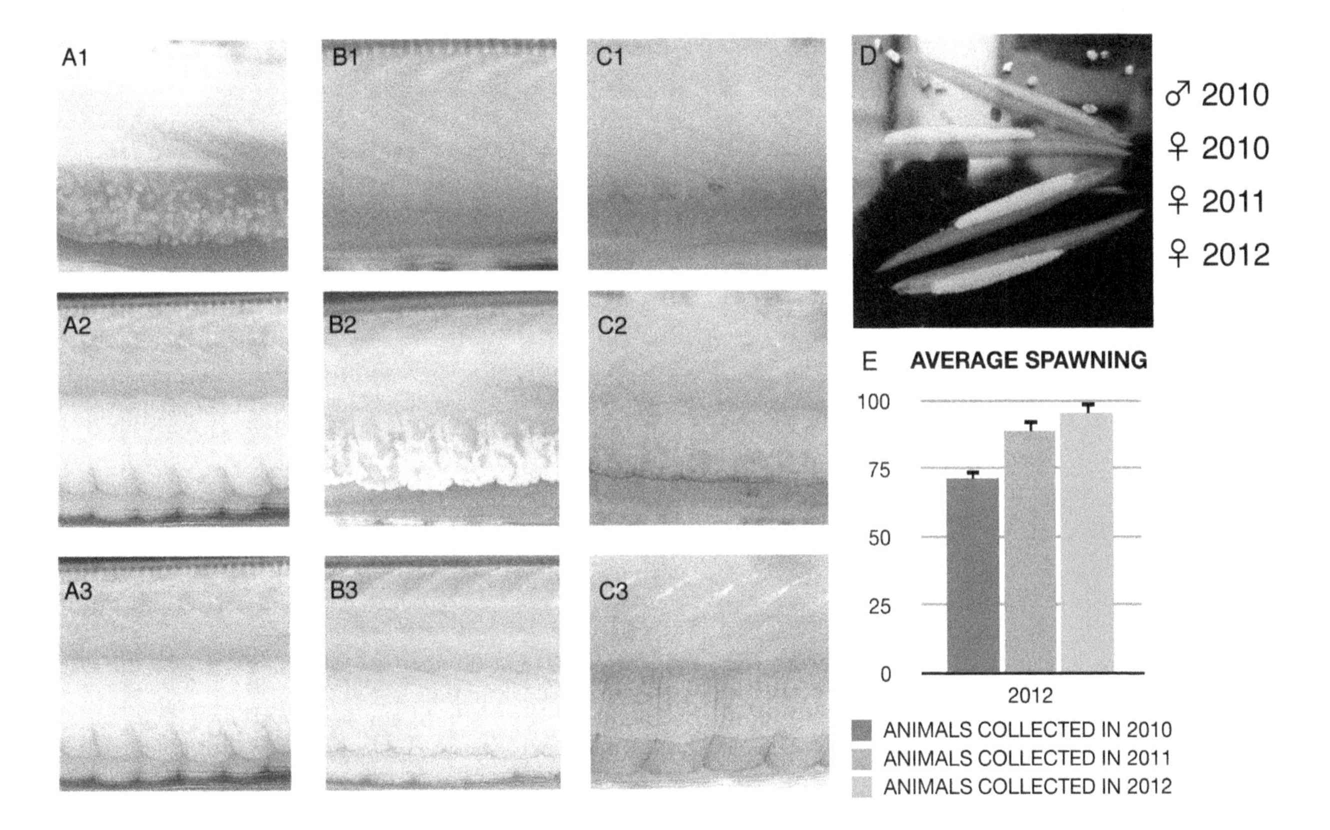

Figure 3. Annual gonad refill in the amphioxus facility. (**A–C**) Gonad status of maturation for three different groups of animals in our optimised amphioxus facility: (**A**) Animals collected with incipient gonads; (**B**) Animals collected with no gonads or animals that spawned in the facility in previous years; (**C**) Animals already spawned in the same breeding season.1 is the initial status of the gonads for a given individual, and its progression (2) until the gonads are fully developed (3). State 3 is achieved in our facility in line with artificial seasonal fluctuations only in combination with feeding on Food Mix C. In state 3 most of our animals are able to spawn after being induced. (**D**) Comparison between animals that refilled their gonads in the facility or in the wild. The picture shows a snapshot of ripe animals in our facility in 2012, with animals collected in 2010 and 2011 versus a female collected in 2012 (with refilled gonads in the wild). By feeding on Food Mix C, no appreciable difference is seen between animals that maturated their gonads in the nature or in the facility. (**E**) Annual spawning percentages of animals collected in 2010 and 2011 that spawned in subsequent years. The spawning in subsequent years appears to improve over the time with a maximum difference of a 30% between 2010 and 2012 animals that spawned in 2012.

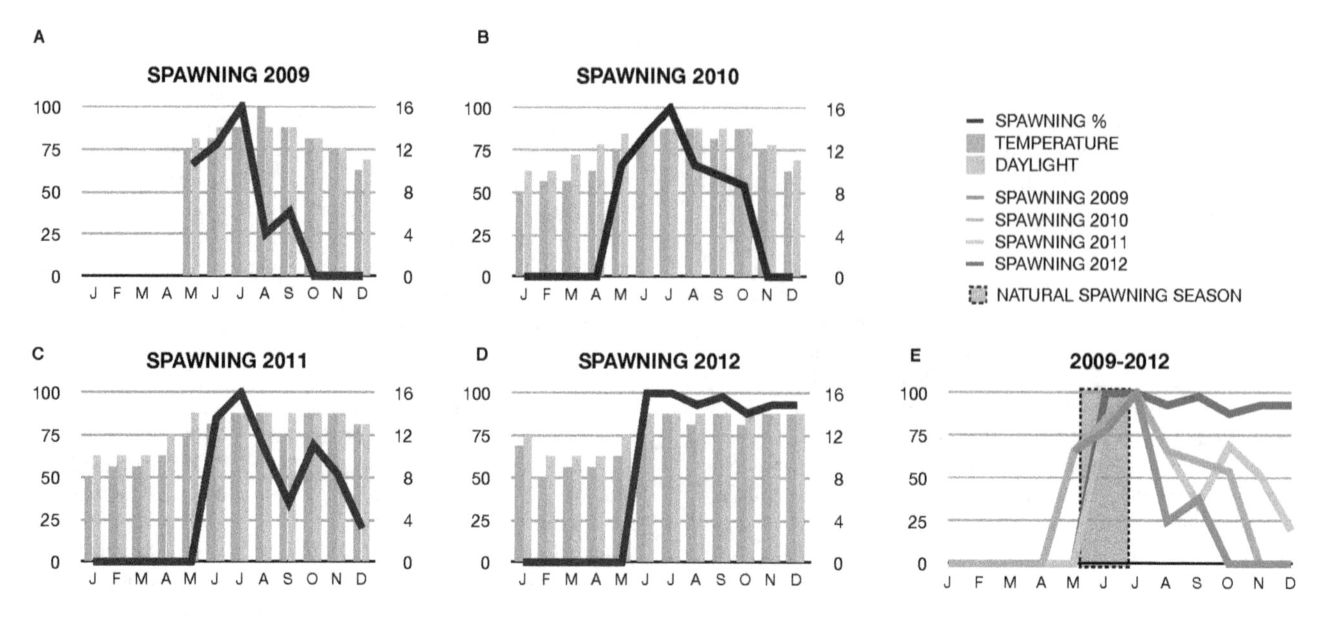

Figure 4. Expansion of the spawning season in the amphioxus facility. (**A–D**) Annual spawning periods and efficiencies in 2009, 2010, 2011 and 2012. The results show that the animals respond to the artificial changes, meaning deliberate fluctuations of temperature and light, which provoked a second peak of spawning, specially pronounced in 2011 (**C**) and maintaining spawning efficiencies no lower that 89% all through the facility spawning season in 2012 (**E**). Progressive expansion of the spawning season in the facility in the last four years and increment of the spawning efficiencies over the time, in-line and off-line the natural spawning season in the wild.

temperature regimen in the facility, which demonstrates that the animals are uncoupled from the natural cycle and fully adopted to the artificial environment. Once food mix C was in place, animals collected in 2010 and 2011 spawned controllably in subsequent years (Fig. 3 E). Among the animals in which spawning was induced, the success rate of spawning was high (62.5% and 71.42% success rate for the 2010 animals in 2011 and 2012; 88.88% for the 2011 animals spawned in 2012). Still, our numbers are small with only a 8% of our long-term cultured population ready to spawn at any time of the year, meaning that further improvement in diet is necessary to get independent from annual collections. Consecutive spawnings have been recently reported for *B. belcheri* [18] but in contrast to our fully controlled system these occur spontaneously in the parental tanks. Therefore, our facility is the first to allow controlled consecutive spawnings in the laboratory, which is critical for exploiting amphioxus in the laboratory.

Artificial expansion of the spawning season

Until now, a major drawback in amphioxus research has been the limited time window during which embryos can be obtained (at present ~2 months per year during the natural breeding season in *B.lanceolatum* [9–10,15,19]). As it was suggested that the raise in temperature during springtime was important for successful spawning [9] we tested the effect of different temperature regimens and daylight/moonlight ratios upon the readiness of the animals to spawn in captivity. To this aim the facility is equipped with a control panel that allows manipulating and recreating natural seasonal fluctuations (See Figure S3). We observed that animals were ready to spawn shortly after entering the artificial summer cycle and, importantly, artificially delaying the start of the autumn program prolonged the breeding season for at least an additional eight months (Fig. 4). Maximal spawning efficiency (100% of animals spawning) was reached in July, followed by a gradual decrease during subsequent months (Fig. 4). However, critically, the spawning efficiency was more than sufficient to enable the collection of embryos for experimental work until late in the year. To further increase spawning efficiency during the entire artificial season we exposed the animals to mild fluctuations in the seasonal temperature regimen. In essence, we generated short periods of colder days before raising the temperature back to standard summer values. This was motivated by the observation that this pattern arises in nature in late spring shortly before the spawning season starts [9]. This enabled a second peak in spawning efficiency to be obtained, which was especially pronounced in 2011, where a spawning efficiency of 71% was observed in October (Fig. 4C). In 2012 mild drops in the temperature allowed us to maintain the spawning efficiency at a high level throughout all eight months, with 93% efficiency in December (Fig. 4D) and 75% efficiency in January 2013 (see Figure S5). Importantly, we have successfully expanded the spawning season in each of the last four years, with increasing efficiency (Fig. 4E). In 2012 spawning efficiency was never lower than 89% and continued into January 2013 (Fig. 4D). Therefore, our artificially expanded and fluctuating seasons generate a pool of responsive animals that can be induced to spawn in a controlled manner (see below) throughout almost the entire year (Fig. 4E).

Controlled spawning at any time of day

As wild amphioxus spawn after sunset, we attempted to control the timing of spawning by varying daylight times. In our facility, the amphioxus shed their gametes almost exactly one hour after sunset, if exposed to heat shock before (see Protocol S2 adapted from 9 and 10). We thus kept the animals in light-proof boxes during the thermal shock and delayed the onset of darkness in incremental steps. By delaying the off time of the bulb inside the light box the animals were induced to spawn at specific times (always one hour after the light was extinguished). Such programmed spawning was effective for up to 10 hours after the normal sunset time in the facility. Since *in vivo* manipulation requires freshly laid eggs [20] our controlled spawning protocol will greatly facilitate experimentation.

It is important to note that our controlled spawning protocol completely eliminates the spontaneous spawning events observed in other systems [9–10,14–15]. For the first time, this ensures the controlled production of offspring from specific animals of interest (e.g., after experimental manipulation).

Concluding Remarks

Our optimized facility provides access to amphioxus adults and embryos throughout the almost entire calendar year, facilitating the establishment and transfer of molecular techniques, the usage of novel imaging techniques and experimental manipulation at much larger scale.

Supporting Information

Figure S1 Environmental Setup of the Facility. (**A**) Environmental setup in the roof of each main module of the facility, showing: water piping (w), oxygen piping (o) and a complete set of night (ml) and day light (dl) bulbs. (**B**) Faucets for manual control of the water flow in the individual amphioxus boxes (arrow) and the main tanks (double arrow). (**C**) Facility in an open (left) and closed (right) configuration via a black-out automatic blind hosted in the upper blind case (bc) at the top of the facility cabinet. (**D**) Filter unit showing three foam mats of different colours according to the pore diameter (black, blue, red), the biological filter, and on the left side the lower reservoir tank containing the protein skimmer and pump for re-circulation of the purified water.

Figure S2 Water Circuitry of the Amphioxus Facility. Schematic representation of the water circuitry of the amphioxus facility. Only one shelf is represented for clarity. The water is cooled in the upper distribution tank. Before being distributed to the individual amphioxus boxes the water is UV sterilised and pumped under the control of magnetic solenoid valves, which flow can be programmed through the operations panel. Through the netted outlets in the amphioxus boxes, water streams out to the main tanks from where it drains down to the filter unit, the protein skimmer and finally the lower reservoir tank from where the water is pumped back to the upper distribution tank. The facility can be also runt in an open configuration, similar to other systems previously shown (e.g. [2]), if the main tanks are drained by opening the outlet as indicated in the diagram. The air circuitry is also indicated, with pipes supplying oxygen to each of the amphioxus boxes. The inlay is a schematic representation of one of our amphioxus tanks showing: a) netted outlet in the front of the tank; b) feeding vent; c) opening to insert the oxygen tubing; d) opening to insert the water jet.

Figure S3 Operations Panel of the Amphioxus Facility. (**A**) Plan of the inner part of the operations panel cabinet. It contains all electronic connections, fuses and digital controllers of the circuits for the different parts of the amphioxus and algae facility. Al circuits are individualised to avoid a shut down of the entire facility in case of short-circuit in any of the components. Accordingly there are individual electric boxes and fuses for: each of the magnetic valves (for the flow-through system), the two pumps, the UV-light, the daylight bulbs for the amphioxus facility, the daylight for the algae facility, the moonlight leds, the skimmer, the air pump for the amphioxus facility, the air pump for the algae facility, the automatic blind and the pH, salinity and temperature controllers. The digital controllers are equipped with programmable timers to mimic seasonal fluctuations, controlling the flow-through and the sterilisation intervals. Accordingly there are timers for the magnetic valves, for the moonlight and daylight to control the length of the day, the UV lamps for controlling the sterilisation time and another timer for controlling the light in the algae facility (Food Light). (**B**) Front door of the operations panel. It allows changing the mode of operation of the facility into automatic, semi-automatic or manual mode. It includes the buttons to operate the facility manually (pairs of green and red for on and off) and light buttons (only in green) informing of the status of the system. H1–H7 indicates the status of the flow-through per shelf; H8 indicates when it is day in the facility; H9 indicates when it is night in the facility; H4 indicates that the main pump is working (this should be permanently on); H10 indicates when the sterilisation device is in operation. The big red button at the bottom of the panel is to cut the power in case of an emergency.

Figure S4 High-Throughput Algae Culturing Facility. To ensure the fresh production of algae in the large quantities necessary to feed the animals on a daily basis we constructed an algae facility. Our algae facility consists of 8 conical funnels with a capacity of 30 litres (http://www.emsustains.co.uk/fish_hatching_jars.htm). Each funnel is oxygenated through a long air tube, which opens at the bottom of the funnel to continuously agitate the culture and impede the sedimentation of the algae. The funnels are illuminated by daylight bulbs and maintained at a constant temperature of 20°C. The funnels are equipped with outlets to dispense saturated solutions of algae (30.000–80.000 cells/ml) whenever necessary. Our algae facility operates in tandem with our amphioxus facility so with a single operations panel we control and program both facilities. The figure shows the accessory algae facility showing different algal cultures. The 30-litre funnels are mounted in a self-constructed scaffold to allow maximal exposure to the light and to facilitate the manipulation. Daylight bulbs are mounted on the wall and the oxygen system is on the upper part of the facility. All components of the algae facility run under the control of the amphioxus facility through the main operation panel.

Figure S5 Raw Spawning Data. The tables show the total numbers of shocked animals per month, in the course of this investigation. The number of spawning animals in relation to the total shocked was used to calculate the percentage of spawning efficiency per month and year (see Figure 4). Over the time the number of animals unable to spawn after the thermal shock diminished with only occasional individuals registered in 2012.

Materials and Methods S1 Materials and Methods S1

Protocol S1 Amphioxus Care

Protocol S2 Time Lapse Spawning

Table S1 Adult Culturing Setups. The table summarizes the major differences between the prototype and the optimised versions of the amphioxus facility assayed. The graphs compare the annual temperature curves in the sea and in the facility. In all cases same natural topology of temperature curves was reproduced annually, though the range applied in the two setups was different. In the prototype we reproduced exactly the same temperature values as observed in nature, whereas in the optimised facility we reduced the overall annual temperature around 3.5°C in average. SSW: Surface Sea Water; DSW: Deep Sea Water (3 meters deep).

Table S2 Raw Survival Data. The table shows the number of animals collected and the number of survivor over the time in the course of this investigation. Recover after quarantine upon arrival of the animals to the facility is also indicated. Death others: indicate recorded deaths after quarantine but before one year of farming.

Acknowledgments

We thank the mechanical and electronics workshop, and building maintenance of the European Molecular Biology Laboratory (EMBL) for customizing both algae and amphioxus facilities and light boxes for time-lapsed controlled spawning; Kresimir Crnokic for help in maintaining the animals; Margret Krüss for access to amphioxus in Helgoland, Hector Escrivà for access to amphioxus in Banyuls-sur-mer. We thank John Marioni for help on the manuscript.

Author Contributions

Conceived and designed the experiments: EBG HW DA. Performed the experiments: EBG. Analyzed the data: EBG DA. Contributed reagents/materials/analysis tools: EBG HW DVB DA. Wrote the paper: EBG DA.

References

1. Delsuc F, Brinkmann H, Chourrout D, Philippe H (2006) Tunicates and not cephalochordates are the closest living relatives of vertebrates. Nature 439: 965–968.
2. Putnam NH, Butts T, Ferrier DEK, Furlong RF, Hellsten U, et al. (2008) The amphioxus genome and the evolution of the chordate karyotype. Nature 453: 1064–1071.
3. Holland LZ, Albalat R, Azumi K, Benito-Gutiérrez E, Blow MJ, et al. (2008) The amphioxus genome illuminates vertebrate origins and cephalochordate biology. Genome Res 18:1100–1111.
4. Yu JK, Holland ND, Holland LZ (2004) Tissue-specific expression of FoxD reporter constructs in amphioxus embryos. Developmental Biology 274: 452–461.
5. Schubert M, Yu JK, Holland ND, Escriva H, Laudet V, et al. (2005) Retinoic acid signaling acts via Hox1 to establish the posterior limit of the pharynx in the chordate amphioxus. Development 132: 61–73
6. Schubert M, Holland ND, Laudet V, Holland LZ (2006) A retinoic acid-Hox hierarchy controls both anterior/posterior patterning and neuronal specification in the developing central nervous system of the cephalochordate amphioxus. Developmental Biology 296: 190–202.
7. Onai T, Yu JK, Blitz IL, Cho KWY, Holland LZ (2010) Opposing Nodal/Vg1 and BMP signals mediate axial patterning in embryos of the basal chordate amphioxus. Developmental Biology 344: 377–398.
8. Zhang SC, Zhu JT, Li GR, Wang R (2001) Reproduction of the laboratory-maintained lancelet *Branchiostoma belcheri tsingtauense*. Ophelia 54:115–118.
9. Fuentes M, Schubert M, Dalfo D, Candiani S, Benito E, et al. (2004) Preliminary observations on the spawning conditions of the European amphioxus (Branchiostoma lanceolatum) in captivity. J Exp Zool Part B: Molec and Dev Evol 302: 384–391.
10. Fuentes M, Benito E, Bertrand S, Paris M, Mignardot A, et al. (2007) Insights into spawning behavior and development of the european amphioxus (Branchiostoma lanceolatum). J Exp Zool Part B: Molec and Dev Evol 308: 484–493.
11. Yasui K, Urata M, Yamaguchi N, Ueda H, Henmi Y (2007) Laboratory Culture of the Oriental Lancelet Branchiostoma belcheri. Zoological Science 24: 514–520.
12. Zhang QJ, Sun Y, Zhong J, Li G, Lü XM, et al. (2007) Continuous culture of two lancelets and production of the second filial generations in the laboratory. J Exp Zool Part B: Molec and Dev Evol 308: 464–472.
13. Somorjai I, Camasses A, Riviere B, Escriva H (2008) Development of a semi-closed aquaculture system for monitoring of individual amphioxus (Branchiostoma lanceolatum), with high survivorship. Aquaculture 281: 145–150.
14. Theodosiou M, Colin A, Schulz J, Laudet V, Peyrieras N, et al. (2011) Amphioxus spawning behavior in an artificial seawater facility. J Exp Zool Part B: Molec and Dev Evol 263–275.
15. Kubokawa K, Mizuta T, Morisawa M, Azuma N (2003) Gonadal State of Wild Amphioxus Populations and Spawning Success in Captive Conditions during the Breeding Period in Japan. Zoological Science 20: 889–895.
16. Mizuta T, Kubokawa K (2004) Non-synchronous spawning behavior in laboratory reared amphioxus *Branchiostoma belcheri* Gray. J Exp Mar Biol Ecol 309:239–251.
17. Yamaguchi T, Henmi Y (2003) Biology of the Amphioxus, Branchiostoma belcheri in the Ariake Sea, Japan II. Reproduction. Zoological Science 20: 907–918.
18. Li G, Yang X, Shu Z, Chen X, Wang Y (2012) Consecutive Spawnings of Chinese amphioxus, Branchiostoma belcheri, in captivity. PLoS One 7:e50838.
19. Holland LZ, Yu JK (2004) Cephalochordate (amphioxus) embryos: procurement, culture, and basic methods. Methods Cell Biol 74:195–215.
20. Holland LZ, Onai T (2011) Analyses of Gene Function in Amphioxus Embryos by Microinjection of mRNAs and Morpholino Oligonucleotides. Methods in Mol Biol 770: 423–438.

10

Bovine Polledness – An Autosomal Dominant Trait with Allelic Heterogeneity

Ivica Medugorac[1]*, Doris Seichter[2], Alexander Graf[3], Ingolf Russ[2], Helmut Blum[3], Karl Heinrich Göpel[4], Sophie Rothammer[1], Martin Förster[1], Stefan Krebs[3]

1 Ludwig-Maximilians-University Munich, Munich, Germany, **2** Tierzuchtforschung e.V. München, Grub, Germany, **3** Laboratory for Functional Genome Analysis (LAFUGA), Gene Center, Ludwig-Maximilians-University Munich, Munich, Germany, **4** Göpel Genetik GmbH, Herleshausen, Germany

Abstract

The persistent horns are an important trait of speciation for the family *Bovidae* with complex morphogenesis taking place briefly after birth. The polledness is highly favourable in modern cattle breeding systems but serious animal welfare issues urge for a solution in the production of hornless cattle other than dehorning. Although the dominant inhibition of horn morphogenesis was discovered more than 70 years ago, and the causative mutation was mapped almost 20 years ago, its molecular nature remained unknown. Here, we report allelic heterogeneity of the *POLLED* locus. First, we mapped the *POLLED* locus to a ~381-kb interval in a multi-breed case-control design. Targeted re-sequencing of an enlarged candidate interval (547 kb) in 16 sires with known *POLLED* genotype did not detect a common allele associated with polled status. In eight sires of Alpine and Scottish origin (four polled versus four horned), we identified a single candidate mutation, a complex 202 bp insertion-deletion event that showed perfect association to the polled phenotype in various European cattle breeds, except Holstein-Friesian. The analysis of the same candidate interval in eight Holsteins identified five candidate variants which segregate as a 260 kb haplotype also perfectly associated with the *POLLED* gene without recombination or interference with the 202 bp insertion-deletion. We further identified bulls which are progeny tested as homozygous polled but bearing both, 202 bp insertion-deletion and Friesian haplotype. The distribution of genotypes of the two putative *POLLED* alleles in large semi-random sample (1,261 animals) supports the hypothesis of two independent mutations.

Editor: Shuhong Zhao, Huazhong Agricultural University, China

Funding: This work was funded by the German Research Foundation (DFG; Germany, ME 3404/2-1) and a grant from the German agriculture network PHÄNOMICS (funded by Federal Ministry of Education and Research funding reference number 0315536F) as well as by substantial budget funds from the Chair of Animal Genetics and Husbandry of the Ludwig-Maximilians-University Munich (TierALL, LMU Munich, Germany), the Laboratory for Functional Genome Analysis (LAFUGA, Gene Center, LMU Munich, Germany) and the Tierzuchtforschung e.V. München (TZF Grub, Germany). The authors IM, SR, MF, SK, AG, HB, DS and IR are employed by TierALL, LAFUGA and TZF which supported this study by substantial budget funds but only the authors and not the funders had a role in study design, data collection and analysis, decision to publish, or preparation of the manuscript.

Competing Interests: The authors have read the journal's policy and declare the following conflicts: Karl Heinrich Göpel, one of the co-authors, is employed by Göpel Genetik GmbH, a commercial company that kindly provided a significant amount of samples and phenotypic information. However no funding was provided by Go¨pel Genetik GmbH or other cattle breeding companies, or cattle breeders. In addition, neither cattle breeding companies nor cattle breeders were involved in the study design, analysis and interpretation of data, writing of the paper, or decision to submit for publication.

* E-mail: ivica.medjugorac@gen.vetmed.uni-muenchen.de

Introduction

Since the beginning of animal husbandry, humans have tended to accumulate particular variations in domesticated animals. Thereby, the variants of practical interest for agriculture were selected as well as phenotypes that visibly distinguished particular animals. This practice was more frequent in pets (e.g. cats and dogs) but led to the fixation of some obvious breed characteristics in cattle, too [1]. The absence of horns (polled phenotype) as well as horn shape diversity represents such evident traits in cattle.

The persistent horn of the *Bovidae* consists of a pneumatised osseous core, which is fused with the frontal bone and covered by a cornified epithelium that grows outward from the skin at the base of the horn, thereby forming the cavernous visible horn. The development of horns is dependent on a number of different tissues and their interaction [2,3]. Before the domestication of cattle, horns were important for the survival of the wild species. Even after domestication, horns were a desired trait (e.g. fixation and use as draught animals) in most cattle breeding areas until recently. Exceptions to that rule are found in some regions (e.g. Scotland and the Nordic countries) where polledness was a desired trait much earlier probably due to dense housing of animals during the long winter. Nowadays, commercial dairy or beef herds are mainly confined to barns or fenced-in enclosures such as pastures or corrals. Under these conditions horns are not only of little value but can lead to considerable economic loss due to a higher risk of injuries and the possible consequences (infection, carcass deterioration etc.). Therefore in modern cattle husbandry removing horns at an early age has become an accepted management practice. However, all used methods are debatable not least because of animal welfare implications [4]. Hence, breeding polled cattle may constitute a non-invasive option to replace the common practice by means of genetic selection.

In cattle, as in other *Bovidae*, horn development and morphology are characterised by a substantial degree of polymorphism. Even if only European cattle breeds (*Bos taurus*) are considered, there are completely hornless cattle breeds like Angus, and breeds with very short horns like Buša as well as breeds with very long and lyra-shaped horns like the Pannonian Podolian cattle [5]. The polled phenotype has been considered to be caused by a dominant inhibitor [3] that is characterised by the complete absence of corneous appendices [6]. In addition to diverse forms of cattle horns, abnormal types of horns called scurs frequently occur, i.e. incompletely developed horns that are not fused to the frontal bone [7]. Scurs can be considered as the intermediate phenotype between polled and horned because the horn core is not an outgrowth of the skull but originates from a separate ossification centre in the tissues located above the periosteum with subsequent fusion to the skull [2,3].

While the inheritance pattern and mapping position of the *POLLED* locus on *Bos Taurus* Autosome 1 (BTA1) [8,9,10] is beyond controversy, the scurs locus displays heterogeneity [2]' [11]' [12]. To avoid complications due to possible interference between *POLLED* locus and type 1 [11] or type 2 [2] scurs syndrome we exclusively focussed on polled and horned animals of 31 cattle breeds of European origin. This material exploited the full capacity of the population structure of European cattle and in combination with high-density SNP genotyping and high-throughput sequencing, was used to identify the genetic variation that causes the polled phenotype in *Bos taurus*.

Results

The *POLLED* Locus Maps to Chromosome 1 in Divergent European Cattle Breeds

We collected DNA samples from twelve polled cattle breeds (Table S1). The applied mapping approach and major part (62%) of the samples has been presented in our previous study [10] and was now completed by additional 61 polled animals as well as by three additional polled breeds: Norwegian Red, Fjall cattle and Braunvieh. Case (homozygous polled, PP) and control (homozygous horned, pp) group each consisted of equal numbers of animals (162) from 18 breeds in total. As some breeds are consolidated either for the polled or horned phenotype uneven phenotype distribution was inevitable within some breeds. An improved case-control design accounting for this uneven distribution did not refine the mapping (Fig. 1). We mapped the *POLLED* locus in the same 381 kb interval (1.668 Mb - 2.049 Mb; UMD3.1 genome build) we already found in our previous study [10]. The genome-wide significance ($P \leq 0.0002$,) was determined by 50,000 random permutations of markers along chromosomes. The applied mapping procedure [13] performs searches for shared segments of homozygosity without the need of an identity by state and thus revealed two haplotypes. The most common haplotype (AGACAAGGA) was found in all but one case animals. Two copies of an alternative haplotype, with a variant allele at the sixth SNP (AGACA**G**GGA) were found in a homozygous polled Braunvieh bull (Fig. S1) that was one of the animals from the current addition to the mapping design. Genotyping of three related bulls revealed the same alternative haplotype associated with the polled phenotype in Braunvieh (Fig. S1). As the origin of polledness in Braunvieh is unknown and recombination within the homozygosity block of the available Braunvieh pedigrees could not be detected, the Braunvieh data did not refine the mapping.

In order to confirm and possibly improve the mapping results we embedded 89 heterozygous polled animals (*POLLED* carriers; Pp) and 62 horned relatives into the analysis. These 151 animals were genotyped with the same Illumina BovineSNP50 assay [14] and have not been used in the case-control design. These 89 Pp bulls comprise ten breeds and are ancestors or descendents of PP bulls included in the case group, except for carriers from the Witrug breed (WTG) (Table S1). The haplotypes were manually traced through the pedigree charts. The result clearly confirmed the mapping result (Fig. S1, S2, S3 and S4), but the absence of any observable recombination within the candidate region hindered further improvement of fine-mapping by linkage analysis.

Sequencing Eight Polled and Eight Horned Sires did not Detected a Common Allele Associated with Polled Status

We selected seven PP, one Pp and eight pp bulls and performed high-throughput sequencing of a 547 kb interval (1.543–2.090 Mb (UMD3.1 genome build)) that nested the most likely location for the *POLLED* mutation [10]. The polled group was represented by four Holstein-Friesian (HF), two Galloway (GLW), one German Angus (DAN) and one Fleckvieh (FV) bull. One HF bull was declared as Pp while remaining seven bulls of polled group were progeny tested as PP. The horned group was represented by four HF, two FV, one Gelbvieh (FGV) and one Murnau-Werdenfelser (MWF) bull.

Paired-end sequencing of the libraries enriched for a 547 kb region on BTA1 resulted in average sequence depth of non-repetitive sequences of ~40× per individual, with 98.9% of the non repeat-masked target region being covered. From these data, we identified 451 putative DNA sequence variants (DSVs), or an average nucleotide diversity of ~0.15%. Further analysis of the sequencing data (on a locally installed instance of GALAXY [15]) revealed almost complete absence of sequence variability in a region of 125 kb (1.697 to 1.822 Mb) between all 16 sequenced animals. Under the assumption that the *POLLED* locus is biallelic with dominant inheritance [7,9] and because the reference sequence (RefSeq) is based on a horned Hereford cow [16], the causative variant has to be homozygous in the seven PP sires, heterozygous in one Pp sire and not present in RefSeq or in the eight pp sires. Applying this filter to the 451 DSVs did not detect a common allele associated with polled status. Also a visual inspection in a genome browser did not yield DSV concordant with the *POLLED* genotypes. These results suggested at least two plausible explanations: (i) the targeted 547 kb interval that was covered to 98.9% for the non repeat region harbors the functional mutation(s) in the remaining 1.1% (ii) some of DSVs are causal but there is allelic heterogeneity at the *POLLED* locus and different alleles have been selected in different geographic regions or breeds.

Dividing Sequenced Animals into Two Groups Suggests Allelic Heterogeneity of the *POLLED* Phenotype

Eight of 16 sequenced sires are Holsteins and in the case of the allelic heterogeneity these could most probably harbour the same *POLLED* mutation. The average sequence depth of non-repetitive sequences in eight HF bulls was ~29× per individual, covering 98.5% of the non repeat-masked target region. From these data, we identified 312 putative DSVs, or an average nucleotide diversity of ~0.11% (Fig. 2). Applying the same consecutive filtering as above reduced the number of potential causal variants to seven. On the proximal end of the region homozygous in polled HF (1.648–2.027 Mb; Fig. 2) we detected a complex InDel event, replacing 7 bp (cgcatca; RefSeq: 1,649,163–1,649,169) by 12 bp (ttctcagaatag) and thus resulting in a 5 bp longer sequence in polled HF animals (allele P_{5ID}). On the distal end of the region of homozygosity in polled HF we detected a duplication of an 80,128 bp (1,909,352–1,989,480 bp) sequence (Fig. 2). This large sequence is seamlessly

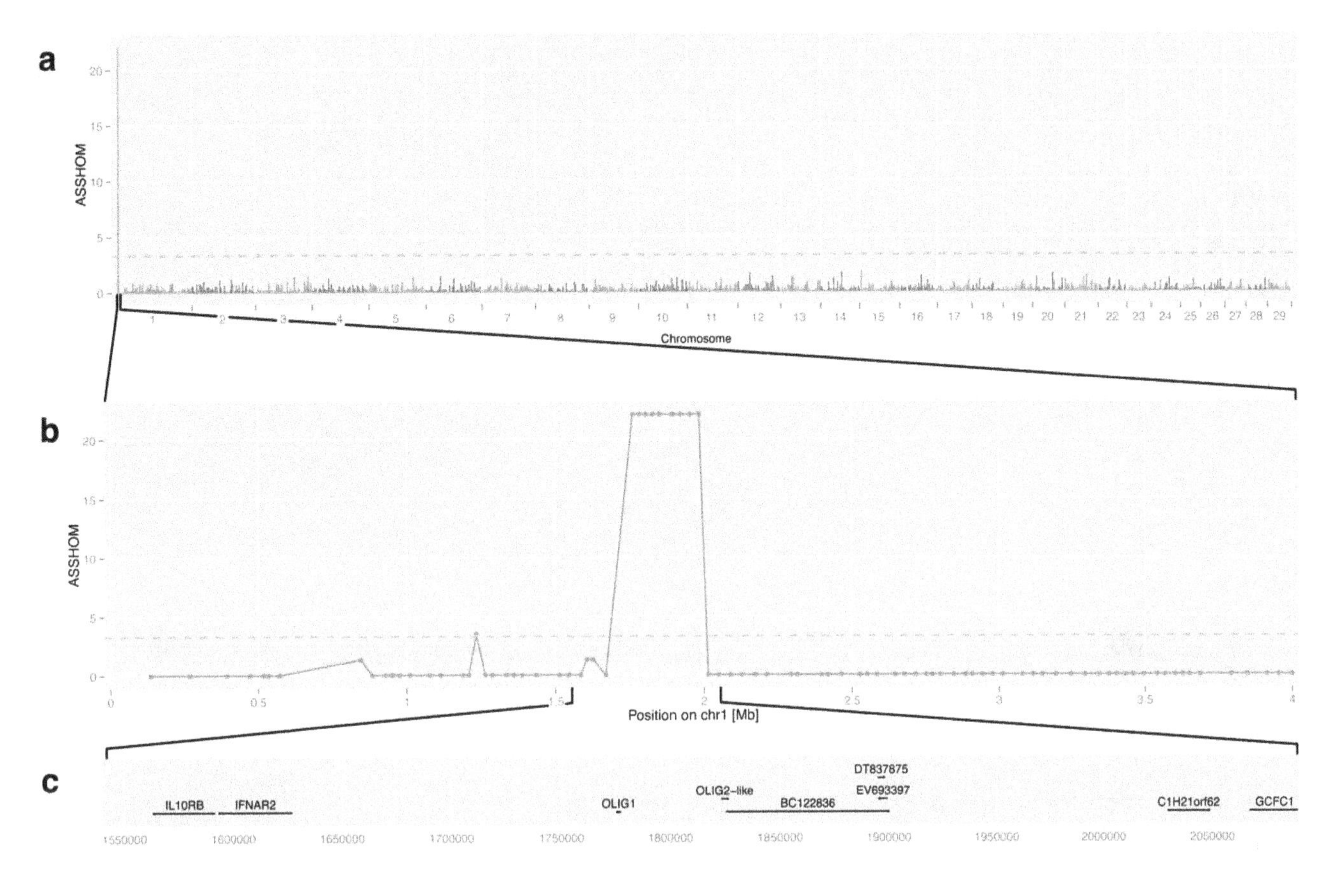

Figure 1. Homozygosity mapping in case-control design of polled and horned cattle animals. (**a**) Genome-wide association mapping of the *POLLED* mutation to the proximal end of bovine chromosome 1 (BTA1) with 162 affected animals and 162 controls (ASSHOM method, Charlier *et al.* 2008). Nominal ASSHOM statistic is presented chromosome by chromosome for all 29 bovine autosomes. Evidence for association (y axis) is measured as nominal ASSHOM statistic with genome-wide significance ($P = 0.0002$, grey line) at 3.315, being determined by 50,000 nominal ASSHOM statistics along chromosomes with randomly permutated markers. (**b**) Association mapping details for the region from 0 to 4 Mb on BTA1. Nine markers homozygous in all 162 cases covered a region from 1,760,113 to 1,983,902. Two informative SNP markers flanking this core (i.e. variable sites in the case group) are *ARS-BFGL-NGS-39992* (1,668,494 bp) and *ARS-BFGL-NGS-29653* (2,049,400 bp). (**c**) Candidate region of 547 kb (1.543–2.090 Mb (UMD3.1 genome build)) chosen for high-throughput sequencing and most likely encompassing the *POLLED* mutation. This interval is larger than homozygosity bracket detected by multi-breed design and nested five genes (IL10RB, IFNAR2, OLIG1, C1H21orf62 and GCFC1), one pseudo gene (OLIG2-like) and some not further annotated ESTs (e.g. BC122836, EV693397 and DT837875).

duplicated in the same direction and only differs from the RefSeq by one T→A transversion at the third position after the beginning of the duplicated sequence ($P_{T1909354A}$) and by a two-base pair (TG) deletion at the 45th position in the duplicated sequence. This 2 bp deletion corresponds to position 1,909,396 in the original sequence and will be denoted as $P_{1909396D2}$. Both variants close to the junction between the original and the duplicated sequence, $P_{T1909354A}$ and $P_{1909396D2}$, allow for a PCR design for the detection of this 80 kb InDel (allele P_{80kbID}). In addition to P_{5ID} and P_{80kbID}, both flanking the homozygosity region in polled HF, we detected five point mutations at the positions 1,654,405 (G→A), 1,655,463 (C→T), 1,671,849 (T→G), 1,680,646 (T→C) and 1,768,587 (C→A), respectively. In these five candidate SNPs ($P_{G1654405A}$, $P_{C1655463T}$, $P_{T1671849G}$, $P_{T1680646C}$, $P_{C1768587A}$) mutant alleles determine the sequence in the polled HF animals. These seven candidate mutations do not include any known coding sequence, or splice site, or intronic region, or any known regulatory elements (Fig. 2). Furthermore, none of these variants were previously reported in dbSNP (http://www.ncbi.nlm.nih.gov/projects/SNP/).

The remaining eight non HF sires (four PP and four pp) belong to spatial and genetic quite differentiated breeds: Angus, Galloway, Fleckvieh, Gelbvieh and Murnau-Werdenfelser. Nevertheless, both PP and pp sires were homozygous for the common haplotype that is always homozygous in PP animals [10]. Considering these eight sires separately paired-end sequencing identified 248 putative DSVs, or an average nucleotide diversity of ~0.08%. Applying the same filter to the 248 DSVs yielded only one complex insertion-deletion (InDel) event as the putative causative variant (Fig. 3), which was confirmed by visual inspection in a genome browser to be the only DSV concordant with the pp and PP genotypes. The sequence of 212 bp (1,705,834–1,706,045 bp) is duplicated and replaces a sequence of 10 bp (1,706,051–1,706,060 bp). This InDel (P_{202ID}) is between the genes *IFNAR2* and *OLIG1* (Fig. 3). The InDel P_{202ID} again does not disrupt any known coding sequence or a splice site, or an intronic region, or any known regulatory regions. All candidate DSVs detected in HF subset were homozygous for RefSeq allele in all eight non HF sires, both in PP as well as in pp.

Target Genotyping of Candidate DSVs in Polled Animals, Random Samples and Diversity Panel Supports Allelic Heterogeneity of the *POLLED* Phenotype

In order to further investigate the presumed allelic heterogeneity of *POLLED*, we chose sperm samples of polled bulls of all of the most important cattle breeds with European origin. This

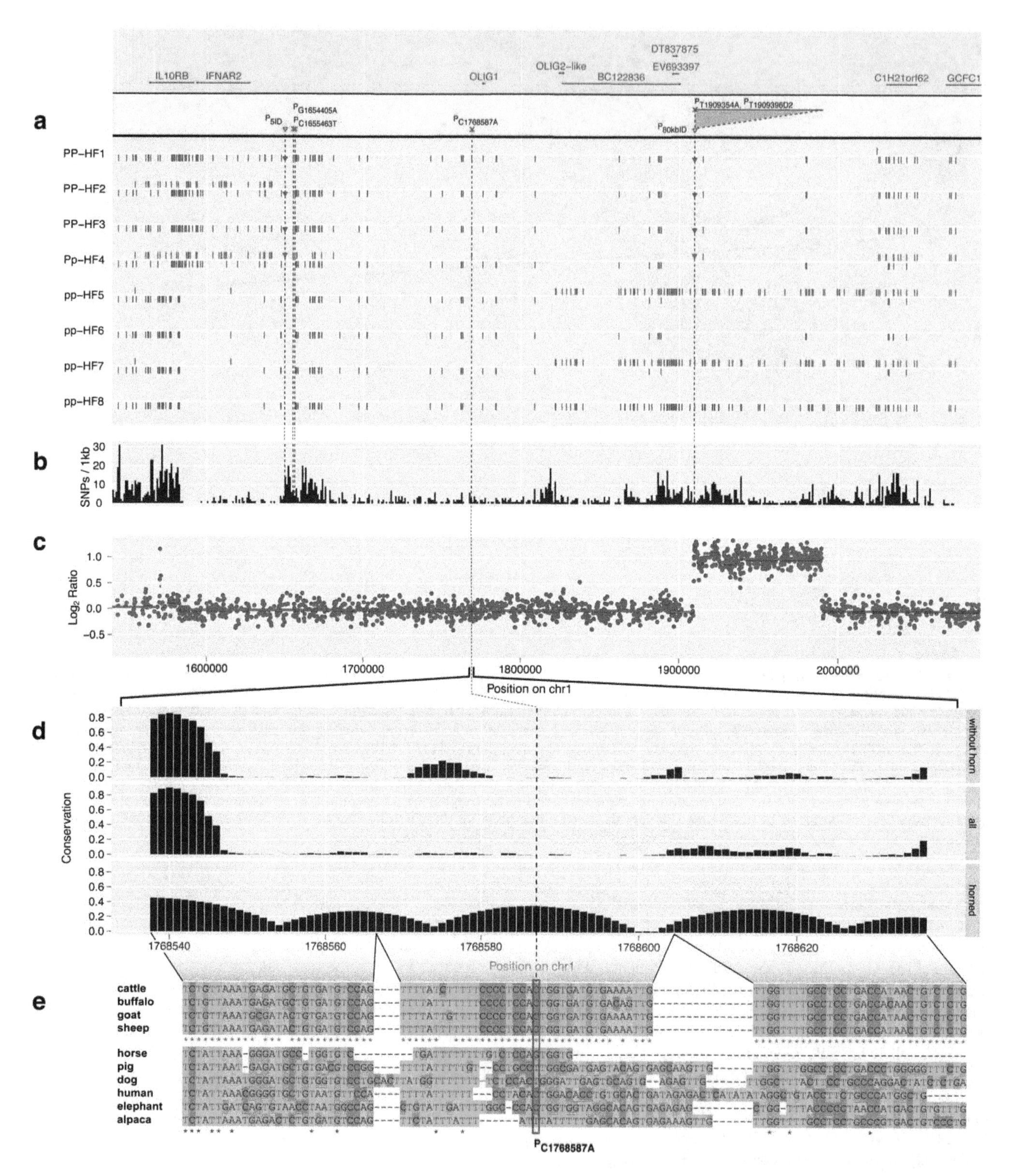

Figure 2. DNA sequence variants in Polled and horned Holstein Friesian. (**a**) Positions of DNA sequence variants (DSV) detected in the eight HF animals are presented by bars for SNPs and inverted triangles for InDels, with red symbols for heterozygous and blue symbols for homozygous differences from the reference sequence. Homozygous polled bulls are PP-HF1, PP-HF2 and PP-HF3. Pp-HF4 is declared as heterozygous polled and pp-HF5, pp-HF6, pp-HF7 and pp-HF8 are horned. Annotated genes and ESTs lying in the re-sequenced region are shown in the green shaded region. The five candidate mutations for polledness are superimposed: three SNPs were outlined by red crosses, 5 bp InDel and the duplicated region P_{80kbID} were highlighted by inverted triangle. The red triangle area above P_{80kbID} marks the duplicated sequence. (**b**) Within-species conservation. The nucleotide diversity in the re-sequenced region is shown by a density plot of bovine variants from dbSNP per kb. (**c**) Copy number variations. The ratio of mapped sequence reads between polled and horned animals is plotted with blue dots. The red line represents the result of segmentation analysis, showing the average ratio in the determined bins. (**d**) Across-species conservation. For each candidate mutation the surrounding base conservation for species without horn, bovid species and all aligned species was determined as *PhastCons* score calculated from MULTIZ alignments. As an example, here we display the base conservation for the candidate mutation $P_{C1768587A}$. (**e**) The underlying multi-species alignment for the across-

species conservation calculation for the candidate region is shown in plot d. The candidate mutation $P_{C1768587A}$ is highlighted within a red frame and stars in red and black indicate identity among bovid and all animals, respectively.

comprised 89 Pp bulls of ten breeds (Table S1). Moreover, we complemented the sample of the homozygous case group by additional PP bulls of Holstein-Friesian (HF, including Red-Holsteins (RH)), German Fleckvieh (FV) and Jersey (JY).

The P_{202ID} allele was in complete LD to the *POLLED* allele in all breeds but HF, JY and WTG which turned out to carry the P_{202ID} allele only sporadically. Polled HF, JY and WTG animals declared as Pp but missing the P_{202ID} allele were carriers of the haplotype block of seven candidates detected by sequencing of HF animals (P_{5ID}, $P_{G1654405A}$, $P_{C1655463T}$, $P_{T1671849G}$, $P_{T1680646C}$, $P_{C1768587A}$ and P_{80kbID}). Furthermore, HF and JY animals declared as PP and missing P_{202ID} were carriers of two copies of the haplotype detected by sequencing of polled HF animals.

To further proof the hypothesis of allelic heterogeneity of *POLLED*, we randomly sampled 400 animals in South-Germany complemented by random samples from target breeds. This semi-random sample comprised sets of 238 random HF/RH, 293 HF-FV crosses, 52 random JY, 50 random PNZ, and 211 animals from our cattle diversity panel (Table S1). These 1,262 DNA samples as well as the case and the carrier group were chosen for genotyping for the candidate variants: P_{202ID}, P_{5ID}, P_{80kbID} $P_{G1654405A}$, $P_{C1655463T}$, $P_{T1671849G}$, $P_{T1680646C}$, and $P_{C1768587A}$. Two SNPs $P_{T1671849G}$ and $P_{T1680646C}$ were excluded due to sporadic and solitary occurrence of the candidate mutation on RefSeq background in horned animals. So far, we do not observe recombination within P_F haplotype block (from P_{5ID} to P_{80kbID}) and consequently do not exclude any candidate mutations by recombination. As soon as one candidate was excluded by confirmed genotyping we cancelled further genotyping of the respective marker. The breeding records, including polled certificates of the genotyped animals and their ancestors as well as their breed affiliation, were used to confirm the congruence

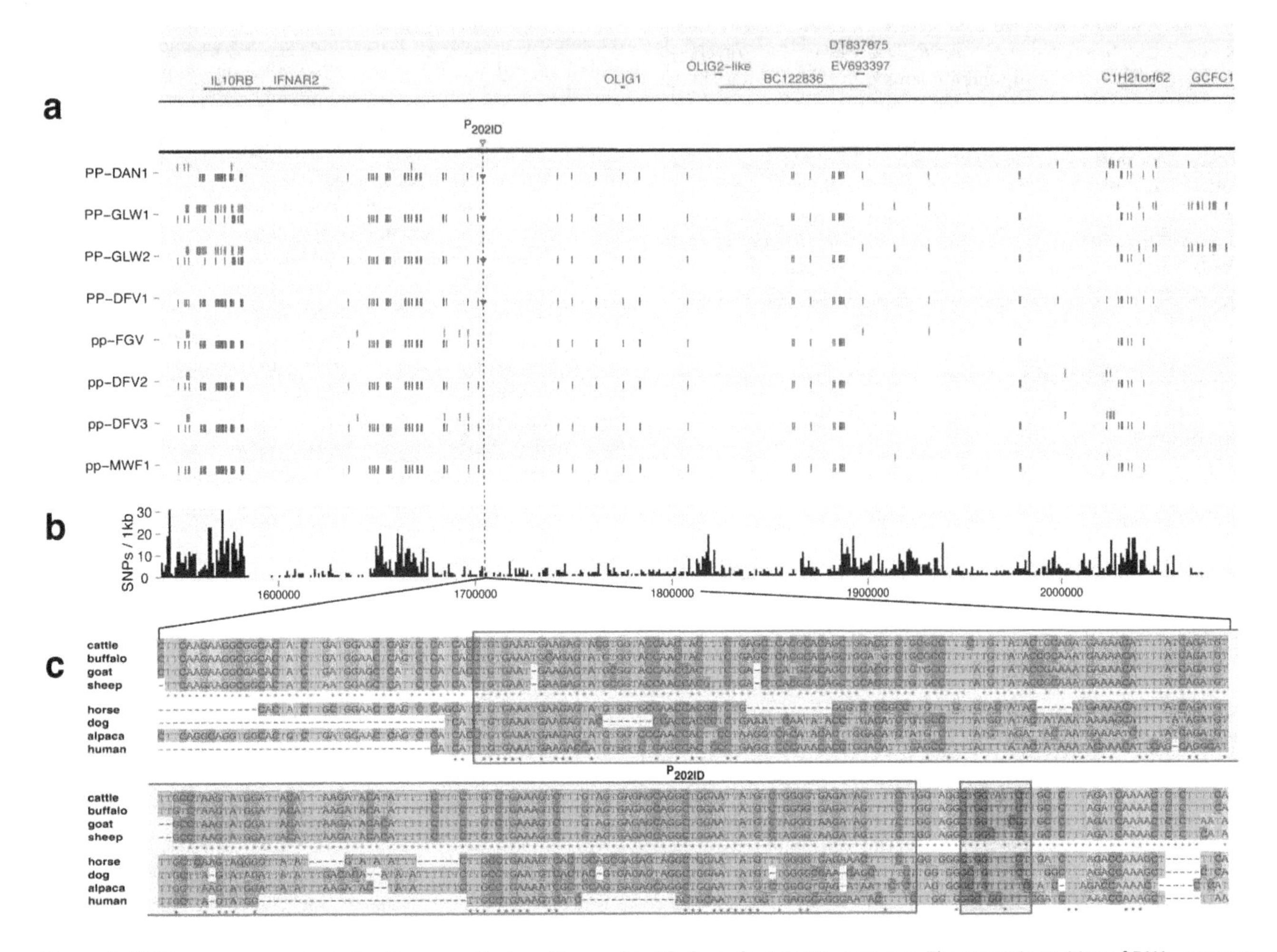

Figure 3. DNA sequence variants in diverse polled and horned cattle breeds. (**a**) SNP genotypes. The genomic position of DNA sequence variants (DSV) called by high-throughput sequencing of the target region are displayed as bars (SNPs) or triangles (InDels), in blue (homozygote) and red (heterozygote). Homozygous polled bulls are PP-DAN1, PP-GLW1, PP-GLW2 and PP-DFV1 and horned bulls are pp-FGV, pp-DFV2, pp-DFV3 and pp-MWF1. The candidate duplication P_{202ID} is highlighted by an inverted triangle. The annotated genes and ESTs in this region are displayed in the green area. (**b**) Within-species conservation. The bovine variant density from dbSNP is shown in the re-sequenced region per kb. (**c**) Multi-species alignment. The genomic multi-species alignment around the candidate duplication P_{202ID} was built with MULTIZ. Sequence identities among bovid animals are displayed with red stars whereas black stars denote the identity among all animals. The duplicated region is highlighted by the red shaded area and replaces the blue shaded nucleotides in polled animals.

between the obtained genotyping results at six remaining candidate mutations and polled status. The results are presented in Fig. 4 and summarised below. The single candidate variant P_{202ID} detected by the sequencing of eight non HF sires and the block of five candidates detected by sequencing of eight HF animals (P_{5ID}, $P_{G1654405A}$, $P_{C1655463T}$, $P_{C1768587A}$ and P_{80kbID}) segregate independently without evidence for any interference or recombination. As all animals carrying one or two copies of the P_{202ID} allele are polled (Pp or PP, respectively) and belong to breeds originating from geographical areas with Celtic culture [17] we hereafter figuratively and provisory called this allele P_C, indicating polledness of Celtic origin (see Fig. 4). Appropriately, the haplotype block (260 kb long) detected by sequencing of HF animals is hereafter called P_F, indicating polledness of Friesian origin. Both P_C and P_F, independently and in combination, were in perfect association with *POLLED* locus. There were two HF and one JY bull declared as PP which we found to be heterogeneous polled at candidate level, i.e. P_C/P_F. These three bulls were progeny-tested and mated to a large number of horned cows, permitting an official declaration as PP by the respective breeder associations. The pedigrees chart and genotype analyses of the P_C/P_F bulls and their sampled relatives (Fig. S3 and S4) clearly support allelic heterogeneity at the *POLLED* locus. Our diversity panel completed the semi-random sample (Table S1) and both together cover the European cattle population from the domestication centre, over the most likely dispersal routes for this species to the western- and northernmost parts of Europe (Fig. 4). This spatial well distributed panel comprises breeds with very small horns as well as extremely large horns [5] and with high genetic diversity between as well as within breeds [18]. In these horned breeds, only wild-type alleles (p_{rs}) were detected for all six candidate mutations, hindering further sieving of causative variants from passenger variants in this design.

Sequence Conservation Among Horned Ruminants and Other Mammals Allows Scoring of Candidate Causal Mutations

The DSVs $P_{G1654405A}$ and P_{5ID} are part of repeat-masked regions (MLT1B and LTR13BT), hence no conservation score (*PHASTCONS* scores) could be computed. Local blat searches against the selected species revealed only short multiple alignments to various genomic locations; in consequence no local alignments could be built. This however makes it unlikely that these two DSVs are linked to horn development as the flanking sequence is not conserved among horned ruminants. Moreover, these DSVs are found in a region with a high density of SNPs (Fig. 2B), indicating little selective pressure and making it less likely to encode for a functional element important for horn development.

The sequence flanking $P_{C1655463T}$ has a low conservation score, even among ruminants, and the local alignment (Fig. S5) shows that the corresponding sequence in the goat genome harbours a T, the same base as the variant allele found in polled Holstein cows. Thus, $P_{C1655463T}$ is less likely being causative for the polled phenotype.

The RefSeq nucleotide C at SNP $P_{C1768587A}$ is conserved only among bovid ruminants, where the variant allele is at the central position of a 24 nucleotide block that is identical in all four available *Bovidae* genomes. Also, the surrounding sequence is highly conserved among *Bovidae* but far less among other mammals (Fig. 2D and E). The SNP density of the whole region (Fig. 3B and 3C) including P_C shows a contiguous area with high sequence conservation within cattle, making it a likely candidate for the presence of an un-annotated functional element.

Both $P_{T1909354A}$, and $P_{1909396D2}$, nested within the 80 kb duplication, affect elements that are conserved only among the bovid ruminants (Fig. S6). However, the 80 kb duplication can be causative itself, and sequence conservation of single elements within these 80 kb does not necessarily reflect the consequence this duplication may induce.

Sequence conservation across and within species thus allows ranking of the candidate mutations within P_F haplotype, resulting in the P_{80kbID} duplication (including the two variations $P_{T1909354A}$ and $P_{1909396D2}$) and the SNP $P_{C1768587A}$ as two most plausible candidates. The candidate $P_{C1655463T}$ is considered as least likely as it coincides with sequence variation within different members of the *Bovidae* family suggesting a lack of functional constraint at this position.

Discussion

The polled phenotype in cattle has been a target of selection and research since long. Large economic value of the phenotype [19], the simple dominant inheritance known since 1936 [6] and the mapping position known since 1993 [9] provided hope for an early and easy detection of the causal variant. This, however, was not the case due to special features of the phenotype, underlying causal mutations and livestock genetics: (i) the identification of functional candidate genes by the usual comparative trajectory with humans or mouse is hardly possible for this phenotype specific to *Bovidae*, (ii) the lack of functional and even positional candidate genes in large fractions of candidate region hindered identification of mutations conferring the polled phenotype and (iii) possible allelic heterogeneity of the phenotype investigated in a highly mobile species (world-wide and across-breeds use of some founders by artificial insemination) can reduce the ability to exploit haplotype diversity between breeds.

In this study, we used approaches beyond LD-based fine mapping to reduce the list of candidate causative genes without relying on any *a priori* assumptions of gene function. According to the contemporary trajectory of fine-mapping by a relatively dense SNP marker panel, high-throughput re-sequencing and association proven in a large panel of individuals sampled across the European continent we detected a single complex insertion-deletion event (P_{202ID}) perfectly associated with the *POLLED* gene in most European cattle breeds. The absence of any other congruent candidate variant in the entire chromosome segment detected by fine-mapping, as well as perfect association with the phenotype, high sequence conservation of the candidate fragment among horned ruminants and absence of the appropriate conservation in un-horned mammals (Fig. 3C), clearly suggest P_{202ID} as most probably being the causal mutation for polledness in most *Bos taurus* breeds. The fact that InDel P_{202ID} resides in a region without known function is neither a clear argument against nor in favour of the candidate but is rather a proof of current rudimentary knowledge about gene function or presumed effects of candidate causal variants. The plausible causal mutation must fall within the limits of the dominant mode of inheritance. One possibility is haploinsufficiency which unlike in type 2 scurs [2] should not result in an enhancement of the effect in homozygotes. The next possibility is a gain-of-function mutation which would be characterized by a complete inhibition of the separate ossification centre in the tissues above the periosteum [3]. Such new regulatory elements with abnormal function may fall within the class of non-coding RNA and micro RNA (miRNA) or promoter sequences. However, their *de novo* prediction from un-annotated sequence of a non-model organism is highly error-prone and was not used for speculation about the mechanisms causing polledness.

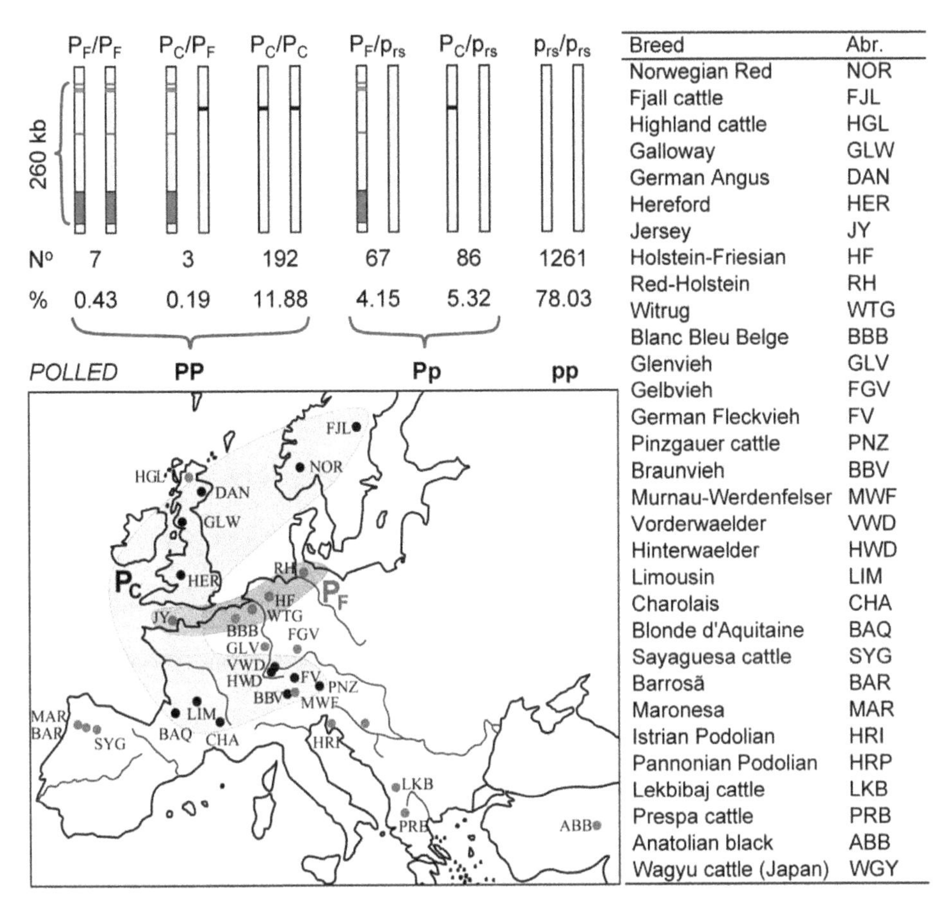

Figure 4. Distribution of *POLLED* candidate variants in European cattle breeds. The largest proportion (shown in the row marked %) of genotyped animals (78.03%) were homozygous for the wild-type allele at all six candidate variants of the *POLLED* locus, i.e. they inherited two copies of the RefSeq haplotype p_{rs} presented by vertical gray bar. Plausibility analyses suggested all these 1261 (shown in the row marked with N°) p_{rs}/p_{rs} animals as being horned, therefore pp at the *POLLED* locus. Five candidate mutations detected by sequencing of polled and horned Holstein-Friesian animals form a haplotype block (P_{5ID}-$P_{G1654405A}$-$P_{C1655463T}$-$P_{C1768587A}$-P_{80kbID}) consisting of three SNPs flanked by two InDels and called P_F signifying polledness of Friesian origin. The five candidate mutations of Friesian polledness were superimposed on RefSeq background (vertical gray bar): three SNPs and a 5 bp InDel were outlined by red horizontal bars and the duplicated region P_{80kbID} by the red area. The P_F block is in perfect association with *POLLED* genotype and segregate only in cattle breeds originating from north-western coast of continental Europe (HF, RH, JY and WTG). The candidate mutation P_{202ID} is represented by black horizontal bar on RefSeq background. All animals carrying one (86) or two copies (192) of the P_{202ID} are polled (Pp or PP, respectively) and belong to breeds originating from Scandinavia, Great Britain, France and South-Germany, hereafter figuratively called polledness of Celtic origin, P_C. Three bulls with PP genotype determined by extensive progeny testing were found to be heterogeneous polled at candidate gene level, i.e. P_C/P_F. The genotyping of their sampled relatives as well as entire experimental design provide no evidence for recombination within P_F haplotype block. The geographic origin of the sampled breeds is outlined by dots on Europe map. Breeds lacking polled samples are represented by gray dots. The breeds with P_{202ID} as only causal variant or variant in perfect association with *POLLED* locus are marked by black dots. The breeds with P_F as predominant variant in perfect association with *POLLED* locus are marked by red dots. The approximate distribution area of Celtic and Friesian polledness are highlighted as gray and red shading respectively.

The biology of horn development has not been studied in detail, partly also due to the obvious lack of model organisms. The development of a new ossification centre should probably engage the well characterized network of bone development with the *RUNX2* master transcription factor and its known regulators including twist proteins [20] and several miRNAs [21]. Indeed, mapping of the type 2 scurs locus has identified a *TWIST1* mutation as likely cause [2] and expression analysis of skin biopsies of the horn bud area in newborn polled and normal calves has shown evidence of endothelial-mesenchymal transition [22], a process involving bone morphogenetic protein signalling and twist protein activity. Genetic polledness is also known in goat and sheep, but the underlying mutations are mapped to chromosomes without synteny to the bovine polled locus investigated, here. Moreover, the polled/horned phenotype shows sexual dimorphism in sheep [23] and in goat is even linked with impaired sexual development [24], so that it is assumed that polledness in these species is caused by different mechanisms.

Including the cattle breeds originating from the North See area –HF, JY and WTG– led (Fig. 4) to the detection of a single haplotype block composed of three SNPs flanked by two InDels, all possible candidate causal variants for an independent mutation causing polledness. This Friesian haplotype, P_F, segregates independently from the previously mentioned Celtic mutation, P_C. Both are complementary and there is no evidence for any interference or recombination. Two Friesian and one Jersey bull were heterogenic at the *POLLED* locus (P_F/P_C) and declared as homozygous polled by progeny testing, i.e. all progeny from mating with horned dams are polled. All three bulls were intensively used for artificial insemination without any particular features and bred hundreds of exclusively polled offspring. Recombinant offspring of these bulls can supply important information for a formal test of allelic heterogeneity of the *POLLED* phenotype.

Not only economical reasons but also increasing serious animal welfare issues urge for a solution in the production of hornless cattle other than by dehorning. Even though the positive effects of genetic dehorning predominate, possible negative aspects should not be dismissed. In spite of a strong selection and serious attempts to improve breeding value of the polled sires these still lag behind in performance. For example, the best progeny tested polled (Pp) Fleckvieh bull currently ranks at position 1,724 of selection index, more than 1.3 standard deviations worse than the top 100 bulls. We find a similar situation in the Holstein breed. The inferior dairy breeding values of polled bulls might be caused simply by long-standing neglect of polled animals within sophisticated breeding programs or by multiple pleiotropic effects. It is known that some domestication traits result from mutations that cause large phenotypic effects but include deleterious pleiotropic effects [25]. Short-termed agricultural interests can obviously overcome the negative pleiotropic effects of some large phenotypic effects like in *myostatin* null mutations in cattle [26]. The recently mapped type 2 scurs syndrome locus with its underlying most probably causal mutation in the *TWIST1* gene [2] represents another breeder-selected trait that is negatively correlated with fitness due to embryonic lethality for the homozygous state. To avoid possible negative long-term impacts, possible deleterious pleiotropic effects should be closely investigated before the massive amplification of the *POLLED* gene in large cattle populations.

In conclusion, we describe the mapping as well as the perfect association of one InDel and one short haplotype causing polledness in cattle and suggest the conservation score as prioritising criterion of the most probable causal mutations. The distribution of the two putative *POLLED* alleles across the European continent supports the hypothesis of two independent mutations. Traits with allelic heterogeneity are no exception but tedious to decipher. We presented research strategies which could be more widely applicable for deciphering the molecular mechanisms of phenotypes with allelic heterogeneity.

Materials and Methods

Ethics Statement

Collection of blood samples was conducted exclusively by local veterinarians during regular health inspection and quality control of breeding records on the farms, so that randomness of the sample was assured and no ethical approval was required for this study. The regular health inspection includes annual IBR diagnostics (BGBl. I Nr. 74 S. 3520 ff) and ruminant metabolisms control at Lehr- und Versuchsgut, Oberschleißheim, Germany (Dec. 55.2-1-54-2531.3-80-10). The regular quality control of breeding records includes paternity testing organised by the respective breeding associations. Paternity testing involves blood, semen and hair root samples. Blood sampling by veterinarians with state examination avoids unnecessary pain, suffering and damage and is in accordance with the German Animal Welfare Act. Semen samples were collected by approved commercial artificial insemination stations as part of their regular breeding and reproduction measures in cattle industry. Hair roots samples were collected by breeders themselves (hair plucking from pinna or tail-tassel) as part of regular sampling for parentage control. As acknowledged below and specified in Table S2 we used these regularly sampled blood, semen and hair roots samples for DNA preparation necessary for this study.

Animals

All animals sampled and genotyped in this study are presented in Table S1 according to their breed affiliations and purpose in the experimental design. These 1,675 animals originate from 31 European cattle breeds belonging to the *Bos taurus* subspecies. The DNA samples of all 162 cases, 162 controls and 89 carriers were genotyped genome-wide with the Illumina BovineSNP50 Bead-Chip [14]. All 162 cases, 89 carriers and 1,262 semi random animals were genotyped for the revealed candidate mutations.

Phenotypes and Declaration of the Underlying *POLLED* Genotype

Polledness or complete absence of horns is a visible phenotype that can be identified at relatively young animals (four to six months). Because the growth of scurs occurs later in life than horns, phenotyping of some scurred animals will not be possible until nine to eighteen months of age [12]. Polled animals, especially breeding bulls and important dams, are declared as polled (name suffix P) in their pedigree certificate and/or other records. One P designates polled animals with the PP or Pp genotype at underlying *POLLED* locus. Breeding animals with 12 to 15 (depending on breeding organisation) consecutively polled offspring originating from horned mates are declared as homozygous polled and get a name suffix PP. One horned offspring with confirmed paternity is sufficient to declare a polled animal as Pp.

Homozygosity Mapping

DNA samples of 162 case and 162 controls were genotyped with the Illumina BovineSNP50 BeadChip [14]. Marker order was based on release UMD3.1 of the *Bos taurus* genome (http://www.cbcb.umd.edu/research/bos_taurus_assembly.shtml). The SNP haplotypes were inferred and missing genotypes imputed using hidden Markov models (software package *BEAGLE* [27]). Three cohorts, namely trios (two parents, one offspring), pairs (one parent, one offspring) and unrelated animals were formed, including those animals that turned out not to be relevant for this study (2,721 animals in total). Genome-wide homozygosity mapping in polled animals and controls was performed using the *ASSHOM* procedure [13]. Case and control genotypes served as input data after completion of haplotype-inference and imputation by the *BEAGLE* package. To determine the statistical significance of each summary score, we permutated (50,000 permutations) the complete markers along the chromosomes [10] and estimated the summary score as harmonic mean across all cases. Thus, the corresponding *P* values were corrected for multiple testing and accounted for the level of inbreeding within the cases [13]. The statistical significance based on 50,000 permutations is comparable to the number of markers used.

Genotyping of the Candidate Causal Mutations

In general, 30–60 ng of genomic DNA were used for genotyping of the candidate mutations. InDel variants P_{202ID} and P_{5ID} were PCR amplified (94°C 30 sec, 58°C 60 sec, 72°C 60 sec for 31 or 35 cycles, respectively) using primer binding sites flanking the InDel events (5′-TCAAGAAGGCGGCACTATCT-3′ and 5′-TGATAAACTGACCCTCTGCCTATA-3′ for P_{202ID} and 5′-FAM-CCTTGTCACGTTAGATGTATGTCC-3′ and 5′-TCAATCTCTAATAAGGAACAGAAGAAA-3′ for P_{5ID}). PCR products were size-separated and visualized by 2% ethidium-bromide stained agarose gel electrophoresis (P_{202ID}) or analysed on an ABI Prism® 3130*XL* DNA sequencer (P_{5ID}).

Genotyping of the P_{80kbID} was performed by use of two primers flanking the variable site $P_{1909396D2}$ (5′-GAAGTCGGTGGTCT-

GAAAGG-3′ and 5′-TGTTCTGTGTGGGTTTGAGG-3′). PCR amplification (32 cycles 94°C 30 sec, 59°C 60 sec, 72°C 60 sec) resulted in a RefSeq related product (p_{rs}) which was obtained in all animals, whether horned or polled. Additionally, a second P_F related product, containing the two-base pair (TG) deletion was observed in all animals bearing one or two copies of the P_F haplotype. Discrimination of the two products differing 2 bp in size was performed on an ABI Prism® 3130*XL* DNA sequencer. Animals bearing P_F/P_F were distinguished from P_F/p_{rs} by quantitative evaluation of the obtained signals. P_F/P_F animals yielded signals of similar peak heights for the P_F and p_{rs} products, while a heterozygous constellation (P_F/p_{rs}) resulted in signal intensities of approximately double height for the p_{rs} product when compared to the P_F specific product.

SNP variants $P_{G1654405A}$ and $P_{C1655463T}$ were genotyped by competitive allele-specific PCR using commercially available kits (KASPar®, KBioscience). PCR was performed as recommended by the manufacturer, while subsequent discrimination of the obtained fluorescent-labelled (FAM and CAL Fluor Orange 560) allele-specific products was performed on an ABI Prism® 3130*XL* DNA sequencer.

SNP variant $P_{C1768587A}$ was analysed by PCR-RFLP. Genomic DNA was amplified (32 cycles 94°C 30 s, 58°C 60 s, 72°C 60 s; 5′-CTGGAACCACGGATTACACAG-3′ and 5′-ACAGT-TATGGTCAGGAGGCAAA-3′). Subsequently, three µl of the PCR product were treated with three units of TspRI (65°C for a minimum of 3.5 h). Obtained fragments were size-separated and visualized by 2% ethidium-bromide stained agarose gel electrophoresis.

Targeted re-sequencing

One µg of Genomic DNA was randomly sheared by sonication (Bioruptor, Diagenode, Liege, Belgium) for 25 cycles (30 sec on/off, "low" intensity). Sheared DNA with a median size of 200–300 bp was rendered blunt-ended and 5′-phosphorylated (NEB-next end repair module, New England Biolabs Inc, Ipswich, USA). After addition of a single non-templated 3′-A (NEBnext A-tailing module) fragments were ligated to Illumina compatible adapters that carried a sample-specific 3 nt barcode and a 3′-T overhang. The ligated library was size-selected on a 2% agarose gel and amplified by PCR with Illumina PE1 and PE2 primers. Equimolar amounts of the eight samples were pooled for array-capture (Agilent 244 k capture Array, Agilent, Santa Clara, USA; custom designed by e-array, repeat-masked, 3 bp tiling). Briefly, the libraries were hybridised for 65 h at 65°C, washed and eluted with nuclease-free water for 10 min at 95°C. The eluted DNA was concentrated in a vacuum centrifuge, amplified by PCR (10 cycles 98°C 15 s, 65°C 30 s, 72°C 30 s) and purified with Ampure XP beads.

Mapping and Variant Calling

Sequence reads from the Illumina Genome Analyzer IIx were aligned to the bovine reference genome (UMD3.1) using *BWA* [28]. The mapped reads were filtered for PCR duplicates with *SAMTOOLS* [29] and only uniquely mapping reads were retained. A pileup of the mapped reads was created for each animal using *SAMTOOLS* and variants were detected with *VARSCAN* (v.2.2.7) [30] at a minimal coverage of 20 and a minimal variant frequency of 0.01.

The applied filters for the detection of potentially causative variants were: Homozygosity in the PP sires with a variant frequency range of 95 to 100%. For the heterozygote carrier Pp they must fall within a range of 40 to 60%, while the resulting candidate variants must be undetected in the pp sires. The putatively causative variants must be different from the UMD3.1 genome sequence, which was derived from a horned Hereford dam and therefore cannot carry the polled allele.

Copy Number Variation

For the identification of large insertion-deletion events between polled and horned animals we calculated the coverage ratio between the two groups in dynamical bin sizes. Each bin size was calculated by iterating over the mapped reads in the reference group until a determined number of reads (400) was reached and the start position of the last read was taken as bin size [31]. This step was repeated until the whole target region was processed. Then, for the target group the number of reads falling in each bin were counted and used for the calculation of the log2 ratios between reference and target group. The log2 ratios were plotted using R [32] and the average ratios were segmented using the circular binary segmentation implementation in the *DNACOPY* package (v1.14.0) from *BIOCONDUCTOR* [33].

Multi-species Alignment and Conservation Scores

For the re-sequenced target region (chr1:1,543,412-2,089,648) we created multiple alignments for bovid ruminants (cattle, sheep and water buffalo), for unhorned mammals (horse, dog, human, mouse, pig, elephant, alpaca and dolphin) and for all selected species. The repeat-masked bovine genome sequence of the target region was obtained from UCSC genome browser [34] and used as reference for each of the pair-wise *LASTZ* genome alignments [35]. The resulting alignments were chained and netted to obtain best matching hits [36] and multi-species alignments were generated with *MULTIZ* [37] projected to the bovine reference. Additionally, short read sequences from the goat genome project (SRA Accession number: SRX016522) were added to the multi-species alignments only in the neighborhood of candidate mutations. To do so we aligned the short read to the bovine reference using *BLAT* [38] and manually selected the best alignments to the bovine reference.

The reference genomes of sheep (*Ovis aries* 1.0), human (hg19), mouse (mm9), dog (canFam2), horse (EquCab2) and pig (Sscrofa9.1) were obtained from the FTP-site at UCSC. The draft genome sequences from alpaca, elephant and dolphin were obtained from the Broad Institute [39] and the water buffalo genome sequence was taken from the Indian Buffalo Genome Project (http://210.212.93.84).

From the multiple species alignments we calculated the base conservation among horned animals, unhorned mammals and for all species using the *PHAST* package. For that purpose, we estimated a phylogenetic tree with *PHYLOFIT* from which the conservation score for each base was calculated with *PHASTCONS*.

Supporting Information

Figure S1 The pedigree chart of most polled Braunvieh. All declared polled Braunvieh bulls are descendents of the well-known American Brown-Swiss bull BS1 which is founder of polledness in Brown-Swiss/Braunvieh cattle population. The case individuals (PP) are represented by solid circles (females) and squares (males); declared carriers by half-filled symbols; not sampled individuals are marked with a diagonal line. The haplotype associated with polledness (red letters) of the four genome-wide genotyped bulls can be traced back to the same carrier bull BS1. This pedigree includes also four polled Braunvieh animals (*) genotyped only for candidate mutations. All sampled PP animals were genotyped as P_C/P_C and all Pp as P_C/p_{rs}.

Figure S2 The pedigree chart of all sampled Fleckvieh bulls composing case and carrier group in Table S1. The case individuals (PP) are represented by solid circles (females) and squares (males); declared carriers by half-filled symbols; not sampled individuals are marked with a diagonal line. To reduce complexity of the pedigree not all relationships were presented. At two positions (A and B) there are relationships to important sires indicated. The inbreeding of the Fleckvieh bull (PP-DFV1) chosen for re-sequencing is superimposed by red lines. All sampled Fleckvieh PP animals bear two copies of the common haplotype (AGACAAGGA) and were genotyped as P_C/P_C. All Pp bear one copy of the common haplotype and were genotyped as P_C/p_{rs}. No recombination was detected in the common haplotype.

Figure S3 The pedigree chart of all sampled Holstein bulls composing case and carrier group in Table S1. The case individuals (PP) are represented by solid circles (females) and squares (males); declared carriers by half-filled symbols; not sampled individuals are marked with a diagonal line. To reduce complexity of the pedigree not all relationships were presented. At positions A, B and C there are relationships to important sires indicated. The founder of the Celtic polledness is indicated by D. The carriers and cases of Friesian and Celtic polledness are filled with gray (genotyped as P_F/p_{rs} and P_F/P_F) and black color (genotyped as P_C/p_{rs} and P_C/P_C), respectively. Two heterogeneous polled bulls were genotyped as P_C/P_F. At the SNP-Chip level all sampled Holstein PP and Pp animals bear two or one copy of the common haplotype (AGACAAGGA) and there was no recombination detected.

Figure S4 The pedigree chart of all sampled case and carrier Jersey bulls. The case individuals (PP) are represented by solid circles (females) and squares (males); declared carriers (Pp) by half-filled symbols; not sampled individuals are marked with a diagonal line. The founder of the Friesian polledness in Jersey breed is indicated by F. The carriers and cases of polledness with Friesian and Celtic origin are filled with gray (genotyped as P_F/p_{rs} and P_F/P_F) and black color (genotyped as P_C/p_{rs} and P_C/P_C), respectively. Two heterogeneous polled bulls were genotyped as P_C/P_F. At the SNP-Chip level all sampled Jersey PP and Pp animals bear two or one copy of the common haplotype (AGACAAGGA) and there was no recombination detected.

Figure S5 Across-species conservation for variant $P_{C1655463T}$. (**a**) The sequence conservation around the position of the candidate mutation $P_{C1655463T}$ among animals without horn, horn-bearing and among all animals are represented as *PhastCons* scores, calculated from the underlying multi-species alignment in plot b. (**b**) Sequence identity among horn-bearing animals is denoted with red stars and black stars are used for identity in all aligned species. The position of the candidate DNA sequence variant is outlined by a dashed black line and is highlighted in the multi-species alignment by a red frame.

Figure S6 Across-species conservation for variants $P_{T1909354A}$ and $P_{1909396D2}$. (**a**) The *PhastCons* conservation scores of the sequence around the two DNA sequence variants $P_{T1909354A}$ and $P_{1909396D2}$ are shown. Both variants are outlined with black dotted lines. (**b**) Sequence identity is marked in the multi-species alignment by a red and black star among horn-bearing and all aligned species, respectively.

Table S1 Breed origin of case-control design, carriers and semi-random samples. Breed names, abbreviations, geographic origin and numbers of genotyped samples are listed for each group within breed. The four groups are homozygous polled cases (PP), heterozygous polled carriers (Pp), horned controls (pp) and the semi-random sample (R) adjusted by the diversity panel.

Table S2 Specification of the samples origin. Name of the samples supplier, affiliation, sampled breed(s), sampled tissue and primary reason for tissue sampling. This table is part of Acknowledgments and Ethics statement.

Acknowledgments

We are grateful for the support of the respective numerous cattle breeders and cattle breeding associations and insemination centres that generously provided the most samples free of charge. In particular, we thank K.H. Göpel (Göpel Genetik GmbH, Germany), Dr. G. Röhrmoser (Arbeitsgemeinschaft Süddt. Rinderzucht-u. Besamungsorganisationen e.V., München, Germany), Dr. A. Scholz (Lehr- und Versuchsgut, Oberschleißheim, Germany), Dr. J. Eder (Hochschule Weihenstephan-Triesdorf, Triesdorf, Germany), Dr. A. Medugorac (Institute for Animal Breeding, TU Munich), Dr. J. Ramljak (Department of Animal Science, University of Zagreb, Croatia), Dr. A. Ivankovic (Department of Animal Science, University of Zagreb, Croatia), Dr. K. Kume (NC of FAnGR, Tirana, Albania), Dr. H.P. Grünenfelder (SAVE Foundation, St. Gallen, Switzerland), Dr. V. Cadavez (Departament of Animal Science, CIMO, Braganca, Portugal), Dr. E.D. Hegemann (Tierarzt Praxis Hegemann, Soest, Germany), Dr. J. Potthast (Rinder-Union West eG, Münster, Germany), Dr. B. Weber (Masterrind GmbH, Verden, Germany), Mrs. B. Brentrup (WWS Germany GmbH, Altenberge, Germany), Mr. M. Schricker (Fleischrinderverband Bayern, Ansbach, Germany), Mr. H. Zuchtriegel (Zweckverband für künstliche Besamung der Haustiere, Greifenberg, Germany), Mr. E. Rosenberger (The Bavarian Gene Reserves, LfL, Grub, Germany), Mr. L. Eule (AgrarCenter Erzgebirge GmbH, Germany), Mr. N. Hartmann (Klöck Hartmann GbR, Bidingen, Germany), Mr. M. Wild (Family Wild, Prittriching, Germany), Mr. S. Rist (AHG Kempten, Germany), Mr. M. Kelz (Rinderbesamungsgenossenschaft Memmingen, Germany) and Mr. P. Lobet (Belgian Blue Group, Ciney). We thank Dr. C.E. Veit (Chair of Animal Genetics and Husbandry, LMU, Munich, Germany) for comments on English grammar. Detailed specification of the samples origin is given in the Supplementary Table S2.

Author Contributions

Conceived and designed the experiments: IM SR MF HB. Performed the experiments: IM SK DS. Analyzed the data: IM SR SK AG DS IR. Contributed reagents/materials/analysis tools: KHG HB MF. Wrote the paper: IM SR SK DS IR.

References

1. Felius M (1995) Cattle breeds : an encyclopedia. Doetinchem: Misset. 799p. p.
2. Capitan A, Grohs C, Weiss B, Rossignol MN, Reverse P, et al. (2011) A newly described bovine type 2 scurs syndrome segregates with a frame-shift mutation in TWIST1. PLoS One 6: e22242.
3. Dove WF (1935) The physiology of horn growth: a study of the morphogenesis, the interaction of tissues and the evolutionary processes of a Mendelian recessive character by means of transplantation of tissues. J Exp Zool 69: 347–405.
4. Graf B, Senn M (1999) Behavioural and physiological responses of calves to dehorning by heat cauterization with or without local anaesthesia. Applied Animal Behaviour Science 62: 153–171.
5. Ramljak J, Ivankovic A, Veit-Kensch CE, Forster M, Medugorac I (2011) Analysis of genetic and cultural conservation value of three indigenous Croatian cattle breeds in a local and global context. J Anim Breed Genet 128: 73–84.

6. White W, Ibsen H (1936) Horn inheritance in Galloway-Holstein cattle crosses. Journal of Genetics 32: 33–49.
7. Long CR, Gregory KE (1978) Inheritance of the horned, scurred, and polled condition in cattle. Journal of Heredity 69: 395–400.
8. Drogemuller C, Wohlke A, Momke S, Distl O (2005) Fine mapping of the polled locus to a 1-Mb region on bovine chromosome 1q12. Mamm Genome 16: 613–620.
9. Georges M, Drinkwater R, King T, Mishra A, Moore SS, et al. (1993) Microsatellite mapping of a gene affecting horn development in Bos taurus. Nat Genet 4: 206–210.
10. Seichter D, Russ I, Rothammer S, Eder J, Förster M, et al. (2012) SNP-based Association mapping of the polled gene in divergent cattle breeds. Anim Genet (in press).
11. Asai M, Berryere TG, Schmutz SM (2004) The scurs locus in cattle maps to bovine chromosome 19. Anim Genet 35: 34–39.
12. Capitan A, Grohs C, Gautier M, Eggen A (2009) The scurs inheritance: new insights from the French Charolais breed. BMC Genet 10: 33.
13. Charlier C, Coppieters W, Rollin F, Desmecht D, Agerholm JS, et al. (2008) Highly effective SNP-based association mapping and management of recessive defects in livestock. Nat Genet 40: 449–454.
14. Matukumalli LK, Lawley CT, Schnabel RD, Taylor JF, Allan MF, et al. (2009) Development and characterization of a high density SNP genotyping assay for cattle. PLoS One 4: e5350.
15. Goecks J, Nekrutenko A, Taylor J (2010) Galaxy: a comprehensive approach for supporting accessible, reproducible, and transparent computational research in the life sciences. Genome Biol 11: R86.
16. Gibbs RA, Taylor JF, Van Tassell CP, Barendse W, Eversole KA, et al. (2009) Genome-wide survey of SNP variation uncovers the genetic structure of cattle breeds. Science 324: 528–532.
17. McEvoy B, Richards M, Forster P, Bradley DG (2004) The Longue Duree of genetic ancestry: multiple genetic marker systems and Celtic origins on the Atlantic facade of Europe. Am J Hum Genet 75: 693–702.
18. Medugorac I, Veit-Kensch CE, Ramljak J, Brka M, Marković B, et al. (2011) Conservation priorities of genetic diversity in domesticated metapopulations: a study in taurine cattle breeds. Ecology and Evolution 1: 408–420.
19. Prayaga KC (2007) Genetic options to replace dehorning in beef cattle–a review*. Australian Journal of Agricultural Research 58: 1–8.
20. Bialek P, Kern B, Yang X, Schrock M, Sosic D, et al. (2004) A twist code determines the onset of osteoblast differentiation. Dev Cell 6: 423–435.
21. Zhang Y, Xie RL, Croce CM, Stein JL, Lian JB, et al. (2011) A program of microRNAs controls osteogenic lineage progression by targeting transcription factor Runx2. Proc Natl Acad Sci U S A 108: 9863–9868.
22. Mariasegaram M, Reverter A, Barris W, Lehnert SA, Dalrymple B, et al. (2010) Transcription profiling provides insights into gene pathways involved in horn and scurs development in cattle. BMC Genomics 11: 370.
23. Johnston SE, McEwan JC, Pickering NK, Kijas JW, Beraldi D, et al. (2011) Genome-wide association mapping identifies the genetic basis of discrete and quantitative variation in sexual weaponry in a wild sheep population. Mol Ecol 20: 2555–2566.
24. Pailhoux E, Vigier B, Chaffaux S, Servel N, Taourit S, et al. (2001) A 11.7-kb deletion triggers intersexuality and polledness in goats. Nat Genet 29: 453–458.
25. Stern DL, Orgogozo V (2009) Is genetic evolution predictable? Science 323: 746–751.
26. Grobet L, Martin LJ, Poncelet D, Pirottin D, Brouwers B, et al. (1997) A deletion in the bovine myostatin gene causes the double-muscled phenotype in cattle. Nat Genet 17: 71–74.
27. Browning SR, Browning BL (2007) Rapid and accurate haplotype phasing and missing-data inference for whole-genome association studies by use of localized haplotype clustering. Am J Hum Genet 81: 1084–1097.
28. Li H, Durbin R (2009) Fast and accurate short read alignment with Burrows-Wheeler transform. Bioinformatics 25: 1754–1760.
29. Li H, Handsaker B, Wysoker A, Fennell T, Ruan J, et al. (2009) The Sequence Alignment/Map format and SAMtools. Bioinformatics 25: 2078–2079.
30. Koboldt DC, Chen K, Wylie T, Larson DE, McLellan MD, et al. (2009) VarScan: variant detection in massively parallel sequencing of individual and pooled samples. Bioinformatics 25: 2283–2285.
31. Wood HM, Belvedere O, Conway C, Daly C, Chalkley R, et al. (2010) Using next-generation sequencing for high resolution multiplex analysis of copy number variation from nanogram quantities of DNA from formalin-fixed paraffin-embedded specimens. Nucleic Acids Res 38: e151.
32. Ihaka R, Gentleman R (1996) R: A Language for Data Analysis and Graphics. J Comput Graph Stat 5: 299–314.
33. Gentleman RC, Carey VJ, Bates DM, Bolstad B, Dettling M, et al. (2004) Bioconductor: open software development for computational biology and bioinformatics. Genome Biol 5: R80.
34. Kent WJ, Sugnet CW, Furey TS, Roskin KM, Pringle TH, et al. (2002) The human genome browser at UCSC. Genome Res 12: 996–1006.
35. Harris RS (2007) Improved pairwise alignment of genomic DNA: The Pennsylvania State University.
36. Kent WJ, Baertsch R, Hinrichs A, Miller W, Haussler D (2003) Evolution's cauldron: duplication, deletion, and rearrangement in the mouse and human genomes. Proc Natl Acad Sci U S A 100: 11484–11489.
37. Blanchette M, Kent WJ, Riemer C, Elnitski L, Smit AF, et al. (2004) Aligning multiple genomic sequences with the threaded blockset aligner. Genome Res 14: 708–715.
38. Kent WJ (2002) BLAT–the BLAST-like alignment tool. Genome Res 12: 656–664.
39. Lindblad-Toh K, Garber M, Zuk O, Lin MF, Parker BJ, et al. (2011) A high-resolution map of human evolutionary constraint using 29 mammals. Nature 478: 476–482.

11

A Genome-Wide Association Study Reveals Loci Influencing Height and Other Conformation Traits in Horses

Heidi Signer-Hasler[1], Christine Flury[1], Bianca Haase[2,3], Dominik Burger[4], Henner Simianer[5], Tosso Leeb[2], Stefan Rieder[4]*

1 School of Agricultural, Forest and Food Sciences, Bern University of Applied Sciences, Zollikofen, Switzerland, **2** Institute of Genetics, Vetsuisse Faculty, University of Bern, Bern, Switzerland, **3** Faculty of Veterinary Science, University of Sydney, New South Whales, Australia, **4** Agroscope Liebefeld-Posieux Research Station Agroscope Liebefeld-Posieux (ALP) Haras, Swiss National Stud Farm (SNSTF), Avenches, Switzerland, **5** Department of Animal Sciences, Georg-August-Universität, Göttingen, Germany

Abstract

The molecular analysis of genes influencing human height has been notoriously difficult. Genome-wide association studies (GWAS) for height in humans based on tens of thousands to hundreds of thousands of samples so far revealed ~200 loci for human height explaining only 20% of the heritability. In domestic animals isolated populations with a greatly reduced genetic heterogeneity facilitate a more efficient analysis of complex traits. We performed a genome-wide association study on 1,077 Franches-Montagnes (FM) horses using ~40,000 SNPs. Our study revealed two QTL for height at withers on chromosomes 3 and 9. The association signal on chromosome 3 is close to the *LCORL/NCAPG* genes. The association signal on chromosome 9 is close to the *ZFAT* gene. Both loci have already been shown to influence height in humans. Interestingly, there are very large intergenic regions at the association signals. The two detected QTL together explain ~18.2% of the heritable variation of height in horses. However, another large fraction of the variance for height in horses results from ECA 1 (11.0%), although the association analysis did not reveal significantly associated SNPs on this chromosome. The QTL region on ECA 3 associated with height at withers was also significantly associated with wither height, conformation of legs, ventral border of mandible, correctness of gaits, and expression of the head. The region on ECA 9 associated with height at withers was also associated with wither height, length of croup and length of back. In addition to these two QTL regions on ECA 3 and ECA 9 we detected another QTL on ECA 6 for correctness of gaits. Our study highlights the value of domestic animal populations for the genetic analysis of complex traits.

Editor: Michael Nicholas Weedon, Peninsula College of Medicine and Dentistry, University of Exeter, United Kingdom

Funding: This study was financed by grants from the Swiss Federal Office for Agriculture, and the Fondation Sur-la-Croix. BH is funded by the Swiss Foundation for Grants in Biology and Medicine. The funders had no role in study design, data collection and analysis, decision to publish, or preparation of the manuscript.

Competing Interests: The authors have declared that no competing interests exist.

* E-mail: stefan.rieder@haras.admin.ch

Introduction

Horse genomics [1] made a tremendous step, when the whole genome sequence of the domestic horse was made publicly available in 2007 [2]. Information from that sequence served as the primary resource for the development of a commercial horse SNP array, and thus high-throughput genotyping. As a result genome-wide association studies (GWAS) became feasible in a so far unprecedented manner. A brief overview on the present state of horse genome research, trait mapping, and breed diversity studies, is given in a special supplementary issue of Animal Genetics from December 2010. To note, that some of the most spectacular findings from GWAS until today were the detection of SNPs on horse chromosome 18 within and proximal to the myostatin gene (MSTN), associated with racing performance in Thoroughbred horses [3–5].

However, progress was also made in the field of decipher mendelian traits, like the detection of a series of allelic variants responsible for different coat colors [6,7] and/or disease traits [8]. So far, less information is available on the genetics of polygenic quantitative traits in the horse such as overall conformation including e.g. height at withers. Morphological traits are key traits in horse breeding since centuries, as they are thought to be related to specific performance (e.g. conformation of a race horse versus a show jumper, a draft horse or a cutting horse) and longevity (e.g. correctness of gaits) [9,10]. A lack of data and phenotypes, and the complex genetic architecture usually underlying quantitative traits are still challenging research efforts.

Human height is a classical quantitative model trait. Its heritability is estimated ~0.8 [11]. This means that 80% of variation in height is explained due to additive genetic factors. Many studies were undertaken to discover association between height and loci using millions of SNPs and large cohorts of many thousand individuals. The GIANT consortium performed a meta-analysis of GWA data comprising 183,727 probands [12]. In this study 180 loci were found to be significantly associated with human adult height, which explained around 10% of the phenotypic variation or $1/8^{th}$ of the heritability of height. For human height and many other complex traits in humans common genetic variants have typically explained only a small part of the phenotypic variation [13]. Potential explanations for the missing heritability are 'rare variants or structural DNA variations that are not well covered by common SNPs' and 'a large number of loci with small effects' [11,13]. However, much more of the phenotypic variance (29%) could be explained when using 5,646 SNPs that are associated at $p<0.01$ with human height and a maximum likelihood method [14]. In another study on human height it was estimated that ~45% of the phenotypic variance can be explained by considering all the SNPs together [15]. GWAS are unable to explain this amount of genetic variation because effect sizes of individual SNPs are too small to reach genome-wide significance level and because SNPs are not in complete LD with causal variants [11,15].

Very recently several independent studies have been published on the detection of quantitative trait loci (QTL) for stature traits in cattle, using different methodological approaches and sample sets [16,17]. These studies identified about 8 loci that were significantly associated with height in cattle and explained up to 20% of the phenotypic variation [17]. The favorable population structure of domestic animals facilitates the identification of the actual underlying causative mutations. Thus a non-coding regulatory mutation in the promoter region of the bovine *PLAG1* gene was found to explain about 1–3.5% of the phenotypic variance in different cattle breeds [18]. This example illustrates that one may often expect fewer functional variants with larger effect sizes when comparing domestic animals to humans [19].

Genomic selection, i.e. the selection based on genomic breeding values, is a rapidly emerging field in plant and animal breeding [20,21]. In many countries genomic selection has been implemented in dairy cattle breeding programs [22,23]. Genomic breeding value estimation is based on linkage disequilibrium between genetic markers and QTL [24], or, equivalently, on the marker-based estimation of the realized relationship in the population [25], using genome-wide SNP marker panels. When analyzing the proportion of black coat color, fat concentration in milk, and conformation in Holstein cattle, many chromosome segments explained <0.1% of the genetic variance [26]. However, taken together these segments with individual small effects explained half of the variance for conformation and for the proportion of black. Few segments explained larger proportions of the genetic variance, e.g. in one case up to 37.5% for fat percent.

We have recently started to establish genomic selection procedures for the Franches-Montagnes (FM) horse breed. FM are a genetically closed and indigenous Swiss horse breed consisting of about 21,000 horses with 2,500 foalings per year [27]. Here, we report the mapping of loci influencing horse conformation traits - in particular height at withers (stature) - in the FM breed.

Results

GWAS for height at withers and other conformation traits

We initially selected a representative sample set of 1,151 FM horses from the active breeding population and obtained their genotypes at 54,602 SNPs. After quality filtering 1,077 horses and 38,124 SNPs remained for the final analysis. We analyzed the association of these data with respect to deregressed estimated breeding values (dEBVs) instead of direct phenotypic measurements for 28 conformation traits including height at withers, which varied between 145 cm and 165 cm on the phenotypic level (Figure 1). The estimated heritability for height at withers is 72% and the corresponding EBVs for height at withers varied between −5.76 and +6.64 (Table 1).

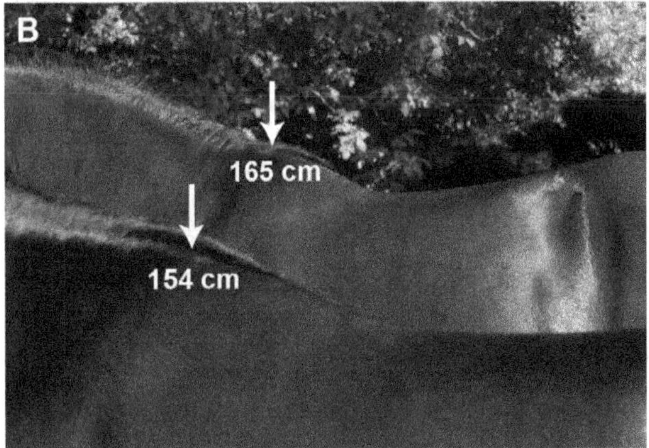

Figure 1. The Franches-Montagnes horse belongs to the type of light draft horse breeds and has its origins in Switzerland (A). Height at withers is one of 28 conformation traits, for which breeding values are estimated once a year (B). The breed standard calls for horses between 150–160 cm in size. The stallion in the background was the tallest horse in our study with a phenotypic height at withers of 165 cm. (Picture: Swiss National Stud Farm).

We analyzed the data using a mixed-model considering the genomic relationships in order to account for population stratification, which resulted in a genomic inflation factor of 1.04 after the correction. Eight SNPs within two QTL regions on ECA 3 and ECA 9 reached the Bonferroni corrected genome-wide significance level (Figure 2, Table 2). After 40,000 permutations the same SNPs were still significantly associated (5% p-level) as if we were using the Bonferroni criterion (Table 2). The two QTL regions for height at withers map near the *LCORL/NCAPG* and *ZFAT* genes, respectively. These results are in accordance with human positions of orthologous genes.

The QTL region on ECA 3 associated with height at withers was also significantly associated with wither height, conformation of legs, ventral border of mandible, correctness of gaits, and expression of the head (Table S1). The region on ECA 9 associated with height at withers was also associated with wither height, length of croup and length of back. In addition to these two QTL regions on ECA 3 and ECA 9 we detected another QTL on

Table 1. Distribution of EBVs and dEBVs for conformation traits in FM horses.

Trait	Min EBV[a]	Max EBV[a]	Min dEBV[b]	Max dEBV[b]
Height at withers	−5.76	+6.64	−8.66	+8.55
Expression of the head	−1.25	+1.10	−2.62	+2.47
Wither height	−0.85	+1.18	−2.10	+2.22
Conformation of legs	−0.57	+0.61	−1.42	+1.32
Ventral border of mandible	−1.25	+0.87	−2.13	+1.88
Correctness of gaits	−0.26	+0.33	−1.20	+1.35
Length of croup	−0.80	+0.68	−2.05	+1.74
Length of back	−0.51	+0.62	−2.26	+2.44

[a]Minimum and maximum estimated breeding values (EBV). The average estimated breeding value for animals born between 1998 and 2000 was set to 0.
[b]Minimum and maximum deregressed breeding values (dEBV). Deregressed EBVs were used for association analysis.

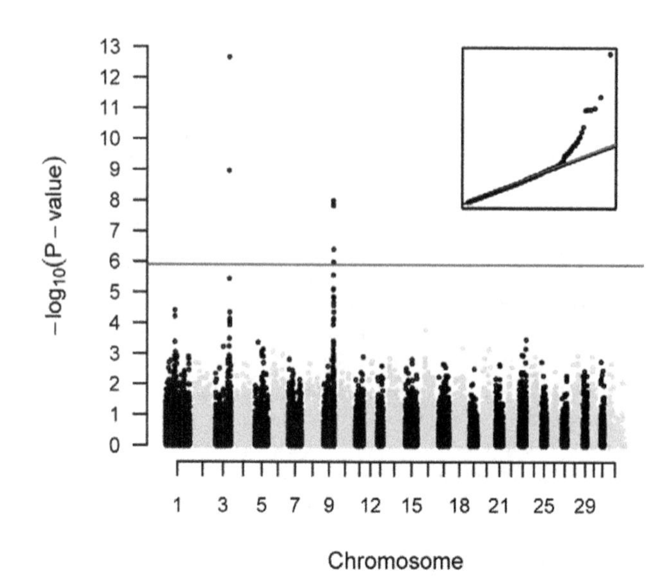

Figure 2. Manhattan plot for height at withers based on dEBV. The red line indicates the Bonferroni-corrected significance level ($p<1.31\times10^{-6}$). The inset shows a quantile-quantile (qq) plot with the observed plotted against the expected p-values. The used mixed-model approach efficiently corrected for the stratification in the sample. The skew at the right edge indicates that these SNPs are stronger associated with height than would be expected by chance. This is consistent with a true association as opposed to a false positive signal due to population stratification.

ECA 6 for correctness of gaits (Table S1). However, the associated SNP was not found within or next to a potential candidate gene.

Effect size on height at withers

We calculated the mean dEBVs for each genotype at the two best associated markers on ECA 3 and ECA 9, respectively. The C-allele for SNP BIEC2-808543 was associated with increased height at withers. The presence of the C-allele at this SNP was found to increase the dEBV for height at withers by ~1.0 cm. The effect at BIEC2-1105377 on ECA 9 was smaller and one copy of the A-allele at this SNP accounted for about ~0.5 cm of increase in height at withers (Table 3).

Taken together, the alleles at these two SNPs are responsible for a total difference of about 3 cm in the dEBV for height at withers. The trait-increasing alleles C for SNP BIEC2-808543 and A for SNP BIEC2-1105377 are the minor alleles at both SNPs and, combining the two SNPs, none of the 1,077 horses in our analysis was found to carry all four of the trait-increasing alleles (Figure 3).

Proportion of the explained variance

The 38,124 autosomal SNPs together account for 70.2% of the dEBV variance (Figure 4). Major fractions of the dEBV variance are attributable to ECA 1 (11.0%), ECA 3 (11.6%) and ECA 9 (7.4%). Interestingly, a large fraction of the dEBV variance results from ECA 1, although the association analysis with the mixed-model approach did not reveal significantly associated SNPs on this chromosome. The major fraction of the dEBV variance on ECA 3 and ECA 9 is attributed to the identified QTL. The two QTL alone explain 18.2% of the variance of the dEBV for height at withers.

Discussion

We carried out an association analysis for conformation traits with 1,077 FM horses and 38,124 SNPs. This analysis led to the identification of two QTL for height at withers. The two identified QTL can be considered as the major determinants for stature since the two QTL account for 18.2% of the dEBV variance. Given the heritability of about 72% we would expect that these two QTL explain together 13.1% of the total phenotypic variance observed. Taking chromosome 1 into account ECA 1, ECA 3 and ECA 9 explain together ~30.0% of dEBV variance. So far we couldn't find significantly associated SNPs on ECA 1, therefore we suppose that either a potential QTL was not detectable or that there are many loci with small effects summing up for the observed variance. Thus, the genetic architecture of the dEBV for height at withers is characterized by a few genes with major effects and a large number of genes with small effects. Such a situation is typical for many complex quantitative traits in domestic animals [28]. In contrast, in studies on human height, so far all detected loci had very small effects [12]. The large effect size of the detected QTL in our horse population will facilitate the identification of the causative variants in the future.

Interestingly both detected QTL for height are located near genes with very large intergenic regions. The QTL on ECA 3 is located near the *LCORL*/*NCAPG* genes. The best-associated SNP is located shortly upstream of the *LCORL* gene, in a 1.7 Mb gene desert. The same locus has already been identified in human and bovine association studies for height [12,17]. In humans no causative mutation has been identified so far. In cattle, a non-synonymous variant in the *NCAPG* gene has been proposed as a potential causative variant for various growth-related traits [29]. However, there is no functional proof for the causality of this variant available. The best-associated SNP for the QTL on ECA 9 is located in the 3′-flanking region of the *ZFAT* gene, which represents 900 kb of intergenic region. For the further follow-up of these QTL one should consider the possibility that these large genomic regions without any coding sequences represent regulatory domains of the chromatin [30]. In this case the QTL would be expected to have an effect on gene regulation, which might extend to genes that are not in the immediate neighborhood of causative nucleotide variants.

In conclusion, we have identified two QTL for height in horses, which explain a substantial fraction of the variance of the trait. The same loci have been previously identified in human, but the

Table 2. Significantly associated SNPs with dEBV for height at withers using a mixed-model approach.

SNP name	Equine position[a]	Human position[b]	Alleles (freq.)[c]	p-value[d]	p-value[e]
BIEC2-808543	ECA3:105,547,002	HSA4:18,084,850	C/T (0.0845)	2.18×10^{-13}	2.50×10^{-5}
BIEC2-808466	ECA3:105,163,077	HSA4:18,595,032	G/A (0.0795)	1.08×10^{-9}	7.50×10^{-5}
BIEC2-1105377	ECA9:74,798,143	HSA8:135,326,638	A/G (0.3908)	9.99×10^{-9}	3.25×10^{-4}
BIEC2-1105370	ECA9:74,795,013	HSA8:135,319,688	T/C (0.3914)	1.30×10^{-8}	4.25×10^{-4}
BIEC2-1105372	ECA9:74,795,089	HSA8:135,319,764	A/C (0.3914)	1.30×10^{-8}	4.25×10^{-4}
BIEC2-1105373	ECA9:74,795,236	HSA8:135,323,667	A/G (0.3907)	1.48×10^{-8}	5.75×10^{-4}
BIEC2-1105840	ECA9:76,254,733	HSA8:~137,138,000	A/G (0.8477)	3.98×10^{-7}	0.0178
BIEC2-1105505	ECA9:75,386,842	HSA8:136,024,172	T/C (0.8695)	1.04×10^{-6}	0.0451

[a]EquCab 2.0 assembly.
[b]corresponding homologous human position, build 37.
[c]trait-increasing allele/trait-decreasing allele; (frequency of the trait-increasing allele).
[d]corresponding list of p-values of 1-d.f. (additive or allelic) test for association between SNP and trait; the Bonferroni-corrected threshold for a 5% genome-wide significance level is $p_{BONF} = 1.31\times10^{-6}$.
[e]corresponding list of empirical p-values derived from permutations with 40,000 replicates.

causative mechanisms remain largely unknown. Interestingly, both QTL are located in very large intergenic regions or gene deserts. The large effect size in horses may facilitate the future identification of the true causative mutations underlying these QTL.

The two QTL on ECA 3 and ECA 9 are also significantly associated with wither height, conformation of legs, ventral border of mandible, correctness of gaits, expression of the head, length of croup and length of back, respectively. These results are not unexpected, as genetic correlations between the 28 conformation traits are known from the analysis of variance components and the routine estimation of breeding values (results not shown). Height at withers shows a negative genetic correlation with expression of the head, ventral border of mandible, conformation of legs and correctness of gaits. Height at withers shows a positive genetic correlation with wither height, length of back and length of croup, to mention just some of the 28 existing genetic and phenotypic correlations. From a biological point of view it makes complete sense that the two QTL-regions on ECA 3 and ECA 9 are also significant for other confirmation traits. In general conformation traits are expected to follow a certain proportionality, i.e. a taller horse has a longer back and a longer croup, compared to a smaller horse. From a biomechanical point of view correctness of gaits is related to size. A horse which is too tall might be less well balanced than a smaller horse, and thus show less correctness of gaits. However, some of our findings might also be due to the structure of our data, and the particular breed we worked with, respectively. Thus, further studies including data from other breeds are needed to completely resolve these questions.

Table 3. Effect of different QTL genotypes on the dEBV height at withers.

Chromosome	SNP name	Genotype	Counts	Mean dEBV[a]	SD
ECA 3	BIEC2-808543	CC	4	2.09	3.29
		CT	170	1.29	2.63
		TT	884	0.00	2.43
ECA 9	BIEC2-1105377	AA	149	1.13	2.51
		AG	523	0.47	2.54
		GG	381	0.00	2.41

[a]The mean dEBV for the homozygous trait-decreasing genotype was arbitrarily set to zero. The values correspond to centimeters.

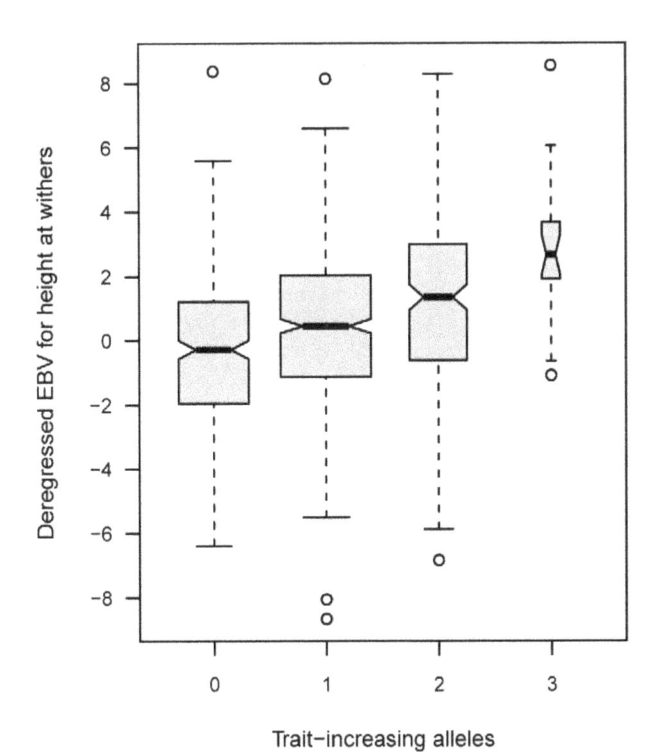

Figure 3. Combined effect size of the two identified QTL on ECA 3 and ECA 9 on the dEBV for height at withers in the FM horse breed. The box plot indicates the median values, 25% and 75% quartiles and the outliers of the distribution. A total of 308, 513, 211, and 21 animals, respectively, represented the classes from zero to three trait-increasing alleles. The difference in the medians between horses with zero and three trait-increasing alleles, respectively, is 2.97 cm. None of the analyzed horses carried four trait-increasing alleles. The association between the number of trait increasing alleles and height at withers is highly significant ($p = 8.4\times10^{-13}$; Kruskal-Wallis test). Medians are significantly different between groups.

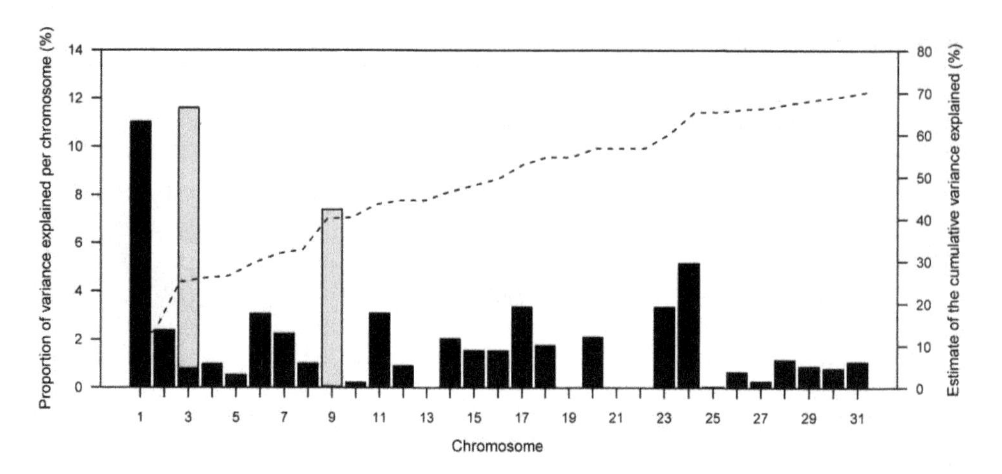

Figure 4. Estimates of height at withers dEBV variance explained by SNPs on 31 autosomes (black bars) and the two identified QTL (green bars). The two QTL combined explain 18.2% of the variance. All chromosomes together explain 70.2% of the genetic variance.

Materials and Methods

Animals

We selected 1,151 FM horses from studbook data based on the following criteria: Each horse had to be an active breeding animal with at least one progeny between 1998 and 2008. We preferably selected stallions with many offspring. However, due to the limited availability of such horses, we also included younger stallions and breeding mares. We also considered the distribution of estimated breeding values (EBVs) for different traits and their accuracy for the selection of the animals (Figure S1). We considered this set of animals to be representative for the active breeding population of the FM breed, which comprises about 3,500 animals in total.

Phenotypes and estimated breeding values (EBV)

Since the year 2006 the FM breeding association runs a breeding program based on estimated breeding values (EBVs). EBVs for 28 conformation traits (linear type traits and measurements such as height at withers) are estimated once a year with a best linear unbiased prediction (BLUP) multiple trait animal model [31] considering all relations between the traits. EBV estimation is based on phenotypic data, which are recorded during the compulsory field tests for stud book horses. Recording of conformation traits is a standardized methodology in livestock breeding, and aims to determine the morphology and general appearance of e.g. a horse [9,10]. Experts from the breeding organisation recorded the linear type conformations and body measurements. The classification of traits is determined by the rules and regulations of the Franches-Montagnes Horse Breeding Association and uses a linear scale from 1 to 9 [32]. Height at withers is the only conformation trait in the FM breed which is measured in centimeters. The average estimated breeding value for animals born between 1998 and 2000 was set to 0. The EBVs were deregressed [33] prior to association analysis.

Ethic statement

All animal work was conducted in accordance with the relevant local guidelines (Swiss law on animal protection and welfare - permit to the Swiss National Stud Farm no. 2227). No experiments with animals have been performed in our study, except of collecting blood samples from horses by a state approved veterinarian.

Genotyping and quality control

We collected EDTA blood samples and isolated genomic DNA from all horses. The DNA samples were genotyped with the illumina equine 50 K SNP beadchip containing 54,602 SNPs. We used the PLINK v1.07 software for pruning of the genotype data set [34]. We removed 48 out of 1,151 genotyped FM horses due to sample duplication. Of the remaining 1,103 FM horses, we removed 10 horses as they had genotype call rates below 90%. Out of the 54,602 markers on the array we removed 12,738 SNPs with minor allele frequencies below 5%, 2,191 SNPs with more than 10% missing genotypes, and 2,730 SNPs strongly deviating from Hardy-Weinberg equilibrium (HWE, $p \leq 0.0001$). We calculated the pairwise identity by descent (IBD) from the remaining SNPs and compared them with the corresponding pedigree numerator relationships calculated with CFC [35]. We excluded further 16 animals due to inconsistencies between the marker-based relationship and the pedigree-derived relationship. Thus, the final data set consisted of 1,077 horses (212 males and 865 females) and 38,124 autosomal SNPs. These horses are direct descendants of 208 sires and 883 dams. The average pedigree completeness index over 10 generations was 97.8% for the 1,077 FM horses. The average inbreeding coefficient and the average numerator relationship calculated was 6.22% and 14.22%, respectively.

Genome-wide association study

We performed a genome-wide association study using a mixed-model approach considering the relatedness of the horses as implemented in the function mmscore in the R package GenABEL [36]. We examined QQ-plots for inflation of small p-values hinting at false positive association signals. After correction for the population stratification the genomic inflation factor was 1.04. We considered SNPs to be genome-wide significantly associated if their p-values were below the 5% Bonferroni-corrected threshold for 38,124 independent tests ($p_{BONF} < 1.31 \times 10^{-6}$). To derive empirical genome-wide significance thresholds permutations with 40,000 replicates were conducted.

Chromosomal partitioning of genetic variance

We used the GCTA software [37] to partition the genetic variance onto different chromosomes and the two identified QTL. For this the genomic relationship matrix was built for the 31 autosomes and the two QTL separately. For each QTL we selected the neighboring SNPs in a 5 Mb interval surrounding the most significantly associated SNP (±2.5 Mb) to build its genomic relationship matrix. All other SNPs were used to build the genomic relationship matrix for the chromosome harbouring the QTL. We used the GCTA command –reml to estimate variance components with the effects of all chromosomes and QTL fitted simultaneously.

Supporting Information

Figure S1 Distribution of genotyped FM horses ranked by the EBV for conformation type and the accuracy for this particular EBV. In practice the EBVs in the FM breed are scaled to a mean of 100 and a standard deviation of 20. The average EBV for animals born between 1998 and 2000 was set to 100.

Acknowledgments

We would like to thank the Franches-Montagnes Horse Breeding Association for providing data. Furthermore, we would like to thank the staff, and in particular Fanny Berruex, at the veterinary clinic at the Swiss National Stud Farm for continuous help during sample collection. We thank Pierre-André Poncet, former director of the Swiss National Stud Farm, for his enthusiasm and support for our study. We also thank the staff of the NCCR Genomics Platform at the University of Geneva Medical School for performing the SNP genotyping experiments. We further thank agn Genetics GmbH and the Syndicate of Swiss Cattle Breeders for their IT and software support.

Author Contributions

Conceived and designed the experiments: SR HSH. Performed the experiments: BH TL. Analyzed the data: HSH CF TL HS SR. Wrote the paper: HSH TL SR. In charge of sample collection and contributed phenotypic information: DB.

References

1. Chowdhary BP, Raudsepp T (2008) The Horse Genome Derby: racing from map to whole genome sequence. Chromosome Res 16: 109–127.
2. Wade CM, Giulotto E, Sigurdsson S, Zoli M, Gnerre S, et al. (2009) Genome Sequence, Comparative Analysis, and Population Genetics of the Domestic Horse. Science 326: 865–867.
3. Binns MM, Boehler DA, Lambert DH (2010) Identification of the myostatin locus (MSTN) as having a major effect on optimum racing distance in the Thoroughbred horse in the USA. Anim Genet 41 Suppl 2: 154–158.
4. Hill EW, McGivney BA, Gu J, Whiston R, MacHugh DE (2010) A genome-wide SNP-association study confirms a sequence variant (g.66493737C>T) in the equine myostatin (MSTN) gene as the most powerful predictor of optimum racing distance for Thoroughbred racehorses. BMC Genomics 11: 552.
5. Tozaki T, Hill EW, Hirota K, Kakoi H, Gawahara H, et al. (2012) A cohort study of racing performance in Japanese Thoroughbred racehorses using genome information on ECA18. Anim Genet 43: 42–52.
6. Rieder S (2009) Molecular tests for coat colours in horses. J Anim Breed Genet 126: 415–424.
7. Bellone RR (2010) Pleiotropic effects of pigmentation genes in horses. Anim Genet 41 Suppl 2: 100–110.
8. Brosnahan MM, Brooks SA, Antczak DF (2010) Equine clinical genomics: A clinician's primer. Equine Vet J 42: 658–670.
9. Saastamoinen MT, Barrey E (2000) Genetics of Conformation, Locomotion and Physiological Traits. In: Bowling AT, Ruvinsky A, eds. The Genetics of the Horse, CABI, Wallingford, UK. pp 439–472.
10. Koenen EPC, van Veldhuizen AE, Brascamp EW (1995) Genetic parameters of linear scored conformation traits and their relation with dressage and show-jumping in the Dutch Warmblood riding horse population. Livest Prod Sci 43: 85–94.
11. Visscher PM, McEvoy B, Yang J (2010) From Galton to GWAS: quantitative genetics of human height. Genet Res (Camb) 92: 371–379.
12. Lango Allen H, Estrada K, Lettre G, Berndt SI, Weedon MN, et al. (2010) Hundreds of variants clustered in genomic loci and biological pathways affect human height. Nature 467: 832–838.
13. Manolio TA, Collins FS, Cox NJ, Goldstein DB, Hindorff LA, et al. (2009) Finding the missing heritability of complex diseases. Nature 461: 747–753.
14. Kutalik Z, Whittaker J, Waterworth D, GIANT consortium, Beckmann JS, et al. (2011) Novel method to estimate the phenotypic variation explained by genome-wide association studies reveals large fraction of the missing heritability. Genet Epidemiol 35: 341–349.
15. Yang J, Benyamin B, McEvoy BP, Gordon S, Henders AK, et al. (2010) Common SNPs explain a large proportion of the heritability for human height. Nat Genet 42: 565–569.
16. Pausch H, Flisikowski K, Jung S, Emmerling R, Edel C, et al. (2011) Genome-wide association study identifies two major loci affecting calving ease and growth-related traits in cattle. Genetics 187: 289–297.
17. Pryce JE, Hayes BJ, Bolormaa S, Goddard ME (2011) Polymorphic regions affecting human height also control stature in cattle. Genetics 187: 981–984.
18. Karim L, Takeda H, Lin L, Druet T, Arias JAC, et al. (2011) Variants modulating the expression of a chromosome domain encompassing PLAG1 influence bovine stature. Nat Genet 43: 405–413.
19. Visscher PM, Goddard ME (2011) Cattle gain stature. Nat Genet 43: 397–398.
20. Habier D (2010) More than a third of the WCGALP presentations on genomic selection. J Anim Breed Genet 127: 336–337.
21. Heffner EL, Lorenz AI, Jannink J-L, Sorrells ME (2010) Plant breeding with genomic selction: potential gain per unit time and cost. Crop Sci 50: 1681–1690.
22. Hayes BJ, Bowman PJ, Charmberlain AJ, Goddard ME (2009) Genomic selection in dairy cattle: Progress and challenges. J Dairy Sci 92: 433–443.
23. Habier D, Tetens J, Seefried F-R, Lichtner P, Thaller G (2010) The impact of genetic relationship information on genomic breeding values in German Holstein cattle. Genet Sel Evol 42: 5.
24. Meuwissen THE, Hayes BJ, Goddard ME (2001) Prediction of total genetic value using genome-wide dense marker maps. Genetics 157: 1819–1829.
25. VanRaden PM (2008) Efficient Methods to Compute Genomic Predictions. J Dariy Sci 91: 4414–4423.
26. Hayes BJ, Pryce J, Chamberlain AJ, Bowman PJ, Goddard ME (2010) Genetic architecture of complex traits and accuracy of genomic prediction: coat colour, milk-fat percentage, and type in Holstein cattle as contrasting model traits. PLoS Genet 6: e1001139.
27. Hasler H, Flury C, Menet S, Haase B, Leeb T, et al. (2011) Genetic diversity in an indigenous horse breed –implications for mating strategies and the control of future inbreeding. J Anim Breed Genet 128: 394–406.
28. Hayes B, Goddard ME (2001) The distribution of the effects of genes affecting quantitative traits in livestock. Genet Sel Evol 33: 209–229.
29. Setoguchi K, Furuta M, Hirano T, Nagao T, Watanabe T, et al. (2009) Cross-breed comparisons identified a critical 591-kb region for bovine carcass weight QTL (CW-2) on chromosome 6 and the Ile-442-Met substitution in NCAPG as a positional candidate. BMC Genet 10: 43.
30. Libioulle C, Louis E, Hansoul S, Sandor C, Farnir F, et al. (2007) Novel Crohn disease locus identified by genome-wide association maps to a gene desert on 5p13.1 and modulates expression of PTGER4. PLoS Genet 3: e58.
31. Henderson CR (1975) Best linear unbiased prediction under a selection model. Biometrics 31: 423–447.
32. Poncet PA, Pfister W, Muntwyler J, Glowatzki-Mullis ML, Gaillard C (2006) Analysis of pedigree and conformation data to explain genetic variability of the horse breed Franches-Montagnes. J Anim Breed Genet 123: 114–121.
33. Garrick DJ, Taylor JF, Fernando RL (2009) Deregressing estimated breeding values and weighting information for genomic regression analyses. Genet Sel Evol 41: 55.
34. Purcell S, Neale B, Todd-Brown K, Thomas L, Ferreira MAR, et al. (2007) PLINK: a toolset for whole-genome association and population-based linkage analysis. Am J Hum Genet 81: 559–575.
35. Sargolzaei M, Iwaisaki H, Colleau JJ (2006) CFC: a tool for monitoring genetic diversity. Proc. 8th World Congr. Genet. Appl. Livest. Prod., CD-ROM Communication no 27–28. Belo Horizonte, Brazil, Aug. 13–18, 2006.
36. Aulchenko YS, Ripke S, Isaacs A, van Duijn CM (2007) GenABEL: An R library for genome-wide association analysis. Bioinformatics 23: 1294–1296.
37. Yang J, Manolio TA, Pasquale LR, Boerwinkle E, Caporaso N, et al. (2011) Genome partitioning of genetic variation for complex traits using common SNPs. Nat Genet 43: 519–525.

12

Field-Isolated Genotypes of *Mycobacterium bovis* Vary in Virulence and Influence Case Pathology but Do Not Affect Outbreak Size

David M. Wright[1*¤], Adrian R. Allen[2], Thomas R. Mallon[2], Stanley W. J. McDowell[2], Stephen C. Bishop[3], Elizabeth J. Glass[3], Mairead L. Bermingham[3], John A. Woolliams[3], Robin A. Skuce[1,2]

1 School of Biological Sciences, Queen's University Belfast, Belfast, Northern Ireland, United Kingdom, **2** Veterinary Sciences Division, Bacteriology Branch, Agri-Food and Biosciences Institute, Belfast, Northern Ireland, United Kingdom, **3** The Roslin Institute and Royal (Dick) School of Veterinary Studies, University of Edinburgh, Midlothian, Scotland, United Kingdom

Abstract

Strains of many infectious agents differ in fundamental epidemiological parameters including transmissibility, virulence and pathology. We investigated whether genotypes of *Mycobacterium bovis* (the causative agent of bovine tuberculosis, bTB) differ significantly in transmissibility and virulence, combining data from a nine-year survey of the genetic structure of the *M. bovis* population in Northern Ireland with detailed records of the cattle population during the same period. We used the size of herd breakdowns as a proxy measure of transmissibility and the proportion of skin test positive animals (reactors) that were visibly lesioned as a measure of virulence. Average breakdown size increased with herd size and varied depending on the manner of detection (routine herd testing or tracing of infectious contacts) but we found no significant variation among *M. bovis* genotypes in breakdown size once these factors had been accounted for. However breakdowns due to some genotypes had a greater proportion of lesioned reactors than others, indicating that there may be variation in virulence among genotypes. These findings indicate that the current bTB control programme may be detecting infected herds sufficiently quickly so that differences in virulence are not manifested in terms of outbreak sizes. We also investigated whether pathology of infected cattle varied according to *M. bovis* genotype, analysing the distribution of lesions recorded at post mortem inspection. We concentrated on the proportion of cases lesioned in the lower respiratory tract, which can indicate the relative importance of the respiratory and alimentary routes of infection. The distribution of lesions varied among genotypes and with cattle age and there were also subtle differences among breeds. Age and breed differences may be related to differences in susceptibility and husbandry, but reasons for variation in lesion distribution among genotypes require further investigation.

Editor: Stephen V. Gordon, University College Dublin, Ireland

Funding: This study was funded by the BBSRC (http://www.bbsrc.ac.uk/) under the CEDFAS initiative, grant numbers BB/E018335/1 and BB/E018335/2; and by the Northern Ireland Department of Agriculture and Rural Development (http://www.dardni.gov.uk/), grant number DARD0407. SCB, JAW, and EJG also acknowledge BBSRC Institute Strategic Programme funding. The funders had no role in study design, data collection and analysis, decision to publish, or preparation of the manuscript.

Competing Interests: The authors have declared that no competing interests exist.

* E-mail: david.m.wright@cantab.net

¤ Current address: Centre for Public Health, Queen's University Belfast, Belfast, Northern Ireland, United Kingdom

Introduction

Bacterial pathogens are frequently classified into distinct strains according to virulence, detectability, host specificity and other parameters that determine the magnitude of their impact on host populations; classifications which may then be used to assist and improve disease management. For example, laboratory trials have found evidence of variation in virulence among clinical strains of *Mycobacterium tuberculosis*, the causative agent of human tuberculosis [1–3]. As genotyping technologies have advanced, classification of strains according to genetic similarity has become more common [4], often followed by efforts to detect phenotypic variation among strains that were originally distinguished using molecular techniques. Variation in immunogenicity, virulence and pathology has been found among the six major lineages of *M. tuberculosis*, along with evidence of host-pathogen coevolution in regions where lineages are long established [5–7].

We investigated whether genotypically-distinct strains of *M. bovis* differ in transmissibility and virulence, and whether an aspect of the pathology of infected cattle varies according to pathogen genotype. Bovine tuberculosis is a chronic disease of farmed cattle and wildlife which may also be transmitted to humans, presenting a public health risk [8,9]. In the UK, a system of regular skin testing followed by compulsory slaughter of infected animals, supported by active abattoir surveillance, is used in an attempt to control bTB incidence in the cattle population [10]. This programme imposes significant costs on the UK cattle industry and government. In England alone, the bTB control programme costs an estimated £91 million annually, comprised mostly of testing costs and compensation for farmers [11].

Since the late 1990s a large number of *M. bovis* isolates from infected cattle in the UK have been genotyped to help trace sources of infection [12,13]. This provides a unique opportunity to assess whether there is phenotypic variation among genotypes and whether knowledge of such variation might be exploited to aid control of the epidemic. A similar approach has been used to compare strains of *M. tuberculosis* infecting human populations, with some strains more likely to be found in clusters of cases, indicating greater virulence or transmissibility [14,15].

In the UK, cattle herds in which *M. bovis* is detected are placed under movement restrictions until all infected animals have been removed (a herd breakdown), and so the national bTB epidemic consists of a series of discrete breakdown events that vary in the number of animals infected. A survey of bTB outbreaks within herds in Great Britain revealed subtle differences among pathogen genotypes in outbreak size and the proportion of cases visibly lesioned [16], leading to speculation that closely related *M. bovis* genotypes might vary in transmissibility. We investigated these effects, conducting a larger scale analysis whilst accounting for variation in the host population structure (especially herd size and mix of cattle breeds) within Northern Ireland, where the *M. bovis* population has been systematically sampled.

Strains of *M. tuberculosis* have also been shown to induce distinctive pathologies in humans, with some lineages associated with a greater proportion of extra-pulmonary cases that carry an increased risk of mortality [6], but no previous studies have searched for *M. bovis* genotype-specific variation in cattle pathology. The site of the initial infection can be deduced from the location of tuberculous lesions, provided that infection has not progressed to multiple sites, and is thought to be indicative of the route of infection [17,18]. In naturally infected cattle in the UK, lesions are most commonly found in lymph nodes draining the respiratory tract, indicating that inhalation is the primary route of infection [19]. However, lesion distribution and severity may be modified by cattle breed and husbandry; Holstein cattle in Ethiopia allowed to graze extensively were shown to have a greater proportion of lesions in the upper respiratory tract and mesenteric lymph nodes than animals kept indoors under intensive conditions, a pattern indicative of infection via ingestion [20]. In the UK, dairy animals are typically managed more intensively than beef animals so we might expect to find variation in lesion distribution among the major beef and dairy breeds.

In this study we used a large population wide survey of *M. bovis* genotypes in Northern Ireland [13] linked with cattle population records, to assess whether there is phenotypic variation among *M. bovis* genotypes. We estimated the relative transmissibility of *M. bovis* genotypes by analysing the distribution of outbreak sizes. We also investigated whether there is variation among *M. bovis* genotypes in virulence, measuring the proportion of those cattle in each breakdown that tested positive (using the single intradermal comparative tuberculin test, henceforth skin test) that were subsequently found to have tuberculous lesions. Finally, we investigated bTB pathology, specifically the influence of *M. bovis* genotype and cattle breed and age on the proportion of infected animals that were lesioned and the distribution of lesions in these animals.

Methods

Outbreak Detection and Genotyping

The bTB control programme in Northern Ireland is based on a regime of annual skin testing of all animals and post-mortem inspection for tuberculous lesions [10,21]. Following detection of bTB by either method, the infected herd is placed under restrictions whereby cattle can only be moved if they are sent directly to slaughter, with skin test positive animals (reactors) being dispatched immediately. All animals in the herd are then subjected to repeated skin tests at sixty day intervals and tissue samples from reactors and lesioned animals are subjected to histopathological tests and laboratory culture to confirm infection with *M. bovis*. If infection is confirmed, all remaining animals in the herd must undergo two successive negative skin tests before restrictions are lifted and a breakdown is deemed over. If infection is not confirmed then a single clear herd test is sufficient for the breakdown to be ended. In 2011 there were approximately 1.6 million cattle in Northern Ireland in ca. 25,000 herds, with the herd incidence of bTB close to 5%.

We combined data from a nine year (2003–2011) survey of the genetic structure of the *M. bovis* population with detailed records of the cattle population during the same period. Beginning in 2003, a single isolate has been genotyped from each newly confirmed herd breakdown, provided that there had been no confirmed cases in the herd during the previous 365 days. Sampling increased to two isolates per breakdown in 2006 and to every confirmed isolate in June 2009. Genotypes were defined using a combination of spoligotyping and VNTR (variable number of tandem repeat) markers selected to provide maximum resolution of the clonal relationships among herd breakdowns (VNTR markers discriminate within spoligotypes) [22]. A total of 23,711 isolates were genotyped during the study period, covering 11,818 herd breakdowns with at least one isolate genotyped. The majority of the 351 genotypes identified were rare (number of breakdowns for each genotype: median = 2, range 1–4443; 289 genotypes were found in less than 10 breakdowns each), and there was pronounced inter-annual variation in relative frequency of occurrence [13]. We extracted corresponding records detailing skin tests, animal life histories and movements among herds from the Animal and Public Health Information System [23], a database administered by the Department of Agriculture and Rural Development.

Breakdown Size

We estimated the average size of breakdowns caused by *M. bovis* genotypes, defining breakdown size as the total number of animals detected with bTB during the period of movement restriction. There were 1892 herd breakdowns between June 2009 and December 2011 in which all lesioned cases were genotyped (5066 isolates). From these we excluded 207 breakdowns in which multiple genotypes were detected because in instances where reactors were not visibly lesioned it would not have been possible to identify the relative contribution of each genotype to the breakdown size. The mean length of herd breakdowns was seven months but a small number of herds remained under restrictions for much longer. Many of these were beef finishing herds that buy in large numbers of cattle from many different sources and which sell directly to abattoirs. Persistent breakdowns in these herds are likely to be the result of multiple imported infections and we therefore excluded 61 outbreaks that lasted longer than fourteen months (the upper 90% quantile of all breakdown durations over the period 1993 to 2012). Following these exclusions 1624 breakdowns remained (89% of fully genotyped breakdowns) with 87 different genotypes represented.

We fitted a series of models to examine the relative influences of pathogen genotype, herd size and the means by which infection was detected on breakdown size. In the UK and Ireland larger herd sizes have been associated with both increased risk of herd breakdown [24,25] and persistent infection within herds [26,27]. In Northern Ireland cattle are housed over winter in large sheds with shared airspace. In addition, although housed cattle are often batched according to age and sex there is likely to be occasional physical contact among batches when animals are moved for veterinary treatment and other routine management. Large herds are typically housed in larger sheds designed to enable mechanized feeding, rather than in an increased number of small units (pers. obs.). Therefore an infected animal in a large herd may have contact with a larger number of susceptible animals and so we expected breakdown size to increase with herd size. The majority of breakdowns were detected either as a result of annual testing and abattoir surveillance but 31% resulted from epidemiological investigations into other breakdowns (tracing of infectious contacts). We expected these breakdowns to differ in size in comparison with routine detections because the different tracing methods give an indication of the probability and timescale of disease presence in the herd (Table 1).

We tested whether herd size or contact tracing had a significant influence on breakdown size by comparing the fit of generalized linear mixed models (GLMMs) incorporating different combinations of these factors as fixed effects. We fitted four models of increasing complexity; M1) no fixed effects (null model) M2) just herd size, M3) just contact tracing and M4) both herd size and contact tracing. In our dataset there were seven different situations by which breakdowns were detected (Table 1) and we estimated a coefficient representing each in models incorporating contact tracing (i.e. varying the intercept). A single additional coefficient (the regression slope) was estimated in models including herd size, representing a linear relationship between the logarithm of herd size and breakdown size (this functional form provided the best fitting models among various approaches tested: ordinary linear models, polynomial fits and treating herd size as a categorical variable). Herd sizes fluctuate throughout the year, peaking in summer months. Therefore, we used total number of animals that had been present in the herd over the previous calendar year as our measure of herd size. In all models genotype effects were included as normally distributed random variables. This mixed modelling approach allowed us to account for uncertainty around estimates for genotypes that were responsible for very few outbreaks; estimates for these are regressed towards the overall mean and have wider confidence intervals. We used a Poisson error distribution with a log link, but noted that the data were overdispersed, with more large outbreaks than expected based on a Poisson model. Following the approach of Elston et al. [28] we explicitly modelled this extra variance by including outbreak effects as normally distributed random variables, nested within genotype effects (i.e. fitting a data level random variable) resulting in a Poisson-lognormal model where each outbreak is associated with variation at both the outbreak and higher hierarchical levels. Models were fitted using the *lme4* package [29] in *R 13.2* [30]. We also searched for variation among genotypes at the coarser (spoligotype) level of discrimination, fitting a similar series of models with effects of VNTR types nested within spoligotypes.

We compared the candidate models using Akaike's Information Criterion (marginal) which scores models according to their complexity and fit to the data (models are penalised for each parameter estimated). In a given set, models with lower AIC are considered to be better supported by the data and those with AIC values that differ by more than two are considered to be significantly different [31]. We then examined the estimated parameters from the best fitting model to see if they supported our predictions about outbreak size and contact tracing (Table 1). Using a likelihood ratio test we also compared the best fitting model with a simpler model with the same fixed effect structure but with no genotype effects.

Virulence

We estimated the proportion of reactors that were found to have tuberculous lesions as a measure of genotype virulence, whilst attempting to control for variation in disease susceptibility among cattle of different ages and breeds. Of the records used for the breakdown size analysis, we selected only those that had at least one reactor (in some breakdowns all cases were detected at abattoir), a subset of 1276 breakdowns with a total of 4706 post mortem records of reactors, 59% of which were lesioned. We modelled the mean proportion of reactors that were lesioned using a logistic GLMM with genotype and breed effects incorporated as normally distributed random variables and with animal age (in months) as a fixed effect. Exploratory analysis showed that the proportion of reactors lesioned depended on the breakdown size. Reactors in breakdowns with one or two reactors were more likely to be lesioned than those in larger outbreaks (Figure 1). This discrepancy is probably related to the ways in which the skin test is interpreted in different sized breakdowns. If there are a large number of reactors detected in a herd, a more severe interpre-

Table 1. Modes of detection for bTB breakdowns in Northern Ireland and predicted effects on breakdown size.

Mode of detection	Predicted effect on breakdown size	Reason
AHT – annual herd test	baseline	
LRS – lesions found at routine slaughter	–	Slaughter usually more frequent than AHT therefore less time for infection to spread.
LCT – lateral check test (herds sharing a boundary with infected herd are tested)	–	Less time for infection to spread since last herd test than under annual testing.
BCT – backward check test (source herds of cattle bought into focal herd are tested)	+	Onward spread to another herd has occurred so likely to be a large number infected in the source herd.
CTS – check test status (individual animals with ID or movement queries tested)	+	Animals bought in with unknown disease status potentially increase risk to the remainder of the herd
FCT – forward check test (destination herds of animals that left focal herd immediately prior to a breakdown)	–	Short period since potentially infectious animal(s) moved in to herd and so little time for large outbreak to develop.
CTT – check test trace (forward trace of individual animal)	–	As above.

tation of the skin test may be applied to try to 'clean' the herd of infected animals. Animals with inconclusive responses to the skin test might be culled, and these may be in very early stages of infection and so be unlikely to have gross lesions. To ensure that this factor did not bias our estimates of genotype virulence we fitted the model twice, first using records from all breakdowns and secondly using only breakdowns with more than two reactors. We then correlated the estimated genotype effects from the two models. We also tested whether the proportion of reactors lesioned varied depending on the manner of disease detection because contact tracing may often disclose outbreaks at an earlier stage than annual skin testing, and hence cases may be less advanced (Table 1). We fitted a second model estimating additional parameters for each form of tracing (i.e. fitting these as fixed effects) but retaining the random genotype effects. The two models were then compared by means of a likelihood ratio test.

Lesion Distribution

We aimed to determine whether the presumed primary site of infection varied depending on *M. bovis* genotype or host age and breed. The distribution of lesions in both experimentally- and naturally-infected animals is considered to be indicative of the primary site of infection, with the majority of lesions found in lymph nodes draining the respiratory tract [17–19]. We selected post mortem records of all reactors from which *M. bovis* had been successfully isolated and genotyped between 2003 and 2011, a total of 16,571 animals. We did not consider animals that were found to be lesioned at routine slaughter because abattoirs vary considerably in the quality of inspection for lesions [32,33]. However, a single abattoir handled 85% of reactors in our sample, minimising bias caused by different inspection regimes. Animals with multiple lesions (approximately 20% of reactors) were also excluded because in these cases it would not have been possible to determine which lesion was closest to the initial site of infection. We also excluded cases where lesions were likely to have resulted from haematogenous spread rather than being close to the site of initial infection (e.g. lesions in the popliteal or prescapular lymph nodes or in the liver). Following these exclusions, a total of 12,633 post mortem records remained. Cases were classified into three groups based on the site at which lesions were found; infection of the upper respiratory tract was indicated by lesions in the head lymph nodes, the lower respiratory tract was represented by lesions in the lungs or the bronchio-mediastinal lymph nodes and the digestive tract was represented by lesions in the intestines or mesenteric lymph nodes.

We modelled the proportion of respiratory tract lesions that were found in the lower section of the tract using a logistic GLMM to determine whether there were differences in lesion distribution among genotypes and cattle of different ages and breeds. Genotype and breed effects were incorporated as normally distributed random variables and animal age (in months) was incorporated as a fixed effect. Finally we correlated the estimated genotype effects from this model with those estimated for virulence.

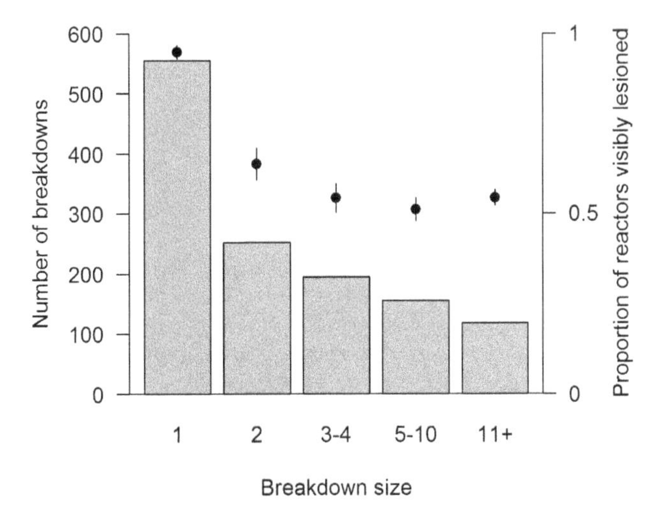

Figure 1. Herd breakdown sizes and proportion of reactors visibly lesioned in Northern Ireland 2009–2011. Distribution of herd breakdown sizes (grey bars). Points and error bars indicate average proportion of reactors found to be visibly lesioned in breakdowns of each size class (mean and 95% CIs).

Results

Breakdown Size

Breakdown size was influenced by both herd size and contact tracing but not significantly by pathogen genotype. Breakdowns ranged in size from one to 73 infected animals but the majority were small with 75% having three cases or fewer. Model M4, incorporating herd size and contact tracing was best supported by the data and was significantly better than the other candidate models (ΔAIC >10 in each case, Table 2). The estimated mean size of breakdowns (number of infected animals) detected at annual herd tests was 2.00 (95% CI: 1.90–2.11) and predicted breakdown size increased significantly with the logarithm of herd size (Table 3). Breakdowns detected by backward or lateral tracing of infected animals were significantly larger than those detected by annual herd tests. There was no significant difference in the mean sizes of breakdowns detected by annual herd tests and forward tracing. Breakdowns detected as a result of abattoir surveillance were significantly smaller than those detected by annual herd tests (Table 2).

There were no significant differences in mean size of breakdowns caused by different *M. bovis* genotypes; in all of the models fitted the variance among genotype estimates was zero. Confirming these findings, there was no significant loss of fit when we compared the best fitting model (M4) with a simpler model that did not incorporate genotype level variation (Likelihood ratio test $X^2 = 0$, $d.f. = 1$, $P = 0.999$, log-likelihood identical for both models). There was considerable residual size variation among breakdowns, with a standard deviation of breakdown sizes estimated with the best fitting model of 0.8. We found no evidence of variation in mean breakdown size among spoligotypes; models separating VNTR and spoligotype effects revealed very similar patterns to those using the compound genotype classification.

Table 2. Comparison of candidate models explaining variation in size of bTB breakdowns in Northern Ireland.

Model	AIC
M4: herd size+contact tracing+genotype	2737
M3:contact tracing+genotype	2827
M2:herd size+genotype	2919
M1: Null model (genotype only)	2983

Models listed in order of decreasing goodness of fit (increasing AIC).

Table 3. Parameter estimates from best fitting linear regression model explaining variation in the size of bTB breakdowns in Northern Ireland.

Effect	Estimate	S.E.	Z	P	n
Intercept (AHT)	−0.45	0.122	−3.66	<0.001	690
LRS	−0.31	0.068	−4.58	<0.001	431
LCT (lateral)	0.49	0.067	7.39	<0.001	334
BCT (backward)	1.15	0.136	8.44	<0.001	49
CTS (check)	1.67	0.330	5.05	<0.001	7
FCT (forward)	−0.04	0.323	−0.11	0.909	11
CTT (forward)	−0.14	0.112	−1.26	0.207	102
Herd size (log)	0.25	0.026	9.65	<0.001	N/A

Detection mode abbreviations: AHT = annual herd test, LRS = lesions detected at routine slaughter, LCT = lateral check test, BCT = backward check test, FCT = forward check test, CTT = check test trace. Herd size parameter represents the increase in breakdown size with increasing log herd size.

Virulence

The mean proportion of reactors visibly lesioned in a breakdown varied among *M. bovis* genotypes. Estimates ranged from 44% of reactors lesioned in breakdowns caused by genotype 19.140 to 73% lesioned for genotype 9.273 (Figure 2A), and there were some significant differences among genotypes (genotypes are named [VNTR type.Spoligotype], e.g. genotype 19.140 = VNTR type 19, spoligotype SB0140). For example, animals infected with genotype 3.140 were significantly less likely to be lesioned than those infected with genotype 11.145 (95% CIs do not overlap, Figure 2A). The variance among genotypes decreased slightly (from 0.128 to 0.115) when breakdowns with only one case were excluded but the order of genotype effects remained very similar (Pearson's $r = 0.94$, $d.f. = 41$). The proportion of reactors lesioned decreased with cattle age (regression coefficient $= -0.66$, $z = -10.6$, $P < 0.001$) and was also influenced by the manner in which an outbreak was detected: those detected by backward or lateral tracing of infected animals had lower proportions of lesioned reactors than those detected at annual herd tests (Likelihood ratio test, contact tracing vs. non-contact tracing model: $X^2 = 16.9$, $d.f. = 6$, $P = 0.010$). Estimated genotype effects were closely correlated across both models (Pearson's $r = 0.99$, $d.f. = 66$). Therefore it is unlikely that differences among outbreaks due to the manner of detection, or the interpretation of the skin test in large and small outbreaks were responsible for the observed inter-genotype variation in proportion of reactors lesioned. We found no systematic differences in the proportion of reactors lesioned among spoligotypes (genotypes with different spoligotypes interspersed throughout the range of responses, Figure 2A).

We found less variation among cattle breeds than among *M. bovis* genotypes in the proportion of reactors lesioned (means ranged from 52% of Friesians to 69% of Aberdeen Angus), and estimates for the majority of breeds did not differ significantly from one another (overlapping CIs, Figure 3A).

Lesion Distribution

Lesion sites varied according to cattle age and pathogen genotype and to a lesser extent with cattle breed. The majority of lesions were found in the respiratory tract indicating that this is the most common route of infection of cattle in Northern Ireland. Only 1.8% of animals were found to have lesions associated with the digestive tract (i.e. the mesenteric lymph nodes). Overall 70% of lesions in the respiratory tract were found in the lower section (i.e. lungs or bronchio-mediastinal lymph nodes), and this proportion increased significantly with animal age. The mean age of animals at slaughter in our dataset was 50 months. Predictions from our fitted regression model indicated that 63% of animals slaughtered at 20 months (lower age quartile) would be lesioned in the lower tract, increasing to 75% at 72 months (upper age quartile).

The proportion of respiratory tract lesions found in the lower tract varied slightly among genotypes, with estimates ranging from 66% for animals infected with genotype 49.140 to 77% for animals with genotype 122.263. Uncertainty around these estimates means that very few of the genotypes effects can be considered to be significantly different from one another, with the majority of 95% CIs overlapping (Figure 2B). In addition, genotype effects were not clustered by spoligotype. There was less variation in lesion site among cattle breeds, with estimated proportions of cases lesioned in the lower tract ranging between 67% and 75% (Figure 3B). The most pronounced difference was between Friesians, which had fewer cases lesioned in the upper tract than the major beef breeds (e.g. Aberdeen Angus, Charolais, Simmental, Limousin).

We found no association between genotype virulence and the lesion distribution. There was only a very weak correlation between genotype estimates of the proportion of reactors visibly lesioned with the proportion of lesioned cases that had lesions in the lower respiratory tract (Pearson $r = 0.15$, $d.f. = 21$, c.f. Figures 2A and 2B). There was a weak negative correlation between these two estimated proportions among cattle of different breeds (Pearson $r = -0.60$, $d.f. = 33$, Figures 3A and 3B).

Discussion

Breakdown Size

We found no differences in the average breakdown size of herds infected with different genotypes of *M. bovis* when the effects of herd size and contact tracing had been accounted for. These results were partially consistent with the single previous study to investigate variation in transmissibility of *M. bovis* (in Great Britain) which indicated that there was no significant variation in outbreak size among VNTR types, although there was subtle variation among spoligotypes (a coarser level of classification) which we did not find [16].

Comparisons of multi-drug resistant with drug susceptible strains of *M. tuberculosis* revealed pronounced differences in transmissibility, although the effect size and direction was highly dependent on the strains compared and the design of the individual study [34]. In contrast, there was no evidence of variation in transmissibility or virulence in a comparison of field isolated *M. tuberculosis* strains that were distinguished by genotype alone rather than by clinical characteristics [35]. However in a population based survey of TB case cluster sizes in Malawi, strains identified using genetic markers were shown to vary in transmissibility. Cluster sizes were also strongly affected by mixing patterns within the host population, with sociable patient groups strongly represented in larger clusters [36].

Herd size and the means by which infection was detected (contact tracing) were related to breakdown size, highlighting the importance of host population structure in determining *M. bovis* transmission rates, as expected based on field studies and mathematical models of *M. tuberculosis* transmission [37–39]. Simulations of human TB transmission indicate that variation in susceptibility within host populations can result in considerable variation in outbreak sizes even when pathogen strains are assumed to be equally transmissible [40]. Multiple risk factors,

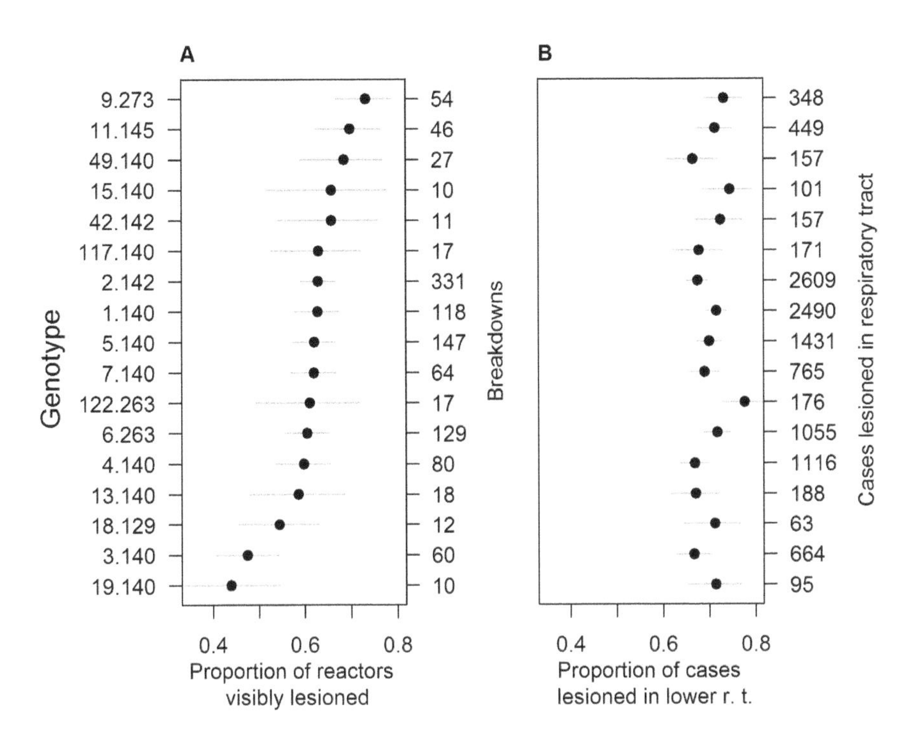

Figure 2. Variation among *M. bovis* genotypes in proportion of reactors visibly lesioned and distribution of lesions. Variation among *M. bovis* genotypes in A) the proportion of reactors found to be visibly lesioned (mean and 95% CIs) and B) the proportion of cases with respiratory tract lesions having lesions in the lower tract. The seventeen most abundant genotypes are plotted.

both genetic and husbandry related have been identified which may modulate cattle susceptibility to bTB at the animal level [9]. Outbreak size can also be affected by the presence of other diseases in the population; patients infected with HIV are at greater risk of TB and an HIV epidemic can significantly increase the size of subsequent TB outbreaks [41]. In cattle herds there is some evidence that liver fluke (*Fasciola hepatica*) can influence host susceptibility and also modify the sensitivity of the skin test, potentially allowing large breakdowns to develop undetected [42]. In some situations herd management may affect breakdown size; *M. bovis* transmission within intensively managed dairy herds in Spain was shown to be faster than that within herds managed for beef or to provide animals for bullfighting, although once detected, breakdowns could be more easily controlled in dairy herds [37].

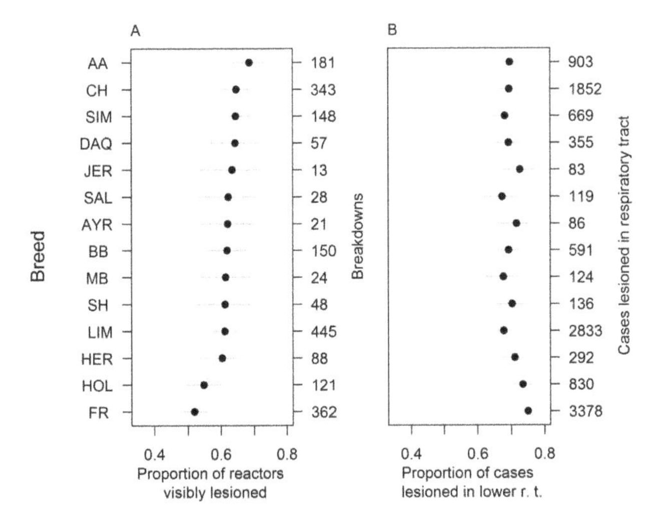

Figure 3. Variation among cattle breeds in proportion of reactors visibly lesioned and distribution of bTB lesions. The fourteen most abundant breeds are plotted. Breeds: AA = Aberdeen Angus, AYR = Ayrshire, BB = Belgian Blue, CH = Charolais, DAQ = Blonde D'Aquitaine, FR = Friesian, HER = Hereford, HOL = Holstein, JER = Jersey, LIM = Limousin, MB = Montbeliarde, SAL = Saler. SH = Shorthorn, SIM = -Simmental.

Having accounted for herd size and contact tracing we still found considerable residual variation in breakdown sizes with a small number of breakdowns much larger than the average. This feature of the breakdown size distribution may be related to variation among herds in the time since infection was introduced but could also be the result of superspreading, whereby a small number of hosts are responsible for a large number of secondary infections and have a large influence on epidemic progression [43,44]. Superspreading may occur when there is contact between an infected individual and a large number of susceptible individuals over a short time period [45], for instance when cattle are mustered for TB testing or for milking.

A similar effect may occur when a few individuals are responsible for shedding a disproportionately large number of infectious particles into the environment (supershedding). For example, a small number of patients in a hospital ward who received inadequate treatment for multi-drug resistant TB were responsible for 90% of transmission to a sentinel animal host [46]. Pathogen strain may also influence supershedding; particular strains of *E. coli* have been shown to induce supershedding in cattle, leading to many more secondary cases than other strains [47]. Shedding in cattle infected with *M. bovis* can be intermittent, with estimates indicating that only 9–19% of animals shed in nasal or tracheal secretions [48], which might explain some of the observed variation in outbreak sizes. Variation among animals in infectious dose received (potentially as a result of supershedding) may also have influenced our measures of virulence and pathology. Cattle experimentally infected with *M. bovis* showed patterns of variation in the size and distribution of lesions that were associated with infectious dose [49]. However infectious doses in natural infection events are likely to be less variable than within these studies (where challenge doses ranged up to 1×10^7 colony forming units in some cases [50]) and so we would not expect this effect to be pronounced in our dataset.

Virulence

Breakdowns associated with some pathogen genotypes had a greater proportion of lesioned reactors than others, indicating that there are subtle differences in the virulence of genotypes, an effect observed in one previous field study of *M. bovis* [16]. Variation in virulence among strains of *M. tuberculosis* has been found in experimental settings on multiple occasions [51,52]. In one such study, strains that were commonly found in large clusters of cases were compared with those that found singly. Clustered strains were found to have a more virulent phenotype, invading human macrophages *in vitro* more rapidly than non-clustered strains and inducing different cytokine responses, giving clues as to the mechanism of invasion [15]. *M. bovis* strains have also been shown to vary in virulence, eliciting strain specific patterns of immune response in mice [53]. A potential line for future investigation would be to compare the genotypes which we identified as differing significantly in virulence (e.g. 9.273 and 19.140), perhaps using transcriptomic techniques to elucidate the mechanisms driving the observed variation [54]. Alternatively, the genes responsible might be identified by mapping virulence traits onto a phylogeny of genotypes. Given the close relationships among genotypes in Northern Ireland, whole genome sequencing of *M. bovis* isolates might be required to construct such a phylogeny [55].

A potential drawback to our choice of virulence measure (i.e. the proportion of reactors lesioned) is that it might have been influenced by variation in detectability of genotypes to the skin test. If certain genotypes were less detectable to the skin test then recently infected animals (probably not lesioned) might remain undetected, decreasing the total number of reactors and thus increasing the observed proportion lesioned. However we consider this possibility unlikely because in a parallel study we found no systematic differences among genotypes using two different measures of skin test detectability [56].

Laboratory studies indicate that more virulent strains of *M. tuberculosis* are also more transmissible than less virulent strains and tend to form larger outbreaks within human populations [57,58]. In contrast, we found evidence of variation among *M. bovis* genotypes in virulence but not in transmissibility. A possible explanation is that the current test and slaughter programme is sufficient to prevent differences in virulence being manifested in terms of outbreak size, with even the most virulent strains being detected prior to large-scale onward spread. Indeed, the majority of cases show limited pathology (1–2 lesions detected) indicating relatively recent infection and cases of generalised bTB (systemic infection with lesions in organs not connected to the respiratory or alimentary tract [50]) are rare in Northern Ireland.

Lesion Distribution

We found variation in pathology induced by different genotypes of *M. bovis*, with subtle differences in lesion location. A study of *M. tuberculosis* infection in humans linked differences in pathology to strain virulence by comparing isolates taken from a group of patients simultaneously infected with two strains, one disseminated and the other localised. Disseminated strains were found to have greater virulence in laboratory assays than localised strains [59]. We found no such relationship (no correlation between proportion of reactors lesioned and proportion of cases lesioned in lower respiratory tract) perhaps because our measure of lesion distribution was derived from cases that were in relatively early stages of infection where only a single lesion was found. The variation in lesion distribution that we observed was therefore more likely to have been linked to processes determining establishment of the initial infection rather than factors governing disease progression, and the former warrants further investigation.

The observed differences in infection site among cattle breeds may be related to variation in animal husbandry. Lesion sites differed in a comparison of groups of cattle kept indoors in Ethiopia with those kept outdoors; animals under intensive management indoors were more likely to be lesioned in the lower respiratory tract, indicating the respiratory route of infection, than those kept outdoors. Animals kept indoors were also more likely to have more severe pathology and an increased risk of acquiring infection [20,60]. We found a similar pattern with Friesian and Holstein cattle (primarily dairy breeds) in particular showing evidence of a greater degree of infection through the respiratory route (a higher proportion of cases lesioned in the lower respiratory tract) than most of the beef breeds, although the differences that we observed among breeds were relatively small. We also found differences between Holstein-Friesian cattle and beef breeds in the proportion of reactors lesioned, with a greater proportion of reactors belonging to beef breeds having visible lesions. A potentially informative area of future research would be to use measures of pathology to investigate links between husbandry (including stocking density) and the risk of infection by the different routes.

Besides differences in husbandry, the observed variation in lesion distribution may also indicate genetic differences in TB susceptibility among breeds. European *Bos taurus* cattle breeds in Ethiopia have been shown to be more susceptible than native *Bos indicus* cattle [61]. There is also evidence of heritable variation in TB susceptibility within breeds in Irish cattle [62] and so our findings are consistent with the view that host genetic variation influences the outcome of exposure to *M. bovis* and that knowledge of this variation may have a role in future disease control programmes [63–65].

The limited genetic diversity of the *M. bovis* population in Northern Ireland may explain the relatively subtle differences that we found in virulence and lesion distribution. Genotypes of *M. bovis* in the UK and Ireland belong almost exclusively to the EU1 clonal complex which has much less diversity at the spoligotype level than the population in continental Europe, where EU1 is relatively rare [66]. Reduced diversity in Great Britain has been attributed to a series of population bottlenecks, the most recent being the introduction of a comprehensive 'test and slaughter' control programme [67]. Diversity is further restricted in Northern Ireland, where 96% of isolates belong to the dominant spoligotype (SB0140) or its derivations [13].

Conclusions

Using a combination of genotyping and epidemiological data we investigated the associations between *M. bovis* genotypes and patterns of outbreak sizes, virulence and pathology in naturally occurring cases across Northern Ireland during a nine year period. Some genotypes were associated with a greater proportion of lesioned cases indicating that genotypes differ in virulence.

However, we found no evidence for systematic variation in breakdown sizes among genotypes, perhaps indicating that the programme of annual skin testing and abattoir surveillance is successfully preventing more virulent and transmissible genotypes from causing large outbreaks. Cases infected with different genotypes also varied in the distribution of lesions and there was variation in lesion distribution among cattle breeds, perhaps indicative of different disease susceptibility and transmission routes in beef and dairy cattle, traits which with further investigation might be exploited to aid disease control.

Acknowledgments

We thank the administrators of APHIS for access to animal level data. We thank the Statutory TB laboratory (AFBI) for specialist mycobacterial culture.

Author Contributions

Conceived and designed the experiments: ARA SWJM RAS JAW EJG SCB. Performed the experiments: TRM ARA RAS SWJM. Analyzed the data: DMW ARA. Wrote the paper: DMW MLB SCB RAS.

References

1. Palanisamy GS, DuTeau N, Eisenach KD, Cave DM, Theus SA, et al. (2009) Clinical strains of *Mycobacterium tuberculosis* display a wide range of virulence in guinea pigs. Tuberculosis 89: 203–209. doi: 10.1016/j.tube.2009.01.005.
2. Manca C, Tsenova L, Bergtold A, Freeman S, Tovey M, et al. (2001) Virulence of a *Mycobacterium tuberculosis* clinical isolate in mice is determined by failure to induce Th1 type immunity and is associated with induction of IFN-α/β. Proceedings of the National Academy of Sciences 98: 5752–5757. doi: 10.1073/pnas.091096998.
3. Valway SE, Sanchez MPC, Shinnick TF, Orme I, Agerton T, et al. (1998) An outbreak involving extensive transmission of a virulent strain of *Mycobacterium tuberculosis*. New England Journal of Medicine 338: 633–639.
4. Li W, Raoult D, Fournier P (2009) Bacterial strain typing in the genomic era. FEMS Microbiology Reviews 33: 892–916. doi: 10.1111/j.1574-6976.2009.00182.x.
5. Gagneux S, DeRiemer K, Van T, Kato-Maeda M, de Jong BC, et al. (2006) Variable host–pathogen compatibility in *Mycobacterium tuberculosis*. Proceedings of the National Academy of Sciences of the United States of America 103: 2869–2873. doi: 10.1073/pnas.0511240103.
6. Click ES, Moonan PK, Winston CA, Cowan LS, Oeltmann JE (2012) Relationship between *Mycobacterium tuberculosis* phylogenetic lineage and clinical site of tuberculosis. Clin Infect Dis 54: 211–219. doi: 10.1093/cid/cir788.
7. Krishnan N, Malaga W, Constant P, Caws M, Thi Hoang Chau T, et al. (2011) *Mycobacterium tuberculosis* lineage influences innate immune response and virulence and is associated with distinct cell envelope lipid profiles. PLoS ONE 6: e23870. doi: 10.1371/journal.pone.0023870.
8. Boukary AR, Thys E, Rigouts L, Matthys F, Berkvens D, et al. (2012) Risk factors associated with bovine tuberculosis and molecular characterization of *Mycobacterium bovis* strains in urban settings in Niger. Transboundary and Emerging Diseases 59: 490–502. doi: 10.1111/j.1865-1682.2011.01302.x.
9. Humblet M, Boschiroli ML, Saegerman C (2009) Classification of worldwide bovine tuberculosis risk factors in cattle: a stratified approach. Veterinary Research 40: 50. doi: 10.1051/vetres/2009033.
10. Abernethy DA, Upton P, Higgins IM, McGrath G, Goodchild AV, et al. (2013) Bovine tuberculosis trends in the UK and the Republic of Ireland, 1995–2010. Veterinary Record. doi: 10.1136/vr.100969.
11. Defra (2011) Bovine TB eradication programme for England.. Available: http://www.defra.gov.uk/publications/files/pb13601-bovinetb-eradication-programme-110719.pdf. via the Internet.
12. Smith NH, Dale J, Inwald J, Palmer S, Gordon SV, et al. (2003) The population structure of *Mycobacterium bovis* in Great Britain: clonal expansion. Proceedings of the National Academy of Sciences 100: 15271–15275. doi: 10.1073/pnas.2036554100.
13. Skuce RA, Mallon TR, McCormick CM, McBride SH, Clarke G, et al. (2010) *Mycobacterium bovis* genotypes in Northern Ireland: herd-level surveillance (2003 to 2008). Veterinary Record 167: 684–689. doi: 10.1136/vr.c5108.
14. Sreevatsan S, Pan XI, Stockbauer KE, Connell ND, Kreiswirth BN, et al. (1997) Restricted structural gene polymorphism in the *Mycobacterium tuberculosis* complex indicates evolutionarily recent global dissemination. Proceedings of the National Academy of Sciences 94: 9869–9874.
15. Theus SA, Cave MD, Eisenach KD (2005) Intracellular macrophage growth rates and cytokine profiles of *Mycobacterium tuberculosis* strains with different transmission dynamics. Journal of Infectious Diseases 191: 453–460. doi: 10.1086/425936.
16. Goodchild AV, de la Rua-Domenech R, Palmer S, Dale J, Gordon SV, et al. (2003) Association between molecular type and the epidemiological features of *Mycobacterium bovis* in cattle. Proceedings of the Annual Conference of the Society for Veterinary Epidemiology and Public Health: 45–59.
17. Cassidy JP (2006) The pathogenesis and pathology of bovine tuberculosis with insights from studies of tuberculosis in humans and laboratory animal models. Veterinary microbiology 112: 151–161.
18. Neill SD, Skuce RA, Pollock JM (2005) Tuberculosis – new light from an old window. Journal of Applied Microbiology 98: 1261–1269. doi: 10.1111/j.1365-2672.2005.02599.x.
19. Liebana E, Johnson L, Gough J, Durr P, Jahans K, et al. (2008) Pathology of naturally occurring bovine tuberculosis in England and Wales. The Veterinary Journal 176: 354–360.
20. Ameni G, Aseffa A, Engers H, Young D, Hewinson G, et al. (2006) Cattle husbandry in Ethiopia is a predominant factor affecting the pathology of bovine tuberculosis and gamma interferon responses to Mycobacterial antigens. Clin. Vaccine Immunol. 13: 1030–1036. doi: 10.1128/cvi.00134-06.
21. Abernethy DA, Denny GO, Menzies FD, McGuckian P, Honhold N, et al. (2006) The Northern Ireland programme for the control and eradication of *Mycobacterium bovis*. Vet Microbiol 112: 231–237. doi: 10.1016/j.vetmic.2005.11.023.
22. Skuce RA, McDowell SW, Mallon TR, Luke B, Breadon EL, et al. (2005) Discrimination of isolates of *Mycobacterium bovis* in Northern Ireland on the basis of variable numbers of tandem repeats (VNTRs). The Veterinary Record 157: 501–504.
23. Houston R (2001) A computerised database system for bovine traceability. Rev - Off Int Epizoot 20: 652–661.
24. Bessell PR, Orton R, White PCL, Hutchings MR, Kao RR (2012) Risk factors for bovine tuberculosis at the national level in Great Britain. BMC Veterinary Research 8: 1. doi: 10.1186/1746-6148-8-51.
25. Carrique-Mas JJ, Medley GF, Green LE (2008) Risks for bovine tuberculosis in British cattle farms restocked after the foot and mouth disease epidemic of 2001. Preventative Veterinary Medicine 84: 85–93. doi: 10.1016/j.prevetmed.2007.11.001.
26. Wolfe DM, Berke O, Kelton DF, White PW, More SJ, et al. (2010) From explanation to prediction: a model for recurrent bovine tuberculosis in Irish cattle herds. Preventative Veterinary Medicine 94: 170–177. doi:10.1016/j.prevetmed.2010.02.010.
27. Brooks-Pollock E, Keeling M (2009) Herd size and bovine tuberculosis persistence in cattle farms in Great Britain. Preventative Veterinary Medicine 92: 360–365. doi: 10.1016/j.prevetmed.2009.08.022.
28. Elston DA, Moss R, Boulinier T, Arrowsmith C, Lambin X (2001) Analysis of aggregation, a worked example: numbers of ticks on red grouse chicks. Parasitology 122: 563–569.
29. Bates D, Maechler M, Bolker B (2012) lme4: Linear mixed-effects models using S4 classes.
30. R Development Core Team (2011) R: A language and environment for statistical computing. Vienna, Austria: R Foundation for Statistical Computing.
31. Burnham KP, Anderson DR (2002) Model selection and multimodel inference: A practical information-theoretic approach: Springer, New York.
32. Frankena K, White PW, O'Keeffe J, Costello E, Martin SW, et al. (2007) Quantification of the relative efficiency of factory surveillance in the disclosure of tuberculosis lesions in attested Irish cattle. Veterinary Record 161: 679–684. doi: 10.1136/vr.161.20.679.
33. Olea-Popelka F, Freeman Z, White P, Costello E, O'Keeffe J, et al. (2012) Relative effectiveness of Irish factories in the surveillance of slaughtered cattle for visible lesions of tuberculosis, 2005–2007. Irish Veterinary Journal 65: 2. doi: 10.1186/2046-0481-65-2.
34. Borrell S, Gagneux S (2009) Infectiousness, reproductive fitness and evolution of drug-resistant *Mycobacterium tuberculosis*. The International Journal of Tuberculosis and Lung Disease 13: 1456–1466.
35. Rhee JT, Piatek AS, Small PM, Harris LM, Chaparro SV, et al. (1999) Molecular epidemiologic evaluation of transmissibility and virulence of *Mycobacterium tuberculosis*. Journal of clinical microbiology 37: 1764–1770.
36. Glynn JR (2008) Determinants of cluster size in large, population-based molecular epidemiology study of tuberculosis, Northern Malawi. Emerging Infectious Diseases 14: 1060–1066. doi: 10.3201/eid1407.060468.
37. Alvarez J, Perez A, Bezos J, Casal C, Romero B, et al. (2012) Eradication of bovine tuberculosis at a herd-level in Madrid, Spain: study of within-herd transmission dynamics over a 12 year period. BMC Veterinary Research 8: 100. doi: 10.1186/1746-6148-8-100.
38. Brooks-Pollock E, Cohen T, Murray M (2010) The impact of realistic age structure in simple models of tuberculosis transmission. PLoS ONE 5: e8479. doi: 10.1371/journal.pone.0008479.
39. Cohen T, Colijn C, Finklea B, Murray M (2007) Exogenous re-infection and the dynamics of tuberculosis epidemics: local effects in a network model of transmission. Journal of The Royal Society Interface 4: 523–531. doi: 10.1098/rsif.2006.0193.

40. Murray M (2002) Determinants of cluster distribution in the molecular epidemiology of tuberculosis. Proceedings of the National Academy of Sciences 99: 1538–1543. doi: 10.1073/pnas.022618299.
41. Porco TC, Small PM, Blower SM (2001) Amplification dynamics: predicting the effect of HIV on tuberculosis outbreaks. JAIDS Journal of Acquired Immune Deficiency Syndromes 28: 437–444.
42. Claridge J, Diggle P, McCann CM, Mulcahy G, Flynn R, et al. (2012) *Fasciola hepatica* is associated with the failure to detect bovine tuberculosis in dairy cattle. Nature Communications 3: 853. doi: 10.1038/ncomms1840.
43. Garske T, Rhodes CJ (2008) The effect of superspreading on epidemic outbreak size distributions. Journal of Theoretical Biology 253: 228–237. doi: 10.1016/j.jtbi.2008.02.038.
44. Lloyd-Smith JO, Schreiber SJ, Getz WM (2006) Moving beyond averages: Individual-level variation in disease transmission. Contemporary Mathematics 410: 235.
45. James A, Pitchford JW, Plank MJ (2007) An event-based model of superspreading in epidemics. Proceedings of the Royal Society B: Biological Sciences 274: 741–747. doi: 10.1098/rspb.2006.0219.
46. Escombe AR, Moore DAJ, Gilman RH, Pan W, Navincopa M, et al. (2008) The infectiousness of tuberculosis patients coinfected with HIV. PLoS medicine 5: e188. doi: 10.1371/journal.pmed.0050188.
47. Matthews L, Reeve R, Woolhouse MEJ, Chase-Topping M, Mellor DJ, et al. (2009) Exploiting strain diversity to expose transmission heterogeneities and predict the impact of targeting supershedding. Epidemics 1: 221–229. doi: 10.1016/j.epidem.2009.10.002.
48. Palmer MV, Waters W (2006) Advances in bovine tuberculosis diagnosis and pathogenesis: What policy makers need to know. Veterinary microbiology 112: 181–190. doi: 10.1016/j.vetmic.2005.11.028.
49. Pollock JM, Neill SD (2002) *Mycobacterium bovis* infection and tuberculosis in cattle. The Veterinary Journal 163: 115–127. doi: 10.1053/tvjl.2001.0655.
50. Neill SD, Bryson DG, Pollock JM (2001) Pathogenesis of tuberculosis in cattle. Tuberculosis 81: 79–86. doi: 10.1054/tube.2000.0279.
51. Caceres N, Llopis I, Marzo E, Prats C, Vilaplana C, et al. (2012) Low dose aerosol fitness at the innate phase of murine infection better predicts virulence amongst clinical strains of *Mycobacterium tuberculosis*. 7: e29010. doi: 10.1371/journal.pone.0029010.
52. Gagneux S, Small PM (2007) Global phylogeography of *Mycobacterium tuberculosis* and implications for tuberculosis product development. The Lancet Infectious Diseases 7: 328–337. doi: 10.1016/s1473-3099(07)70108-1.
53. Aguilar León D, Zumárraga MJ, Jiménez Oropeza R, Gioffré AK, Bernardelli A, et al. (2009) *Mycobacterium bovis* with different genotypes and from different hosts induce dissimilar immunopathological lesions in a mouse model of tuberculosis. Clinical & Experimental Immunology 157: 139–147. doi: 10.1111/j.1365-2249.2009.03923.x.
54. Blanco FC, Nunez-García J, García-Pelayo C, Soria M, Bianco MV, et al. (2009) Differential transcriptome profiles of attenuated and hypervirulent strains of *Mycobacterium bovis*. Microbes and Infection 11: 956–963. doi: 10.1016/j.micinf.2009.06.006.
55. Biek R, O'Hare A, Wright D, Mallon T, McCormick C, et al. (2012) Whole genome sequencing reveals local transmission patterns of *Mycobacterium bovis* in sympatric cattle and badger populations. PLoS Pathog 8: e1003008. doi: 10.1371/journal.ppat.1003008.
56. Wright DM, Allen AR, Mallon TR, McDowell SWJ, Bishop SC, et al. (2013) Detectability of bovine TB using the tuberculin skin test does not vary significantly according to pathogen genotype within Northern Ireland. Infection, Genetics and Evolution 19: 15–22. doi: 10.1016/j.meegid.2013.05.011.
57. Aguilar LD, Hanekom M, Mata D, Gey van Pittius NC, van Helden PD, et al. (2010) *Mycobacterium tuberculosis* strains with the Beijing genotype demonstrate variability in virulence associated with transmission. Tuberculosis 90: 319–325. doi: 10.1016/j.tube.2010.08.004.
58. Hernández-Pando R, Marquina-Castillo B, Barrios-Payán J, Mata-Espinosa D (2012) Use of mouse models to study the variability in virulence associated with specific genotypic lineages of *Mycobacterium tuberculosis*. Infection, Genetics and Evolution 12: 725–731. doi: 10.1016/j.meegid.2012.02.013.
59. Garcia de Viedma D, Lorenzo G, Cardona P, Rodriguez NA, Gordillo S, et al. (2005) Association between the infectivity of *Mycobacterium tuberculosis* strains and their efficiency for extrarespiratory infection. J Infect Dis 192: 2059–2065. doi: 10.1086/498245.
60. Biffa D, Bogale A, Godfroid J, Skjerve E (2012) Factors associated with severity of bovine tuberculosis in Ethiopian cattle. Tropical Animal Health and Production 44: 991–998. doi: 10.1007/s11250-011-0031-y.
61. Vordermeier M, Ameni G, Berg S, Bishop R, Robertson BD, et al. (2012) The influence of cattle breed on susceptibility to bovine tuberculosis in Ethiopia. Comparative Immunology, Microbiology and Infectious Diseases 35: 227–232. doi: 10.1016/j.cimid.2012.01.003.
62. Bermingham ML, More SJ, Good M, Cromie AR, Higgins IM, et al. (2009) Genetics of tuberculosis in Irish Holstein-Friesian dairy herds. Journal of Dairy Science 92: 3447–3456. doi: 10.3168/jds.2008-1848.
63. Allen AR, Minozzi G, Glass EJ, Skuce RA, McDowell SWJ, et al. (2010) Bovine tuberculosis: the genetic basis of host susceptibility. Proceedings of the Royal Society B: Biological Sciences 277: 2737–2745. doi: 10.1098/rspb.2010.0830.
64. Driscoll EE, Hoffman JI, Green LE, Medley GF, Amos W (2011) A preliminary study of genetic factors that influence susceptibility to bovine tuberculosis in the British cattle herd. PLoS ONE 6: e18806. doi: 10.1371/journal.pone.0018806.
65. Amos W, Brooks-Pollock E, Blackwell R, Driscoll E, Nelson-Flower M, et al. (2013) Genetic predisposition to pass the standard SICCT test for bovine tuberculosis in British cattle. PLoS ONE 8: e58245. doi: 10.1371/journal.-pone.0058245.
66. Smith NH, Berg S, Dale J, Allen A, Rodriguez S, et al. (2011) European 1: A globally important clonal complex of *Mycobacterium bovis*. Infection, Genetics and Evolution 11: 1340–1351. doi: 10.1016/j.meegid.2011.04.027.
67. Smith NH, Gordon SV, de la Rua-Domenech R, Clifton-Hadley RS, Hewinson RG (2006) Bottlenecks and broomsticks: the molecular evolution of *Mycobacterium bovis*. Nat Rev Micro 4: 670–681. doi: 10.1038/nrmicro1472.

Allometry of Sexual Size Dimorphism in Domestic Dog

Daniel Frynta[1], Jana Baudyšová[1], Petra Hradcová[1], Kateřina Faltusová[1], Lukáš Kratochvíl[2]*

1 Department of Zoology, Faculty of Science, Charles University in Prague, Prague, Czech Republic, 2 Department of Ecology, Faculty of Science, Charles University in Prague, Prague, Czech Republic

Abstract

Background: The tendency for male-larger sexual size dimorphism (SSD) to scale with body size – a pattern termed Rensch's rule – has been empirically supported in many animal lineages. Nevertheless, its theoretical elucidation is a subject of debate. Here, we exploited the extreme morphological variability of domestic dog (*Canis familiaris*) to gain insights into evolutionary causes of this rule.

Methodology/Principal Findings: We studied SSD and its allometry among 74 breeds ranging in height from less than 19 cm in Chihuahua to about 84 cm in Irish wolfhound. In total, the dataset included 6,221 individuals. We demonstrate that most dog breeds are male-larger, and SSD in large breeds is comparable to SSD of their wolf ancestor. Among breeds, SSD becomes smaller with decreasing body size. The smallest breeds are nearly monomorphic.

Conclusions/Significance: SSD among dog breeds follows the pattern consistent with Rensch's rule. The variability of body size and corresponding changes in SSD among breeds of a domestic animal shaped by artificial selection can help to better understand processes leading to emergence of Rensch's rule.

Editor: Vincent Laudet, Ecole Normale Supérieure de Lyon, France

Funding: The project was supported by the Grant Agency of the Czech Academy of Sciences (project No. IAA6111410). The personal costs of K.F. were provided by the Czech Science Foundation (project No. 206-05-H012). L.K. was supported by the Czech Science Foundation (project No. 206/09/0895). The funders had no role in study design, data collection and analysis, decision to publish, or preparation of the manuscript.

Competing Interests: The authors have declared that no competing interests exist.

* E-mail: lukas.kratochvil@natur.cuni.cz

Introduction

Animal species that have undergone domestication processes usually exhibit rapid phenotypic change [1–4]. The extraordinary ability of domestic species to radiate into numerous morphologically and behaviorally distinct breeds within a few generations is explained by episodes of strong artificial selection [1,5]. Domestic dog (*Canis familiaris*) is the most morphologically variable mammalian species [6]. Although dog is a product of multiple domestication events [7], its ancestors belonged to a single, morphologically rather uniform species, the wolf (*Canis lupus*). Differences in size are especially apparent among dog breeds, ranging from about two kilograms in the miniature Chihuahua to about 100 kilograms in the giant mastiff. This range encompasses the range of body mass reported for all other canids: from 1.3 kilograms in the fennec fox (*Fennecus zerda*) to 36.3 kilograms in the wolf [8]. Size differences, including those concerning differences between males and females, i.e., sexual size dimorphism (SSD), have many direct and indirect consequences [9,10]. Exceptional magnitude of size variation and emerging knowledge on proximate mechanisms of body size changes [11–17] predetermines domestic dog to be a proper model for the assessment of allometric relationships.

In this study, we focus on allometry in SSD. Sexes differ in body size in many animal lineages [18,19], but the relationship between female and male body size increase is usually not isometric. Males in larger species within an animal lineage tend to be larger relative to females than are males in smaller species. This empirically documented trend is currently referred to as Rensch's rule ([20], cf. [21,22]). In recent years, this rule has attracted considerable research effort, and conforming patterns have been reported by interspecific comparisons in various animal taxa [20,23–29], especially [30] or exclusively [31,32] in taxa exhibiting male-larger SSD. The pattern consistent with Rensch's rule also has been demonstrated at an intraspecific level [33–36]. Several hypotheses have been formulated to explain Rensch's rule. Most popular has probably been the sexual selection hypothesis, being considered a general explanation for SSD allometry [27,36,37]. The sexual selection hypothesis suggests that Rensch's rule is driven by a correlated evolutionary change in female body size to directional sexual selection on increased body size in males.

Modern breeds of the domestic dog present a unique opportunity to test this hypothesis in the unusual situation where we know the selective agent responsible for size variation (artificial selection). The sexual selection hypothesis thus predicts that the pattern of SSD among dog breeds should not follow Rensch's rule. The aim of this paper is to examine the allometry of SSD and to test Rensch's rule among breeds of the domestic dog.

Results

The Lovich-Gibbons ratios computed from the FCI standards [38] range from 1.00 up to 1.17 (mean = 1.05), showing male-larger SSD in most breeds. SSD was positively correlated with female size irrespective of the trait used for its calculation (Spearman correlation coefficients $r_s = 0.29$, $p < .0001$, $N = 311$

for shoulder height, and $r_s = 0.46$, $p < .0001$, $N = 159$ for body mass).

In some breeds, FCI standards do not provide different values for stud dogs and bitches, and these theoretically monomorphic breeds tend to be smaller. It cannot be excluded that the absence of different size standards for males and females reflect just oversimplification during the process of standard formation. Therefore, we excluded these theoretically monomorphic breeds and computed the correlation coefficients again. The relationships did not remain significant after exclusion of the breeds with monomorphic standards ($r_s = -0.05$, $p < .48$, $N = 231$ for shoulder height, and $r_s = 0.16$, $p = .14$, $N = 85$ for body mass).

Next, we compared shoulder height of males and females based on our original data (Table S1). Mean shoulder height in males was larger than in females in 73 out of 74 breeds. ANOVA revealed that the variability in shoulder height can be explained by breed ($F_{73,6073} = 5{,}449$, $p < .0001$), sex ($F_{1,6073} = 1{,}867$, $p < .0001$), and their interaction ($F_{73,6073} = 4$, $p < .0001$). The Lovich-Gibbons ratios for shoulder height (mean = 1.067, median = 1.071) ranged from 0.995 in caniche toy to 1.099 in Slovakian hound. Size differences between the sexes were statistically significant (t-tests, $p < 0.05$) in 69 breeds. Twenty-four breeds monomorphic according to the FCI standards appeared significantly dimorphic according to the original data. The sexual differences in shoulder height were not statistically significant in five extremely small breeds (Chihuahua, Prague ratter, papillon, miniature spitz, caniche toy). Moreover, t-tests failed to prove significant SSD for body mass in three of them (Chihuahua, Prague ratter, papillon) where sample size was larger than 17 individuals of each sex.

Lovich-Gibbons ratios expressing SSD in shoulder height correlate with female shoulder height among 74 breeds of the domestic dog ($r = 0.47$; $p < .0001$; Figure 1) proving conformity with Rensch's rule among dog breeds. An alternative, but mathematically equivalent computation confirmed this trend. We found a clear linear relationship between log-transformed mean female shoulder height against log-transformed mean male shoulder height (Figure 2). The RMA slope of the line was 0.971 (95% confidence interval 0.956–0.982), which significantly deviates from the slope 1.0 expecting under isometry. The significant deviation from isometry remained unchanged when breeds represented by sample sizes smaller than 20, 30, and 50 individuals in at least one sex were excluded from the analyses; the corresponding values of the allometric slopes were 0.965, 0.956 and 0.961, respectively.

Discussion

Our analysis confirms the pattern of SSD conforming to Rensch's rule among modern breeds of domestic dog. We were able to demonstrate significant allometry in SSD among breeds in our original dataset (Table S1). The results for data obtained from the FCI standards [38] confirmed Rensch's rule among breeds, as well, but many breeds treated in the standards as monomorphic proved to be in fact dimorphic based upon the real data. Our findings contradict the conclusions by Sutter *et al.* [39] that the breed standards rigorously adhered to morphological variation for domestic dog breeds. Sutter *et al.* [39] also concluded that the proportional size difference between males and females of small and large breeds in shoulder height is more or less the same, i.e., that breeds do not follow Rensch's rule (cf. to our Figures 1,2). The discrepancies between the studies can be explained by the different rules for including a particular breed into the respective dataset and different total number of studied breeds. Sutter *et al.* [39] included data for 53 breeds for which at least three males and three females were measured, while we analyzed data for 74 breeds with at least 10 males and 10 females measured.

Comparative analyses have long recognized the need to control for the non-independence of data that arises through patterns of shared common ancestry (phylogenetic non-independence). Ideally, the phylogenetic relationships among breeds should be taken into account to control for this effect in our test of Rensch's rule. Contemporary pedigree breeds have relatively short histories. Most of them were officially established in the 19th or even 20th century, and separation of their gene pools was thus completed only recently [40]. Consequently, mitochondrial [7,41–45] and Y chromosome [46] haplotypes are distributed rather erratically among modern breeds. Microsatellite markers provide better resolution and allow genetic characterization of individual breeds [47], however, neither they can recover any reliable phylogenetic structure among the vast majority of modern breeds originating in Europe [40,48]. Recently, an extensive SNP dataset allowed defining main genetic clusters of these breeds that fairly correspond to phenotypic/functional groups of breeds [49]. Nevertheless, these results are not in contradiction with earlier studies suggesting that the phylogenetic component represents only a small portion of genetic variation among modern dog breeds. Therefore, we assume that phylogenetic non-independence among dog breeds does not largely bias our conclusions.

The pattern consistent with Rensch's rule, i.e. larger evolutionary plasticity in body size in males than in females, among domestic dog breeds has emerged as a result of rapid radiation in size during domestication processes and subsequent formation of breeds. Interestingly, the domestication itself had no marked effect on the magnitude of SSD in domestic dog. Wolf, the direct ancestor of domestic dogs, is the largest species of the family Canidae and at the same time belongs to the most sexually dimorphic canids with male to female ratio in body mass about 1.28 [8]. This value fits well with male-to-female body mass ratios in dog breeds of comparable body size, such as hovawart (1.26), beauceron (1.25), giant schnauzer (1.22), or leonberger (1.21). Because the basal clades of canids comprise exclusively species of small, fox-like body size, the lineage leading to the genus *Canis* has been previously subject to substantial evolutionary increase in body size (cf. canid phylogenies in [50,51]). On the other hand, according to the archaeozoological evidence, there was probably a great variability of size in dogs from the very beginning of the domestication process [52,53]. Recent evolutionary trajectories of contemporary breeds probably involve changes in body size in both directions. Nevertheless, as the largest breeds of domestic dogs (e.g., great dane, Irish wolfhound) are only slightly larger than the largest subspecies of the wolf [54], the predominant direction in the evolution of body size in domestic dog was reduction rather than enlargement of the body size.

Thus, it seems that the pattern of SSD consistent with Rensch's rule among dog breeds evolved due to more pronounced size reduction in males in comparison to females. One can thus argue that the SSD allometry can be attributed to stronger selection on male than female size, and indeed, it seems that male dogs rather than bitches were the primary target of intentional artificial selection (cf. genetic evidence in [55]). Nevertheless, assuming that the size is primarily controlled by loci without sex-biased expression pattern, the genetic correlation between male and female body size should quickly eliminate the size difference caused by the higher selection pressure on a single sex [56]. Thus, it seems that Rensch's rule among domestic dog breeds can be rather attributed to more constrained female than male body size, especially in small breeds. There is some evidence that the presence of functional constraints represented mainly by size of an

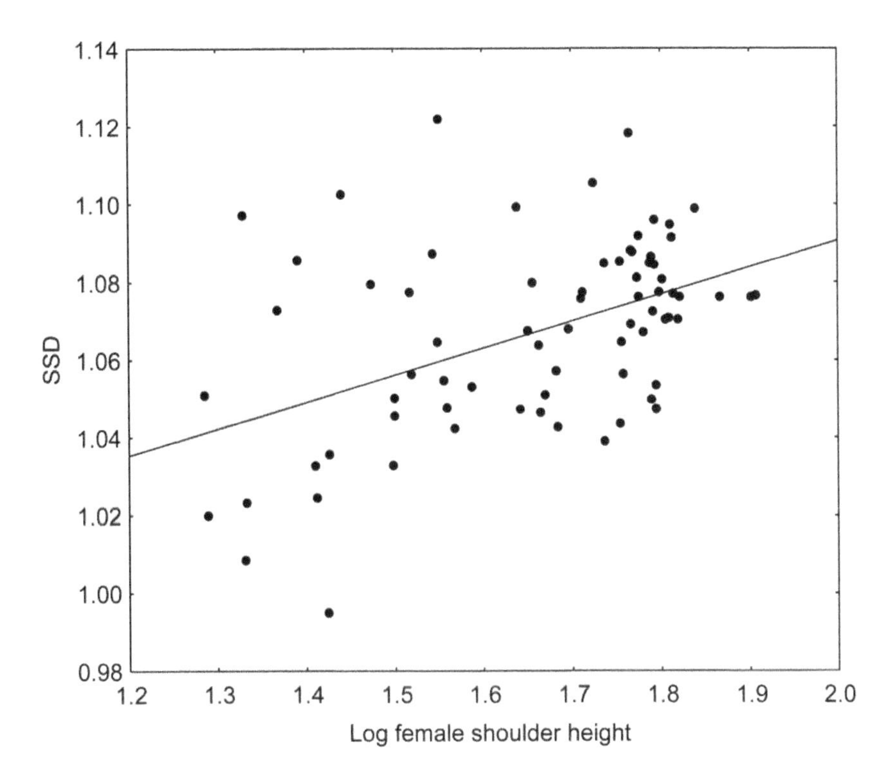

Figure 1. SSD increases with body size among dog breeds. Lovich-Gibbons ratios expressing SSD in shoulder height correlate with female shoulder height among 74 breeds of the domestic dog ($r = 0.47$; $p < .0001$).

individual neonate limits further miniaturization of female size in small breeds. In large breeds such as hovawart, for example, the neonate mass relatively to female mass (1.45%) is about 75% smaller and the relative litter mass (12.2%) about 33% smaller than in Chihuahua (J. Baudyšová & D. Frynta, unpublished data). Moreover, breeders point out that bitches of the smallest breeds frequently suffer from complicated parturition.

The actual processes leading to pattern consistent with Rensch's rule among domestic dog breeds remain unclear. Nevertheless, the hypothesis on the functional constraint imposed by neonate size as a cause of SSD allometry consistent with Rensch's rule obtains support from the comparison between SSD allometry and reproductive allometry of domestic dog breeds and species of wild canids. The Lovich-Gibbons ratios computed from the data provided by Moehlmann and Hofer [8] do not correlate with female body mass (Spearman $r_s = 0.075$, $p = .688$, $N = 31$ species,) among species of wild canids. Wild canids thus do not follow Rensch's rule, and, in accord with the hypothesis, small wild canids have much smaller neonate mass relatively to female body mass than do small dog breeds of the same size category. For example, neonate mass of 123 grams in Chihuahua ($N = 62$ newborns from 20 litters; J. Baudyšová, unpublished data) sharply contrast with 80 grams in *Vulpes corsac* [8]. The neonate mass relative to female body mass is about 1.5 times larger in Chihuahuas (5.9) than in corsac fox (3.9), while values obtained for hovawart (i.e., a breed resembling wolf in body mass) are well comparable to those in wolves [8,57]. Differences in neonate mass scaling between wild canids and domestic dog breeds mirror different scaling in gestation length. Unlike in wild canids, which show a positive relationship between gestation length and body size [8,57], gestation length is nearly the same across all domestic dog breeds irrespective of body mass [11,58]. In view of that, it seems that during their evolution particular species of wild canids were able to adjust reproductive characteristics (gestation length, neonate size) according to female body size, while breeds of domestic dogs still possess values more similar to their relatively recent ancestor. These differences probably affect evolutionary plasticity in female body size and consequently also scaling of SSD. The difference in consistency with Rensch's rule between wild canids and domestic dogs are also concordant with the previously suggested scenario that Rensch's rule should be followed in lineages experiencing miniaturization in body size [24]. The common ancestor of wild canids was probably relatively small and the largest forms possess derived body size [8,50,51]. On the other hand, miniaturization was the predominant trend during formation of recent variability in body size among dog breeds from their wolf ancestor. Nevertheless, we should keep in mind that artificial selection and formation of breeds in domesticated animals is a different process involving for instance different genetic changes than speciation [59], and the comparison between body size radiations among dog breeds and in wild canids should be taken with caution.

In conclusion, we have demonstrated that the SSD pattern of domestic dogs follows Rensch's rule. Recently, SSD patterns consistent with Rensch's rule were demonstrated also in domestic cattle [60], goats and sheep [61], but not in domestic chicken breeds [37]. The evidence of Rensch's rule among breeds of a domestic animal shaped mainly by artificial selection supports a view that sexual selection cannot be considered as a general explanation for Rensch's rule. Rather, that evidence contributes to a more pluralistic view as to its formation. A pattern consistent with Rensch's rule seems to emerge in some clades exhibiting considerable evolutionary changes in body size, no matter whether they were driven by sexual, artificial or other type of selection.

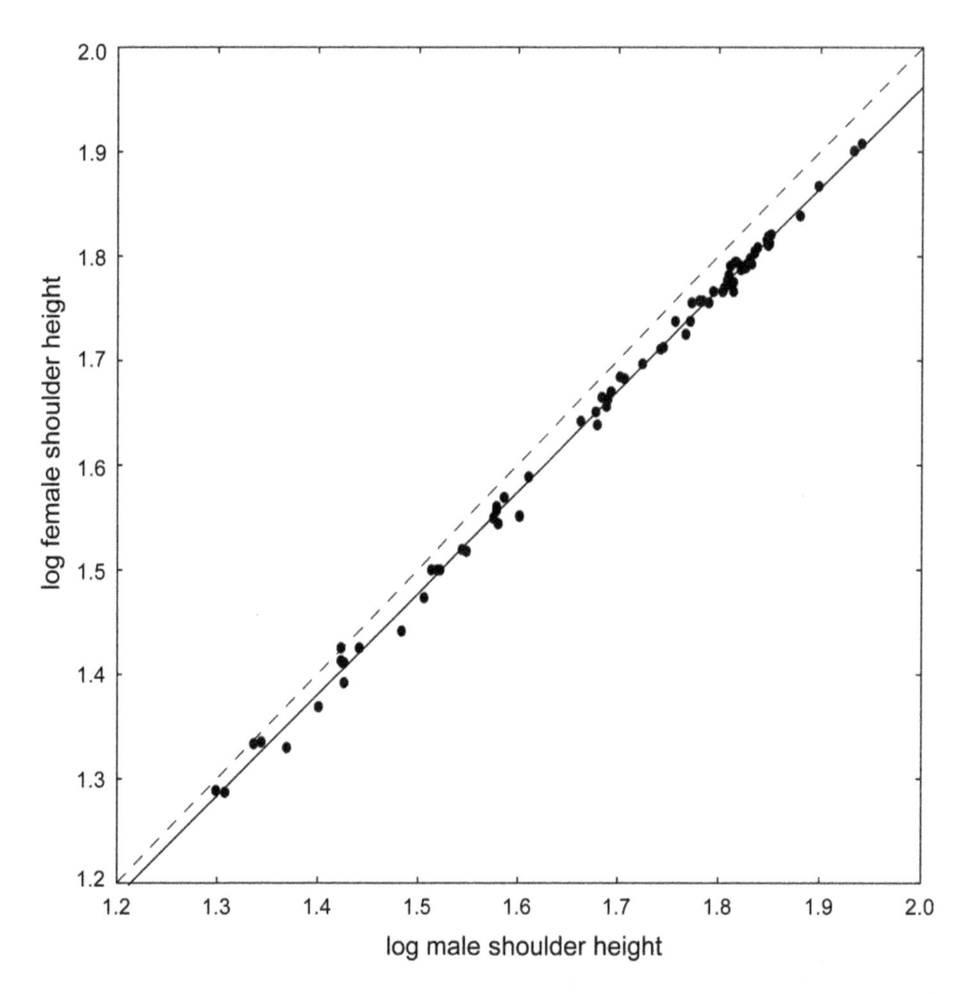

Figure 2. Relationship between mean male and female shoulder heights among 74 domestic dog breeds. Each point represents means computed from at least 10 males and 10 females. The data are naturally log-transformed. The slope of the fitted line is significantly less than 1, thus conforming to Rensch's rule. The broken line illustrates a 1:1 relationship.

Materials and Methods

Male and female shoulder height and body mass were excerpted from the Féderation Cynologique Internationale (FCI) standards for particular breeds [38]. The standards provide data on 311 and 140 breeds for shoulder height and body mass, respectively. When the standards provide ranges instead of means or idealized/typical values, we used midpoints of minimum and maximum values.

Because the standards can be imprecise or biased, we also collected data on shoulder height in 6,221 dogs (2,714 males and 3,507 females) from 74 pedigree breeds. Each breed was represented by at least 10 individuals of each sex. All studied animals were included in studbooks of the Czech and Moravian Cynological Unit. They were measured in adulthood (i.e., at an age exceeding one year). The measurements were taken by the authors and/or authorities of particular breeding clubs (exhibition judges or breed advisors). The shoulder height was selected because (1) it is easily measurable and with high repeatability, (2) it does not depend on body condition, and (3) this measurement is the most correlated to the first principal component computed from a series of external measurements separately in three minutely studied breeds (briard, giant schnauzer; P. Hradcová, unpublished data; hovawart, J. Baudyšová & D. Frynta, unpublished manuscript). In morphometric studies, the first principal component often represents generalized body size. When available, data on body mass were recorded as well. Collection of the measurements was performed in accordance with Czech law implementing all corresponding European Union regulations and were approved by the institutional animal care and use committee (the Czech Ministry of Education, Youth and Sports No. 21384/2011- 30).

The SSD was expressed as the Lovich-Gibbons ratio [62] computed as M/F in male-larger breeds and 2-F/M in female-larger breeds. This ratio assures both linearity and proportional symmetry of SSD index (for details, see [63]). Significance of SSD was tested by t-tests and analysis of variance (ANOVA) with sex and breed as factors. Relationships between SSD and body size were tested by nonparametric or parametric correlations between the Lovich-Gibbons ratios and means of naturally log-transformed expressions of female size. Nonparametric tests were applied in the case of FCI standards, where distribution of SSD index deviated from normality. Means of naturally log-transformed data were further used in the estimation of interbreed allometries of female body size on male body size following [30]. Briefly, we estimated the slope of a regression of log-transformed mean female shoulder height against log-transformed mean male shoulder height. The slope 1.0 is expected under isometric increase of male and female size, while slope < 1.0 represents an increase in male-biased SSD with body size and thus the pattern correspondent with Rensch's rule. For the regression slope estimations, we employed the reduced major axis regression (RMA) model, which accounts for error in both dependent and independent variables [64]. Deviations from isometric relationship were considered significant when the expected isometric slope (1.0) fell outside the 95% confidence interval of the estimated slope. The calculations were performed using STATISTICA, version 6.0 [65].

Supporting Information

Table S1 Descriptive statistics for male and female size in dog breeds.

Acknowledgments

We thank all breeders of dogs measured during the work for their patience, and exhibition judges or breed advisors for sharing their measurements. The manuscript was improved thanks to comments of two anonymous reviewers.

Author Contributions

Conceived and designed the experiments: DF LK. Performed the experiments: KF PH JB. Analyzed the data: DF JB. Wrote the paper: DF JB LK.

References

1. Darwin CR (1859) On the origin of species by means of natural selection, or the preservation of favoured races in the struggle for life. London: John Murray. 432 p.
2. Darwin CR (1868) The variation of animals and plants under domestication. London: John Murray. 2 vols.
3. Herre W (1980) Grundfragen zoologischer Domestikations-forschung. Nova Acta Leopold 241: 1–16.
4. Trut LN (1999) Early canid domestication: The farm-fox experiment. Am Sci 87: 160–169.
5. Sol D (2008) Artificial selection, naturalization, and fitness: Darwin's pigeons revisited. Biol J Linn Soc 93: 657–665.
6. Drake AG, Klingenberg CP (2010) Large-scale diversification of skull shape in domestic dogs: disparity and modularity. Am Nat 175: 289-301.
7. Savolainen P, Leitner T, Wilton AN, Matisoo-Smith E, Lundeberg J (2004) A detailed picture of the origin of the Australian dingo, obtained from the study of mitochondrial DNA. Proc Natl Acad Sci USA 101: 12387–12390.
8. Moehlman PD, Hofer H (1997) Cooperative breeding, reproductive suppression, and body mass in canids. In: Solomon NG, French JA, editors. Cooperative breeding in mammals. Cambridge: Cambridge University Press. 76–127.
9. Schmidt-Nielsen K (1984) Scaling: Why is animal size so important? Cambridge: Cambridge University Press. 241 p.
10. Isaac JL (2005) Potential causes and life-history consequences of sexual size dimorphism in mammals. Mamm Rev 35: 101–115.
11. Wayne RK (1986) Cranial morphology of domestic and wild canids: the influence of development on morphological change. Evolution 40: 243–261.
12. Chase K, Carrier DR, Adler FR, Ostrander EA, Lark KG (2005) Interaction between the X chromosome and an autosome regulates size sexual dimorphism in Portuguese water dogs. Gen Res 15: 1820–1824.
13. Wang W, Ewen F, Kirkness EF (2005) Short interspersed elements (SINEs) are a major source of canine genomic diversity. Gen Res 15: 1798–1808.
14. Sutter NB, Bustamante CD, Chase K, Gray MM, Zhao K, et al. (2007) A single IGF1 allele is a major determinant of small size in dogs. Science 316: 112–115.
15. Gray MM, Sutter NB, Ostrander EA, Wayne RK (2010) The IGF1 small dog haplotype is derived from Middle Eastern grey wolves. BMC Biol 8: 16.
16. Boyko AR, Quignon P, Li L, Schoenebeck JJ, Degenhardt JD, et al. (2010) A simple genetic architecture underlies morphological variation in dogs. PLOS Biology 8: e1000451.
17. Shearin AL, Ostrander EA (2010) Canine morphology: hunting for genes and tracking mutations. PLoS Biol 8: e1000310.
18. Darwin CR (1871) The descent of man, and selection in relation to sex. London: John Murray. 399 p.
19. Andersson M (1994) Sexual selection. Princeton University Press. 336 p.
20. Abouheif E, Fairbairn DJ (1997) A comparative analysis of allometry for sexual size dimorphism: assessing Rensch's rule. Am Nat 149: 540–562.
21. Rensch B (1950) Die Abhangigkeit der relativen Sexualdifferenz von der Körpergröße. Bonn Zool Beitr 1: 58–69.
22. Rensch B (1959) Evolution above the species level. London: Methuen and Co. Ltd. 419 p.
23. Colwell RK (2000) Rensch's rule crosses the line: convergent allometry of sexual size dimorphism in hummingbirds and flower mites. Am Nat 156: 495–510.
24. Kratochvíl L, Frynta D (2002) Body size, male combat and the evolution of sexual dimorphism in eublepharid lizards (Squamata: Eublepharidae). Biol J Linn Soc 76: 303–314.
25. Kratochvíl L, Frynta D (2007) Phylogenetic analysis of sexual dimorphism in eye-lid geckos (Eublepharidae): the effects of male combat, courtship behaviour, egg size, and body size. In: Fairbairn D, Székely T, Blanckenhorn W, editors.Sex, size and gender roles. Oxford University Press. 154–162.
26. Cox RM, Butler MA, John-Alder HB (2007) Sexual size dimorphism in reptiles. In: Fairbairn D, Székely T, Blanckenhorn W, editors. Sex, size and gender roles. Oxford University Press. 38–49.
27. Dale J, Dunn PO, Figuerola J, Lislevand T, Székely T, et al. (2007) Sexual selection explains Rensch's rule of allometry for sexual size dimorphism. Proc R Soc B 274: 2971–2979.
28. Székely T, Lislevand T, Figuerola J (2007) Sexual size dimorphism in birds. In: Fairbairn D, Székely T, Blanckenhorn W, editors. Sex, size and gender roles. Oxford University Press. 27–37.
29. Lindenfors P, Gittleman JL, Jones KE (2007) Sexual dimorphism in mammals. In: Fairbairn D, Székely T, Blanckenhorn W, editors. Sex, size and gender roles. Oxford University Press. 16–26.
30. Fairbairn DJ (1997) Allometry for sexual size dimorphism: pattern and process in the coevolution of body size in males and females. Ann Rev Ecol Syst 28: 659–687.
31. Webb TJ, Freckleton RP (2007) Only half right: species with female-biased dimorphism consistently break Rensch's rule. PlosOne 2007: 1–10.
32. Stephens PR, Wiens JJ (2009) Evolution of sexual size dimorphisms in emydid turtles: ecological dimorphism, Rensch's rule, and sympatric divergence. Evolution 63: 910–25.
33. Fairbairn DJ, Preziosi RF (1994) Sexual selection and the evolution of allometry for sexual size dimorphism. Am Nat 144: 101–118.
34. Blanckenhorn WU, Stillwell RC, Young KA, Fox CW, Ashton KG (2006) When Rensch meets Bergmann: Does sexual size dimorphism change systematically with latitude? Evolution 60: 2004–2011.
35. Starostová Z, Kubička L, Kratochvíl L (2010) Macroevolutionary pattern of sexual size dimorphism in geckos corresponds to intraspecific temperature-induced variation. J Evol Biol 23: 670-677.
36. Walker SPW, McCormick MI (2009) Sexual selection explains sex-specific growth plasticity and positive allometry for sexual size dimorphism in a reef fish. Proc R Soc B 276, 3335–3343.
37. Remeš V, Székely T (2010) Domestic chickens defy Rensch's rule: sexual size dimorphism in chicken breeds. J Evol Biol 23: 2754-2759.
38. Féderation Cynologique Internationale (2008) Nomenclature et Standards. Available: http://www.fci.be. Accessed 2008, Aug 15.
39. Sutter NB, Mosher DM, Gray MM, Ostrander EA (2008) Morphometrics within dog breeds are highly reproducible and dispute Rensch's rule. Mamm Gen 19: 713–723.
40. Lindblad-Toh K, Wade CM, Mikkelsen TS, Karlsson EK, Jaffe DB, et al. (2005) Genome sequence, comparative analysis and haplotype structure of the domestic dog. Nature 438: 803–819.
41. Vilá C, Maldonado JE, Wayne RK (1999) Phylogenetic relationships, evolution, and genetic diversity of the domestic dog. J Hered 90: 71–77.
42. Vilá C, Savolainen P, Maldonado JE, Amorim IR, Rice JE, et al. (1997) Multiple and ancient origins of the domestic dog. Science 276: 1687–1689.
43. Leonard JA, Wayne RK, Wheeler J, Valadez R, Guillén S, et al. (2002) Ancient DNA evidence for Old World origin of New World dogs. Science 298: 1613–1616.
44. Savolainen P, Zhang Y, Luo J, Lundeberg J, Leitner T (2002) Genetic evidence for an east Asian origin of domestic dogs. Science 298, 1610–1613.
45. Gundry RL, Allard MW, Moretti TR, Honeycutt RL, Wilson MR, et al. (2007) Mitochondrial DNA analysis of the domestic dog: control region variation within and among breeds. J Forens Sci 52: 562–572.
46. Bannasch DL, Bannasch MJ, Ryun JR, Famula TR, Pedersen NC (2005) Y chromosome haplotype analysis in purebred dogs. Mamm Gen 16: 273–280.
47. Parker HG, Kim LV, Sutter NB, Carlson S, Lorentzen TD, et al. (2004) Genetic structure of the purebred domestic dog. Science 304: 1160–1164.
48. Irion DN, Schaffer AL, Famula TR, Eggleston ML, Hughes SS, et al. (2003) Analysis of genetic variation in 28 dog breed populations with 100 microsatellite markers. J Hered 94: 81–87.
49. vonHoldt BM, Pollinger JP, Lohmueller KE, Han E, Parker HG, et al. (2010) Genome-wide SNP and haplotype analyses reveal a rich history underlying dog domestication. Nature 464: 898-903.
50. Řičánková V, Zrzavý J (2004) Phylogeny of Canidae (Mammalia) and evolution of reproductive, developmental and socio-ecological traits: inferences from morphological, behavioural and molecular data. Zool Scri 33: 311–333.
51. Bardeleben C, Moore RL, Wayne RK (2005) A molecular phylogeny of the Canidae based on six nuclear loci. Mol Phyl Evol 37: 815–831.
52. Pionnier-Capitan M, Bemilli C, Bodu P, Célérier G, Ferrié J-G, et al. (2011) New evidence for Upper Palaeolithic small domestic dogs in South-Western Europe. J Archaeol Sci 38: 2123-2140.

53. Morey DF (1994) The early evolution of the domestic dog. Am Sci 82: 336–347.
54. MacDonald DW, Sillero-Zubiri Z (2004) The biology and conservation of wild canids. Oxford University Press. 464 p.
55. Sundquist AK, Björnerfeldt K, Leonard JA, Hailer F, Hedhammar A, et al. (2006) Unequal contribution of sexes in the origin of dog breeds. Genetics 172: 1121–1128.
56. Lande R (1980) Sexual dimorphism, sexual selection, and adaptation in polygenic characters. Evolution 34: 292–307.
57. Geffen E, Gompper ME, Gittleman JL, Hang-Kwang L, MacDonald DW, et al. (1996) Size, life-history traits, and social organization in the Canidae: A reevaluation. Am Nat 147: 140–160.
58. Kutzler MA, Yeager AE, Mohammed HO, Meyers-Wallen VN (2003) Accuracy of canine parturition date prediction using fetal measurements obtained by ultrasonography. Theriogenology 60: 1309–1317.
59. Stern DL, Orgogozo V (2009) Is genetic evolution predictable? Science 323: 746-751.
60. Polák J, Frynta D (2010) Sexual size dimorphism in domestic cattle supports Rensch's rule. Evol Ecol 24: 1255-1266.
61. Polák J, Frynta D (2009) Sexual size dimorphism in domestic goats, sheep and their wild relatives. Biol J Linn Soc 98: 872-883.
62. Lovich JE, Gibbons JW (1992) A review of techniques for quantifying sexual size dimorphism. Growth, Devel Aging 56: 269–281.
63. Smith RJ (1999) Statistics of sexual size dimorphism. J Hum Evol 36: 423–459.
64. McArdle BH (1988) The structural relationship: regression in biology. Can J Zool 66: 2329–2339.
65. StatSoft Inc. (accessed 2001) STATISTICA, Version 6.0. http://www.statsoft.com.

A Quasi-Exclusive European Ancestry in the Senepol Tropical Cattle Breed Highlights the Importance of the *slick* Locus in Tropical Adaptation

Laurence Flori[1,2]*, Mary Isabel Gonzatti[3], Sophie Thevenon[2], Isabelle Chantal[2], Joar Pinto[3], David Berthier[2], Pedro M. Aso[3], Mathieu Gautier[4]

1 INRA, UMR 1313 GABI, F-78350 Jouy-en-Josas, France, **2** CIRAD, UMR INTERTRYP, Montpellier, France, **3** Departamento de Biología Celular, Universidad Simón Bolívar, Caracas, Venezuela, **4** INRA, UMR CBGP, Montferrier-sur-Lez, France

Abstract

Background: The Senepol cattle breed (SEN) was created in the early XXth century from a presumed cross between a European (EUT) breed (Red Poll) and a West African taurine (AFT) breed (N'Dama). Well adapted to tropical conditions, it is also believed trypanotolerant according to its putative AFT ancestry. However, such origins needed to be verified to define relevant husbandry practices and the genetic background underlying such adaptation needed to be characterized.

Methodology/Principal Findings: We genotyped 153 SEN individuals on 47,365 SNPs and combined the resulting data with those available on 18 other populations representative of EUT, AFT and Zebu (ZEB) cattle. We found on average 89% EUT, 10.4% ZEB and 0.6% AFT ancestries in the SEN genome. We further looked for footprints of recent selection using standard tests based on the extent of haplotype homozygosity. We underlined i) three footprints on chromosome (BTA) 01, two of which are within or close to the *polled* locus underlying the absence of horns and ii) one footprint on BTA20 within the *slick* hair coat locus, involved in thermotolerance. Annotation of these regions allowed us to propose three candidate genes to explain the observed signals (TIAM1, GRIK1 and RAI14).

Conclusions/Significance: Our results do not support the accepted concept about the AFT origin of SEN breed. Initial AFT ancestry (if any) might have been counter-selected in early generations due to breeding objectives oriented in particular toward meat production and hornless phenotype. Therefore, SEN animals are likely susceptible to African trypanosomes which questions the importation of SEN within the West African tsetse belt, as promoted by some breeding societies. Besides, our results revealed that SEN breed is predominantly a EUT breed well adapted to tropical conditions and confirmed the importance in thermotolerance of the *slick* locus.

Editor: David Caramelli, University of Florence, Italy

Funding: This work was supported by an INRA (Institut National de la Recherche Agronomique)-Animal Genetics Department Grant (TROPSENEPOL project), an ECOS (Évaluation-orientation de la COopération Scientifique) Nord No. PI-2008002104 grant and by FONACIT (FOnds NAtional pour les Sciences, les Technologies et l'Innovation) projects G98-3462 and 2007-1425. MG acknowledges partial funding by the Agence Nationale de la Recherche programme BLANC EMILE 09-BLAN-0145-01. The funders had no role in study design, data collection and analysis, decision to publish, or preparation of the manuscript.

Competing Interests: The authors have declared that no competing interests exist.

* E-mail: laurence.flori@jouy.inra.fr

Introduction

Over the past decades, population genetic approaches associated with archeological evidence have provided better insights into the origin of modern livestock breeds including domestication processes, migration routes and their relationships [1–5]. The recent release of complete genome sequences and the development of high density SNP genotyping assays (e.g. [6]) in most livestock species have greatly increased available genomic information to refine characterization of breed origins and to efficiently detect footprints of selection in animal genomes. As a result, assembling large genetic datasets from livestock breeds, it is now possible to evaluate the accuracy of the reports on the origin of breeds which are widespread in the scientific literature. For instance, several publications referred to the Italian Piemontese cattle breed as a hybrid between the extinct aurochs and Indo-Pakistani Zebu. This myth was only recently refuted by genetic fingerprinting which showed that Italian Piemontese (as might have been expected) is a mixture of several European taurine (EUT) breeds [7]. Less fanciful, among the tropical cattle, the Kuri cattle breed has long been reported as an African taurine (AFT) breed in particular because these animals are humpless [8], characterized by a submetacentric Y chromosome from a taurine origin and devoid of Zebu allele on the Y chromosome [9]. However, nuclear markers such as microsatellites and SNPs showed that Kuri is actually a hybrid between AFT and Zebu (ZEB) breeds [10–15]. Apart from being of historical interest, such clarification of beliefs could be essential in the management and diffusion of some cattle breeds.

In addition, high density SNP dataset has highlighted chromosomal regions and genes targeted by artificial selection in dairy and/or beef cattle and by natural selection in West African cattle

[14,16–20]. Recently, by combining genetic structure analysis, examination of extended haplotype homozygosities (EHH,[21–23]) and detection of local excess or deficiency of a given ancestry relative to the average genome admixture level [24], we characterized the genetic origins of an admixed New World Creole cattle breed and detected footprints of selection in the genome.

Such a global approach is applied hereby on the Senepol cattle breed (SEN), a completely polled breed originating from the Virgin Islands, now widespread in several tropical regions. The SEN breed is particularly well adapted to tropical conditions, SEN animals being able for instance to maintain a normal body temperature during heat stress and appearing as heat tolerant as Brahman Zebu (BRA) cattle [25–27]. It is hypothesized that their sleek and shiny hair coat is responsible for or participates in their thermotolerance.

According to the St. Croix (US Virgin Island) breed society, SEN was created by Henry Nelthropp's family at the beginning of 1900s on the St. Croix Island [28]. Nelthropp's declared objective was to develop a hornless, early maturing breed, with gentle disposition, that combined the adaptation to tropical conditions of AFT cattle with meat production abilities of EUT cattle. He thus chose to cross supposed N'Dama (NDA) cows originating from Senegal with bulls belonging to the Red Poll cattle breed bought in Trinidad island and originating from Great Britain [29]. Since 1977, SEN have been spread in tropical areas around the world, in particular to some states of the US mainland, Australia and several countries of Latin America. The SEN Cattle Breeders Association (SCBA) recognized in 1999 over 500 breeders and more than 14,000 SEN records [30].

As an important consequence, many technical documents about SEN breed report that SEN individuals are considered to have inherited from their presumed NDA ancestry trypanotolerance characteristics, i.e. the ability displayed by West AFT breeds such as NDA to "survive, reproduce and remain productive under trypanosomiasis risk without trypanocidal drug" [31,32]. SEN has thus been proposed by some breeding societies as a good candidate to improve production performances of local western African cattle living in areas infested by tsetse flies, the biological vectors of *Trypanosoma* (Associação Brasileira dos Criadores de Bovinos Senepol, http://senepol.org.br). To our knowledge, trypanotolerance has never been rigorously assessed in SEN, although failure to verify it could have dramatic consequences both for local breeders and for local breed diversity if importation of SEN animals were carried out towards West Africa. Moreover, the official version of the SEN origins, which represents the only argument to suggest a possible trypanotolerance of the SEN breed, has been disputed [33,34]. De Alba early questioned the veracity of pure NDA importation in St. Croix and also noticed that supposed NDA animals harbored phenotypic characteristics of ZEB cattle [33]. He finally suggested that SEN animals could be derived from Creole cattle imported from the island of Viquez in St. Croix before the creation of the SEN breed. Although the official version of origin is routinely repeated by breed societies, the true SEN origin remains uncertain, but needs to be clarified before the breed is widely disseminated into tsetse infested areas.

Our study was primarily aimed at providing a detailed assessment of the genetic origins of the SEN using high density SNP data. To that purpose, a total of 153 SEN cattle were genotyped on the Illumina Bovine SNP50 chip and the resulting data were combined with those already available on 18 other breeds [6,14–16]. In order to better characterize the origin of the adaptation to tropical conditions of the SEN breed, we further examined footprints of selection within this population using standard Extended Haplotype Homozygosity (EHH) based tests (e.g. [24]).

Results

The Genetic Characterization of SEN Cattle Breed Excludes AFT Ancestry

In order to characterize the SEN genetic history, we first combined the SNP data obtained on the 147 SEN animals (from the 153 genotyped SEN animals, see Materials and Methods) with those available on 18 worldwide cattle populations [14–16] corresponding to six European taurine (EUT), four AFT, four ZEB and four admixed breeds (one AFTxEUT, two AFTxZEB and one EUTxZEB), respectively. The combined data set consists of 623 individuals genotyped for 47,365 SNPs (see Materials and Methods).

The neighbor joining (NJ) tree based on allele sharing distance (ASD) separated individuals according to their population of origin (Figure S1). Breeds can also be grouped according to the three main populations (EUT, AFT and ZEB). Admixed breeds (OUL, KUR, BOR and SGT) branched at intermediate positions between their populations of origin. SEN animals from different origins (different Venezuelan states, St. Croix Island and USA) appear as a homogeneous group and branched at the same position than the SGT (EUT×ZEB).

An individual Principal Component Analysis (PCA) was then carried out using all available SNP information (Figure 1A). The first and the second components (PC1 and PC2) explained 7.91 and 6.22% of the variation, respectively. Focusing on this first factorial plan, we obtained the previously described triangle-like 2-Dimensional global organization of cattle genetic diversity [14,15]. This triangle is shaped by EUT (in blue), AFT (in red) and ZEB (in green) at the three apexes. Hybrid breeds (in orange) such as SGT lay at their intermediated position in agreement with the NJ tree. Unexpectedly, SEN (in black) is positioned between SGT (EUTxZEB) and EUT on the side of the triangle limited by ZEB and EUT, at the opposite position from AFT. The SEN position suggests an influence of EUT and ZEB ancestries, with a greater EUT ancestry. Interestingly, PC2 also discriminates EUT populations according to a North/South gradient and SEN appeared closer on this axis to ANG and HFD (northern Europe breeds) than MON and SAL (southern Europe breeds).Using a model-based unsupervised hierarchical clustering of the individuals considering different K numbers of predefined clusters (Figures 1B and S2), we quantified the different SEN ancestry proportions. Results obtained with $K = 3$ were in agreement with PCA ones (Figure 1.B). The three clusters in dark blue (K1), red (K2) and green (K3) could be broadly interpreted as EUT, AFT and ZEB ancestry, respectively. SEN individuals had on average 89% (ranging from 66 and to 95%), 10.4% (ranging from 4 and to 33%) and 0.6% (ranging from 0 and to 4.1%) of EUT, ZEB and AFT ancestries, respectively. Hence, AFT ancestry in SEN individuals was even lower than that observed in individuals belonging to Northern-Europe breeds (Table S2). Increasing the number of clusters ($K = 4$) led to a similar overall picture for non-SEN individuals and the additional cluster (in green) isolated SEN individuals. Such a clustering of SEN individuals into a single group when $K = 4$ is essentially a consequence of overrepresentation of SEN individuals ($n = 147$) compared to other breeds (Table S1; Figure S2). Indeed, when a new analysis was performed, replacing the 147 initial SEN individuals with a sample of 30 randomly chosen SEN individuals (Figure S3), this trend was not observed. Note that the analysis on the reduced data sets essentially confirmed our previous observation for $K = 3$ and

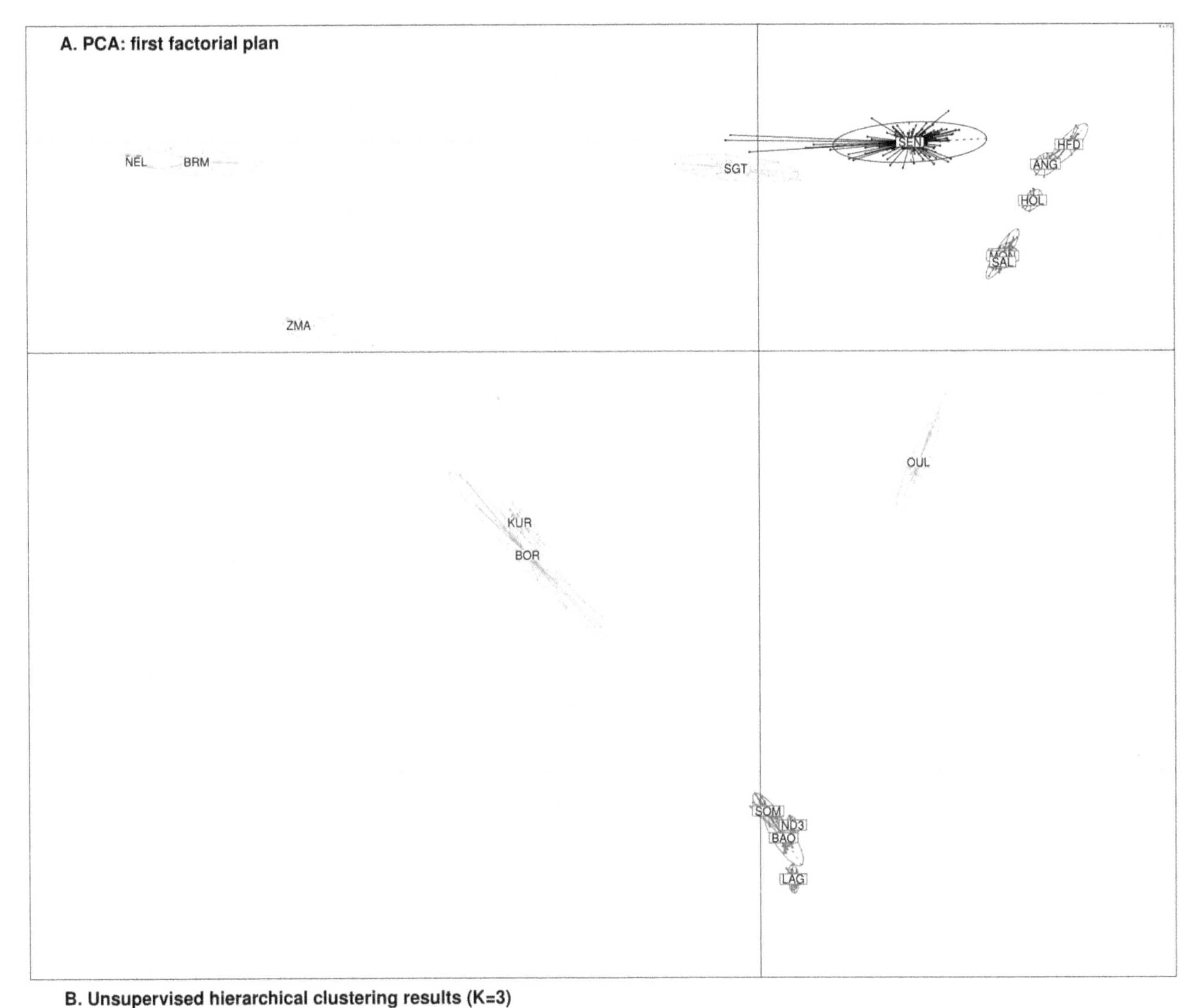

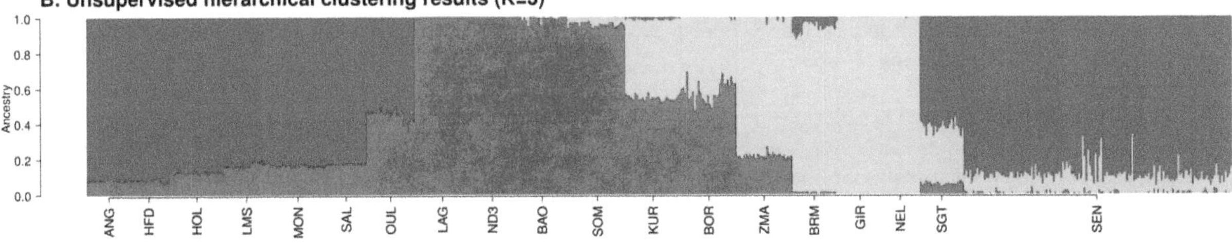

Figure 1. First factorial plan resulting from the PCA and unsupervised hierarchical clustering results. A. PCA was performed on 629 animals and 47365 SNPs. The first factorial plan is composed by PC1 and PC2 with corresponding eigenvalues equal to 7.91 and 6.22%. Ellipses characterize the dispersion of each population around its center of gravity. SEN individuals are plotted in black and individuals from the EUT, AFT, ZEB and hybrid (BOR, KUR, OUL and SGT) populations are plotted in blue, red, green and orange respectively. B. An unsupervised hierarchical clustering based on 629 individuals genotyped for 47365 SNPs is presented with an inferred number of clusters K = 3. For each individual, the proportion of each cluster (y) which were interpreted as representative of EUT, AFT and ZEB ancestries are plotted in blue, red and green, respectively.

K = 4 and SEN individuals displayed a high percentage of Northern-Europe breed ancestry for K = 4.

Focusing now on the different SEN population origins, we observed i) a low level of differentiation among populations from the four different Venezuelan states represented (F_{ST} <0.021), and ii) no inbreeding within the whole SEN population ($|F_{IS}|$ <0.01) (Table 1). This was consistent with the recent introduction of SEN in Venezuela and exchanges of animals. It justified in turn considering SEN populations as one single population.

As expected, regarding the individual-based analyses (v Figures 1, S1 and S2), pairwise F_{ST} between SEN and EUT breeds were lower than those between SEN and ZEB or AFT breeds (Table 1) and close to F_{ST} between EUT breeds (Table S3). As shown in Table 1, SNP average heterozygosity for the SEN breed (0.29) was within the range observed in EUT breeds (from 0.28 to 0.30) and above those observed in AFT (from 0.18 to 0.22) or ZEB (from 0.15 to 0.19) populations.

Table 1. Heterozygosity (HZ), inbreeding coefficient (F_{IS}) for each breed and differentiation (F_{ST}) of SEN vs other breeds.

Origin	Breed	HZ	F_{IS}	F_{ST}
AFT	BAO	0.2080	−0.0123	0.2168
	LAG	0.1822	0.0193	0.2479
	NDA	0.2053	0.0085	0.2191
	SOM	0.2232	0.0500	0.2000
EUT	ANG	0.2998	0.0212	0.1428
	HFD	0.3077	0.0274	0.1558
	HOL	0.3050	0.0003	0.1385
	LMS	0.3026	0.0002	0.1113
	MON	0.2805	−0.0356	0.1417
	SAL	0.2808	0.0198	0.1363
HYB	BOR	0.2596	0.0042	0.1612
	KUR	0.2575	0.0058	0.1632
	OUL	0.2703	−0.0342	0.1431
	SEN	0.2912	0.0058	
	SGT	0.3039	−0.0028	0.1123
ZEB	BRM	0.1939	0.0289	0.2530
	GIR	0.1596	0.0093	0.2870
	NEL	0.1571	−0.0171	0.2869
	ZMA	0.1958	0.0174	0.2453

Identification of Footprints of Selection

Following Gautier & Naves (2011) [24], we i) computed *iHS* scores [23] for each SNPs over the whole genome using the 294 SEN phased haplotypes [35] for each bovine autosome and ii) calculated *Rsb* statistics [22] to compare the local extent of haplotype homozygoties between SEN and its different ancestries (EUT and ZEB). For each autosome, haplotypes representative of EUT and ZEB ancestry, respectively, were pooled. Figure 2 represents the plots of the three different scores (*iHS* for SEN and *Rsb* for both EUT/SEN and ZEB/SEN comparisons).

Table 2 summarizes characteristics of the four chromosomal regions displaying significant *iHS* and/or significant *Rsb* scores, three of which (#1, #2 and #3) map to BTA01 and one (#4) to BTA20. According to our somewhat stringent criteria, region #4 was significant for the three tests performed (*iHS*, $Rsb_{EUT/SEN}$ and $Rsb_{ZEB/SEN}$), region #1 was significant for two tests (*iHS* and $Rsb_{ZEB/SEN}$) and both regions #2 and #3 were found only significant with the $Rsb_{EUT/SEN}$ statistic. In order to illustrate the haplotype structure around the ancestral and derived alleles for each of the four SNPs at the peak position, we drew i) the decay of site-specific EHH or EHHS (weighted average of both EHH used for the computation of *Rsb*) within SEN, EUT and ZEB populations, ii) the EHH decay from the core SNP at both alleles within the SEN population and iii) haplotype bifurcation diagrams for both ancestral and derived allele in each region under selection within the SEN population (Figure 3). For Region#1, haplotypes containing the derived core SNP allele harbored a clear long-range Linkage Disequilibrium (LD) *i.e.* a thick branch in both directions from the core SNP. This suggests that the underlying causal variant is associated to this allele. Conversely, for regions #2 and #3, the underlying selected variant might be associated to the ancestral allele at the corresponding peak core SNP. For region #4, the signal is less clear since haplotype diversity is reduced at both the ancestral and derived alleles suggesting several favorable variants associated to various haplotype background might be under segregation (Figure 3). Following the UCSC UMD3 assembly, candidate genes could be identified at or close to the peaks for regions #2 (TIAM1: T-cell Lymphoma Invasion and Metastasis 1), # 3 (GRIK1: Glutamate Receptor, Ionotropic, Kainate 1) and #4 (RAI14: Retinoic Acid Induced 14). No candidate gene could be found for region #1.

Discussion

Revisiting the Origin of SEN Cattle Breed: Consequences for Breeding Strategies

Our study provides a first fine-scale genetic characterization of SEN through a comparison with other breeds representative of major breed groups of cattle distributed world-wide. The SEN individuals are representative of the whole population since they were sampled in different regions of Venezuela, and included some animals brought directly from the USA and the St. Croix Island, where the breed originated. No clear sub-structure was observed among the SEN individuals, suggesting an homogeneity of the population. In addition, SEN individuals displayed on average 89% of EUT ancestry meaning that SEN is predominantly a EUT breed. The origin of the EUT ancestry from North European breeds thus remains in agreement with historical and SEN breed association reports indicating that Red Poll individuals were initially used to create SEN. Although limited to 10.4% on average in SEN individuals, a ZEB ancestry was unexpectedly observed. More strikingly, given the very recent creation of the SEN breed (beginning of the XXth century), our results clearly challenge the official historical version since no significant AFT ancestry was found in the SEN population. In order to explain such discrepancies, two hypotheses might be proposed. First, NDA breeding individuals were actually not used to create the SEN breed. As mentioned above, some authors raised doubts about a possible NDA ancestry [33,34]. Hence, De Alba indicated that individuals participating in the breed creation in St.Croix had ZEB phenotypic characteristics suggesting Creole cattle (present in the St. Croix Island before the SEN breed creation) contributed to the SEN breed. Alternatively, AFT genome might not have been retained in the first individuals because AFT background had been counter-selected. We might for instance speculate that descendants from EUTxNDA crosses were not chosen as reproducers because of poor zootechnical characteristics or because they did not fit breeding objectives, one of them being hornless. Indeed, the inheritance of horns is mainly controlled by two locus: i) the *polled* locus located on BTA01 in *Bos Taurus* and ii) the *African horn* locus in *Bos taurus* and especially *Bos indicus*, which has an epistatic effect on the *polled* locus [36]. Based on the genetic determinism of horn development, NDA (horned) x RedPoll (completely polled) crosses result in polled F1 females and horned F1 males probably not retained as reproducers, leading to a counter-selection of individual carrying AFT ancestry.

Importantly, the absence of AFT ancestry makes it unlikely that SEN animals have inherited trypanotolerance trait from their ancestries although such ability represents the main argument of some breeder associations for the diffusion of SEN to West-African countries. Indeed, both EUT and ZEB cattle are recognized as trypanosusceptible. Similarly, SEN might not have acquired trypanotolerance during their recent history since Caribbean islands such as St. Croix are *Trypanosoma* free. Moreover, both *Trypanosoma congolense (T.congolense)* which is responsible for severe outcome of the

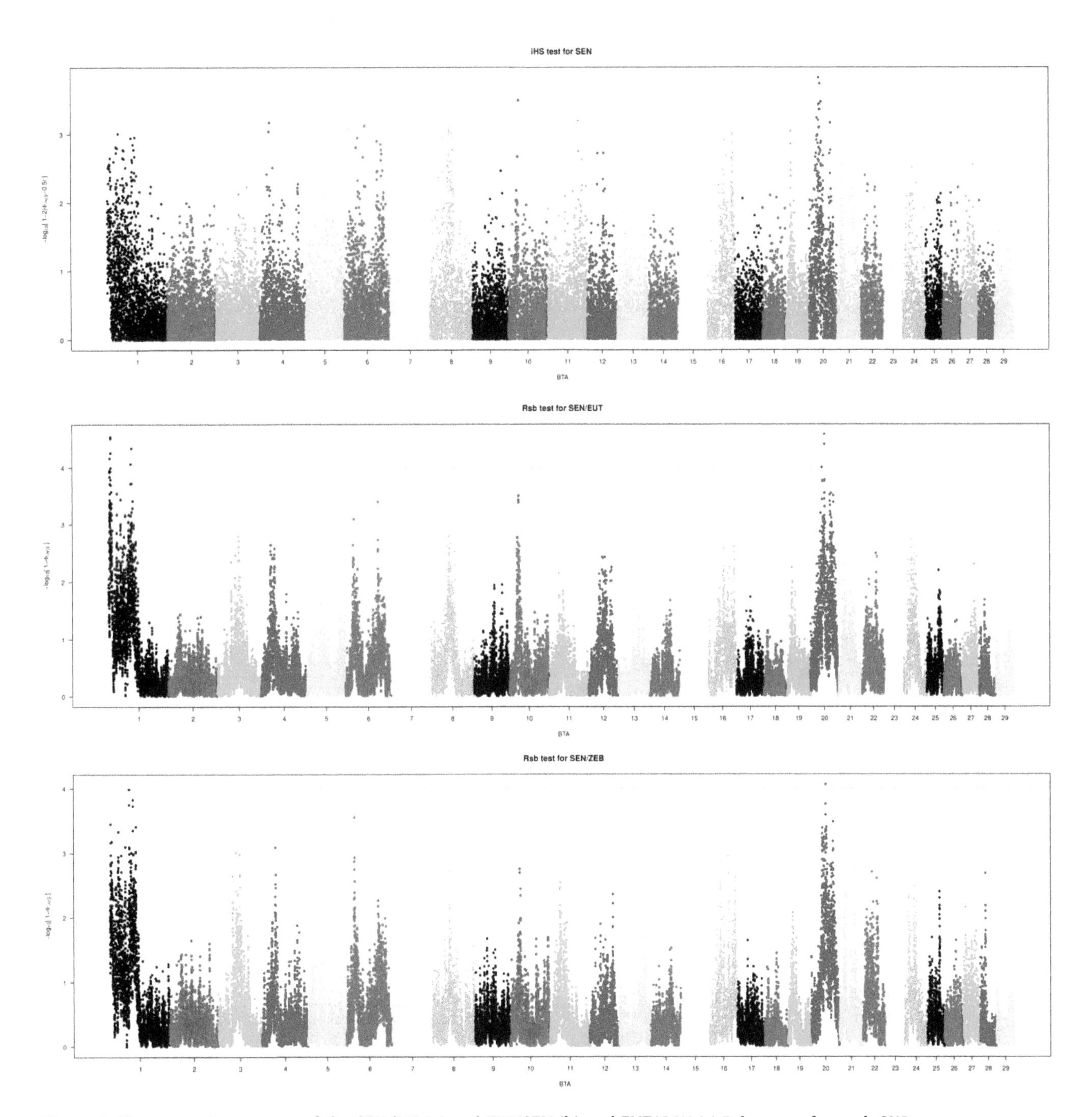

Figure 2. Plots over the genome of the SEN iHS (a) and ZEB/SEN (b) and EUT/CGU (c) Rsb scores for each SNP.

Table 2. Chromosomal regions under selection.

Region	BTA	Position (Mb)	Peak position (Mb)	iHS_{SEN}	$Rsb_{EUT/SEN}$	$Rsb_{ZEB/SEN}$	Nb of significant SNPs	Gene closest to the maximum
#1	1	52.6–53.6	52.9	3.984	NS	3.984	2	No gene found
#2	1	2.4–3.4	3.2	NS	4.161	NS	6	TIAM1
#3	1	4.7–5.8	5.5	NS	4.538	NS	8	GRIK1
#4	20	38.6–39.6	39.5	4.055	4.576	4.055	3–6	RAI14

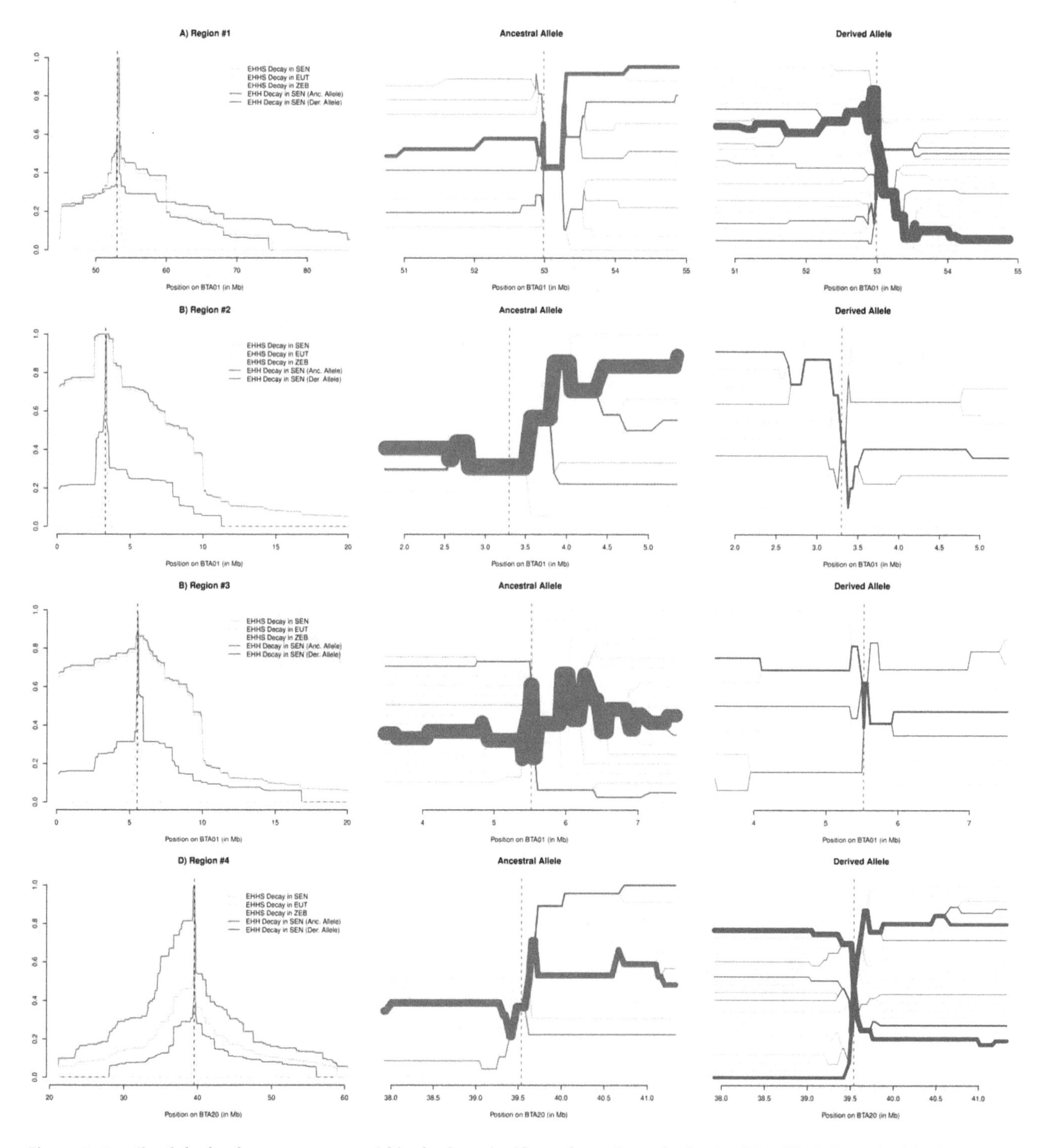

Figure 3. Details of the haplotype structure within the four significant footprints of selection identified. For each of the four regions, the decays of EHH (SEN ancestral and derived allele) and EHHS (for SEN, EUT and ZEB populations) from the core SNP located under the peak position are plotted in the leftmost panel. Haplotype bifurcation diagrams for both the ancestral (in red in the center panel) and derived allele (in blue in the rightmost panel) are also represented.

disease in African livestock and its obligate vector (the tsetse fly) are absent in South America. SEN cattle are already present in South Africa, Botswana, Namibia and Zimbawe and some other African countries are planning to import SEN animals because they are more productive under tropical conditions than local breeds in a tsetse free context. It might be highly advisable to limit SEN importation to countries or areas outside of the tsetse belt.

Detection and Characterization of Footprints of Selection within SEN

Such new insights into SEN origins raised another interesting question. Indeed, how can a breed with almost 90% of European cattle ancestry be well adapted to tropical conditions? Due to the recent origin of the SEN breed, it is tempting to speculate that such adaptation was inherited from the ZEB ancestry. ZEB, that are widespread in African, Asian and South America tropical areas, are indeed recognized for their thermotolerance, their ability to survive during food shortage and their resistance to several tick-borne diseases [37,38].

In such a case, SEN represents an attractive model to identify those genomic regions from ZEB origin involved in its adaptation to tropical conditions. Alternatively, some favorable alleles might have segregated in EUT ancestry of the SEN breed in which theyhave subsequently been selected. To unravel loci potentially involved in adaptation of SEN to tropical conditions, we then searched for footprints of selection using previously described EHH-related tests (based on *iHS* and *Rsb* statistics). *iHS* based tests allow to detect loci carrying favorable variants subjected to strong recent or ongoing selection and display maximal power when they have not yet reached fixation [23]. Conversely, *Rsb* based tests show higher power to detect selective sweeps that have resulted in near or complete fixation of the favorable variant in one of the compared populations [22]. Four footprints of selection were then identified using *iHS* and *Rsb* statistics, three located on BTA01 (#1, #2 and #3) and one located on BTA20 (#4).

Interestingly, this latter region (#4) is within the previously described *slick* hair locus responsible for the very short, sleek hair coat [27,39]. The slick phenotype was shown to be controlled by a single dominant unknown gene [27] located on a 4.4 cM region on BTA20 [39]. The resulting particular hair coat plays an important role in thermotolerance since hair coat thickness and hair weight per unit surface, are involved in adaptability to warm climate by limiting evaporative heat [40]. More precisely, Olson and colleagues showed that vaginal temperature, skin temperature and respiration rate of animals classified as slick was lower than normal-haired animals in ¾ Holstein ¼ Senepol crossbred cattle during acute heat stress [41]. This difference is partly due to an increase sweating rate [41], but could also result from an effective reflexion of solar radiations and an increase of heat loss via convection and conduction [42]. In addition, Olson and colleagues observed that the slick-haired dairy cows in Venezuela have an increased milk yield compared to normal haired animals [27]. Hence the *slick* locus appears as a strong candidate to explain the observed footprint of selection underlining thereby its importance in SEN adaptation to tropical conditions. Under this assumption, our study greatly refines the mapping interval of the underlying gene and allows us to propose a relevant positional and functional candidate gene. Indeed, the peak of the signal is located within the Retinoic Acid induced 14 gene (RAI14 or NORPEG), a developmentally regulated gene induced by all trans retinoic acid [43]. Retinoic acids are able to modulate several biological functions and are particularly involved in the control of hair follicule morphogenesis and cycling. Indeed, the prototypic retinoid all-trans retinoic acid is able to impair hair shaft elongation *in vitro* and can induce hair loss through premature hair follicule regression [44]. Topical treatment of human patients suffering from baldness, with retinoid all-trans retinoic acid in association with minoxidil enhances the hair follicule anagen-prolonging effects of the latter [45].

Among regions located on BTA01, no candidate genes could be found for region #1. However and interestingly, regions #2 and #3 mapped within (or close to in the case of region #3) the *polled* locus [46]. The polled phenotype corresponding to the absence of horns is caused by a single dominant mutation which has been mapped on BTA01 in several independent studies but has not yet been identified [36,46–48]. Focusing on cattle breeds artificially selected for dairy production, other authors reported also a footprint of selection located in the *polled* locus defined in Drogemüller et al. and Wunderlich et al. studies [46,48,49]. To date, the most refined *polled* critical interval published extends from microsatellite BM6438 and RP42-218J17_MS1 on BTA01 [48] and is contained in a 2.5 Mb contig spanning the interval between SLC5A3 and SOD1 [46]. Assuming that the *polled* allele is responsible for the observed signals, our study allows us to propose two candidate genes lying under these peaks. TIAM1 encoding the T-cell lymphoma invasion and metastasis protein and GRIK1, encoding a ionotropic glutamate receptor, which are located in regions #2 and #3, respectively. TIAM1 modulates the activity of Rho-like proteins and connects extracellular signals to cytoskeletal activities [50]. This protein is crucial for the integrity of adherens junctions and cell matrix interactions [51]. The Glutamate receptor family, which comprises GRIK1, is involved in the mediation of excitatory synaptic transmission and plays a key role in cognitive function. Glutamate signaling is also involved in modifications of differentiation and osteoclasts and osteoblasts activities in bone [52]. Recently, Goto and colleagues showed that two ionotropic glutamate receptor genes (GRIN2C and GRIN3A) are the most divergent glutamate receptor genes between humans and chimpanzees [53]. GRIN3A showed also significant dN/dS acceleration in the human branch based on the analysis of six mammalian species [54]. A cluster of keratoproteins (KRTAP) is located between regions #2 and #3. In our previous study in West African cattle breeds, we also highlighted a region under selection near the *polled* region, with a peak near the KRTAP8-1 gene, located 0.5 Mb from TIAM1 and 1.38 Mb from GRIK1 [14].

For regions #2 and #3, examination of EHH decay and haplotype bifurcation diagrams suggested that the ancestral allele at the corresponding peak SNPs is associated to the underlying selected variant. For region #4, both ancestral and derived allele seem associated to an underlying selected variant. Due to the high EUT ancestry, ancestral alleles are thus more likely to be of EUT origin (see Materials and Methods). However, it remains difficult based on such observations to infer the origin of the selected variants. Nevertheless, a suggestive excess of local EUT ancestry [24] was observed in the footprints of selection observed in regions #2, #3 and #4 (data not shown). For regions #2 and #3, such observations are thus in agreement with a EUT origin of the hornless phenotype. Interestingly, for region #4 located within the *slick* hair locus (BTA20), the underlying favorable variant might be associated to several haplotype backgrounds from EUT ancestry. Discovered in SEN, the *slick* allele has also been described in other Creole breeds such as the Carora [27], Romosinuano (which is also polled) and Criollo Limonero which all have supposedly an important EUT ancestry. Fine scale genetic characterization of the SEN cattle breed allowed us to revisit the SEN breed origin. Since no AFT ancestry could be found in this population, the SEN breed might be viewed as a EUT breed well adapted to tropical conditions making it a relevant biological model to study adaptation to tropical climates. In addition, such results may help in the definition of management and breeding practices, namely to avoid importation of this breed to countries within the tsetse belt. Finally, we identified several footprints of selection presumably driven by artificial (selection toward the hornless phenotype) and natural (heat) pressures.

Materials and Methods

Ethics Statement

No ethics statement was required for the collection of DNA samples. DNA was extracted either from commercial AI bull semen straws or from blood samples obtained from different veterinary practitioners visiting farms with the permission of the owners.

Selection of Animals, Blood Sampling and Genotyping

A total of 153 SEN individuals from Venezuela were included in our genetic study. In Venezuela, the first importation of SEN cattle was done in 1988, in the form of 250 frozen embryos from American Senepol Limited (Harrogate, Tenessee), which were implanted in Zebuin and Holstein breeds between 1988 and 1992. In 1991, the first bull was imported and 30,000 doses of its semen were sold. In May 1993 a total of 200 animals were imported from Saint Croix to seven different farms in Venezuela. Up to 2011, 540 animals have been imported from St. Croix or USA, for a total of 1,774 fullblood and purebred Senepol animals and 4,974 crossbred animals that are participating in the upgrade program in Venezuela. The Venezuelan Association of Senepol Cattle Breeders (ASOSENEPOL) was created in 1998 and has 25 active members spread in the territory and 13 veterinarians that classify the cattle in the whole country. They have a collaborative program with the international Senepol Cattle Breeders Association.

These 153 SEN animals (Table S1) belong to four Venezuelan states (39, 20, 36 and 58 animals collected in Guarico, Anzoategui, Bolivar and Monagas, respectively) and 10 different farms (two, one, three and four breeding units located in Guarico, Anzoategui, Bolivar and Monagas, respectively). Among them, 134, 13 and five are from Venezuelan, St. Croix and USA origin. Special care was taken to choose pure SEN animals belonging to the herdbook and to limit animal relationships based on available pedigree information.

Blood samples were collected in EDTA Vacutainer tubes and DNA was extracted using the Wizard® Genomic DNA Purification Kit (Promega, France). DNA quantity and quality were evaluated on Nanodrop and on agarose gels (1% stained with SYBR safe, Invitrogen).

The 153 animals (Table S1) were genotyped on the second version of the Illumina BovineSNP50 chip (v2) at INRA Labogena plateform (Jouy-en-Josas, France) using standard procedures (http://www.illumina.com) [6]. Six SEN individuals were further removed from subsequent analyses because they were genotyped for less than 95% of SNPs. We completed the SEN genotyping data with SNPs genotypes already obtained using the first version of the Illumina BovineSNP50 chip (v1) for 21 to 30 individuals belonging to18 other breeds (Table S1) representative of i) EUT: SAL, LMS, HOL, MON, ANG, HFD, ii) AFT : BAO, SOM, LAG, ND3, iii) ZEB: NEL, GIR, BRM, ZMA and iv) four crossbreed populations : OUL (EUTxAFT), BOR (AFTxZEB), KUR (AFTxZEB), SGT (EUTxZEB) [14,15]. The complete data set finally consists of genotypes for 623 individuals including 147 SEN individuals. Both versions of the Illumina BovineSNP50 chip have 52340 SNPs in common. Among these SNPs, we discarded from the analyses SNPs genotyped for less than 75% of the individuals per breed.

Following Gautier et al (2010) an exact test for Hardy-Weinberg Equilibrium (HWE) was subsequently carried out within each breed separately on the remaining SNPs [55] and q-values were estimated for each SNP using the R package qvalue (http://cran.r-project.org/web/packages/qvalue/index.html) [56]. SNP with a q-value below 0.1 in at least one population were discarded leading to a total of 47,365 SNPs in the final data set.

Analyses of Population Structure

F-statistics F_{IT}, F_{ST} and F_{IS}, and the diversity estimation for each locus and population both within and among individuals within a population were estimated using the program GENEPOP 4.0 [57].The within breed F_{IS} was derived from the average of these two quantities over all the SNPs.

A neighbor-joining tree [58] based on Allele Sharing Distance (ASD) was computed with APE R package [59]. For a given pair of individuals i and j, ASD was defined as 1-x_{ij} where x_{ij} represents the proportion of alleles alike in state averaged over all genotyped SNPs.

Principal Component Analysis (PCA) and unsupervised hierarchical clustering of individuals based on SNP genotyping data were performed using the smartpca software package and the program Admixture 1.04, respectively [60,61]. PCA results were visualized using the R package ade4 [62,63].

Detection of Footprints of Selection

Haplotypes were estimated using fastphase 1.4 for SEN and each AFT, ZEB and EUT populations [35]. We further computed *iHS* and *Rsb* score [22,23] using haplotype information. To compute *iHS*, the SEN ancestral allelic state was defined as the highest ancestral frequency estimated as $fa = w_{EUT} f_{EUT} + w_{ZEB} f_{ZEB}$, were w_{EUT} and w_{ZEB} represent the average proportions of EUT and ZEB ancestries in the SEN genome and f_{EUT} and f_{ZEB} the allele frequency of EUT and ZEB clusters [24]. The different quantities were estimated using Admixture 1.04 [60] as described above. For convenience SNP *iHS* scores were further transformed into $p_{iHS} = -\log[1-2|\Phi(iHS)-0.5|]$ where $\Phi(x)$ represents the Gaussian cumulative function. Assuming *iHS* scores are normally distributed under neutrality (Figure S4), p_{iHS} might thus be interpreted as $log_{10}(1/P)$ where P is the two-sided *p-value* associated to the neutral hypothesis (no selection). All the above analyses and plots (including bifurcation diagrams) were performed using the newly developed R package rehh [64].

Rsb was computed for two comparisons (EUT/SEN and ZEB/SEN) by standardizing the ratio of the corresponding ancestral cluster (EUT or ZEB) *iES* and the SEN *iES* [22]. SNP scores were transformed into $p_{Rsb} = -\log[\Phi(Rsb)]$. As above, assuming *Rsb* are normally distributed (under neutrality), p_{Rsb} might be interpreted as $\log_{10}(1/P)$ where P is the one-sided *p-value* associated to the neutral hypothesis (no selection).

Identification and Annotation of Candidate Regions

For each 1 Mb window over the genome (with a 0.5 Mb overlap), the candidate regions were identified by counting the number of SNP with $p_{iHS} > 4$ ($P<0.0001$) and similarly the number of SNP with $p_{Rsb} > 4$ for each of the two comparisons (EUT/SEN and ZEB/SEN). Regions containing at least two SNPs exceeding these thresholds for at least one test were considered as candidate. When several overlapping and contiguous windows were candidates, the chosen one contained the highest proportion of significant SNPs and also the peak. The candidate regions were then annotated using the UCSC Genome browser (http://genome.ucsc.edu) and the *Bos taurus* UMD 3.1/bosTau6 assembly. A gene was considered a candidate if the peak position was located <25 kb from its boundaries.

Supporting Information

Figure S1 Neighbor-Joining tree relating the 629 individuals. Among the 623 individuals, 147 SEN from different origins and states and 476 animals from 18 other breeds were analysed. The tree was constructed using allele sharing distances averaged over 47365 SNPs. Edges are colored according to the individual breed of origin.

Figure S2 Unsupervised hierarchical clustering results with different number of clusters of 623 individuals genotyped for 47365 SNPs. The 623 animals, comprising 147 SEN individuals, were analyzed with an inferred number of clusters K = 2 (A), K = 3 (B), K = 4 (C), K = 5 (D) and K = 6 (E). For each individual, the proportion of each cluster (y) which were interpreted as representative of EUT, AFT and ZEB ancestries for K = 3.

Figure S3 Unsupervised hierarchical clustering results with different number of clusters of 506 individuals genotyped for 47365 SNPs. The 506 individuals, comprising 30 SEN individuals randomly chosen, were analyzed with an inferred number of clusters K = 2 (A), K = 3 (B) and K = 4 (C). For each individual, the proportion of each cluster (y) which were interpreted as representative of EUT, AFT and ZEB ancestries are plotted in blue, red and green, respectively.

Figure S4 Empirical distribution of SEN, EUT and ZEB *iHS* scores.

Table S1 List of population symbols, names and number of individuals.

Table S2 Percentage of the corresponding AFT, ZEB and EUT ancestries for each breed.

Table S3 Differentiation score (*Fst*) for all pairs of breeds.

Acknowledgments

Sincere thanks are due to the Venezuelan Association of Senepol Cattle Breeders (ASOSENEPOL Venezuela) and Senepol breeders for their contribution.

The authors also wish to thank Céline Chantry-Darmon and Guy Noe (Labogena, INRA, France) for providing support in genotyping.

Author Contributions

Conceived and designed the experiments: MG LF PMA. Performed the experiments: MIG PMA LF ST DB IC JP. Analyzed the data: LF MG. Wrote the paper: LF MG.

References

1. Hanotte O, Bradley DG, Ochieng JW, Verjee Y, Hill EW, et al. (2002) African pastoralism: genetic imprints of origins and migrations. Science 296: 336–339.
2. Bruford MW, Bradley DG, Luikart G (2003) DNA markers reveal the complexity of livestock domestication. Nat Rev Genet 4: 900–910.
3. Cymbron T, Freeman AR, Isabel Malheiro M, Vigne JD, Bradley DG (2005) Microsatellite diversity suggests different histories for Mediterranean and Northern European cattle populations. Proc Biol Sci 272: 1837–1843.
4. Freeman AR, Bradley DG, Nagda S, Gibson JP, Hanotte O (2006) Combination of multiple microsatellite data sets to investigate genetic diversity and admixture of domestic cattle. Anim Genet 37: 1–9.
5. Teasdale MD, Bradley DG (2012) The origins of cattle. In: Womack JE, ed. Bovine Genomics. Ames: Wiley-Blackwell. 284 p.
6. Matukumalli LK, Lawley CT, Schnabel RD, Taylor JF, Allan MF, et al. (2009) Development and characterization of a high density SNP genotyping assay for cattle. PLoS One 4: e5350.
7. Felius M, Koolmees PA, B. T, Consortium ECGD, Lenstra JA (2011) On the Breeds of Cattle-Historic and Current Classifications. Diversity 3: 660–692.
8. Quéval R, Petit JP, Tacher G, Provost A, Pagot J (1971) Le Kouri: race bovine du lac Tchad, I. introduction générale à son étude zootechnique et biochimique: origine et écologie de la race. Revue d'Elevage et de Médecine Véterinaire Des Pays Tropicaux 24: 667–687.
9. Petit JP, Quéval R (1973) Le Kouri: race bovine du lac Tchad. II.Etude biochimique: les hémoglobines et les constituants du sérum. Revue d'Elevage et de Médecine Véterinaire Des Pays Tropicaux 26: 97–104.
10. Souvenir P, Zeuh V, Moazami-Goudarzi K, Laloë D, Bourzat D, et al. (1999) Etude du statut phylogénétique du bovin Kouri du lac Tchad à l'aide de marqueurs moléculaires. Revue d'Elevage et de Médecine Véterinaire Des Pays Tropicaux 52: 155–162.
11. Hanotte O, Tawah CL, Bradley DG, Okomo M, Verjee Y, et al. (2000) Geographic distribution and frequency of a taurine Bos taurus and an indicine Bos indicus Y specific allele amongst sub-saharan African cattle breeds. Mol Ecol 9: 387–396.
12. Freeman AR, Meghen CM, MacHugh DE, Loftus RT, Achukwi MD, et al. (2004) Admixture and diversity in West African cattle populations. Mol Ecol 13: 3477–3487.
13. Dayo GK, Thevenon S, Berthier D, Moazami-Goudarzi K, Denis C, et al. (2009) Detection of selection signatures within candidate regions underlying trypanotolerance in outbred cattle populations. Mol Ecol 18: 1801–1813.
14. Gautier M, Flori L, Riebler A, Jaffrezic F, Laloe D, et al. (2009) A whole genome Bayesian scan for adaptive genetic divergence in West African cattle. BMC Genomics 10: 550.
15. Gautier M, Laloe D, Moazami-Goudarzi K (2010) Insights into the genetic history of French cattle from dense SNP data on 47 worldwide breeds. PLoS One 5: e13038.
16. Flori L, Fritz S, Jaffrezic F, Boussaha M, Gut I, et al. (2009) The genome response to artificial selection: a case study in dairy cattle. PLoS One 4: e6595.
17. Gibbs RA, Taylor JF, Van Tassell CP, Barendse W, Eversole KA, et al. (2009) Genome-wide survey of SNP variation uncovers the genetic structure of cattle breeds. Science 324: 528–532.
18. Hayes BJ, Chamberlain AJ, Maceachern S, Savin K, McPartlan H, et al. (2009) A genome map of divergent artificial selection between Bos taurus dairy cattle and Bos taurus beef cattle. Anim Genet 40: 176–184.
19. Qanbari S, Pimentel EC, Tetens J, Thaller G, Lichtner P, et al. (2010) A genome-wide scan for signatures of recent selection in Holstein cattle. Anim Genet 41: 377–389.
20. Qanbari S, Gianola D, Hayes B, Schenkel F, Miller S, et al. (2011) Application of site and haplotype-frequency based approaches for detecting selection signatures in cattle. BMC Genomics 12: 318.
21. Sabeti PC, Reich DE, Higgins JM, Levine HZ, Richter DJ, et al. (2002) Detecting recent positive selection in the human genome from haplotype structure. Nature 419: 832–837.
22. Tang K, Thornton KR, Stoneking M (2007) A new approach for using genome scans to detect recent positive selection in the human genome. PLoS Biol 5: e171.
23. Voight BF, Kudaravalli S, Wen X, Pritchard JK (2006) A map of recent positive selection in the human genome. PLoS Biol 4: e72.
24. Gautier M, Naves M (2011) Footprints of selection in the ancestral admixture of a New World Creole cattle breed. Mol Ecol 20: 3128–3143.
25. Hammond AC, Olson TA, Chase CC, Jr., Bowers EJ, Randel RD, et al. (1996) Heat tolerance in two tropically adapted Bos taurus breeds, Senepol and Romosinuano, compared with Brahman, Angus, and Hereford cattle in Florida. J Anim Sci 74: 295–303.
26. Hammond AC, Chase CC, Jr., Bowers EJ, Olson TA, Randel RD (1998) Heat tolerance in Tuli-, Senepol-, and Brahman-sired F1 Angus heifers in Florida. J Anim Sci 76: 1568–1577.
27. Olson TA, Lucena C, Chase CC, Jr., Hammond AC (2003) Evidence of a major gene influencing hair length and heat tolerance in Bos taurus cattle. J Anim Sci 81: 80–90.
28. Padda DS (1999) Development of Senepol cattle: A collaborative research story. In: Sterns R, ed. Proceedings International Senepol Research Symposium Kingshill.
29. Fleming CB (1999) Early stages of the breed: the farmers' perspective. In: Sterns R, ed. International Senepol Research Symposium St. Croix.
30. Godfrey RW (1999) Foreword for the 1999 edition. In: Sterns R, ed. International Senepol Research Symposium St. Croix.
31. Murray M, Trail JC, Davis CE, Black SJ (1984) Genetic resistance to African Trypanosomiasis. J Infect Dis 149: 311–319.
32. d'Ieteren GD, Authie E, Wissocq N, Murray M (1998) Trypanotolerance, an option for sustainable livestock production in areas at risk from trypanosomosis. Rev Sci Tech 17: 154–175.
33. De Alba J (1987) Criolo cattle of Latin America. In: Hodges J, ed. Animal Genetic Resources: Strategies for Improved Use and Conservation FAO Animal Production and Health Paper 66 Rome, Italy: Food and Agriculture Organization of the United Nations. pp 17–40.
34. Payne WJA, Hodges J (1997) Tropical Cattle: Origins, Breeds and Breeding Policies: Wiley-Blackwell. 336 p.
35. Scheet P, Stephens M (2006) A fast and flexible statistical model for large-scale population genotype data: applications to inferring missing genotypes and haplotypic phase. Am J Hum Genet 78: 629–644.
36. Georges M, Drinkwater R, King T, Mishra A, Moore SS, et al. (1993) Microsatellite mapping of a gene affecting horn development in Bos taurus. Nat Genet 4: 206–210.
37. Hansen PJ (2004) Physiological and cellular adaptations of zebu cattle to thermal stress. Anim Reprod Sci 82–83: 349–360.
38. Porto Neto LR, Jonsson NN, D'Occhio MJ, Barendse W (2011) Molecular genetic approaches for identifying the basis of variation in resistance to tick infestation in cattle. Vet Parasitol 180: 165–172.
39. Mariasegaram M, Chase CC, Jr., Chaparro JX, Olson TA, Brenneman RA, et al (2007) The slick hair coat locus maps to chromosome 20 in Senepol-derived cattle. Anim Genet 38: 54–59.

40. Bennett JW (1964) Thermal insulation of cattle coats. Proceedings of the Australian Society Animal Production 5: 160–166.
41. Dikmen S, Alava E, Pontes E, Fear JM, Dikmen BY, et al. (2008) Differences in thermoregulatory ability between slick-haired and wild-type lactating Holstein cows in response to acute heat stress. J Dairy Sci 91: 3395–3402.
42. Berman A (2004) Tissue and external insulation estimates and their effects on prediction of energy requirements and of heat stress. J Dairy Sci 87: 1400–1412.
43. Kutty RK, Kutty G, Samuel W, Duncan T, Bridges CC, et al. (2001) Molecular characterization and developmental expression of NORPEG, a novel gene induced by retinoic acid. J Biol Chem 276: 2831–2840.
44. Foitzik K, Spexard T, Nakamura M, Halsner U, Paus R (2005) Towards dissecting the pathogenesis of retinoid-induced hair loss: all-trans retinoic acid induces premature hair follicle regression (catagen) by upregulation of transforming growth factor-beta2 in the dermal papilla. J Invest Dermatol 124: 1119–1126.
45. Bergfeld WF (1998) Retinoids and hair growth. J Am Acad Dermatol 39: S86–89.
46. Wunderlich KR, Abbey CA, Clayton DR, Song Y, Schein JE, et al. (2006) A 2.5-Mb contig constructed from Angus, Longhorn and horned Hereford DNA spanning the polled interval on bovine chromosome 1. Anim Genet 37: 592–594.
47. Brenneman RA, Davis SK, Sanders JO, Burns BM, Wheeler TC, et al. (1996) The polled locus maps to BTA1 in a Bos indicus x Bos taurus cross. J Hered 87: 156–161.
48. Drogemuller C, Wohlke A, Momke S, Distl O (2005) Fine mapping of the polled locus to a 1-Mb region on bovine chromosome 1q12. Mamm Genome 16: 613–620.
49. Stella A, Ajmone-Marsan P, Lazzari B, Boettcher P (2010) Identification of selection signatures in cattle breeds selected for dairy production. Genetics 185: 1451–1461.
50. Lambert JM, Lambert QT, Reuther GW, Malliri A, Siderovski DP, et al. (2002) Tiam1 mediates Ras activation of Rac by a PI(3)K-independent mechanism. Nat Cell Biol 4: 621–625.
51. Shepherd TR, Klaus SM, Liu X, Ramaswamy S, DeMali KA, et al. (2010) The Tiam1 PDZ domain couples to Syndecan1 and promotes cell-matrix adhesion. J Mol Biol 398: 730–746.
52. Seidlitz EP, Sharma MK, Singh G (2010) Extracellular glutamate alters mature osteoclast and osteoblast functions. Can J Physiol Pharmacol 88: 929–936.
53. Goto H, Watanabe K, Araragi N, Kageyama R, Tanaka K, et al. (2009) The identification and functional implications of human-specific "fixed" amino acid substitutions in the glutamate receptor family. BMC Evol Biol 9: 224.
54. Toll-Riera M, Laurie S, Alba MM (2011) Lineage-specific variation in intensity of natural selection in mammals. Mol Biol Evol 28: 383–398.
55. Wigginton JE, Cutler DJ, Abecasis GR (2005) A note on exact tests of Hardy-Weinberg equilibrium. Am J Hum Genet 76: 887–893.
56. Storey JD, Tibshirani R (2003) Statistical significance for genomewide studies. Proc Natl Acad Sci U S A 100: 9440–9445.
57. Rousset F (2008) GenePop'007: a complete re-implementation of the HenePop software for Windows and Linux. Molecular Ecology Resources 8: 103–106.
58. Saitou N, Nei M (1987) The neighbor-joining method: a new method for reconstructing phylogenetic trees. Mol Biol Evol 4: 406–425.
59. Paradis E, Claude J, Strimmer K (2004) APE: Analyses of Phylogenetics and Evolution in R language. Bioinformatics 20: 289–290.
60. Alexander DH, Novembre J, Lange K (2009) Fast model-based estimation of ancestry in unrelated individuals. Genome Res 19: 1655–1664.
61. Patterson N, Price AL, Reich D (2006) Population structure and eigenanalysis. PLoS Genet 2: e190.
62. Chessel D (2004) The ade4 package. I. One-table methods. R News 4: 5–10.
63. Jombart T (2008) adegenet: a R package for the multivariate analysis of genetic markers. Bioinformatics 24: 1403–1405.
64. Gautier M, Vitalis R (2012) rehh: an R package to detect footprints of selection in genome-wide SNP data from haplotype structure. Bioinformatics 28: 1176–1177.

15

Evaluation of Animal Genetic and Physiological Factors That Affect the Prevalence of *Escherichia coli* O157 in Cattle

Soo Jin Jeon[1,2], Mauricio Elzo[1], Nicolas DiLorenzo[1,3], G. Cliff Lamb[1,3], Kwang Cheol Jeong[1,2]*

1 Department of Animal Sciences, Institute of Food and Agricultural Sciences, University of Florida, Gainesville, Florida, United States of America, **2** Emerging Pathogens Institute, University of Florida, Gainesville, Florida, United States of America, **3** North Florida Research and Education Center, Institute of Food and Agricultural Sciences, University of Florida, Marianna, Florida, United States of America

Abstract

Controlling the prevalence of *Escherichia coli* O157 in cattle at the pre-harvest level is critical to reduce outbreaks of this pathogen in humans. Multilayers of factors including the environmental and bacterial factors modulate the colonization and persistence of *E. coli* O157 in cattle that serve as a reservoir of this pathogen. Here, we report animal factors contributing to the prevalence of *E. coli* O157 in cattle. We observe the lowest number of *E. coli* O157 in Brahman breed when compared with other crosses in an Angus-Brahman multibreed herd, and bulls excrete more *E. coli* O157 than steers in the pens where cattle were housed together. The presence of super-shedders, cattle excreting $>10^5$ CFU/rectal anal swab, increases the concentration of *E. coli* O157 in the pens; thereby super-shedders enhance transmission of this pathogen among cattle. Molecular subtyping analysis reveal only one subtype of *E. coli* O157 in the multibreed herd, indicating the variance in the levels of *E. coli* O157 in cattle is influenced by animal factors. Furthermore, strain tracking after relocation of the cattle to a commercial feedlot reveals farm-to-farm transmission of *E. coli* O157, likely via super-shedders. Our results reveal high risk factors in the prevalence of *E. coli* O157 in cattle whereby animal genetic and physiological factors influence whether this pathogen can persist in cattle at high concentration, providing insights to intervene this pathogen at the pre-harvest level.

Editor: A. Mark Ibekwe, U. S. Salinity Lab, United States of America

Funding: This work was supported by the Institute of Food and Agricultural Sciences, and Emerging Pathogens Institute at University of Florida, grants to KCJ. The funders had no role in study design, data collection and analysis, decision to publish, or preparation of the manuscript.

Competing Interests: The authors have declared that no competing interests exist.

* E-mail: kcjeong@ufl.edu

Introduction

The prevalence of *Escherichia coli* O157 in cattle herds (ranging 0–61%) [1,2,3,4] is positively correlated to outbreaks of this pathogen causing severe diseases including hemorrhagic colitis (HC) and hemolytic uremic syndrome (HUS), which can cause kidney failure and be fatal [3,5]. Despite the implementation of government regulations and development of process interventions, food recalls and human illness related to *E. coli* O157 remain concerns around the world. Reducing the prevalence of this pathogen in cattle at the pre-harvest level has been highlighted recently as a critical control point to decrease the number of *E. coli* O157 entering the food chain [6,7,8,9,10]. Awareness of the risk factors that may increase the prevalence of *E. coli* O157 at the pre-harvest level can provide insights to develop intervention technologies to reduce its prevalence. Although risk factors on farms have been extensively studied from the bacterial perspectives, information regarding animal factors that may contribute to the prevalence of this pathogen is lacking. Here we identify high risk factors that significantly affect the prevalence of this pathogen in cattle.

Cattle are the primary reservoir of *E. coli* O157, and ground beef remains a significant source of foodborne transmission with other sources such as fresh vegetables [11]. Cattle that excrete more than 10^4 colony forming unit (CFU)/g of cattle feces have been defined as super-shedders [9,12]. The super-shedders are responsible for about 90% of the total number of bacteria in the cattle herd [9,12] and raise the prevalence of cattle infected with this pathogen on farms, making them a high risk factor at the pre-harvest level [6,7,13]. However, colonization of this pathogen in cattle is usually asymptomatic due to the lack of Shiga toxin receptor, globotriaosylceramide (Gb3), in cattle endothelial cells [14] that prevents elimination of super-shedding cattle contaminated with this pathogen at farms.

Several aspects influence the prevalence of *E. coli* O157 in cattle. The colonization of *E. coli* O157 at the rectal anal junction (RAJ) likely allows this pathogen to persist and shed high levels of bacteria for weeks or months [15,16]. *E. coli* O157 primarily colonizes the mucosal epithelium at the RAJ [17], although it can be isolated from the gall bladder and along the gastrointestinal tract [18,19]. More than 100 genes are involved in the colonization of the bovine intestine identified by genetic and biochemical analyses [20], including the *E. coli* O157 type III secretion system that enables the translocation of effector proteins into host cells and is required for colonization of cattle [21,22,23,24,25]. Besides the bacterial factors, environmental factors are believed to contribute to the prevalence of *E. coli* O157 in cattle. The prevalence of *E. coli* O157 in feces fluctuates by season with the peak between late spring and early fall [26,27,28].

Water, soil, wild animals, insects, and dirty equipment are important vectors for spreading and transmission of *E. coli* O157 [29,30,31,32]. Transmission of *E. coli* O157 by environmental sources is one of the challenges to the development of pre-harvest interventions.

Even though similar *E. coli* O157 strains are flourishing in the same herds with identical cattle husbandry practices applied, a portion of animals are considered super-shedders [9,12], suggesting that the phenomenon of super-shedders are controlled by multilayers of factors. However, underling mechanisms by which certain animals (2–5% in herds) become super-shedders are not clearly understood. In early studies of animal inoculation with *E. coli* O157, inoculated orally with water, some animals were never colonized with this pathogen, indicating that some animals were resistant to *E. coli* O157 [10,32]. On the basis of these observations, we hypothesized that, in addition to bacterial factors, animals play a critical role in modulating the colonization of this pathogen in animals. This study was designed to address the impact of animal genetic factors on *E. coli* O157 prevalence, as well as animal husbandry. Here, we present our findings that genetic and physiological factors of animals and animal husbandry practices significantly affect the prevalence of *E. coli* O157, suggesting potential intervention practices to reduce this pathogen entering to the food production chain.

Materials and Methods

Ethics Statement

Standard practices of animal care and use were applied to animals used in this project. Research protocols were approved by the University of Florida Institutional Animal Care and Use Committee (IACUC number 201003744).

Animal Genetic Background

Cattle belonged to the Angus-Brahman multibreed herd at the University of Florida. The herd was established in 1988 to conduct long-term genetic studies in beef cattle under subtropical environmental conditions. Cattle were assigned to six breed groups according to the following breed composition ranges: calf breed group 1 = 100% to 80% of Angus, 0% to 20% of Brahman; calf breed group 2 = 79% to 60% of Angus, 21% to 40% of Brahman; calf breed group 3 = 62.5% of Angus, 37.5% of Brahman, calf breed group 4 = 59% to 40% of Angus, 41% to 60% of Brahman, calf breed group 5 = 39% to 20% of Angus, 61% to 80% of Brahman, and calf breed group 6:19% to 0% of Angus, 81% to 100% of Brahman. Mating was diallel, i.e., sires from the 6 breed groups defined above were mated to dams of these same 6 breed groups [33]. Calves (n = 91) were kept at the Beef Research Unit of the University of Florida before weaning and were moved to the University of Florida Feed Efficiency Facility (UFEF) after weaning.

Cattle Maintenance

Angus, Brahman, and Angus × Brahman crossbred steers (n = 80, 253±38 kg) and bulls (n = 11, 345±29 kg) were housed in the UFEF at the North Florida Research and Education Center in Marianna, Florida for 97 days (From October 19, 2011 until January 24, 2012). Upon arrival (day 0) to the UFEF, cattle were fitted with electronic identification tags (Allflex USA Inc., Dallas-Fort Worth, TX) and were randomly allocated to 5 concrete floor pens of 108 m^2 each with wood shavings bedding for a total of 18 animals per pen with the exception of one pen which contained 19 animals. The 11 bulls were allocated to only 2 of the 5 pens (one pen with 5 and another pen with 6 bulls). Cattle had ad libitum access to feed and water at all times and intake was monitored continuously via a GrowSafe system (GrowSafe Systems Ltd., Airdrie, Alberta, Canada). Each pen contained two GrowSafe nods, thus the mean stocking rate per node was 9 cattle. Cattle received a diet comprised of (DM basis) 34% ground bahiagrass hay, 31% corn gluten feed pellets, 31% soybean hulls pellets, 4% of supplement containing molasses, urea, vitamin and minerals. The diet was formulated to have 14.7% crude protein and 0.97 Mcal of Net Energy of gain/kg of diet dry matter.

Detection and Enumeration of *E. coli* O157

The presence or absence of *E. coli* O157 in swabs of the RAJ was determined as previously described with minor modifications [10]. Rectal anal junction swab samples were collected from 91 cattle at the UFEF and a commercial feedlot. Samples were taken to the lab on ice within 4 h of collection to minimize bacterial growth and further tested for microbiological identification. Swab samples were resuspended in 2 ml of Tryptic soy broth (Difco) and serially diluted in 0.1% (w/v) peptone. Two hundred microliters of diluted samples (neat, 10^{-1}, 10^{-2}, 10^{-3}, and 10^{-4}) were then plated on MacConkey sorbitol agar (Difco) supplemented with cefixime (50 μg/liter; Lederle Labs, Pearl River, N.Y.) and potassium tellurite (2.5 mg/liter; Sigma) (CT-SMAC) to determine the number of *E. coli* O157 [34]. Plates were incubated at 37°C for 18–24 h and typical *E. coli* O157 colonies (i.e., sorbitol-negative colonies and multiplex PCR positive, described below) were enumerated. The minimum detection limit of the direct plating procedure was approximately 10 CFU/swab. Enrichment was used for the presence or absence determinations on samples that did not yield *E. coli* O157 by direct plating. For this purpose, samples were enriched in TSB supplemented with novobiocin (20 μg/ml; Sigma) for 18 to 24 h at 37°C with shaking, and *E. coli* O157 was detected by direct plating after serial dilution (neat, 10^{-1}, 10^{-2}, 10^{-3}, and 10^{-4}) with 0.1% (w/v) peptone. Sorbitol-negative colonies were confirmed by multiplex PCR.

Multiplex PCR

Multiplex PCR was conducted to confirm *E. coli* O157. Bacterial strains used in this paper are listed (Table 1). *E. coli* O157:H7 EDL933 and DH5α were used as a positive and negative control of multiplex PCR, respectively. Primers were designed to detect *stx1*, *stx2*, *hly*, and *rbfE* (Table 2). *stx1* and *stx2* primers detected subunit A of Stx1 and Stx2, respectively. Each PCR wells contained 25 μl of reaction mix, comprised of 2.5 μl of 10X buffer, 0.5 μl of dNTP, 1 μl of Taq polymerase and a mixture of the 8 primers. PCR cycling conditions were as follows: 94°C for 5 min for pre-denature, 94°C for 30 sec, 55°C for 30 sec, 72°C for 1 min each 30 cycles, and 72°C for 10 min for a final extension. PCR products were visualized on 1.5% agarose gel in Tris-EDTA buffer after electrophoresis.

Subtyping of *E. coli* O157 Isolates Using Pulsed-field Gel Electrophoresis

Pulsed-Field Gel Electrophoresis (PFGE) was performed to subtype farm isolates in accordance with PulseNet standardized laboratory protocol. A colony purified on CT-SMAC was grown overnight in a shaker in Luria Broth (LB) at 37°C. Concentration of cell suspension was adjusted to an optical density of 1.0 at 600 nm. Cells (400 μl) were mixed with 20 μl of proteinase K (20 mg/ml stock, Fisher Scientific) and 1% Sekem Gold agarose (Lonza). The mixture was placed into a well of disposable plug molds (Bio-Rad Laboratories). Agarose plugs were lysed in cell lysis buffer (50 mM Tris, 50 mM EDTA, pH 8.0 and 1%

Table 1. Strains used in this study.

Strain name	Description	references
E. coli O157:H7 EDL933	ATCC43895	[37]
E. coli DH5α	Lab collection	
KCJ1220	Isolated from medium-shedding animal in pen 17	This study
KCJ1225	Isolated from medium-shedding animal in pen 13	This study
KCJ1231	Isolated from super-shedding animal in pen 17	This study
KCJ1237	Isolated from medium-shedding animal in pen 15	This study
KCJ1238	Isolated from medium-shedding animal in pen 16	This study
KCJ1242	Isolated from low-shedding animal in pen 15	This study
KCJ1244	Isolated from low-shedding animal in pen 16	This study
KCJ1252	Isolated from super-shedding animal in pen 14	This study
KCJ1254	Isolated from medium-shedding animal in pen 14	This study
KCJ1265	Isolated from low-shedding animal in pen 13	This study
KCJ1266	Isolated from super-shedding animal in pen 13	This study
KCJ1268	Isolated from low-shedding animal in pen 14	This study
KCJ1430	Isolated from a commercial feedlot	This study
KCJ1432	Isolated from a commercial feedlot	This study

Sarcosyl) for 2 h at 55°C with constant shaking at 170 rpm. Lysed plugs were washed one time with sterile distilled-water, followed by TE buffer (10 mM Tris, 1 mM EDTA, pH 8.0) twice at 55°C. Plugs were digested with 10 units of *Xba*I (New England BioLabs), and then electrophoresed using a CHEF Mapper (Bio-Rad Laboratories) under the following condition: 0.5 X Tris-Borate EDTA buffer at 14°C, 6 V/cm for 19 h, and an initial switch time from 2.16 s to 54.17 s. Lambda Ladder PFG Marker (New England BioLabs) was run as a size marker. The PFGE patterns were visualized by using GelDoc™ XR+ with Image Lab™ software (Bio-Rad Laboratories). The banding patterns were analyzed with GelCompar II software (Applied Maths, Kortrijk, Belgium).

Table 2. Oligonucleotides used in this study.

Target	Name	Sequence (5′- 3′)	Orientation	Size (bp)
rfbE[a]	KCP57	CGGACATCCA TGTGATATGG	F	259
	KCP58	TTGCCTATGTA CAGCTAATCC	R	
stx1[b]	KCP11	TGTCGCATAGT GGAACCTCA	F	655
	KCP12	TGCGCACTGAG AAGAAGAGA	R	
stx2[b]	KCP13	CCATGACAACG GACAGCAGTT	F	477
	KCP14	TGTCGCCAGTTA TCTGACATTC	R	
hlyA[b]	KCP19	GCGAGCTAAGC AGCTTGAAT	F	199
	KCP20	CTGGAGGCTGC ACTAACTCC	R	

[a]Referenced by Valadez *et al.* [38].
[b]Referenced by Bai et al. [39].

Statistical Analysis

Statistical analyses were performed by GraphPad InStat version 3.10. Differences among pens were analyzed by the one-way analysis of variance (ANOVA) test followed by Tukey's test. Proportions of the super-shedders between bulls and steers were compared by Fisher's exact test. All data were expressed as mean ± standard error. A P value of <0.05 was considered significant.

Results

Prevalence of *E. coli* O157 in the Multibreed Cattle

As a first step to understand the animal factors that may affect the prevalence of *E. coli* O157 in cattle, we enumerated this pathogen at the RAJ. A total of 91 animals were produced by the diallel mating system described previously [33] and assigned to six groups according to their estimated breed composition. As shown in figure 1A, calf breed group 1 contained the largest expected portion of genetic traits from Angus (100–80%) and the smallest expected portion from Brahman (0–20%). As the calf breed group numbers increase, the Angus portion decreases and Brahman portion increases. Thus, breed group 1 is more closely related to Angus while breed group 6 is more closely related to Brahman. Six different breed groups were randomly housed in 5 pens, but bulls were housed only in pen 1 and 2 to test if castration may affect the levels of this pathogen in the groups (Fig. 1B).

The total number of *E. coli* O157 from the RAJ swab samples was enumerated by a direct plating method without enrichment to monitor the real number of bacteria colonized on the RAJ. Swab samples were serial diluted before plated on CT-SMAC plate and incubated for 24 hours. Colonies with characteristic sorbitol negative color were picked and confirmed by using multiplex PCR. Out of 91 samples, 37 samples (40.66%) were found to be positive for *E. coli* O157 with a detection limit of 10 CFU/swab (Fig. 2). Samples that were negative for the direct plating method were enriched overnight followed by direct plating on CT-SMAC after serial dilution, but O157 was not detected from these samples (data not shown). The total number of *E. coli* O157 from the RAJ varied between animals, ranging from 10^1 to more than 10^6

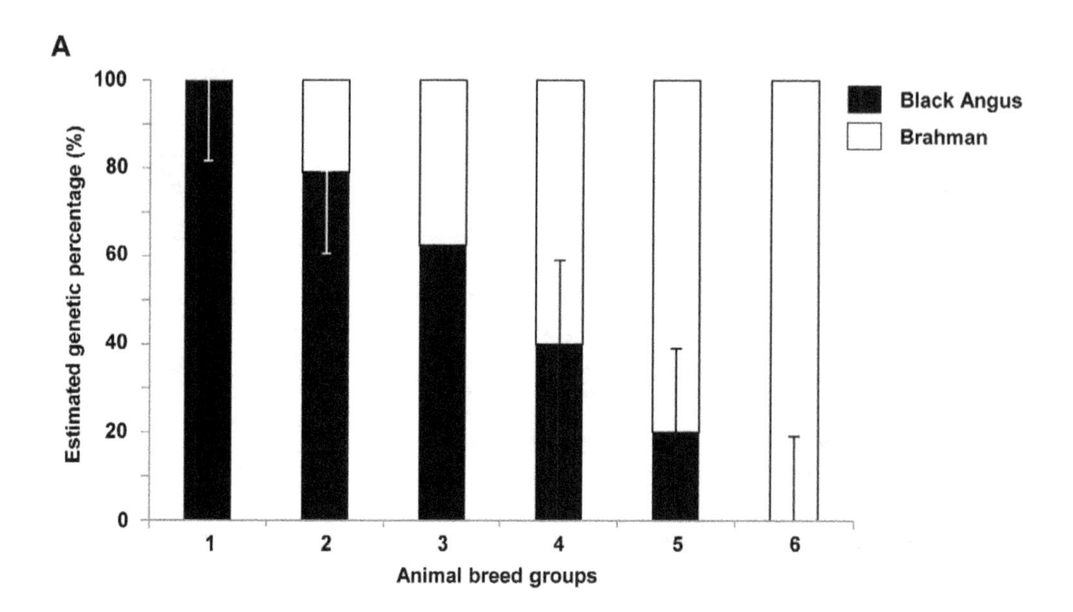

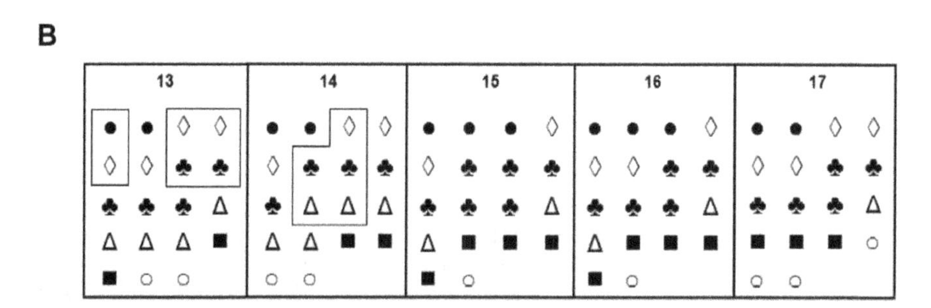

Figure 1. Estimated genetic backgrounds of animals used and housing in five pens. (A) Cattle consisted of six Angus-Brahman multibred groups. Group 1 = 100% to 80% Angus, 0% to 20% Brahman; Group 2 = 79% to 60% Angus, 21% to 40% Brahman; Group 3 = 62.5% Angus, 37.5% Brahman, Group 4 = 59% to 40% Angus, 41% to 60% Brahman, Group 5 = 39% to 20% Angus, 61% to 80% Brahman, Group 6:19% to 0% Angus, 81% to 100% Brahman. Bars indicate the expected portion of Angus and Brahman genetic traits in each breed group. (B) Animals were systemically housed in five pens according to their genetic background and sex. ● is Calf breed group 1, ◇ is Calf breed group 2, ♣ is Calf breed group 3, Δ is Calf breed group 4, ■ is Calf breed group 5, and ○ is Calf breed group 6. Bulls in pen 13 and 14 are boxed.

(Fig. 2). The majority of positive samples contained this pathogen at 10^2–10^5 CFU/swab (n = 31; 83.78% of positive samples) and 6 cattle (16.21% of positive samples) contained more than 10^5 CFU/swab (Table 3). Cattle were categorized into four groups depending on the number of *E. coli* O157 shedding, non-shedder (<10^1/swab; n = 54; 59.34% of cattle), low-shedder (10^1–10^3 CFU/swab; n = 16; 17.59% of cattle), medium shedder (10^3–10^5 CFU/swab, n = 15; 16.48% of cattle), and super-shedder (>10^5 CFU, n = 6; 6.6% of cattle). Previously, super-shedder was defined as an animal that excretes *E. coli* O157 at more than 10^4 CFU/g of feces. However, in this study we defined a super shedder at excretions of more than 10^5 CFU/swab because previous results showed that RAJ swab samples were 10 fold higher than fecal samples [10].

Presence of Super-shedders in Pens Increases the Total Number of Bacteria in Herds

It has been shown that the presence of super-shedders increases the prevalence of *E. coli* O157 in hides and carcass by enhancing transmission of this pathogen [6,7,13]. However, it is not well known if the super-shedders increase the total number of this pathogen in herds. We determined if there was a correlation between the presence of super-shedders and the level of *E. coli* O157 in different pens. Pen 13, 14, and 17 had 3, 2, and 1 super-shedder in the group of cattle, respectively (Fig. 3A), resulting in a higher number of *E. coli* O157 in the pens compared to the pens without super-shedders (pen 15 and 16). To understand the role of super-shedders in the transmission of *E. coli* O157, the total number of *E. coli* O157 bacteria excreted from super-shedders was calculated to determine what percentage of this pathogen was shed from super-shedders. *E. coli* O157 excreted from super-shedder accounted for more than 95% of total bacteria in the herds (Fig. 3B). These data confirm that super-shedders serve as a high risk factor for the transmission of this pathogen to other animals (Fig. 2 or Table 3). In addition, it suggests that removing super-shedders in cattle herds can be an effective method to reduce the number of this pathogen in herds, thus identification of super-shedders is a critical control point to reduce potential outbreaks caused by this pathogen.

Bulls are More Susceptible to *E. coli* O157 than Steers

To determine whether castration may affect the prevalence of *E. coli* O157, the number of this pathogen was counted in bulls and steers at the RAJ. The number of super-shedders was significantly different ($P = 0.022$) between bulls (27.27%, n = 11) and steers (3.75%, n = 80), indicating bulls are more susceptible to become

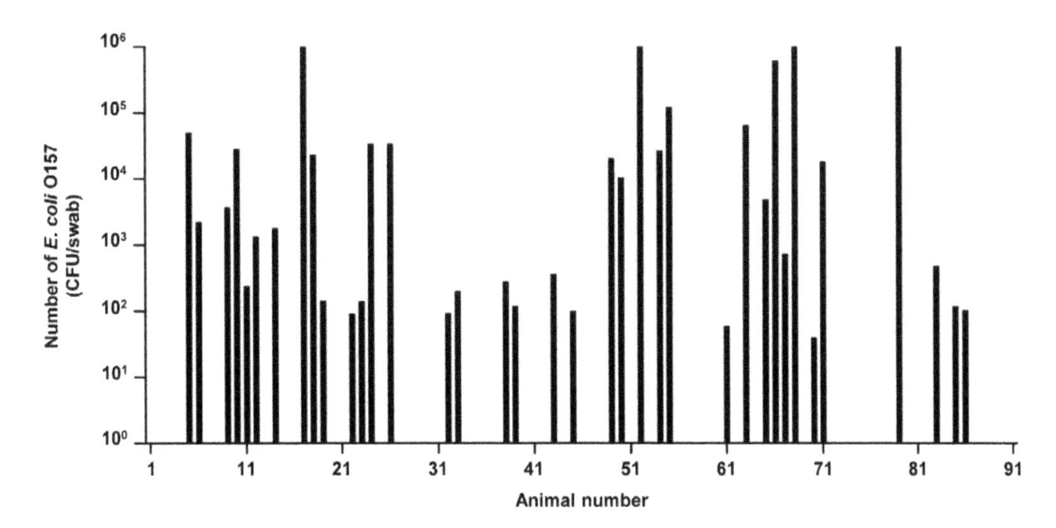

Figure 2. The number of *E. coli* O157 isolated from rectal anal junction in cattle. Rectal swabs were enumerated by direct plating on CT-SMAC and further identified by multiplex PCR detecting the *stx1, stx2, rbfE,* and *hlyA* genes. Cattle were designate as low-shedders (10^1–10^2 CFU), medium-shedders (10^3–10^4 CFU), and super-shedders (>10^5 CFU). Six out of 91 cattle were super-shedders, accounting for 6.6% of all calves. Limit of detection was 10 CFU in this direct plating method.

super-shedders. However, it was remarkable that the prevalence of *E. coli* O157 in both cattle groups was similar at 36.36% and 41.25% respectively (Table 3). Taken together, even though steers and bulls are exposed to this pathogen in the same environment, probably by transmission from super-shedders, bulls are more likely to develop into super-shedders. This data suggest that steers may need to be separated from bulls to reduce total number of this pathogen in herds.

Brahman Cattle are More Resistant to *E. coli* O157

Results from previous inoculation studies of *E. coli* O157 with steers showed that certain cattle were resistant to this pathogen, despite repeated inoculation [32], suggesting animal factors were critical for colonization at the RAJ. We hypothesized that genetic factors play a significant role in the prevalence of *E. coli* O157 in animals. To assess the role of the genetic factors, we examined six different breed groups to identify if there is a correlation between breed groups and resistance to *E. coli* O157. The breed group 6 excreted the lowest number of *E. coli* O157 among the groups. As shown in Fig. 4A, the level of *E. coli* O157 was the lowest in breed group 6 compared to other groups, indicating that Brahman calves were more resistant to *E. coli* O157 than others. When calves were progeny of dams or sires from group 6, the calves had the lowest number of *E. coli* O157 compared to other groups (Fig. 4B and 4C). Notably, we could not observe a linear relationship between genetic composition and *E. coli* O157 resistance in calves, suggesting the resistance against pathogens may not be additive, but existing only in 100% Brahman cattle.

Table 3. Distribution of cattle based on the level of *E. coli* O157 counts.

Type of shedder	CFU/swab	Number of cattle (%) Total	Bull	Steer
Non-shedder	0–10^1	54 (59.34)	7 (63.64)	47 (58.75)
Low-shedder	10^1–10^2	4 (4.40)	0 (0)	4 (5.00)
	10^2–10^3	12 (13.19)	1 (9.09)	11 (13.75)
Medium-shedder	10^3–10^4	5 (5.49)	0 (0)	5 (6.25)
	10^4–10^5	10 (10.99)	0 (0)	10 (12.50)
Super-shedder	10^5–10^6	2 (2.20)	1 (9.09)	1 (1.25)
	>10^6	4 (4.40)	2 (18.18)	2 (2.50)

One Dominant *E. coli* O157 Strain was Prevalent Among Cattle

Although we observed that animal factors play key roles in determining the levels of *E. coli* O157 in cattle, we could not remove the possibility that a different level of this pathogen among cattle may have been mediated by different *E. coli* O157 strains. Bacterial factors are critical for prevalence and persistence of this pathogen in hosts; therefore, the different levels of *E. coli* O157 among cattle could have resulted by difference of bacterial strains rather than animal factors. Thus, we conducted molecular subtyping to eliminate the possibility that super-shedders carried a well-adapted *E. coli* O157 strain to cattle, while low shedders carried not-well-fitted strains. Strains isolated from low, medium, and super shedders of each pen were analyzed by multiplex PCR to compare strains. Multiplex PCR amplified the *rfbE* gene, which is specific for the O157 serotype, and three virulence genes, *stx1, stx2,* and *hlyA*. Unlike the EDL933 strain, all of the *E. coli* O157 isolates contained only *stx2* and *hlyA* genes without *stx1* (Fig. 5A). Furthermore, PFGE analysis with *XbaI* digestion identified only one type of PFGE pattern in the pens (Fig. 5B). Strains isolated from different cattle shedding low, medium, or super number of *E. coli* O157 displayed 100% similarity, indicating they are the same strains. Only one homogenous *E. coli* O157 strain was dominant among the cattle, indicating one specific strain containing only the *stx2* gene was originated from one common source. These results support that bacterial factors were not major factors making super-shedding cattle in this experiment.

Farm-to-farm Transmission of *E. coli* O157 via Animals

After the pen study at the NFREC, 80 steers were transported to a commercial feedlot X, and we traced the dominant *E. coli* O157

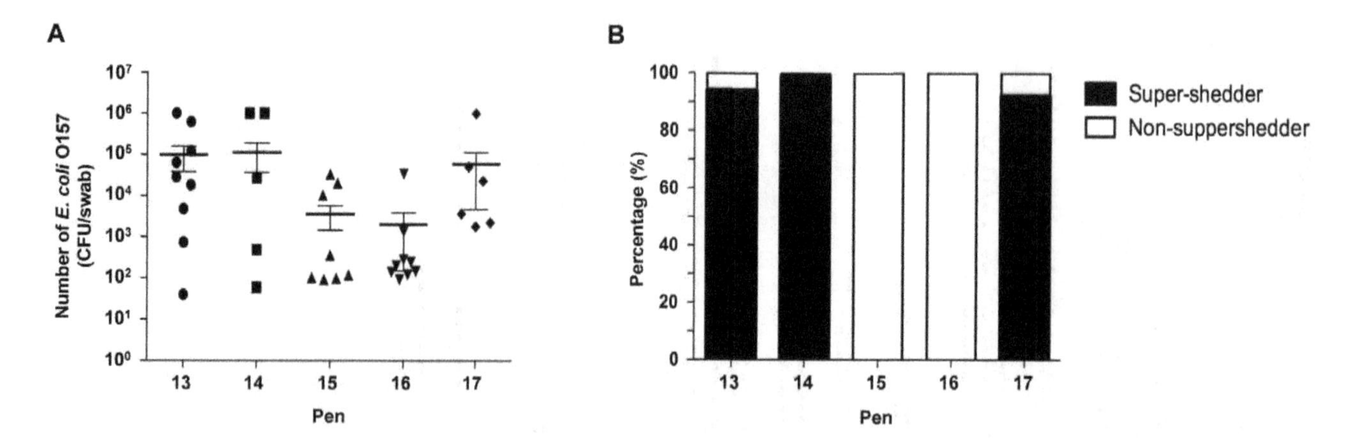

Figure 3. Super-shedders increase the total number of *E. coli* O157 in pens with high density. (A) The total number of *E. coli* O157 was calculated from individual animals in the five pens. Super-shedders increased the total number of *E. coli* O157 in the pens (pen 13, 14, and 17), and the average number of *E. coli* O157 was 100 fold less when a super-shedder was not identified in the pens (pen 15 and 16). The results are represented as mean ± SEM. (B) The percentage of *E. coli* O157 shed from super-shedders in the pens. Black bars represent the percentage of *E. coli* O157 isolated from super-shedders and white bars represent the percentage of *E. coli* O157 from non super-shedders.

strain, which was first identified at the NFREC, to study if farm-to-farm transmission is mediated by cattle. After six months in feedlot X, we collected rectal swab samples (n = 44) from cattle used in the pen study and additional rectal swab samples (n = 19) were collected from random cattle raised in the feedlot. Swab samples were directly, without enrichment, plated on CT-SMAC plate and *E. coli* O157 strains were isolated. Sorbitol negative colonies were further identified by multiplex PCR and PFGE was conducted to evaluate similarity among strains. Unexpectedly, we identified only 2 steers that carried *E. coli* O157 strains, which were positive for *rbfE, stx1, stx2,* and *hlyA*. This is an unexpected result because normally the prevalence of *E. coli* O157 is high at feedlots compared to cattle on pasture. It is not known at this time point why the prevalence of *E. coli* O157 in feedlot X was very low compared to previous studies. However, we isolated the same *E. coli* O157 strain, which was dominant at NFREC, from the *E. coli* O157 positive cattle in feedlot X. As shown in Fig. 5C, one strain had 96.6% similarity, suggesting this strain probably originated from cattle transported from NFREC. It is not known at this time whether the strain originated from super-shedders or not, although it is likely that the strain originated from the super-shedders because they accounted for 90% of total bacteria at NFREC. It is noteworthy that we also isolated one strain showing 70.2% similarity compared to the dominant NFREC strain that may have originated from other sources, probably from cattle in feedlot X because it was not existing at NFREC. Therefore, these data strongly support our hypothesis that cattle transmit pathogens between farms and can be a critical control point to intervene *E. coli* O157 at the pre-harvest level.

Discussion

We report here the role of animal and environmental factors in the prevalence of *E. coli* O157. In addition to the bacterial factors, animal genetic and physiological factors contribute to the prevalence of this pathogen in cattle. Brahman breed among the Angus-Brahman multibreed excreted the lowest level of *E. coli* O157, suggesting this breed is less prone to colonization of this pathogen. Bulls were more susceptible for colonization of this pathogen when compared with steers. These data indicate animal factors significantly contribute to the generation of super-shedders. The presence of super-shedders increased the number of bacteria in the pens and cattle. Super-shedders found in 6.6% of total cattle were responsible for more than 90% of the total number of

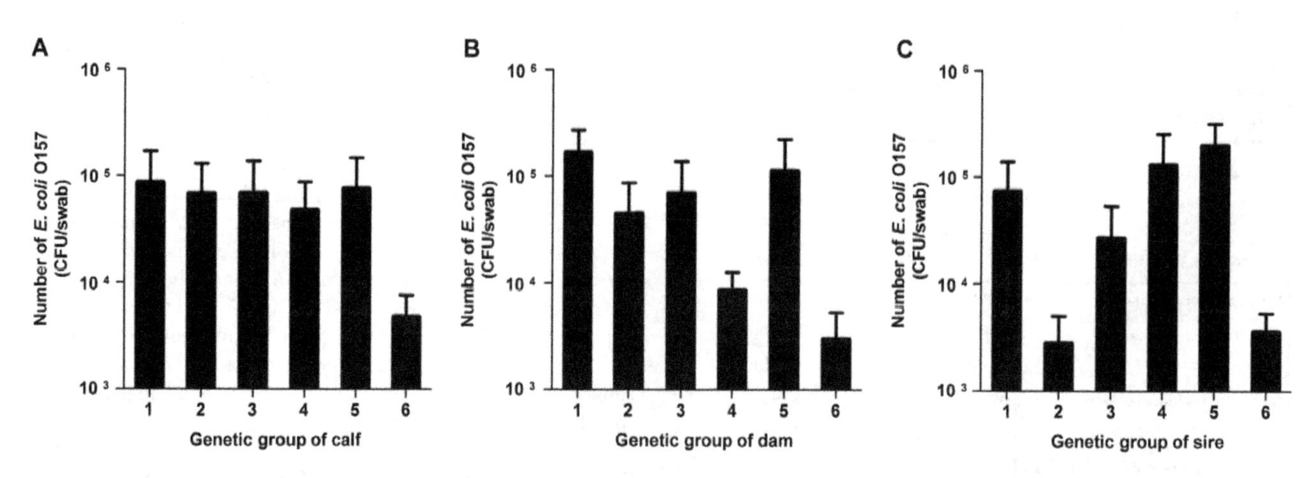

Figure 4. Brahmans are more resistant to *E. coli* O157 than other animals containing different genetic background. All calves (A), dams (B), and sires (C) were classified into the same breed groups. Overall, resistance to O157 was shown in breed group 6.

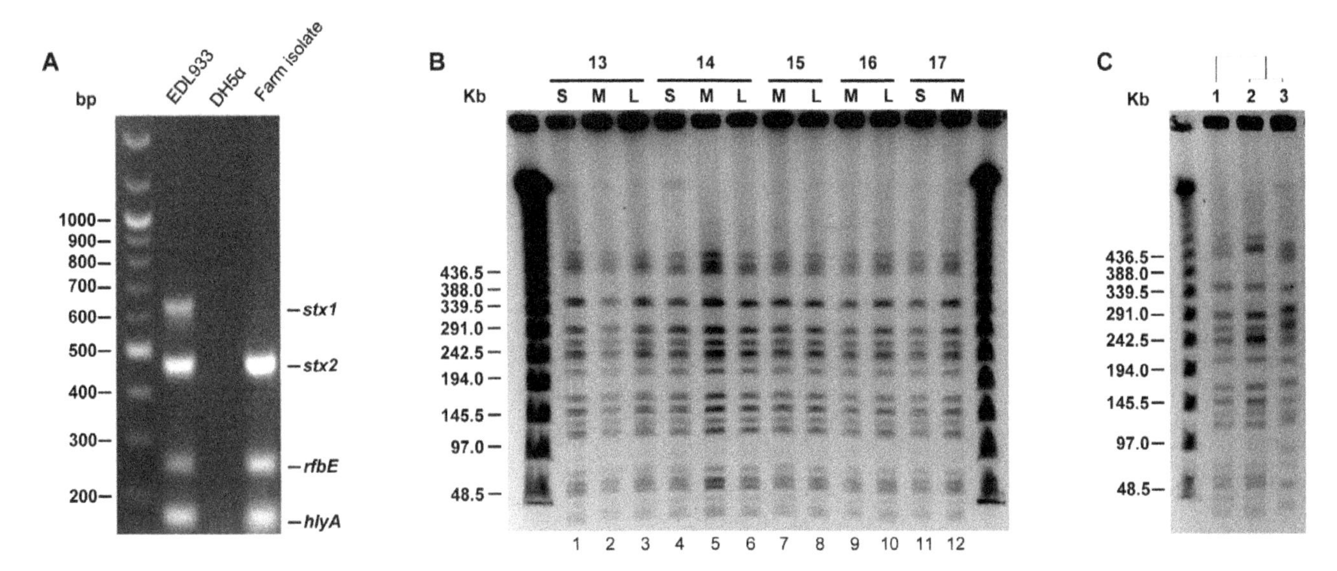

Figure 5. All of the *E. coli* O157 isolates were identical, indicating the strain originated from one common source. (A) Multiplex PCR was conducted using primers amplifying *stx1, stx2, rbfE*, and *hlyA*. Strains used were *E. coli* O157:H7 EDL933 (lane 1), DH5α (lane 2), and KCJ1266 (lane 3). (B) *Xba*I-digested PFGE analysis. Strains used were isolated from the pens indicated on top of the line from animals shed low, medium, and super number of *E. coli* O157. Strains used; lane 1: KCJ1266, lane 2: KCJ1225, lane 3: KCJ1165, lane 4: KCJ1152, lane 5: KCJ1154, lane 6: KCJ1168, lane 7: KCJ1137, lane 8: KCJ1142, lane 9: KCJ1138, lane 10: KCJ1144, lane 11: KCJ1131, lane 12: KCJ1120 (C) The *E. coli* O157 strain isolated from animals raised at a commercial feedlot facility was probably transmitted from super-shedders. *Xba*I-digested PFGE analysis was conducted using the strains isolated from a super-shedder at NFREC and two animals at a commercial feedlot facility. Strains used; lane 1: KCJ1266, lane 2: KCJ1430, lane 3: KCJ1432.

pathogen in herd, increasing the chance of animal-to-animal and farm-to-farm transmission by either direct or indirect contact.

PFGE results from the commercial feedlot X isolates were in agreement with the data acquired from the pen isolates at the NFREC (Fig. 5C). The *E. coli* O157 strains isolated from the commercial feedlot X had high similarity, 96.6%, compared to the strains isolated from the NFREC dominant strain, indicating the commercial feedlot isolates were clonal variants of NFREC strain. Interestingly, we also isolated *E. coli* O157 showing lower similarity (70.2%) compared to the NFREC strains, indicating cattle likely acquired this strain from other cattle at the commercial feedlot X. Although it is plausible that this strain originated from NFREC, we believe that this strain was introduced from the commercial feedlot X. Taken together our data demonstrate the farm-to-farm transmission of *E. coli* O157 via cattle.

In addition to our data showing the transmission of *E. coli* O157 strains between farms, we observed a profound effect on animal-to-animal transmission of *E. coli* O157 strains in the pens. *E. coli* O157 is present in cattle at the prevalence ranging from 0% to 61% [3]; however, the prevalence of this pathogen at NEFREC farm was 40.66% (37 of 91 cattle). Furthermore, when we investigated the prevalence of *E. coli* O157 from grazing cattle at the same farm (i.e., NFREC), prevalence of this pathogen was 1.1% (1 out of 91 animals; Oh and Jeong's unpublished data). Although the high prevalence of *E. coli* O157 in the pens could be mediated by other risk factors such as transmission vectors (i.e., wild animals, insects, and soil) and diet, it is unlikely because the two groups of cattle were raised in the same environmental conditions, except cattle density (pen vs. pasture). Previous studies [10,32] showed that contaminated water could be a major source for *E. coli* O157 contamination in the confined environment; however, water was negative for *E. coli* O157 at NFREC (data not shown), eliminating the possibility that the cattle in the pans were contaminated via water. Thus, this unusual high prevalence of *E. coli* O157 was probably caused by the high density of cattle in the pens that may increase animal-to-animal transmission. Taken together cattle were probably the major source of *E. coli* O157 contamination via animal-to-animal transmission, and we suggest that maintaining cattle at the high density in the pens likely in part increased the prevalence of *E. coli* O157.

Based on an extensive *E. coli* O157 enumeration analysis, we have identified heterogeneous levels of this pathogen among cattle. Six cattle carried high level of *E. coli* O157 (>10^5 CFU/swab) that corresponds to the top 6.6% and 59.34% of cattle did not carry this pathogen even though they were housed in the same pens. Consistent with these data, previous research has shown that some cattle harbor and shed *E. coli* O157 at higher concentration than others [2,6,12]. In this study, we observed the role of super-shedders at high cattle density. Super-shedders were present in 3 pens, and the total number of bacteria in those pens was about 55 fold higher than pens without super-shedders. Super-shedders were responsible for more than 90% of the total number of pathogen in the pens. These data suggest that a super-shedder could increase not only the mean level of O157 among cattle but also the risk of *E. coli* O157 transmission to other cattle. Thus, identification and segregation of super-shedders from uninfected cattle may be a practical strategy to reduce the *E. coli* O157 prevalence in cattle and human disease.

Previous research has shown that bacterial factors may determine super-shedders by showing that diverse *E. coli* O157 strains were observed, but a few of them with particular phage types such as PT 21/28 are more likely to associate with super-shedding cattle via alteration in gene expression of *E. coli* O157 [9]. Unlike the previous findings, our data indicated that only one type of *E. coli* O157 strains was predominant among cattle in 37 cattle without bacterial strain preference (Fig. 5B), demonstrating that animal factors likely determine the colonization or shedding of this pathogen, but not bacterial factors.

We obtained multiple pieces of data indicating that the levels of *E. coli* O157 in cattle were modulated by multiple factors. These data include genetic factors (Fig. 4), physiological characteristics (i.e., castration, Table 3), and cattle density. Among the six breed groups examined, calves in the breed group 6 excreted the lowest number of *E. coli* O157. In addition, super-shedders were not identified in the group 6. As the breed group 6 was composed of Brahman and high percent Brahman cattle, it is reasonable to conclude that Brahman carry genetic factors that confer resistance against *E. coli* O157. Our findings are supported by a previous study where Riley *et al.* [35] found that 17.3% of Angus beef cattle (n = 52) were positive for *E. coli* O157 while 10.1% of Brahman (n = 109) beef cattle were positive with this pathogen by using an overnight enrichment method. The finding suggested that breed difference might affect the prevalence of *E. coli* O157 in the herd. Further studies regarding identification of genetic loci that confer resistance to *E. coli* O157 shedding and colonization will give us insights to develop intervention technologies to reduce *E. coli* O157 at the pre-harvest levels.

In addition to the genetic factors, castration decreased the susceptibility of male calves against this pathogen (Table 3). This was supported by data showing that 27.27% of bulls were super-shedders, whereas only 3.75% of steers were super-shedders. However, prevalence and colonization of *E. coli* O157 is probably influenced by many factors including rumen microflora [36], age [2], immune stress, the amount of feed intake, and unknown animal factors; thus, genetic variation and castration are likely associated with the prevalence of *E. coli* O157 directly or indirectly. Further studies, including identification of genetic loci, will be necessary to identify animal factors that directly govern the prevalence of pathogens. Our principal finding from this study is that animal factors, including genetic factors and castration, influence the prevalence of *E. coli* O157 in cattle. Super-shedders play a critical role in animal-to-animal and farm-to-farm transmission of *E. coli* O157. Identification of animal factors and controlling super-shedders will undoubtedly contribute to the development of intervention technologies to reduce outbreaks caused by this pathogen.

Acknowledgments

We are grateful to Mara Brueck, Matthew Taylor, and Drs. Je Chul Lee, Won Sik Yeo, Man Hwan Oh, and Sung Cheon Hong for the technical assistance.

Author Contributions

Conceived and designed the experiments: SJJ MAE ND KCJ. Performed the experiments: SJJ KCJ. Analyzed the data: SJJ MAE KCJ. Contributed reagents/materials/analysis tools: MAE ND GCL. Wrote the paper: SJJ MAE LD GCL KCJ.

References

1. Besser RE, Lett SM, Weber JT, Doyle MP, Barrett TJ, et al. (1993) An outbreak of diarrhea and hemolytic uremic syndrome from *Escherichia coli* O157:H7 in fresh-pressed apple cider. Jama 269: 2217–2220.
2. Cray WC, Jr., Moon HW (1995) Experimental infection of calves and adult cattle with *Escherichia coli* O157:H7. Applied and environmental microbiology 61: 1586–1590.
3. Elder RO, Keen JE, Siragusa GR, Barkocy-Gallagher GA, Koohmaraie M, et al. (2000) Correlation of enterohemorrhagic *Escherichia coli* O157 prevalence in feces, hides, and carcasses of beef cattle during processing. Proc Natl Acad Sci U S A 97: 2999–3003.
4. Shere JA, Bartlett KJ, Kaspar CW (1998) Longitudinal study of *Escherichia coli* O157:H7 dissemination on four dairy farms in Wisconsin. Appl Environ Microbiol 64: 1390–1399.
5. Nataro JP, Kaper JB (1998) Diarrheagenic *Escherichia coli*. Clinical microbiology reviews 11: 142–201.
6. Arthur TM, Brichta-Harhay DM, Bosilevac JM, Kalchayanand N, Shackelford SD, et al. (2010) Super shedding of *Escherichia coli* O157:H7 by cattle and the impact on beef carcass contamination. Meat science 86: 32–37.
7. Arthur TM, Nou X, Kalchayanand N, Bosilevac JM, Wheeler T, et al. (2011) Survival of *Escherichia coli* O157:H7 on cattle hides. Applied and environmental microbiology 77: 3002–3008.
8. Callaway TR, Carr MA, Edrington TS, Anderson RC, Nisbet DJ (2009) Diet, *Escherichia coli* O157:H7, and cattle: a review after 10 years. Curr Issues Mol Biol 11: 67–79.
9. Chase-Topping ME, McKendrick IJ, Pearce MC, MacDonald P, Matthews L, et al. (2007) Risk factors for the presence of high-level shedders of *Escherichia coli* O157 on Scottish farms. J Clin Microbiol 45: 1594–1603.
10. Jeong KC, Kang MY, Kang J, Baumler DJ, Kaspar CW (2011) Reduction of *Escherichia coli* O157:H7 shedding in cattle by addition of chitosan microparticles to feed. Applied and environmental microbiology 77: 2611–2616.
11. Rangel JM, Sparling PH, Crowe C, Griffin PM, Swerdlow DL (2005) Epidemiology of *Escherichia coli* O157:H7 outbreaks, United States, 1982–2002. Emerging infectious diseases 11: 603–609.
12. Matthews L, Low JC, Gally DL, Pearce MC, Mellor DJ, et al. (2006) Heterogeneous shedding of *Escherichia coli* O157 in cattle and its implications for control. Proceedings of the National Academy of Sciences of the United States of America 103: 547–552.
13. Arthur TM, Keen JE, Bosilevac JM, Brichta-Harhay DM, Kalchayanand N, et al. (2009) Longitudinal study of *Escherichia coli* O157:H7 in a beef cattle feedlot and role of high-level shedders in hide contamination. Applied and environmental microbiology 75: 6515–6523.
14. Lingwood CA (1996) Role of verotoxin receptors in pathogenesis. Trends in microbiology 4: 147–153.
15. Lim JY, Li J, Sheng H, Besser TE, Potter K, et al. (2007) *Escherichia coli* O157:H7 colonization at the rectoanal junction of long-duration culture-positive cattle. Appl Environ Microbiol 73: 1380–1382.
16. Davis MA, Rice DH, Sheng H, Hancock DD, Besser TE, et al. (2006) Comparison of cultures from rectoanal-junction mucosal swabs and feces for detection of *Escherichia coli* O157 in dairy heifers. Appl Environ Microbiol 72: 3766–3770.
17. Naylor SW, Low JC, Besser TE, Mahajan A, Gunn GJ, et al. (2003) Lymphoid follicle-dense mucosa at the terminal rectum is the principal site of colonization of enterohemorrhagic *Escherichia coli* O157:H7 in the bovine host. Infection and immunity 71: 1505–1512.
18. Jeong KC, Kang MY, Heimke C, Shere JA, Erol I, et al. (2007) Isolation of *Escherichia coli* O157:H7 from the gall bladder of inoculated and naturally-infected cattle. Vet Microbiol 119: 339–345.
19. Keen JE, Laegreid WW, Chitko-McKown CG, Durso LM, Bono JL (2010) Distribution of Shiga-toxigenic *Escherichia coli* O157 in the gastrointestinal tract of naturally O157-shedding cattle at necropsy. Applied and environmental microbiology 76: 5278–5281.
20. Dziva F, van Diemen PM, Stevens MP, Smith AJ, Wallis TS (2004) Identification of *Escherichia coli* O157 : H7 genes influencing colonization of the bovine gastrointestinal tract using signature-tagged mutagenesis. Microbiology 150: 3631–3645.
21. Abe A, Heczko U, Hegele RG, Brett Finlay B (1998) Two enteropathogenic *Escherichia coli* type III secreted proteins, EspA and EspB, are virulence factors. The Journal of experimental medicine 188: 1907–1916.
22. Bretschneider G, Berberov EM, Moxley RA (2007) Reduced intestinal colonization of adult beef cattle by *Escherichia coli* O157:H7 tir deletion and nalidixic-acid-resistant mutants lacking flagellar expression. Veterinary microbiology 125: 381–386.
23. Dean-Nystrom EA, Bosworth BT, Moon HW, O'Brien AD (1998) *Escherichia coli* O157:H7 requires intimin for enteropathogenicity in calves. Infection and immunity 66: 4560–4563.
24. Sharma VK, Sacco RE, Kunkle RA, Bearson SM, Palmquist DE (2012) Correlating levels of type III secretion and secreted proteins with fecal shedding of *Escherichia coli* O157:H7 in cattle. Infection and immunity 80: 1333–1342.
25. Sheng H, Lim JY, Knecht HJ, Li J, Hovde CJ (2006) Role of *Escherichia coli* O157:H7 virulence factors in colonization at the bovine terminal rectal mucosa. Infection and immunity 74: 4685–4693.
26. Barkocy-Gallagher GA, Arthur TM, Rivera-Betancourt M, Nou X, Shackelford SD, et al. (2003) Seasonal prevalence of Shiga toxin-producing *Escherichia coli*, including O157:H7 and non-O157 serotypes, and Salmonella in commercial beef processing plants. J Food Prot 66: 1978–1986.
27. Laegreid WW, Elder RO, Keen JE (1999) Prevalence of *Escherichia coli* O157:H7 in range beef calves at weaning. Epidemiol Infect 123: 291–298.
28. Heuvelink AE, van den Biggelaar FL, de Boer E, Herbes RG, Melchers WJ, et al. (1998) Isolation and characterization of verocytotoxin-producing *Escherichia coli* O157 strains from Dutch cattle and sheep. J Clin Microbiol 36: 878–882.
29. Ferens WA, Hovde CJ (2011) *Escherichia coli* O157:H7: animal reservoir and sources of human infection. Foodborne pathogens and disease 8: 465–487.
30. Franz E, Semenov AV, van Bruggen AH (2008) Modelling the contamination of lettuce with *Escherichia coli* O157:H7 from manure-amended soil and the effect of intervention strategies. Journal of applied microbiology 105: 1569–1584.

31. Franz E, van Bruggen AH (2008) Ecology of *E. coli* O157:H7 and *Salmonella enterica* in the primary vegetable production chain. Critical reviews in microbiology 34: 143–161.
32. Shere JA, Kaspar CW, Bartlett KJ, Linden SE, Norell B, et al. (2002) Shedding of *Escherichia coli* O157:H7 in dairy cattle housed in a confined environment following waterborne inoculation. Appl Environ Microbiol 68: 1947–1954.
33. Elzo MA, West RL, Johnson DD, Wakeman DL (1998) Genetic variation and prediction of additive and nonadditive genetic effects for six carcass traits in an Angus-Brahman multibreed herd. J Anim Sci 76: 1810–1823.
34. Zadik PM, Chapman PA, Siddons CA (1993) Use of tellurite for the selection of verocytotoxigenic *Escherichia coli* O157. J Med Microbiol 39: 155–158.
35. Riley DG, Gray JT, Loneragan GH, Barling KS, Chase CC, Jr. (2003) *Escherichia coli* O157:H7 prevalence in fecal samples of cattle from a southeastern beef cow-calf herd. Journal of food protection 66: 1778–1782.
36. Zhao T, Doyle MP, Harmon BG, Brown CA, Mueller PO, et al. (1998) Reduction of carriage of enterohemorrhagic *Escherichia coli* O157:H7 in cattle by inoculation with probiotic bacteria. Journal of clinical microbiology 36: 641–647.
37. Perna NT, Plunkett G, 3rd, Burland V, Mau B, Glasner JD, et al. (2001) Genome sequence of enterohaemorrhagic *Escherichia coli* O157:H7. Nature 409: 529–533.
38. Valadez AM, Debroy C, Dudley E, Cutter CN (2011) Multiplex PCR detection of Shiga toxin-producing *Escherichia coli* strains belonging to serogroups O157, O103, O91, O113, O145, O111, and O26 experimentally inoculated in beef carcass swabs, beef trim, and ground beef. Journal of food protection 74: 228–239.
39. Bai J, Shi X, Nagaraja TG (2010) A multiplex PCR procedure for the detection of six major virulence genes in *Escherichia coli* O157:H7. Journal of microbiological methods 82: 85–89.

Genetics of Microenvironmental Sensitivity of Body Weight in Rainbow Trout (*Oncorhynchus mykiss*) Selected for Improved Growth

Matti Janhunen[1]*, Antti Kause[1], Harri Vehviläinen[1], Otso Järvisalo[2]

1 MTT Agrifood Research Finland, Biometrical Genetics, Jokioinen, Finland, **2** Finnish Game and Fisheries Research Institute, Laukaa, Finland

Abstract

Microenvironmental sensitivity of a genotype refers to the ability to buffer against non-specific environmental factors, and it can be quantified by the amount of residual variation in a trait expressed by the genotype's offspring within a (macro)environment. Due to the high degree of polymorphism in behavioral, growth and life-history traits, both farmed and wild salmonids are highly susceptible to microenvironmental variation, yet the heritable basis of this characteristic remains unknown. We estimated the genetic (co)variance of body weight and its residual variation in 2-year-old rainbow trout (*Oncorhynchus mykiss*) using a multigenerational data of 45,900 individuals from the Finnish national breeding programme. We also tested whether or not microenvironmental sensitivity has been changed as a correlated genetic response when genetic improvement for growth has been practiced over five generations. The animal model analysis revealed the presence of genetic heterogeneity both in body weight and its residual variation. Heritability of residual variation was remarkably lower (0.02) than that for body weight (0.35). However, genetic coefficient of variation was notable in both body weight (14%) and its residual variation (37%), suggesting a substantial potential for selection responses in both traits. Furthermore, a significant negative genetic correlation (−0.16) was found between body weight and its residual variation, i.e., rapidly growing genotypes are also more tolerant to perturbations in microenvironment. The genetic trends showed that fish growth was successfully increased by selective breeding (an average of 6% per generation), whereas no genetic change occurred in residual variation during the same period. The results imply that genetic improvement for body weight does not cause a concomitant increase in microenvironmental sensitivity. For commercial production, however, there may be high potential to simultaneously improve weight gain and increase its uniformity if both criteria are included in a selection index.

Editor: Stephen Moore, University of Queensland, Australia

Funding: MTT Agrifood Research Finland (url: www.mtt.fi). The funders had no role in study design, data collection and analysis, decision to publish, or preparation of the manuscript.

Competing Interests: The authors have declared that no competing interests exist.

* E-mail: matti.janhunen@mtt.fi

Introduction

Early phases of selective breeding can generate rapid genetic responses in farmed animals. This typically involves genetic improvement of mean performance in the direction of selection. It is well established that many concurrent improvements in animal husbandry, including nutrition, housing and veterinary practices, accompany the genetic enhancement in animal performance. Additionally, trait heterogeneity can evolve over time, for example via increased or reduced susceptibility of individuals to variable and unmeasured microenvironmental factors. Understanding the genetic basis of such concurrent changes in quantitative traits reveals how selection influences the ability of individuals to respond to unpredictably fluctuating environmental conditions via developmental mechanisms, and helps us to explain the persistence of phenotypic variability within populations.

Microenvironmental sensitivity refers to an individual's ability to be buffered against local non-specific environmental factors (e.g., fluctuating weather, light conditions and food supply, and competitive social interactions) and subtle developmental noise, and it is considered synonymous to developmental instability [1–3]. Microenvironmental sensitivity of a genotype can be quantified by the amount of residual variation in a trait expressed by the genotype's offspring within a (macro-)environment the offspring share. In modern quantitative genetic analysis, residual variance can be best estimated using an animal model which partitions a phenotype of an individual into its additive genetic and residual components, the latter being the part left unexplained by genetics and systematic fixed effects such as gender, age and management treatments [4,5]. In farm animal husbandry, increased residual and thus phenotypic variation is disadvantageous because it hampers the efficiency of production throughout the supply chain from producers to consumers [6,7]. Moreover, large size variation in rearing groups promotes the formation of behavioral dominance hierarchies which reduce animal welfare and elevate mortality [8–10]. This can be partly avoided by active size sorting and grouping of animals. Currently, there is increasing interest to investigate to what extent residual variation can be genetically reduced by animal breeding programmes.

Permanent changes in microenvironmental sensitivity are possible only when there is additive genetic variation for residual variation. In other words, different genotypes should produce differently variable progenies. The recent evidence from both wild and farmed animals imply that genotypes indeed differ in their amount of residual variation of traits [11]. Even though heritability of residual variation is generally low, it can be exploited to increase uniformity by direct selection [12–16]. Further, it has been suggested that intense directional selection for a trait (mean) value can lead to increased residual (and thus phenotypic) variation because the extreme individuals with a higher selection probability are also the genotypes passing down high variability [14,17,18]. This would be worrisome because selection would make individuals more sensitive to their environment. The counter hypothesis is that during adaptation to an environment, either in the wild or in human-controlled conditions of farmed species, microenvironmental sensitivity is decreased due to the adaption to a focal environment [19,20].

Previous work has concentrated on terrestrial vertebrates and laboratory model species, which greatly differ from aquatic species, and from salmonids in particular. Salmonids have a multitude of characteristics that make the genetic analysis of microenvironmental sensitivity in growth important. In aquaculture production, new populations and species are constantly introduced in intensive captive breeding, providing an opportunity to investigate the genetic effects of artificial selection (or domestication process [21]) on both the trait mean value and its underlying variation. Furthermore, salmonids exhibit an extraordinary polymorphism and diversity in morphological, behavioral and life-history traits, including alternative growth, migration and reproduction strategies expressed across and within single populations [22–25]. Some of these responses are adaptive responses to the highly stochastic natural conditions. Salmonids also display strong dominance hierarchies, especially within farmed populations, in which few individuals can defend food resources, increasing phenotypic variation in growth [26–28]. Given that fish as ectotherms are particularly sensitive to varying ambient conditions that can influence ontogenetic trajectories, individual differences in growth are more pronounced in fish compared to farmed terrestrial animals. For example, in cultured salmonids, phenotypic coefficient of variation (CV) of body weight varies between 20–40% [29], whereas in chicken and pigs it is around 10–15% [30–32]. Finally, an additional strength of using salmonids to study genetic architecture of microenvironmental sensitivity is that the established breeding programmes generate large number of families in successive generations, and due to their high fecundity, high family sizes can be produced, both factors needed for an effective genetic analysis of residual variation.

To investigate the inheritance of microenvironmental sensitivity and its genetic responses across generations when directional selection is performed for improved growth, we analyzed multigenerational pedigreed data covering ten year classes and 46 546 individuals from the Finnish breeding programme for rainbow trout, *Oncorhynchus mykiss* (Walbaum). We first estimated the proportion of genetic variation in residual variation for body weight in fish being maintained in the same location. By providing a common macroenvironment across year classes and by using the animal model, we ensured that residual variation can be regarded as microenvironmental sensitivity (or developmental stability) that results from non-systematic environmental factors and internal developmental noise. Second, we estimated the genetic correlation between the additive genetic effects for body weight and its residual variation. Finally, by estimating genetic trends that quantify genetic responses across multiple generations, we investigated the effects of selective breeding for body weight on the genetic change in microenvironmental sensitivity.

Methods

Ethics Statement

All procedures involving animals were approved by the animal care committee of the Finnish Game and Fisheries Research Institute (FGFRI).

Data Source

The data originated from the Finnish national rainbow trout breeding programme maintained by the FGFRI and MTT Agrifood Research Finland. The breeding nucleus is held at the Tervo Fisheries Research and Aquaculture station in Central Finland (63°1′ N, 26°39′E).

The phenotypic data included 45 900 records of body weight from individuals born during 1992–2002 and reared at the same freshwater nucleus station. The fish represented eight year classes and belonged to two subpopulations with four successive generations (Pop I and Pop IIa) [33,34]. Each year class consisted of 94–270 full-sib families established from matings of 37–90 sires with 92–270 dams. The subpopulations share a common genetic base from which the founding individuals were sampled in 1989 for PopI and in 1990 for PopII. Even though the base population was preceded by a long-term cultivation background, only the studied generations belong to a systematic breeding programme in which intensive genetic selection based on estimated breeding values has been practiced. The pedigree information extended over the five generations and comprised 46 546 individuals, including the 364 base population animals without phenotypic observations.

The generation interval of the study population was 3–4 years. Annual selection of breeding candidates was made using a multitrait selection index with main emphasis on improved growth The selection index has consisted of best linear unbiased predictions of breeding values for body weight measured at the age of 2 and 3 years (since 1992), maturity age (since 2001) [35], and body shape, skin color and its spottiness (since 2001) [36]. Parental fish were mated in spring using either nested paternal hierarchical or partial factorial designs [33].

Full-sib egg batches were incubated separately, and at the eyed-egg stage, they were transferred to one or two 150-liter indoor family tanks. Hatching of eggs occurred in June. During the following winter, after six months of growing in the family tanks, equal amount of fingerlings (of 50–100 g body weight) from each family tank were haphazardly sampled and individually tagged with passive integrated transponders (Trovan, Ltd., Ulm, Germany) and then transferred to a flow-through earth-bottomed raceway at the Tervo station. The fish were fed with commercial dry feed. In Finland, year is highly seasonal and the effective growing season lasts from early May to late October.

After the second growing season, the two-year-old fish were individually weighed to the nearest 1 g (mean 1020±315 (SD) g, n = 45 900). The number of individuals within each year class ranged between 2 518–10 753. The proportion of sexually matured (2+) males in the entire data-set was 14.9%, whereas no mature females were found.

To improve the reliability of genetic parameters for residual variation, only sire families with at least 35 offspring (n = 457 sires) were selected for the analysis. Large family sizes are needed to obtain accurate and unbiased genetic parameters and estimated breeding values (EBVs) for residual variation [14].

Genetic Analysis

The estimation of genetic parameters and genetic trends was conducted using a bivariate animal model [31]. The ASReml 3.0

software applying restricted maximum likelihood (REML) was used [37]. The first trait was body weight for which a linear mixed 'mean model' was fitted:

$$y_{ijk} = \mu + year_j + \tan k_k + A_i + e_{ijk} \quad (1)$$

where y_{ijk} is body weight of an individual i, μ is the overall population mean, $year_j$ is the fixed effect of birth year (j = 8 years), $\tan \imath k_k$ is the random interaction effect between birth year and common environment shared by full-sibs before tagging (k = family tan k × year number), A_i is the random genetic animal effect with a pedigree (i = number of animals), and e_{ijk} is the residual error term with separate variance $\sigma^2_{e_{sf}}$ for each sire family sf. The common environment effect is modeled without the pedigree information. The values for Akaike's Information Criteria (AIC) [38] and Bayesian Information Criteria (BIC) [39] were lower for the model with heterogeneous residual variance structure, suggesting a better fit to the data compared to the model with homogeneous residual variance (AIC: 539554 and 542160; BIC: 541235 and 542171, respectively).

The second trait was microenvironmental sensitivity which was quantified by the log-transformed squared residual values, $\ln(e^2_{ijk})$, obtained from the mean model (1) and used as new observations in the 'variance model'. Log-transformed squared residual values quantify the contribution of each individual to population's residual variation [15,31,32,40]. In contrast to sire-dam models, the residuals of an animal model include only unexplained environmental and developmental noise, and they are not confounded by the additive genetic Mendelian sampling term. The animal 'variance model' was:

$$\ln(e^2_{ijk}) = \mu + year_j + A_{res_i} + e_{res_{ij}} \quad (2)$$

where A_{res_i} is the genetic effect of animal i for $\ln(e^2_{ijk})$ and $e_{res_{ij}}$ is the random residual effect. For the random effects of $\ln(e^2_{ijk})$, the assumptions were $A_{res} \sim N(0, A\sigma^2_{Ares})$ and $e_{res} \sim N(0, I\sigma^2_{eres})$, where $\mathbf{A}$ is the additive genetic relationship matrix with additive genetic variance σ^2_{Ares} and $\mathbf{I}$ is the identity matrix with homogeneous residual variance σ^2_{eres}. The random effect for common environment × birth year was omitted from the variance model because its variance explained less than 2% of the total phenotypic variance and it did not significantly differ from zero.

Because the residuals of the model 1 are used as an input variable for the model 2, the model for the mean and the residual variation was iteratively solved by conducting 30 consecutive bivariate analyses. At each iterative round, $\ln(e^2_{ijk})$ for the variance model were updated with residuals from the previous round's mean model. The residuals e_{ijk} and $e_{res_{ijk}}$ were assumed to follow a bivariate normal distribution and be uncorrelated (i.e., their residual covariance was set to zero). The convergence criteria within separate runs were fulfilled when the REML log-likelihood changed less than 0.002 × iteration number and the individual variance parameter estimates changed less than 1% between successive iterations [37].

Calculation of Genetic Parameters and Genetic Trends

Heritability of weight mean was calculated as $h^2 = \sigma^2_A/\sigma^2_P$ and the common environment effect ratio as $c^2 = \sigma^2_{\tan k}/\sigma^2_P$ using the variance components from model 1. Here $\sigma^2_P = \sigma^2_{\tan k} + \sigma^2_A + \overline{\sigma^2_e}$, where $\overline{\sigma^2_e}$ is the average residual variance of sire families. In addition to common environment effects of full sibs, $\sigma^2_{\tan k}$ may include parts of non-additive genetic and maternal variance. Genetic coefficient of variation was calculated as $GCV = \sigma_A/\mu$, where μ is the phenotypic mean of the population. GCV describes the propensity of the trait to respond to selection, that is, its evolvability [41].

Heritability of residual variation was calculated as $h^2_v = \sigma^2_{A_V}/(2\sigma^4_P + 3\sigma^2_{A_v})$, where $\sigma^2_{A_v}$ is the transformed additive genetic variance of residual variation from model 2 and σ^2_P is the phenotypic variance of body weight obtained from model 1 [14]. The genetic variance $\sigma^2_{A_v}$ was calculated as $\sigma^2_{A_v} = h^2_{res} 2(\overline{\sigma^2_e})^2$, where $h^2_{res} = \sigma^2_{A_{res}}/(\sigma^2_{A_{res}} + \sigma^2_{e_{res}})$ is the heritability of $\ln(e^2_{ijk})$ and $\overline{\sigma^2_e}$ is the average residual variance obtained from model 1. Genetic coefficient of variation for residual variation was calculated as: $GCV_E = \sigma_{A_v}/\overline{\sigma^2_e}$. An estimate of genetic correlation between the additive genetic effects for body weight and its residual variance was obtained from the bivariate analysis where direct estimation of co-variance between the two traits is possible.

The approximate standard errors for estimated variance components and variance ratios were calculated using ASReml. The standard error of h^2_v was approximated according to Mulder *et al.* [31].

REML log-likelihood values and the parameter estimates for body weight were found to remain relatively stable across the 30 iterative rounds, whereas $\sigma^2_{A_{res}}$ oscillated. Therefore, the results from bivariate analysis are presented as averages of all ASReml runs ($n = 30$ rounds). The observed oscillation is inherent to the statistical model used and is mainly due to an interplay between A_i, A_{res_i} and the residual e_{ijk}. An increase in A_i causes a decrease in the residual and thereby lowers A_{res_i} (and *vice versa*).

To investigate whether or not genetic changes in mean body weight and its mircroenvironmental sensitivity occurred during selective breeding, genetic trends were determined for both traits and for both subpopulations separately. The genetic trends were obtained by plotting the average estimated breeding values (i.e., the predicted genetic levels for y_{ijk} and $\ln(e^2_{ijk})$ obtained from individuals' averages across the 30 iterative rounds) against the birth year of fish.

Results

Genetic Variation

Heritability for body weight was moderate (0.35), whereas the common environment ratio was low (0.05) (Table 1). Genetic coefficient of variation for body weight was slight (0.14). Heritability estimate of residual variation was low (0.02), though it was greater than its standard error (Table 2). Yet, the moderately high genetic coefficient of variance for residual variation (GCV_E = 0.37) suggests that there is notable genetic potential in microenvironmental sensitivity of body weight.

Genetic Correlation between Body Weight and its Residual Variation

There was a slight but significant negative genetic correlation between body weight and its residual variation ($r_G = -0.157 \pm 0.039$ (S.E.)), indicating that high body weight was genetically associated with decreased microenvironmental sensitivity.

Genetic Trends

Body weight showed a clear genetic improvement during the study period. Over the four generations of selection, the

Table 1. Estimates of variance components and variance ratios (± approximate standard errors) for body weight.

Parameter[a]	Estimate
σ^2_A	20 888 (1515)
σ^2_{tank}	3 089 (286)
$\overline{\sigma^2_e}$	35 674 (7444)
σ^2_P	59 652 (7439)
h^2	0.350 (0.051)
c^2	0.052 (0.009)
GCV	0.142

[a]additive genetic variance; σ^2_{tank} common environment variance; $\overline{\sigma^2_e}$ the average residual variance of sire families; σ^2_P phenotypic variance; h^2– heritability, $h^2=\sigma^2_A/\sigma^2_P$; c^2– common environment effect ratio, $c^2=\sigma^2_{\text{tank}}/\sigma^2_P$; GCV– coefficient of genetic variation, $GCV=\sigma_A/\mu$.

Table 2. Estimated variance components and variance ratios (± approximate standard errors) for microenvironmental sensitivity of body weight.

Parameter[a]	Estimate
$\sigma^2_{A_{res}}$	0.374 (0.028)
$\sigma^2_{A_v}$	1.81 E + 08
h^2_v	0.024 (0.006)
GCV_E	0.376

[a]additive genetic variance in ln(e^2) (model 2);
$\sigma^2_{A_v}$ – transformed genetic variance in the quantitative genetic model for genetic heterogeneity of residual variation [13], $\sigma^2_{A_v}=h^2_{res}2(\overline{\sigma^2_e})^2$; h^2_v – heritability, $h^2_v=\sigma^2_{A_V}/(2\sigma^4_P+3\sigma^2_{A_v})$;
GCV_E– genetic coefficient of variation, $GCV_E=\sigma_{A_v}/\overline{\sigma^2_e}$.

cumulative genetic gains in the two sub-populations were 199 g to 208 g, corresponding to an average of 0.83 increase in phenotypic standard deviation or 5.5% per year (Fig. 1a). In contrast, mean estimated breeding values for microenvironmental sensitivity remained stable across the year classes (Fig. 1b).

Discussion

Low Heritability but Moderate Evolvability for Microenvironmental Sensitivity

We found a low 0.02 heritability estimate for residual variation of body weight (i.e., microenvironmental sensitivity) in 2-year-old rainbow trout. This is in marked contrast with the moderate heritability of 0.35 for body weight. The low heritability estimate of microenvironmental sensitivity is somewhat surprising as large within- and between-family variation in fish growth is created by multiple factors, some of them presumably exhibiting substantial genetic variation. However, the notable genetic coefficient of variation (37%, when genetic variation for residual variance is scaled by average residual variance) indicates the presence of substantial additive genetic variation for microenvironmental sensitivity. Regarding most life-history traits, the low heritabilities yet paradoxically high evolvability are attributed to the high residual variation accumulating from the variable underlying physiological and behavioral traits [41,42]. Similarly, body weight and its variation can be influenced by many underlying component traits such as feeding behavior, feed utilization and metabolism [43].

There are several factors that can maintain genetic variation in microenvironmental sensitivity in the population under study. First, high initial growth rates and energy resources are related to increased probability of early maturation in salmonids [44,45]. Likewise, in rainbow trout, rapid growth is genetically and phenotypically correlated to early maturity age [35,46,47]. In our population, male fish primarily mature at ages of 2 to 3 years. Maturity age in males has a moderate heritability of 0.23–0.34, and thus there are family differences in the frequency of maturing individuals [33,35]. This alone may create genetic variation in microenvironmental sensitivity: high residual variation would be found in families with both early and late maturing individuals, and low residual variation in families with either only early or only late maturing individuals. Accordingly, it was logical not to include maturation as a fixed factor in the statistical model because sexual development itself captures part of the within-family variation we were interested in. Second, following the former reasoning, the genetic variation observed for resistance and/or tolerance to parasite-mediated cataract (*Diplostomum* spp.) in our population may create genetic variation in microenvironmental sensitivity. Some families remain uninfected while others have both infected and uninfected individuals, and the infected individuals exhibit reduced growth [48]. Third, social interactions associated with behavior and growth differences have also been found to create additional genetic variation in chicken and pigs [49,50], and presumably in fish as well [51]. A large proportion of the genetic variation underlying socially affected traits remains hidden, i.e., is not accounted for by the direct heritability estimates, and can thus only be revealed by unexplained residual variation. Last, it is important to recall that even though the genetic characteristics of farmed fish populations are influenced by life histories originating from their wild ancestors, the results from a genetic analysis of farmed populations cannot be extrapolated back to wild populations [52]. Nevertheless, the estimates of genetic parameters obtained from farmed populations help us to understand biologically meaningful phenomena and also advance general knowledge of the factors underlying phenotypic variation in quantitative traits [53].

Although rainbow trout, among other salmonids, possess a capacity of considerable growth and life-history strategy variation both across and within families [33,35,54], the observed heritability estimate for residual variation in body weight is of similar low magnitude that has been reported for less variable terrestrial animals [11,14]. Correspondingly, GCV_E was in the range of those found in chickens, mice, pigs and rabbits (25–50%) [11]. Fluctuating asymmetry, the degree of random non-directional deviations between morphological characteristics measured from left and right hand side of individuals, is an alternative measure of developmental instability. In accordance with the original idea by Lerner [55], increased heterozygosity has been found to reduce fluctuating asymmetry in bilateral traits of both wild and farmed rainbow trout [56,57]. However, the estimated low heritability for fluctuating asymmetry led the authors to conclude that dominance effects have a major contribution to the control of developmental stability [58]. Developmental instability is often assumed to be selectively disadvantageous due to the increased risk of drift from the phenotypic optimum [3,59,60], but empirical support for this view is largely inconclusive [61]. It is probable that in some cases, such as the morphological traits of plants, selection favors increased sensitivity as a bet-hedging strategy [62].

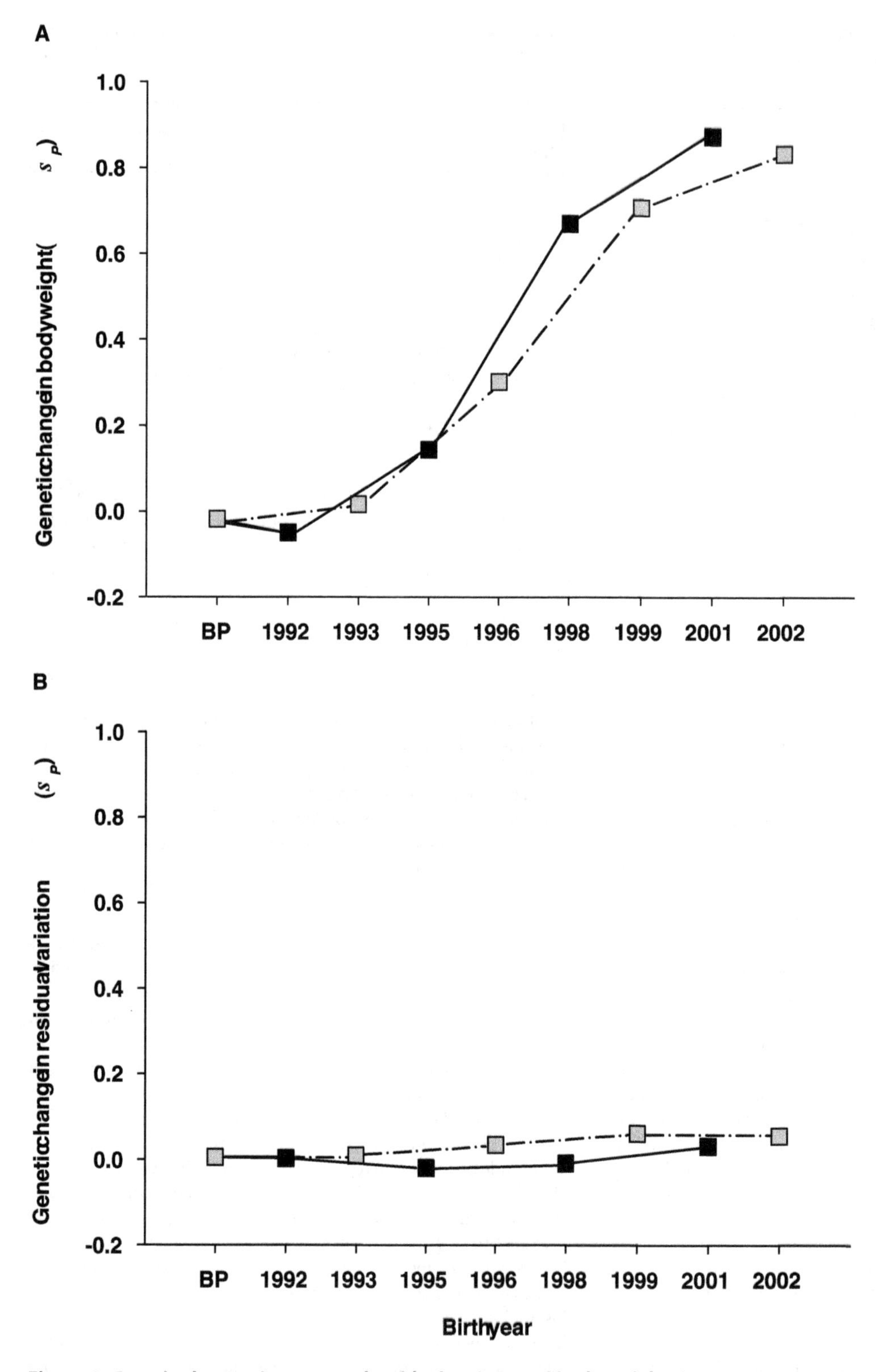

Figure 1. Genetic changes in mean and residual variation of body weight. Average genetic changes for A) body weight mean and B) its microenvironmental sensitivity in two subpopulations (black and grey box) of rainbow trout. The averages are given in the units of phenotypic standard deviation (σ_P).

It is possible that similar to life-history traits [63–66], developmental stability is inherently an important fitness correlate, and the strong directional selection during the long history of animals has led to its low heritability [58]. Meanwhile, many underlying environmental and genetic factors affecting microenvironmental sensitivity retain its genetic coefficient of variation at a moderate level. Nevertheless, further analyses are needed to test whether the genetic parameters show similar values in wild fish populations or when fish populations are in their first generations of domestication. The methods developed by animal breeders and also used here [31,32] can be applied to wild populations when pedigree information is established using molecular genetic markers.

Direct and Correlated Responses to Selection in Microenvironmental Sensitivity

The low heritability estimate observed here does not necessarily indicate that microenvironmental sensitivity would be weakly responsive to selection. Heritability, the ratio of additive genetic variance to phenotypic variance, is one predictor of genetic potential to selection responses, though in this context, genetic coefficient of variation provides a more reasonable measure of evolvability, similar to *GCV* for trait means [41,67]. In our study, GCV_E was over two times higher than *GCV* for body weight, suggesting a good opportunity to obtain reduction in random environmental variation by selection.

Some selection experiments and breeding programmes have obtained considerable genetic responses in traits with low heritability (e.g., developmental stability in *Drosophila* [68,69]; piglet survival [70]), supporting the idea that also the amount of residual variation can be modified by selection. Similarly, residual variation is expected to be reduced by 10% after one generation of selection when it is included in a selection index along with the phenotypic trait value [71]. To effectively breed for a trait with a low heritability, phenotypic records from a large number of relatives are required. Controlled matings and large family sizes inherent to rainbow trout and many other aquaculture species enhance the estimation of breeding values with moderate accuracy [14,72].

To our knowledge, this study is the first multigenerational breeding experiment on aquatic organisms to assess the correlated genetic effect of strong directional selection on microenvironmental sensitivity of a trait. The genetic correlation between body weight and its residual variation was negative, implying that a high trait value was linked to a slightly reduced microenvironmental sensitivity. This combined with the low heritability of residual variation predicts only a weak decreasing microenvironmental sensitivity across successive generations in response to selection for rapid growth. However, the genetic trend for microenvironmental sensitivity remained stable or slightly elevated over the course of the selection period, while body weight mean displayed a 6% genetic increase per generation. These results together indicate that genetic improvement of body weight does not make rainbow trout more sensitive to microenvironmental perturbations. This is important animal welfare issue, because increase in size heterogeneity would lead to serious challenges in animal husbandry.

Intense mass selection based on individuals' own phenotype is expected to increase phenotypic variation within a population even when there is no additive genetic correlation between trait and its residual variation [14,18]. Moreover, studies on salmonid fish suggest that selection for rapid growth may indirectly select for competitive ability and aggressiveness, thus increasing the likelihood for increased size variation in farmed fish during the breeding process [73–76]. The observed patterns in genetic trends do not conform to these assumptions. Referring to the former proposition, however, a multitrait selection method in our study population was not only based on the phenotypic information of an individual itself but also the performance of its all relatives was taken into account. This makes the predictions concerning responses in environmental variation more difficult. Nevertheless, the negative genetic correlation between the body weight and its microenvironmental sensitivity could be expected to counterbalance, to some extent, the rate of increase in growth variation due to scale effects. Previous studies on terrestrial animals have shown that the genetic correlation between quantitative traits and their residual variations can vary from negative to positive, depending on the species and trait analyzed [31,40,77–79]. Similar inconsistent results have been found in selection experiments. For example, Ibáñez-Escriche *et al.* [16] demonstrated a decrease in phenotypic CV of body weight traits in mice selected for increased growth. In contrast, long-term selection experiments on *Drosophila* fruitfly showed that phenotypic variation can be substantially higher in the lines selected for high and low abdominal bristle number relative to the unselected base population [80,81].

In conclusion, heterogeneity of residual variation in rainbow trout growth was found to be partly under genetic control. This implies the possibility for selection to favor genotypes with low variability when constancy across microenvironmental conditions is important. The negative genetic relationship between body weight and its microenvironmental sensitivity presumably facilitates improving weight gain and simultaneously increasing its uniformity/robustness if both objectives are incorporated into a selection index. In addition, increasing the growth potential of fish does not seem to cause a concomitant change in the trait's microenvironmental sensitivity.

Acknowledgments

We express our gratitude to the staff at the Tervo Fisheries Research and Aquaculture station for the solid data collection and managing the fish.

Author Contributions

Analyzed the data: MJ. Mainly wrote the manuscript: MJ. Took part in data analysis and writing: AK. Took part in writing the manuscript: HV. Contributed materials: OJ. Responsible for the maintenance of the experiment: OJ.

References

1. Thoday JM (1955) Balance, heterozygosity and development stability. Cold Spring Harbor Symp Quant Biol 20: 318–326.
2. Falconer DS, Mackay TFC (1996) Introduction to Quantitative Genetics, 4th edn. Essex: Logman Group Ltd. 464 p.
3. Gavrilets S, Hastings A (1994) A quantitative-genetic model for selection on developmental noise. Evolution 48: 1478–1486.
4. Henderson CR (1986) Estimation of variances in animal model and reduced animal model for single traits and single records. J Dairy Sci 69: 1394–1402.
5. Kruuk LEB (2004) Estimating genetic parameters in natural populations using the 'animal model'. Phil Trans R Soc B 359: 873–890.
6. Hovenier R, Brascamp EW, Kanis E, van der Werf JHJ, Wassenberg APAM (1993) Economic values of optimum traits: the example of meat quality in pigs. J Anim Sci 71: 1429–1433.
7. Dekkers JCM, Birke PV, Gibson JP (1995) Optimum linear selection indexes for multiple generation objectives with non-linear profit functions. Anim Sci 61: 165–175.
8. Poignier J, Szendrö ZS, Levai A, Radnai I, Biro-Nemeth E (2000) Effect of birth weight and litter size on growth and mortality in rabbit. World Rabbit Sci 8: 103–109.
9. Milligan BN, Fraser D, Kramer DL (2002) Within-litter birth weight variation in the domestic pig and its relation to pre-weaning survival, weight gain, and variation in weaning weights. Livest Prod Sci 76: 181–191.
10. Gilmour KM, DiBattista JD, Thomas JB (2005) Physiological causes and consequences of social status in salmonid fish. Integr Comp Biol 45: 363–273.
11. Hill WG, Mulder HA (2010). Genetic analysis of environmental variation. Genet Res 92: 381–395.
12. Rendel JM, Sheldon BL, Finlay DE (1966) Selection for canalization of the scute phenotype. II. Amer Nat 100: 13–31.
13. Cardin S, Minvielle F (1986) Selection on phenotypic variation of pupa weight in *Tribolium castaneum*. Can J Genet Cytol 28: 856–861.
14. Mulder HA, Bijma P, Hill WG (2007) Prediction of breeding values and selection responses with genetic heterogeneity of environmental variance. Genetics 175: 1895–1910.

15. Garreau H, Bolet G, Larzul C, Robert-Granie C, Safeil G, et al. (2008) Results on four generations of a canalizing selection for rabbit birth weight. Livest Sci 119: 55–62.
16. Ibáñez-Escriche N, Moreno A., Nieto B, Piquears P, Salgado C, et al. (2008) Genetic parameters related to environmental variability of weight traits in a selection experiment for weight gain in mice: signs of correlated canalised response. Genet Select Evol 40: 279–293.
17. Hill WG (1984) On selection among groups with heterogeneous variance. Anim Prod 39: 473–477.
18. Hill WG, Zhang X-S (2004) Effects on phenotypic variability of directional selection arising through genetic differences in residual variability. Genet Res 83: 121–131.
19. Waddington CH (1960) Experiments of canalizing selection. Genet Res 1: 140–150.
20. Lewontin RC (1957) The adaptations of populations to varying environments. Cold Spring Harb Symp Quant Biol 22: 395–408.
21. Price EO, King JA (1968) Domestication and Adaptation. In: Hazef ESE, editor. Adaptation of Domestic Animals. Philadelphia: Lea and Febiger. 34–45.
22. Gross MR (1985) Disruptive selection for alternative life histories in salmon. Nature 313: 47–48.
23. Hendry AP, Stearns SC (2004) (eds.). Evolution Illuminated: Salmon and their Relatives. New York: Oxford University Press. 510 p.
24. Mangel M, Satterthwaite WH (2008) Combining proximate and ultimate approaches to understand life history variation in salmonids with application to fisheries, conservation, and aquaculture. Bull Mar Sci 83: 107–130.
25. Páez DJ, Brisson-Bonenfant C, Rossignol O, Guderley HE, Bernatchez L, et al. (2011) Alternative developmental pathways and the propensity to migrate: a case study in the Atlantic salmon. J Evol Biol 24: 245–255.
26. Abbott JC, Dunbrack RL, Orr CD (1985) The interaction of size and experience in dominance relationships of juvenile steelhead trout (*Salmo gairdneri*). Behaviour 92: 241–253.
27. Jobling M (1985) Physiological and social constraints on growth of fish with special reference to Arctic charr, *Salvelinus alpinus* L. Aquaculture 44: 83–90.
28. McCarthy ID, Carter CG, Houlihan DF (1992) The effect of feeding hierarchy on individual variability in daily feeding of rainbow trout, *Oncorhynchus mykiss* (Walbaum). J Fish Biol 41: 257–263.
29. Gjedrem T (2000) Genetic improvement of cold-water fish species. Aquaculture Research 31: 25–33.
30. Damgaard LH, Rydhmer L, Lovendahl P, Grandinson K (2003) Genetic parameters for within-litter variation in piglet birth weight and change in within-litter variation during suckling. *J* Anim Sci 81: 604–610.
31. Mulder HA, Hill WG, Vereijken A, Veerkamp RF (2009) Estimation of genetic variation in residual variance in female and male broilers. Animal 3: 1673–1680.
32. Wolc A, White IMS, Avendano S, Hill WG (2009) Genetic variability in residual variation of body weight and conformation scores in broiler chickens. Poult Sci 88: 1156–1161.
33. Kause A, Ritola O, Paananen T, Wahlroos H, Mäntysaari EA (2005) Genetic trends in growth, sexual maturity and skeletal deformations, and rate of inbreeding in a breeding programme for rainbow trout (*Oncorhynchus mykiss*). Aquaculture 247: 177–187.
34. Vehviläinen H, Kause A, Quinton C, Koskinen H, Paananen T (2008) Survival of the currently fittest: genetics of rainbow trout survival across time and space. Genetics 180: 507–516.
35. Kause A, Ritola O, Paananen T, Mäntysaari E, Eskelinen U (2003) Selection against early maturity in large rainbow trout *Oncorhynchus mykiss*: the quantitative genetics of sexual dimorphism and genotype-by-environment interactions. Aquaculture 228: 53–68.
36. Kause A, Ritola O, Paananen T, Eskelinen U, Mäntysaari E (2003) Big and beautiful? Quantitative genetic parameters for appearance of large rainbow trout. J Fish Biol 62: 610–622.
37. Gilmour AR, Gogel BJ, Cullis BR, Thompson R (2009) ASReml User Guide Release 3.0. VSN International Ltd, Hernel Hempstead, HP1 1ES, UK.
38. Akaike H (1973) Information theory and an extension of the maximum likelihood principle. In: Petrov BN, Csaki F, editors. Proceedings of the 2nd International Symposium on Information Theory. Budapest: Akademiai Kiado. 267–281.
39. Schwarz G (1978) Estimating the dimension of a model. Ann Statist 6: 461–464.
40. Neves HHR, Carvalheiro R, Roso VM, Queiroz SA (2011) Genetic variability on residual variance of production traits in Nellore beef cattle. Livest Sci 142: 164–169.
41. Houle D (1992) Comparing evolvability and variability of quantitative traits. Genetics 130: 195–204.
42. Price T, Schluter D (1991) On the low heritability of life-history traits. Evolution 45: 853–861.
43. Kause A, Saloniemi I, Haukioja E, Hanhimäki S (1999) How to become large quicky: quantitative genetics of growth and foraging in a flush feeding lepidopteran larva. J Evol Biol 12: 471–482.
44. Rowe DK, Thorpe JE, Shanks AM (1991) The role of fat stores in the maturation of male Atlantic salmon (*Salmo salar*) parr. Can J Fish Aquat Sci 48: 405–413.
45. Shearer KD, Swanson P (2000) The effect of whole body lipid on early maturation of 1+ age male Chinook salmon (*Oncorhynchus tshawytscha*). Aquaculture 190: 343–367.
46. Crandell PA, Gall GAE (1993) The genetics of body weight and its effect on early maturity based on individually tagged rainbow trout (*Oncorhynchus mykiss*). Aquaculture 117: 77–93.
47. Martyniuk CJ, Perry GMI, Mogahadam HK, Ferguson MM, Danzmann RG (2003) The genetic architecture of correlations among growth-related traits and male age at maturation in rainbow trout. J Fish Biol 63: 746–764.
48. Kuukka-Anttila H, Peuhkuri N, Kolari I, Paananen T, Kause A (2010) Quantitative genetic architecture of parasite-induced cataract in rainbow trout, *Oncorhynchus mykiss*. Heredity 104: 20–27.
49. Bijma P, Muir WM, van Arendonk JAM (2007) Multilevel selection 2: Estimating the genetic parameters determining inheritance and response to selection. Genetics 175: 277–288.
50. Bergsma R, Kanis E, Knol EF, Bijma P (2008) The contribution of social effects to heritable variation in finishing traits of domestic pigs (*Sus scrofa*). Genetics 178: 1559–1570.
51. Monsen BB, Ødegård J, Arnesen KR, Toften H, Nielsen HM, Damsgård B, Bijma P, Olesen I (2010) Genetics of social interactions in Atlantic cod (*Gadus morhua*). 9th World Congr Genet Appl Livest Prod, August 1–6, 2010, Leipzig, Germany.
52. Carlson AM, Seamons TR (2008) A review of quantitative genetic components of fitness in salmonids: implications for adaptation to future change. Evol Appl 1: 222–238.
53. Weigensberg I, Roff DA (1996). Natural heritabilities: can they be reliably estimated in the laboratory? Evolution 50: 2149–2157.
54. Rasmussen RS, Ostenfeld T (2010) Intraspecific growth variation among rainbow trout and brook trout: impact of initial body weight and feeding level. Aquacult Int 18: 933–941.
55. Lerner IM (1954) Genetic Homeostasis. Edinburg: Oliver and Boyd. 134 p.
56. Leary RF, Allendorf FW, Knudsen KL (1983) Developmental stability and enzyme heterozygosity in rainbow trout. Nature 301: 71–72.
57. Leary RF, Allendorf FW, Knudsen KL (1984) Superior developmental stability of heterozygotes at enzyme loci in salmonid fishes. Am Nat 124: 540–551.
58. Leary RF, Allendorf FW, Knudsen KL (1985) Inheritance of meristic variation and the evolution of developmental stability in rainbow trout. Evolution 39: 308–314.
59. Møller AP (1997) Developmental stability and fitness: a review. Am Nat 149: 916–932.
60. Wagner GP, Booth G, Bagheri-Chaichian H (1997) A population genetic theory of canalization. Evolution 51: 329–347.
61. Clarke GM (1998) Developmental stability and fitness: the evidence is not quite so clear. Am Nat 152: 762–766.
62. Hall MC, Dworkin I, Ungerer MC, Purugganan M (2007) Genetics of microenvironmental canalization in *Arabidopsis thaliana*. Proc Natl Acad Sci U S A 104: 13717–13722.
63. Mousseau TA, Roff DA (1987) Natural selection and the heritability of fitness components. Heredity 59: 181–197.
64. Houle D (1998) How should we explain variation in the genetic variance of traits? Genetica 102–103: 241–253.
65. Merilä J, Sheldon BC (1999) Genetic architecture of fitness and nonfitness traits: empirical patterns and development of ideas. Heredity 83: 103–109.
66. Merilä J, Sheldon BC (2000) Lifetime reproductive success and heritability in nature. Am Nat 155: 301–310.
67. Hansen TF, Pélabon C, Houle D (2011) Heritability is not evolvability. Evol Biol 38: 258–277.
68. Mather K (1953) Genetical control of stability in development. Heredity 7: 297–336.
69. Reeve ECR (1960) Some genetic tests on asymmetry of sternopleural chaeta in *Drosophila*. Ge- net Res 1: 151–172.
70. Knol EF (2003) Quantitative selection for piglet survival as a safe way to reduce the cost of weaners. Adv Pork Prod 14: 59–65.
71. Mulder HA, Bijma P, Hill WG (2008) Selection for uniformity in livestock by exploiting genetic heterogeneity of residual variance. Genet Selec Evol 40: 37–59.
72. Sae-Lim P, Komen H, Kause A (2010) Bias and precision of estimates of genotype-by-environment interaction: A simulation study. Aquaculture 310: 66–73.
73. Fenderson OC, Everhart WH, Muth KM (1968) Comparative agonistic and feeding behaviour of hatcheryreared and wild salmon in aquaria. J Fish Res Bd Can 25: 1–14.
74. Ruzzante DE (1994) Domestication effects on aggressive and schooling behaviour in fish. Aqaculture 120: 1–24.
75. Johnsson JI, Petersson E, Jonsson E, Björnsson BT, Järvi T (1996) Domestication and growth hormone alter antipredator behaviour and growth patterns in juvenile brown trout Salmo trutta. Can J Fish Aquat Sci 53: 1546–1554.
76. Sundström LF, Petersson E, Höjesjö J, Johnsson JI, Järvi T (2004) Hatchery selection promotes boldness in newly hatched brown trout (*Salmo trutta*): implications for dominance. Behav Ecol 15: 192–198.
77. Ros M, Sorensen D, Waagepetersen R, Dupont-Nivet M, SanCristobal M, et al. (2004) Evidence for genetic control of adult weight plasticity in the snail *Helix aspersa*. Genetics 168: 2089–2097.
78. Gutierrez JP, Nieto B, Piqueras P, Ibáñez N, Salgado C (2006) Genetic parameters for canalization analysis of litter size and litter weight traits at birth in mice. Genet Select Evol 38: 445–462.

79. Wolc A, Lisowski M, Hill WG, White IMS (2011) Genetic heterogeneity of variance in production traits of laying hens. Br Poult Sci 52: 537–540.
80. Clayton GA, Robertson A (1957) An experimental check on quantitative genetical theory. II. The long-term effects of selection. J Genet 55: 152–170.
81. Mackay TFC, Fry JD, Lyman RF, Nuzhdin SV (1994) Polygenic mutation in *Drosophila melanogaster*: estimates from response to selection of inbred strains. Genetics 136: 937–951.

17

Could Dromedary Camels Develop Stereotypy? The First Description of Stereotypical Behaviour in Housed Male Dromedary Camels and How It Is Affected by Different Management Systems

Barbara Padalino[1]*, Lydiane Aubé[2], Meriem Fatnassi[3], Davide Monaco[4], Touhami Khorchani[3], Mohamed Hammadi[3], Giovanni Michele Lacalandra[4]

1 Department of Veterinary Medicine, University of Bari, Valenzano (Bari), Italy, 2 Laboratoires d'éthologie animale et humaine EthoS -University of Rennes, Rennes, France, 3 Livestock and Wildlife Laboratory, Arid Lands Institute, Médenine, Tunisia, 4 Department of Emergency and Organ Transplantation (D.E.T.O.), Veterinary Clinics and Animal Production Section, University of Bari, Valenzano (Bari), Italy

Abstract

Dromedary camel husbandry has recently been evolving towards a semi-intensive system, due to the changes in use of the animal and the settlement of nomadic populations. Captivity could restrict its social activities, limiting the expression of various behavioural needs and causing the manifestation of stereotypy. The aims of this trial were, firstly, to identify and describe some stereotypical behaviours in captive male dromedary camels used for artificial insemination and, secondly, to study the effects on them of the following husbandry management systems: i) housing in single boxes for 24 hours (H24), ii) housing in single boxes for 23 hours with one hour free in the paddock (H23), and iii) housing in single boxes for 22 hours 30 min with 1 h of paddock time and 30 min exposure to a female camel herd (ExF). Every day, the camels were filmed in their single box in the morning for 30 minutes to record their behavioural activities and a focal animal sampling ethogram was filled in. In this study, male camels showed both oral and locomotor stereotypy most frequently when the bulls were reared in H24. Overall, this preliminary study is a starting point in the identification of stereotypies in male camels, reporting the positive effects of spending one hour outdoor and of social interaction with females.

Editor: Cédric Sueur, Institut Pluridisciplinaire Hubert Curien, France

Funding: This document has been produced with the financial assistance of the European Union through the "PROCAMED" Project: Promotion des systèmes camelins innovants et des filières locales pour une gestion durable des territoires saharienne (reference number: I.B/1.1/493). The contents of this document are the sole responsibility of Livestock and Wildlife Laboratory, IRA Medenine, Tunisia and Veterinary Clinics and Animal Productions Unit D.E.T.O. Bari, Italy, and can under no circumstances be regarded as reflecting the position of the European Union. The funders had no role in study design, data collection and analysis, decision to publish, or preparation of the manuscript.

Competing Interests: The authors have declared that no competing interests exist.

* E-mail: barbara.padalino@uniba.it

Introduction

Animal behaviour is influenced by the prevailing environment, and behavioural modifications are used to assess the impact of different kinds of management on animal welfare [1]. Animals housed in artificial habitats are confronted by a wide range of potentially provocative environmental challenges, and animals in captivity can develop stereotypical behaviours [2], *i.e.* repetitive, unvarying and apparently functionless behaviour patterns [3]. Since these behaviours have usually been associated with sub-optimal living conditions [4], they have often been used to assess animal welfare in different species (*e.g.* [5–8]). Thus, Mason & Latham [6] suggested that stereotypy should always be taken seriously as a warning sign of potential suffering. Stereotypy can take a wide range of different forms (e.g. locomotor or oral; [9]) and the causes of these abnormal behaviours have been the subject of much discussion [3,5,10]. The animal's lack of control over its environment, frustration, threat, fear, and lack of stimulation have all been mentioned as the main causes leading to the development of abnormal behaviour [3,5]. One of the reasons why animals develop stereotypies is that endorphins are released when performing them, producing a form of pleasure that can help the animal to cope with the various captivity stressors, which in turn may positively reinforce the behaviour (in sows [11]; in macaques [12]).

Therefore, intensive management systems which do not allow the animal to express its behavioural needs could lead to the development of repetitive and functionless behaviours. Cooper and McGreevy [13] reported that, in horses, stereotypies were related to a number of management factors, such as concentrate feeding or social isolation, and that the form of stereotypy usually depended on the constraints to which the animals were exposed. Nicol [14] also suggested that oral stereotypies in horses (*e.g.* crib-biting, wood-chewing) may develop in response to a low-forage diet, because these behaviours may increase salivary flow, reducing gastric tract acidity and speeding up the transit of ingested feed. Locomotor stereotypy in horses (*e.g.* weaving) may derive from some frustrated attempt to move or escape from the stable [14].

Moreover, different studies have shown negative correlations between enclosure size and the prevalence of pacing in different species (red deer [15]; giraffe and okapi [16]).

The husbandry of male camels has been changing recently to more intensive management systems where they are kept isolated in a box or pen and used for programmed mating [17] or artificial insemination [18,19]. One of the major problems of male camel rearing is that during the breeding season bulls can become very aggressive towards other males or humans and for this reason they are kept in a single box or tethered with ropes [20]. In an individual box, the animals are isolated and it is known that social isolation can create stress [21] and may lead to stereotypical behaviour. McGreevy *et al.* [22] have shown that the time spent in a single box was positively correlated with an increased risk of abnormal behaviour in horses. Camels are also social animals and in feral conditions usually live in herds and spend most of the day walking to pasture [23], so captivity could affect their behaviour, as already reported in other feral animals housed in artificial habitats [24]. Thus, our hypothesis was that, as in other species, confinement stressors such as restricted movement, reduced retreat space, forced proximity to humans, reduced feeding opportunities and maintenance in abnormal social groups, could also lead to the development of stereotypical behaviour in male dromedary camels. Since stereotypical behaviours have not yet been reported in dromedary camels, the aim of this study was to identify and describe them for the first time in males housed in single boxes.

While there are several studies on the effects of different housing systems on the behaviour of cattle, horses, hens, pigs and other domestic animals [2,22,25], few studies have been carried out to assess the effects of captivity in camels. Therefore, it was thought of interest to study what effects different forms of husbandry could have on their behaviour, in an attempt to suggest how to optimize camel breeding techniques in the future.

Materials and Methods

Animals and management systems

Four clinically healthy male dromedary camels (*Camelus dromedarius*), ranging in age from 5 to 8 years, with a mean body weight of 526±25 kg and good body condition score (3.5±0.25 arbitrary units; from 0 to 5 accordingly with Faye et al. [26]), were used for this study. All animals were identified by ear tags (#808, #514, #515, #504). The camels had been reared at the Arid Lands Institute's experimental station in Médenine, Tunisia (33°30′N, 10°40′E), 18 m above sea level.

In summer, the bulls are kept in a single open-air paddock shaded by trees whereas, starting from October, they were put into single boxes (Height = 3 m, Length = 5 m and Width = 3 m) with sand floors. They were tethered with a rope on the fetlock of the foreleg and were able to walk around inside the box. The boxes were located far from the females' pen, preventing them from seeing and touching any dams; the gates of the stable pointed eastwards, facing an open-air paddock and with a small window on the opposite wall. The gates of the stable were made by bars; camels were able to put their head outside the box through the bars or the window.

The male dromedary camels were tested in three different management systems: i) housed in single boxes for the whole day (H24; their usual and traditional method), ii) housed in the same box for 23 hours, adding 1 hour free in a paddock from 2 p.m. to 3 p.m. (H23) or iii) housed in the same box for 22 hours and 30 min with 1 hour of freedom again in the paddock from 2 p.m. to 3 p.m. and 30 minutes from 8.00 a.m. to 8.30 a.m. in boxes placed in a little pen adjacent to the female herd's pen (ExF). The paddock lies in front of the stable where the boxes are located and measures 250 square metres. Female herd's pen is bordered by a 130 cm-high wall dividing the two pens, but females were free to move and reach the males.

Each experimental condition lasted 7 days and was preceded by a habituation week, so the whole trial took six weeks (three weeks for the habituation period and three weeks of experimental situations) from February to March 2013, from the middle to the end of the breeding season, starting with the traditional husbandry form (H24) and ending with the exposure to the female rearing system (ExF).

The camels were fed with 5 kg oat hay at 9.00 a.m., and 3 kg concentrate supplement based on barley (60%), wheat bran (17.5%), olive cake (17.5%) and a mineral and vitamin complex (5%) at 3.00 p.m. The chemical composition of the oat hay was: Dry Matter (DM) = 90%, Crude Protein (CP) = 6.81%, Ash = 7.9%. Dry matter content of the concentrate was 90.9% and its chemical composition was CP = 11.4%; Acid-detergent fibre (ADF) = 13.2%; Neutral-detergent fibre (NDF) = 31.6% and Ash = 8.1%. The feeding quantity and quality remained constant during the experiment. The diet met the maintenance requirements as set by Laudadio et al. [27], and water was available once every two days.

During the trial, the bulls were used for semen collection twice weekly. They were well accustomed to this practice and to the traditional husbandry system, so we changed only the management system in accordance with the experimental protocol.

Ethics Statement

The experiments were conducted according to the protocols approved by the Italian Ministry for Scientific Research in accordance with EC regulations. No special permission for behavioural research on wild animals such as this study is required in Italy.

Behavioural parameters

In each management system, the four males were filmed in their single box by a video-camera (Sony Camcorder digital video) from 8.00 to 8.30 a.m. every morning for 7 days in each experimental condition, without being disturbed by the operator. The videos were analysed by an expert ethologist, who filled out a focal animal sampling ethogram, defined as the sampling method whereby the recorder chooses one individual and records all behaviours performed by the individual in a specified time window (one bull in his single box, located far or adjacent to the females' paddock for 30 minutes) [28].

The duration of the subsequent behavioural states was noted down: rumination, resting, standing, walking, looking outside and stereotypy. On the basis of the ethogram, the average time spent on these behavioural activities during the 30-min observation periods was calculated for each management system.

The videos were then studied again and after accurate analysis the presence of the stereotypical behaviours was identified and on the basis of their nature, were split into two categories:

Locomotor stereotypes. Head-shaking: the camel raised his head to the vertical with a very fast movement (this behaviour included a movement of the head by up to 90°). This stereotypy was considered as punctual behaviour because it lasted only about one second.

Pacing in a circle: the camel walked to the other side of his box (stopped and tried to look through a small window in the wall), and walked back to his initial position (in doing so, the camel always followed the same path which described a circle). The camel repeated this movement several times without any clear motivation:

this stereotypy was considered as a state, because it always lasted more than 10 seconds.

Oral stereotypy. Self-biting or self-mutilation: the camel bit different parts of his own forelegs (right or left) from the shoulders to the feet. This stereotypical behaviour was considered as a state - indeed the camel could bite his legs for a variable length of time, ranging from just a few seconds to several minutes.

Bar-mouthing: licking, biting or playing with the lips on the bars of box's gate. This stereotypy was considered as punctual behaviour because it lasted only a few seconds.

Thus, a behavioural sampling ethogram, in which the observer notes all the durations and frequencies of a specific behaviour [28]) was filled out. The duration of the following behavioural states were calculated: locomotor and oral stereotypical behaviour; the total duration of stereotypical behaviours was calculated as the sum of the duration of the two categories (locomotor+oral). The frequency of the following behavioural events (punctual behaviours) was also recorded: locomotor and oral stereotypical behaviour; the total frequency of stereotypical events was also calculated as the sum of locomotor+oral. Moreover, the frequencies of putting the head outside the box and of scratching were recorded, so as to measure how many times they were stimulated by the situation outside their box and how many times they scratched, which could be a sign of boredom in captivity.

Statistical Analyses

All behavioural parameters were subjected to repeated-measures analysis of variance using the Generalized Linear Model procedure (SAS, version 9, 1999). Independent variables were the management system (H24, H23 and ExF), the periodof observation (from Monday to Sunday), and the interaction between those variables. Data were normally distributed. Tukey's post hoc test was used to perform statistical multiple comparisons. The p-level was set at 0.05. All data were expressed as quadratic mean and mean standard error.

Results

The average time spent in behavioural activities during 30-min observation periods while in the single box in the three different management systems (H24, H23, ExF) is reported in Fig. 1.

Three of the four male camels showed stereotypical behaviours, each differing from the others', while one of the males showed two types of locomotor stereotypy (Table 1).

The effect of the management system was significant (df = 2; $F_{(2,6)} = 3.86$; P = 0.02) on the frequency and the duration of the stereotypical behaviours, whereas no significant difference was observed in period (from the first to the seventh day of the week) (df = 6; $F_{(2,6)} = 0.99$; P = 0.44) nor in the interaction between management system and period(df = 12; $F_{(2,6)} = 0.80$; P = 0.64). Consequently, only the effect of the three different management systems on the behavioural parameters was considered.

The duration (in sec) of stereotypical behaviours recorded during the thirty-minute observation period every morning was highest in H24 and decreased progressively from H24 to ExF. Figure 2 shows that total stereotypical duration decreased during the weeks with one hour free in the paddock, and was significantly lower when camels were in the box adjacent to the female herd (H24 vs. ExF: 186.8±49.9 vs. 0.1±4.9 s/30 min; P = 0.03). The number of times these behavioural patterns occurred followed the same trend. When the camels were housed in H24, there was a stereotypical behaviour frequency of 12.7±1.4 in 30 min, significantly higher than for camels in systems H23 (P = 0.002) and ExF (P<0.0001); in addition, the value for H23 was also significantly higher than for ExF (P = 0.038) (Fig. 3).

The frequency of camels putting their heads outside their box was higher in the third housing system, when the camels were stimulated by the female herd, compared with the other two management systems (P<0.001). By contrast, the frequency of scratching behaviour was very low when they were in the pen adjacent to the female herd 0.6±0.5, and was significantly lower than for those allowed to roam free in the paddock for one hour (H23 3.0±0.9; P = 0.02) or kept in a box (H24 3.3±0.5; P = 0.003) (Fig. 4).

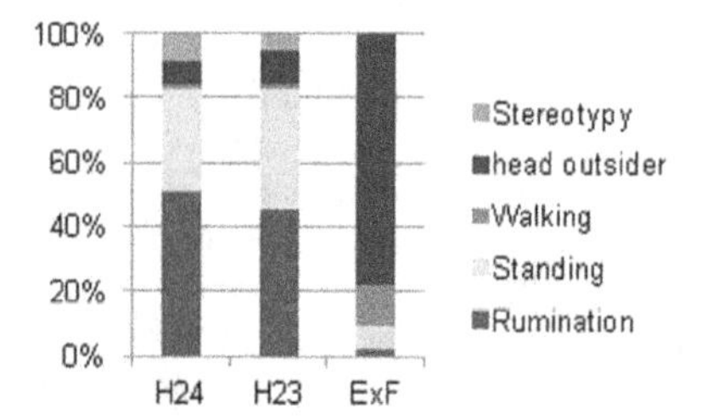

Figure 1. Average percentages of time spent in behavioural activities during 30 min observation periods of camels while in their single box in the three different management system: housed in single box for 24 hours (H24), housed in single box for 23 hours and one hour in paddock (H23), housed in single box for 22 hours and 30 minutes, one hour in paddock and 30 minutes of female exposition (ExF).

Discussion

Three out of four male camels showed abnormal repeated behaviours corresponding to the general definition of stereotypical behaviours (*i.e.* repetitive, unvarying and apparently functionless behaviour patterns, [3]). As different kinds of stereotypical behaviour were observed in the three camels, a distinction was made between oral and locomotor stereotypies were distinguished as in other species [16,9,29,30]. Mason et al. [31] assessed that animals could develop different repertoires of stereotypical behaviour and broke down stereotypy by taxon, to show that different orders of mammals typically favour different types of abnormal repetitive behaviour (locomotor, oral or non-locomotor body movements). This analysis revealed that stereotypical carnivora systematically prefer locomotor movements, while ungulates display oral forms. Accordingly with the data reported for ungulates, these camels developed both oral and locomotor stereotypies, showing a preference for oral ones. Indeed, two of the four camels performed different oral stereotypies: self-biting and bar-mouthing; while one of the four males performed two different locomotor stereotypies: head-shaking and pacing in a circle. The camel which exhibited head-shaking also exhibited circling, so this individual developed two different kinds of locomotor stereotypy. This finding agrees with previous observations in horses [32]: a horse already showing one locomotor stereotypy is more likely to develop a second than horses either performing oral stereotypy or expressing no form of stereotypy.

The camels in this experiment spent about 10% of the observation period stereotyping in H24; the range reported for cows is from 1% to 38% of a 24-h period before, during and after grazing [33], but in another study, where animals were reared in better conditions, this figure dropped to 1–2% [34]. A horse housed in a single box can spend up to 8 h crib-biting each day [35], whereas one female captive giraffe could spend more than 40% of the night licking and tongue-playing [36]. The latter

Table 1. Description of the stereotypy shown by each camel while in their single box.

Camel	Stereotypy
808	"Bar-mouthing": Licking, biting or playing with the lips on the bars
514	None
515	"Self-biting": the camel bit different parts of his own forelegs (right or left) from the shoulders to the feet.
504	"Head-shaking": the camel raised his head to the vertical with a very fast movement (this behaviour included a movement of the head up to 90°). "Pacing in a circle": The camel walked to the other side of his box (and sometimes stopped and looked through the window), and walked again until he was back in his initial position (in doing so, the camel always followed the same path which described a circle)

stereotypical behaviours were related to poor management, i.e. diets with low fibre, thus confirming the effect of husbandry on the prevalence of stereotypy.

It could therefore be supposed that the traditional housing system (H24) in which camels showed the greatest incidence of stereotypies was a sub-optimal management system for this species, in agreement with studies carried out by Mills in horses [4], and that the presence of stereotypical behaviour in these individuals was a sign of poor welfare, as inferred by Mason & Latham [6] who suggested that stereotypy could be a sign of suffering. Thus, in H24 the camels were probably frustrated, lacking stimulation and control of their environment and could not therefore exhibit natural behaviours (*e.g.* it would be impossible for them to perform any social interaction), because this housing system did not satisfy the behavioural needs of this species.

These four stereotypies have already been reported in other species and different explanations have been suggested concerning the cause of these abnormal behaviours [13,14,16,22]. The three major constraints were limited space, lack of stimulation (especially social contact) and controlled feeding. The development of stereotypy in these camels could also be explained by one of these three constraints or by their cumulative effect.

One of the four camels exhibited an unusual kind of oral stereotypy, *i.e.* bar-mouthing, which consisted in biting, licking or playing with the lips on the bars of the cage. The development of this stereotypy in a camel housed in a box was not surprising, indeed, these kinds of abnormal oral behaviours have also been reported in other captive animals (bank voles, [37]; pigs, [38]). Rebdo [33] suggested that feeding frustration could facilitate oral stereotypy; in our study, the camels were fed with 5 kg of oat hay and 3 kg of concentrate and did not have the opportunity to forage on pasture. Therefore, these camels may have felt feeding frustration, which could explain why one individual had developed this kind of stereotypy. This hypothesis agrees with previous studies in gilts where the time spent performing oral stereotypies (*e.g.* bar-chewing) was negatively correlated with their feed allowance (review by Lawrence & Terlouw, [38]). Nicol [14] also suggested that low-forage diets could be the main cause of the onset of oral stereotypy in horses. Moreover, Rebdo [33] has shown that heifers exhibited no abnormal behaviours when at pasture. Camels are also herbivorous animals and in natural conditions usually graze for 8–12 hours per day [23]; therefore, as has been proposed for other species, the lack of pasture may have been the trigger for this oral stereotypy.

Self-biting or self-mutilation was performed by one of the four camels during our observations. In captive-reared rhesus monkeys, the absence of physical contact with conspecifics negatively affected their behaviour and the prevalence of self-biting was positively correlated to the number of years spent in a single stall [39]. Camels are social animals and, while old males can occasionally be solitary, camels usually live in herds made up of males, females and young, or females and young without a male, or males and females without young or only one male, with an average of 25 individuals per herd [23]. Therefore, it could be supposed that social deprivation in this species may lead to the development of self-biting. McDonnell [40] suggested that social and/or feeding distraction could reduce the prevalence of self-mutilation in horses.

Pacing has been reported in a wide range of captive animals (cats, dogs, hens and horses, review by Dallaire, [30]; okapi and

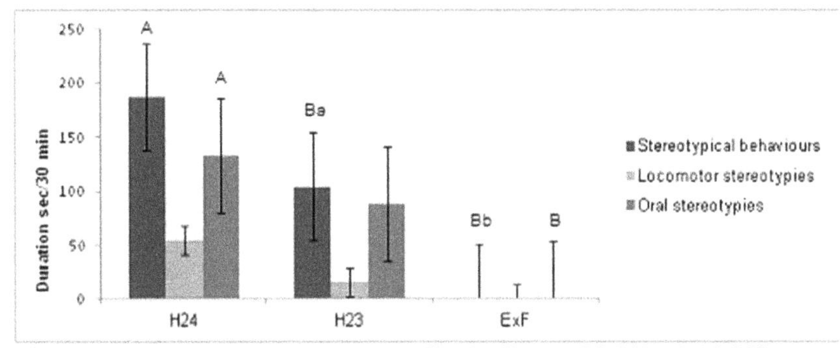

Figure 2. Effect of three different management systems (housed in single box for 24 hours (H24), housed in single box for 23 hours and one hour in paddock (H23), housed in single box for 22 hours and 30 minutes, one hour in paddock and 30 minutes of female exposition (ExF) on the duration (s/30 min) of stereotypical behaviour shown by male dromedary camels while in their single box.
Oral stereotypies: self-biting or self-mutilation and bar-mouthing; Locomotor stereotypies: pacing in a circle; Stereotypical behaviour: sum of oral and locomotor stereotypies.

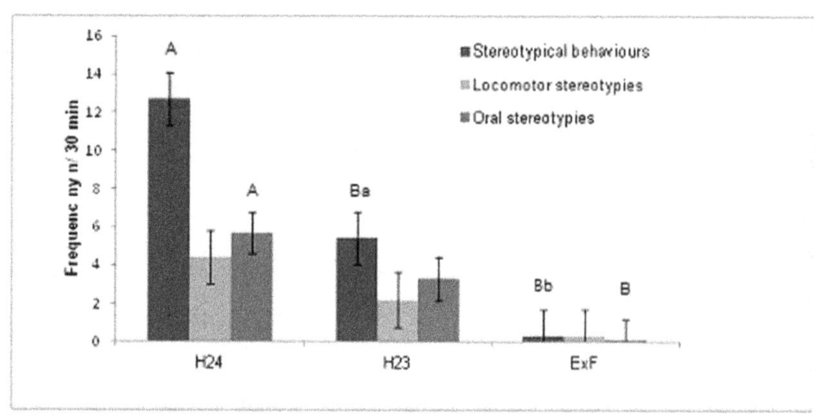

A, B,a,b, : columns with different superscripts are significantly different A, B P<0.01 and a,b P<0.05

Figure 3. Effect of three different management systems (housed in single box for 24 hours (H24), housed in single box for 23 hours and one hour in paddock (H23), housed in single box for 22 hours and 30 minutes, one hour in paddock and 30 minutes of female exposition (ExF) on the frequency (n/30 min) of stereotypical behaviour shown by male dromedary camels while in their single box. Oral stereotypies: self-biting or self-mutilation and bar-mouthing; Locomotor stereotypies: head-shaking and pacing in a circle; Stereotypical behaviour: sum of oral and locomotor stereotypies.

giraffe, [16]; red deer stags, [15]; arctic fox, polar bear, American mink, and lion, [41]; bears, [42]) and has been related to confinement-specific stressors [24]. In natural conditions, camels usually walk a lot during the day, grazing 8–12 hours daily and walking at an average speed of 2 km/h, but if necessary, they can walk 150 km per day in the desert [23]. In our study, the camels were housed in single boxes, so it is to be presumed that this area was unable to fulfil the camels' needs to walk as much during the day as they would do under natural conditions, which is probably why one camel developed this locomotor stereotypy. This hypothesis is in agreement with different studies showing a negative correlation between enclosure size and the prevalence of pacing in different species (red deer, [15]; giraffe and okapi, [16], monkeys, [3], carnivores, [41]). Moreover, Nicol [14] suggested that locomotor stereotypy in horses may derive from some frustrated attempt to move or escape from the stable, and Lawrence & Terlouw [38] supposed that the development of pacing could be based on escape behaviour. Therefore, as has been proposed for other species, providing camels with a bigger enclosure would help improve their welfare. However, Morgan and Tromborg [24] concluded that increasing the space available to the animal did not always have a positive effect on its welfare, particularly for prey animals, because it may well be that it is not the quantity of space available to the animal which is important but rather its quality, and what it gives the animals in the way of behavioural opportunity.

Head-shaking was observed in one individual. This or similar forms of stereotypical behaviour (head-tossing, bobbing, nodding or shaking) has been reported in a broad range of species (in horses, [43]; okapi and giraffe, [16]; bears, [42]; humans, [44]; rats, [45], cats, [30]; elephant, [46]). The causes of such stereotypical behaviours have been poorly investigated. According to Crowell-Davis [47], head-shaking can have a great variety of causes. As for the other stereotypies, we can suppose that the lack of stimulus, space and social contact may have led to the development of this stereotypy. Cooper et al. [48] found that

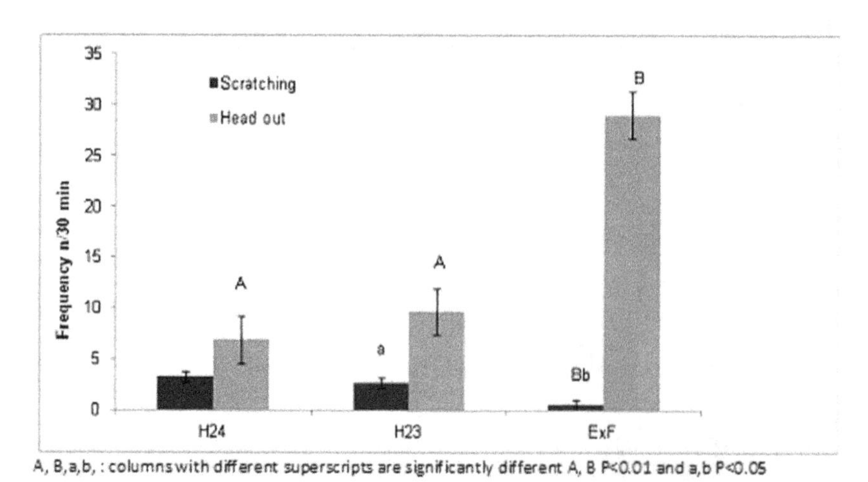

A, B,a,b, : columns with different superscripts are significantly different A, B P<0.01 and a,b P<0.05

Figure 4. Effect of three different management systems (housed in single box for 24 hours (H24), housed in single box for 23 hours and one hour in paddock (H23), housed in single box for 22 hours and 30 minutes, one hour in paddock and 30 minutes of female exposition (ExF) on the frequency (n/30 min) of scratching and putting the head outside the box shown by male dromedary camels while in their single box.

increasing the visual horizon significantly decreased the prevalence of head-shaking in horses housed in single boxes (i.e. increasing visual contact between neighbours and towards the environment allows horses to monitor the environment and to interact with other horses).

Our study showed the impact of the different kinds of management on the duration and frequency of stereotypy, with more frequent stereotypical behaviour among camels kept in a box for 24 h than among those allowed 1 hour free. Therefore, we could suggest that allowing camels to walk for 1 hour daily would be a good way of improving their living conditions, rather than keeping them in a single box around the clock. Time spent stereotyping also tended to decrease between camels in groups H24 and H23, which could suggest that 1 hour free has a positive impact (or at least not a negative one) on camel welfare but it would seem to be insufficient. The duration of locomotor stereotypy decreased between H24 and ExF and this could be explained by the fact that these animals had 30 minutes more to spend in an area adjacent to the female herd where the dams could walk around as much as they wanted, stimulating the bulls.

Overall, the frequency and duration of stereotypical behaviours were higher in H24 and H23 than ExF. This suggests that exposure to females in the pen could be a better environment for male camels because it more closely matches the needs of this species (i.e. more time to walk around and more chance for social contact).

The frequency of oral stereotypy (bar-mouthing and self-biting) decreased from H24 and H23 to ExF. This is not surprising and is in accordance with our hypothesis as well as with that of McDonnell [40]: self-injuries decreased among stallions when they were placed in pasture with mares because it provided plenty of distraction and allowed the animals to perform social behaviour. In our study, camels were not placed directly with females but they could interact with and touch them (with an average of 35.9 touching events in 30 minutes), if females came near the wall and put their neck and head in the male's area, allowing contacts. Consequently, in ExF, the males put their heads outside the box (through the window or between the bars of the gate) more often than in the other groups, showing that they were monitoring their environment more when the females were close by. This could be explained by the fact that during ExF, they had a larger area of view (a wider horizon) of a more interesting environment around them than in their box, where they could only look at an empty space (poor of stimuli). Scratching was also influenced by the management system; indeed, its frequency was higher in H24 and H23 compared with ExF. According to Maestripieri et al. [49] scratching could be a sign of stress, frustration or anxiety. Moreover, Basset et al. [50] measured the frequency of self-scratching as an indicator of stress in the common marmoset (*Callithrix jacchus*) and considered an increase in this behaviour as a sign of reduced welfare. Thus, in this study, the lower frequency of scratching in ExF compared to boxed conditions could be interpreted as an improvement in their welfare needs, i.e. exposure to females could be a good way of providing male camels with stimulation and the opportunity to perform social and sexual behaviours instead of stereotypical ones.

On the basis of our preliminary findings, the traditional husbandry system of male dromedary camels reared under intensive management systems should be changed, by integrating it with at least one hour thirty minutes daily of walking around in paddocks, spending more time feeding (decreasing the concentrate/forage ratio), opening a window between the boxes which would allow the camels to have visual contact with their neighbours and spend some time near females.

Conclusion

Male dromedary camels may develop abnormal behaviour, just as other animals do, if they live in sub-optimal conditions, and this trial was the first step in identifying locomotor and oral stereotypies in male dromedary camels housed in single boxes. Overall, this preliminary study suggests that the traditional husbandry method could be improved by allowing free movement and social contact, both of which had positive impacts on the incidence of stereotypy. Further studies are needed to identify the behavioural needs of camels reared under intensive management systems and to optimize dromedary camel welfare and breeding techniques.

Acknowledgments

The authors are grateful to the Tunisian authorities and thank all of the technical staff at the Laboratory of Livestock and Wildlife for their valuable help during this study. The authors would also like to thank Dr. Julian Skidmore and Dr. Anthony Green for kindly suggesting stylistic improvements to the text and CIRAD (Centre de Coopération Internationale en Recherche Agronomique pour le Développement) for coordination and administrative support.

Author Contributions

Conceived and designed the experiments: BP LA. Performed the experiments: BP LA MF DM. Analyzed the data: BP MH. Contributed reagents/materials/analysis tools: TK GML. Wrote the paper: BP LA DM MH.

References

1. Inglis IR, Langton S (2006) How an animal's behavioural repertoire changes in response to a changing environment: a stochastic model. Behaviour 143: 1563–1596.
2. Haskell MJ, Rennie LJ, Bowell VA Wemelsfelder F, Lawrence AB (2003) On-farm assessment of the effect of management and housing type on behaviour and welfare in dairy cattle. Animal welfare 12: 553–556.
3. Mason GJ (1991) Stereotypy: a critical review. Animal Behaviour 41: 1015–1038.
4. Mills DS (2005) Repetitive movement problems in the horse, in: McDonnell editor, The Domestic Horse, The Origins, Development and Management of its Behaviour, Cambridge University Press, Cambridge, pp 212–227.
5. Broom DM (1991) Animal welfare: concepts and measurement. Journal of Animal Science 69: 4167–4175.
6. Mason G, Latham N (2004) Can't stop, won't stop: is stereotypy a reliable animal welfare indicator? Animal Welfare 13: 57–69.
7. Normando S, Meers L, Samuels WE, Faustini M, Ödberg FO (2011) Variables affecting the prevalence of behavioural problems in horses. Can riding style and other management factors be significant? Applied Animal Behaviour Science 133: 186–198.
8. Hausberger M, Gautier E, Biquand V, Lunel C, Jégo P (2009) Could Work Be a Source of Behavioural Disorders? A Study in Horses. PLoS ONE 4: 1–7.
9. Rushen J, Mason G (2006) A decade-or-more's progress in understanding stereotypic behaviour Stereotypic Animal Behaviours. In: Rushen J, Mason G editors. Stereotypic animal behaviour: fundamentals and applications to welfare. CABI. pp. 1–17.
10. Ödberg F (1978) Abnormal behaviours: stereotypies. Proceedings of the First World Congress on Ethology applied to Zootechnics, Madrid, 475–480.
11. Cronin GM, Wiepkema PR, Van Ree JM (1985) Endogenous opioids are involved in abnormal stereotyped behaviours of tethered sows. Neuropeptides 6: 527–530.
12. Crockett CM, Sackett GP, Sandman CA, Chicz-DeMet A, Bentson KL (2007) Beta-endorphin levels in longtailed and pigtailed macaques vary by abnormal behaviour rating and sex. Peptides 28: 1987–1997.
13. Cooper J, McGreevy P (2007) Stereotypic Behaviour in the Stabled Horse: Causes, Effects and Prevention without Compromising Horse Welfare In: The Welfare of Horses, Animal Welfare (Ed by N . Waran) pp 99–124.
14. Nicol C (1999) Understanding equine stereotypies. Equine Veterinary Journal 28: 20–25.

15. Pollard JC, Littlejohn RP (1996) The effects of pen size on the behaviour of farmed red deer stags confined in yards. Applied Animal Behaviour Science 47: 247–253.
16. Bashaw MJ, Tarou LR, Maki TS, Maple TL (2001) A survey assessment of variables related to stereotypy in captive giraffe and okapi. Applied Animal Behaviour Science 73: 235–247.
17. Rahim S, El Nazier AT (1992) Studies on the sexual behaviour of the dromedary camel Proc 1st Int Camel Conf, 115–118.
18. Skidmore JA, Morton KM (2013) Artificial insemination in dromedary camels. Animal Reproduction Science 136: 178–186.
19. Monaco D, Fatnassi M, Padalino B, Kchira B, El Bahrawy K, et al. (2013) The experimental semen collection centers for dromedary camels in Egypt and Tunisia: Current situation and future developments Proc.11th Congress of the Italian Society of Animal Reproduction vol Ustica, Italy, 1: 132–136.
20. El-Whishy AB (1988) Reproduction in the male Dromedary (*Camelus dromedarius*): a review. Animal Reproduction Science 17, 217–241.
21. Kim JW, Kirkpatrick B (1996) Social isolation in animal models of relevance to neuropsychiatric disorders. Biological Psychiatry 40: 918–922.
22. McGreevy P, French N, Nicol C (1995) The prevalence of abnormal behaviours in dressage, eventing and endurance horses in relation to stabling. Veterinary Record 137: 36–37.
23. Gauthier-Pilters H, Dagg AI (1981) The camel. Its evolution, ecology, behaviour, and relationship to man. The University of Chicago Press. 208 p.
24. Morgan KN, Tromborg CT (2007) Sources of stress in captivity. Applied Animal Behaviour Science 102: 262–302.
25. Lay DC, Fulton RM, Hester PY, Karcher DM, Kjaer JB, et al. (2011) Hen welfare in different housing systems. Poultry Science 90: 278–294.
26. Faye B, Bengoumi M, Cleradin A, Tabarani A, Chilliard Y (2001) Body condition score in dromedary camel: A tool for management of reproduction. Emirates Journal of Agricultural Science 13: 01–06.
27. Laudadio V, Dario M, Hammadi M, Tufarelli V (2009) Nutritional composition of three fodder species browsed by camels (*Camelus dromedarius*) on arid area of Tunisia. Tropical Animal Health and Production 41: 1219–1224.
28. Altmann J (1974) Observational study of behaviour: sampling methods. Behaviour 49: 227–265.
29. McGreevy P (2004) Equine behaviour: a guide for veterinarians and equine scientists. The University of Wisconsin - Madison. 369 p.
30. Dallaire A (1993) Stress and behaviour in domestic animals: Temperament as a predisposing factor to stereotypies. Annals of the New York Academy of Sciences 697: 269–274.
31. Mason G, Clubb R, Latham N, Vickery S (2007) Why and how should we use enviromental enrichment to tackle stereotypic behaviour? Applied Animal Behaviour Science 102: 163–188.
32. Mills DS, Alston RD, Rogers V, Longford NT (2002) Factors associated with the prevalence of stereotypic behaviour amongst Thoroughbred horses passing through auctioneer sales. Applied Animal Behaviour Science 78: 115–124.
33. Redbo I (1990) Changes in duration and frequency of stereotypies and their adjoining behaviours in heifers, before, during and after the grazing period. Applied Animal Behaviour Science 26: 57–67.
34. Bolinger DJ, Albright JL, Morrow-Tesch J, Kenyon SJ, Cunningham MD (1997) The effects of restraint using self locking stanchions on relation to behavior, feed intake, physiological parameters, health, and milk yield. Journal of Dairy Science 80: 2411–2417.
35. McGreevy PD, Nicol CJ (1998) Prevention of crib-biting: a review: Equine Veterinary Journal Supplement 27:35–38.
36. Baxter E, Plowman AB (2001) The effect of increasing dietary fibre on feeding, rumination and oral stereotypies in captive giraffe (*Giraffa camelopardalis*). Animal Welfare 10:281–290.
37. Garner JP, Mason GJ (2002) Evidence for a relationship between cage stereotypies and behavioural disinhibition in laboratory rodents. Behavioural Brain Research 136: 83–92.
38. Lawrence AB, Terlouw E (1993) A review of behavioural factors involved in the development and continued performance of stereotypic behaviours in pigs. Journal of Animal Science 71: 2815–2825.
39. Lutz C, Well A, Novak M (2003) Stereotypic and self-injurious behaviour in rhesus macaques: A survey and retrospective analysis of environment and early experience. American Journal of Primatology 60: 1–15.
40. McDonnell SM (2008) Practical review of self-mutilation in horses. Animal Reproduction Science 107: 219–228.
41. Clubb R, Mason G (2003) Animal welfare: captivity effects on wide-ranging carnivores. Nature 425: 473–474.
42. Vickery S, Mason G (2004) Stereotypic behaviour in Asiatic black and Malayan sun bears. Zoo Biology 23: 409–430.
43. Cook WR (2003) Bit-induced pain: a cause of fear, flight, fight and facial neuralgia in the horse. Pferdeheilkunde 19: 75–82.
44. Singer HS (2009) Motor stereotypies. In: Seminars in Pediatric Neurology 16: 77–81.
45. Holmgren B, Urbá-Holmgren R, Valdés M (1976) Spontaneous and amphetamine induced head-shaking in infant rats. Pharmacology Biochemistry and Behavior 5: 23–28.
46. Friend TH (1999) Behaviour of picketed circus elephants. Applied Animal Behaviour Science 62: 73–88.
47. Crowell-Davis SL (2008) Understanding Behavior: Head Shaking. Compendium Equine 466–474.
48. Cooper JJ, McDonald L, Mills DS (2000) The effect of increasing visual horizons on stereotypic weaving: implications for the social housing of stabled horses. Applied Animal Behaviour Science 69: 67–83.
49. Maestripieri D, Schino G, Aureli F, Troisi A (1992) A modest proposal: Displacement activities as an indicator of emotions in primates. Animal Behaviour 44: 967–979.
50. Bassett L, Buchanan-Smith HM, McKinley J, Smith TE (2003) Effects of Training on Stress-Related Behavior of the Common Marmoset (*Callithrix jacchus*) in Relation to Coping With Routine Husbandry Procedures. Journal of Applied Animal Welfare Science 6: 221–233.

18

Making Friends: Social Attraction in Larval Green and Golden Bell Frogs, *Litoria aurea*

Stephan T. Leu[1*¤], Martin J. Whiting[2], Michael J. Mahony[1]

1 School of Environmental and Life Sciences, University of Newcastle, Newcastle, New South Wales, Australia, **2** Department of Biological Sciences, Macquarie University, Sydney, New South Wales, Australia

Abstract

Socio-ecological models combine environmental and social factors to explain the formation of animal groups. In anurans, tadpole aggregations have been reported in numerous species, but the factors driving this behaviour remain unclear. We conducted controlled choice experiments in the lab to determine whether green and golden bell frog (*Litoria aurea*) tadpoles are directly attracted to conspecifics (social factors) in the absence of environmental cues. Using repeated measures, we found that individual tadpoles strongly preferred associating with conspecifics compared to being alone. Furthermore, this preference was body size dependent, and associating tadpoles were significantly smaller than non-associating tadpoles. We suggest that small tadpoles are more vulnerable to predation and therefore more likely to form aggregations as an anti-predator behaviour. We demonstrate that tadpoles present an ideal model system for investigating how social and ecological factors influence group formation in vertebrates.

Editor: Brock Fenton, University of Western Ontario, Canada

Funding: STL was funded by a Port Waratah Coal Services Limited (PWCS) grant while conducting the study. The funders had no role in study design, data collection and analysis, decision to publish, or preparation of the manuscript.

Competing Interests: The authors have the following competing interest: STL was funded by a Port Waratah Coal Services Limited (PWCS) grant while conducting the study. There are no patents, products in development or marketed products to declare.

* E-mail: stephan.leu@flinders.edu.au

¤ Current address: School of Biological Sciences, Flinders University, Adelaide, South Australia, Australia

Introduction

Group formation in animals has evolved many times independently in response to a wide range of selective pressures [1]. Understanding the ultimate factors and proximate mechanisms influencing group formation and ultimately sociality, is a major challenge in evolutionary biology. Both environmental and social factors have been identified to confer a fitness advantage and to influence the formation of animal aggregations and have been combined into socio-ecological models. Important environmental factors include the abundance, distribution, and quality of resources, such as discrete food patches [2], nest sites [3] and refuges [4].

The most commonly recognized social factors and benefits of being in a group include dilution of predation risk [5], increased vigilance and early predator detection [6,7], thermoregulation [8] and enhanced feeding through the use of social information [9,10]. Nevertheless, there are also costs to group formation which are often proportional to group size and include increased within-group competition [11], home range size and travel distances [12], and susceptibility to parasites and pathogens [13,14]. These costs and benefits, in a species-specific combination, selected for the independent evolution of group living in numerous taxa [15].

Tadpole schooling has evolved multiple times in distantly related taxa and occurs in at least 10 families [16]. Much of our understanding of the causes and consequences of tadpole schooling is based on observations in the field and a number of laboratory studies [16–19]. A key finding suggesting the importance of social cues for group formation is that tadpoles of *Bufo americanus* and several other species are capable of kin recognition and preferentially school with siblings [19–22]. Interestingly, the majority of aggregating species breed in ephemeral ponds [16], indicating that tadpoles may aggregate at scarce food resources. Conversely, it has been shown that daylight induces group formation [23], and water clarity increases group size and decreases inter-individual distances [24], which both suggest a spacing pattern that enhances anti-predator behaviour. Social foraging [25] and the use of social information to find food [26] have also been suggested as important social factors selecting for the formation of tadpole aggregations.

Taken collectively, the relative importance of ecological and/or social factors for the formation of tadpole aggregations is as yet unclear. The green and golden bell frog (*Litoria aurea*) is a large endangered Australian frog (adult snout-vent length 55–90 mm) [27]. It is adapted to breeding in permanent ponds [28] and its tadpoles (snout-vent length 3–30 mm from hatching to metamorphosis) [27] can be under strong predation pressure, for example by the mosquitofish (*Gambusia holbrooki*) [29,30]. *Litoria aurea* tadpoles have occasionally been found in aggregations in the wild [27]. We tested whether tadpoles form groups on the basis of social conspecific attraction in the absence of environmental resources.

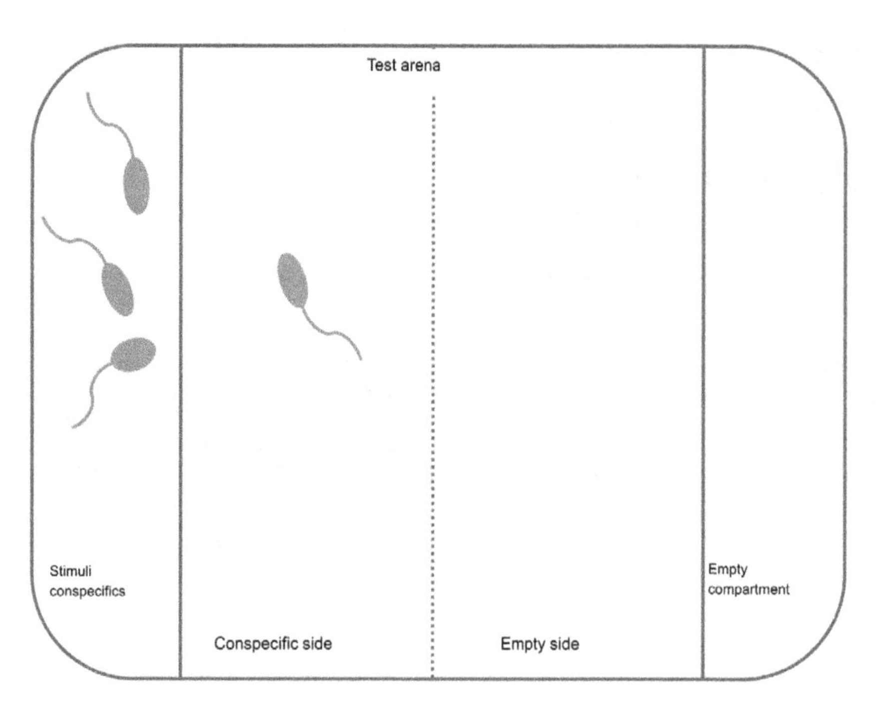

Figure 1. Test arena design. Arena with adjoining compartments on either side to hold conspecifics (test stimulus). A line on the base of the chamber (here dashed) divided the arena into halves.

Materials and Methods

Ethics statement

All methods and procedures were formally approved by the University of Newcastle Animal Care and Ethics Committee (approval no A-2011-150) in compliance with the Australian Code of Practice for the use of animals for scientific purposes.

Tadpole husbandry

Adult green and golden bell frogs were collected from Kooragang Island in the Hunter Estuary in New South Wales (−32°51′48″/151°43′55″) during 2010 as part of an independent breeding project. For the experiment reported here, we randomly selected tadpoles from two clutches laid by two different females in captivity, in December 2011. Tadpoles were housed in two mixed sibship groups, with both group sizes varying repeatedly over time between 25–100 individuals (approximately 5–20 tadpoles/L water). Social experience through raising tadpoles in different group sizes does not influence their propensity to aggregate [31]. Tadpoles were maintained in aged carbon-filtered water at room temperature on the natural light cycle (light-dark cycle of 14:10 h) and they were fed three times/week with sprinkles of Goldfish Flake Plus (Marine Master). Water was changed as required, on average every four days (range 1–7, $N=9$). Tadpoles were returned to the breeding project after the experiment.

Tests of social attraction

We tested a total of 48 tadpoles. Each tadpole was tested once, and all were at Gosner developmental stage 25 [32,33]. The study group was divided into two groups of 24 tadpoles, which were tested on consecutive days at similar times when tadpoles were active. Tadpoles were fed within 24 hours prior to the experiment to minimize the influence of the nutritional state on grouping behaviour. We used six test chambers at a time, which were visually isolated from each other. Each test chamber (figure 1) was partitioned into three compartments, the rectangular test arena (9.0×9.2×5.0 cm) in the middle and one adjoining compartment (2.8 cm×9.2×5.0 cm) on opposite sides. The test arena was divided in half again by a line on the base of the arena. We used three conspecifics as stimuli in each apparatus. Different stimuli tadpoles were used during the second experimental day. On each day focal and stimuli tadpoles were taken from one of the two mixed sibship groups. Focal and stimuli tadpoles were not size matched. Partitioning walls were transparent to allow visual contact between stimuli and focal tadpoles. Each day, we filled the six test chambers with water from a container that held all 42 tadpoles (24 focal tadpoles + 6×3 conspecific stimuli tadpoles) for 15 minutes, to minimize differences in tadpole chemical cues between trials. The partitioning walls were not watertight and tadpole-borne chemical cues could diffuse from the conspecific compartment into the arena. Visual and chemical cues were available to the focal tadpoles. We controlled for spatial preference in focal tadpoles by haphazardly assigning conspecifics to three left and three right compartments in the six test chambers used at a time. The water level in the test chambers was 2 cm. *L. aurea* tadpoles can occur in shallow water and often swim near the water surface [27]. Furthermore, following similar experiments [17,20,26], we chose this relatively low water level to allow us to clearly determine the position and distance of the focal individual relative to conspecifics. After introducing a focal tadpole into each arena we allowed them two minutes to acclimate, inspect the arena with its two adjoining compartments, and make a decision. We then took four photographs at two minute intervals from behind a blind to establish the tadpole's position. At the end of each experimental day we measured the body size (snout-vent length) of each tadpole to the nearest 0.1 mm.

Data analyses

We used the photographs to score whether tadpoles were present on the half of the test arena adjacent to the compartment with conspecifics or the empty compartment. We discarded five trials because they could not be accurately scored, leaving a total of 43 trials. Tadpoles were active and usually moved during the two minute interval between photographs. We used Cochran's Q test (an extension to the McNemar test) to ensure that repeated testing did not affect the decision making process in a systematic way [34], as the precision of decision making may improve with repetition [35,36]. We then allocated trials to one of two categories: (1) focal tadpole present on the conspecific side three or four times out of the four measures; and (2) focal tadpole present on the empty side two, three or four times. We used a chi-square test to determine whether option (1) occurred more frequently than expected by chance and hence whether tadpoles were attracted to conspecifics. Expected frequencies were calculated using the probability mass function for binominal distributions. Finally, we used a two-sample *t*-test to investigate whether social attraction strength was a function of body size. Data were reflect square root transformed due to negative skew to meet the assumption of normality. We used an alpha level of 0.05 to denote significant results. Statistical analyses were performed in SPSS 17.0.2.

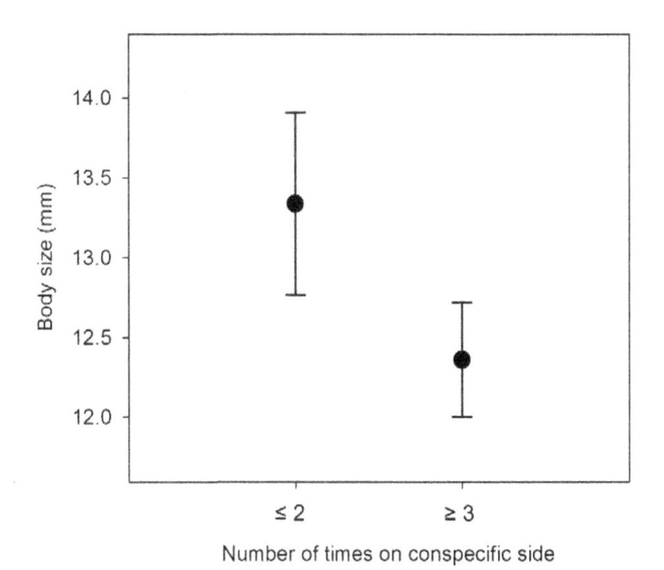

Figure 3. Mean (± SE) tadpole body size in relation to social preference. Body sizes (snout-vent lengths) are shown as untransformed data.

Results

Whether tadpoles chose to be alone or to associate with conspecifics was not significantly different among trials (Cochran's Q test: $Q_3 = 6.66$, $P = 0.08$), and neither increased nor decreased through the course of multiple trials. This indicated a high repeatability among trials. Significantly more tadpoles than expected by chance chose the compartment containing conspecifics at least three out of four times (category 1;Chi-square test: $\chi^2_1 = 29.69$, $P<0.0001$; figure 2). Furthermore, social attraction was a function of body size. Tadpoles that were attracted to their conspecifics (category 1) were significantly smaller than tadpoles that showed no social attraction (category 2, unpaired *t*-test: $t_{41} = -2.84$, $P = 0.007$; figure 3).

Discussion

We conducted a relatively simple choice experiment in which isolated tadpoles had the option of spatially associating with an empty compartment or a compartment containing a group of three conspecific tadpoles. We found a very strong preference for the compartment containing conspecifics, which suggests that *Litoria aurea* tadpoles prefer to aggregate with conspecifics through social attraction. Whether visual or olfactory cues were used to locate the conspecifics remains to be determined. Alternatively, tadpoles may have avoided the empty compartment. However this is unlikely, because *L. aurea* tadpoles occupy structured habitats in the wild, yet have been observed to school [27].

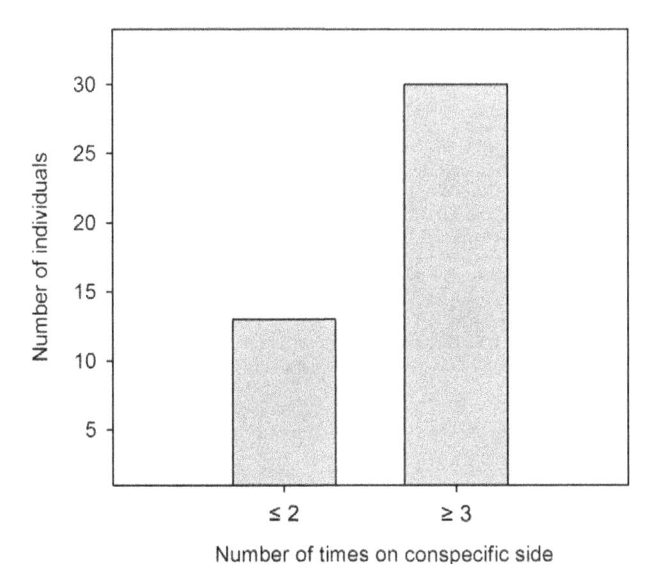

Figure 2. Social preference in *Litoria aurea* tadpoles. Number of individuals that repeatedly chose the side with conspecifics over the empty side.

Both environmental and social factors have been identified to select for the formation of animal aggregations. By conducting controlled experiments in the lab, we were able to exclude environmental factors such as food, temperature and light gradients that might influence tadpole grouping behaviour [37,38]. This indicates that aggregation behaviour in *L. aurea* may be driven by social factors. One such social factor is the benefit of schooling in the presence of a predator [39] and/or when tadpoles detect chemical cues from a predator [40]. Predators may include larger conspecifics [41] and therefore represent another form of natural selection. This may result in relatively higher predation pressures on smaller individuals. Small tadpoles are most vulnerable to predation [42], and tadpoles generally become less vulnerable as they grow [43], because larger tadpoles are able to escape from predators more readily [44] or are too large for canibalistic conspecifics. However, during metamorphosis tadpoles are once again vulnerable because their emerging limbs may compromise locomotor ability [44]. Interestingly, we found a relationship between social attraction and tadpole size in *L. aurea* tadpoles. Individuals that were attracted to conspecifics were significantly smaller than tadpoles that showed no social attraction. This may indicate that schooling in *L. aurea* could be an anti-predator behaviour. This hypothesis requires empirical testing, especially since the tadpoles in our study were born in captivity and not exposed to predatory cues.

Another social route to tadpole schooling is the benefit of social foraging and the transfer of social information to obtain food. For example, tadpoles of the spadefoot toad *Spea multiplicata* have been shown to form a vortex in response to introduced food particles in feeding experiments [25]. It has been suggested that this collective behaviour allows access to food resources through agitating the pond substrate that otherwise would be inaccessible [25]. Similarly, *Bufo americanus* tadpoles use social information through the presence of conspecifics, to locate food patches, rather than chemical cues from the food source itself [26].

A third social route to schooling appears to be relatedness. The tadpoles of several species preferentially school with kin and remain in schools even in the absence of an immediate predatory threat [19,20]. However, the mechanisms driving kin recognition and preference for siblings appear to be varied [16] and there is even debate that kin recognition might simply be a by-product of species and/or group member recognition and therefore not true kin recognition [45]. Nevertheless, a popular argument in support of kin-based schooling is that if distasteful tadpoles are preyed upon, this cost is balanced against the improved survival of siblings [20]. Hence, relatedness could be an alternative explanation for the observed association behaviour in *L. aurea*, which would indicate a genetic determination. However, we did not distinguish tadpoles on the basis of their clutch, so the effects of relatedness on schooling behaviour are difficult to gauge in our study.

There has been considerable debate and speculation about the factors and cues that drive group formation in tadpoles [16]. While environmental drivers appear to be important for some species, social reasons are equally important for other species. Our study establishes the existence of tadpole schooling in *L. aurea* and indicates the importance of social factors, but more research is required to distinguish between the possible functions of this behaviour. For example, whether *L. aurea* tadpoles are attracted to conspecifics as an anti-predator behaviour or because conspecifics signal the presence of food remains to be determined. Nevertheless, the observed inverse relationship between tadpole size and aggregation preference supports the notion of an adaptive anti-predator behaviour.

Given the significance of the tadpole phase for successful reproduction in frogs, we need more studies examining social and environmental factors that potentially drive schooling behaviour, in order to construct a general theory for the evolution of social aggregations in tadpoles. Tadpoles represent an ideal model system for the understanding of socio-ecological factors that drive group formation during a vulnerable life stage.

Acknowledgments

We would like to thank J. Newson for his help during data collection. Three reviewers provided valuable feedback on the manuscript.

Author Contributions

Conceived and designed the experiments: STL MJM. Performed the experiments: STL. Analyzed the data: STL MJW. Contributed reagents/materials/analysis tools: MJM. Wrote the paper: STL MJW.

References

1. Wilson EO (1975) Sociobiology: the new synthesis: Harvard University Press.
2. Schradin C (2005) When to live alone and when to live in groups: ecological determinants of sociality in the African striped mouse (*Rhabdomys pumilio*, Sparrman, 1784). Belg J Zool 135: 77–82.
3. Komdeur J (1992) Importance of habitat saturation and territory quality for evolution of cooperative breeding in the Seychelles warbler. Nature 358: 493–495.
4. Visagie L, Mouton PlFN, Bauwens D (2005) Experimental analysis of grouping behaviour in cordylid lizards. The Herpetological Journal 15: 91–96.
5. Hamilton WD (1971) Geometry for the selfish herd. J Theor Biol 31: 295–311.
6. Lanham EJ, Bull CM (2004) Enhanced vigilance in groups in *Egernia stokesii*, a lizard with stable social aggregations. J Zool 263: 95–99.
7. Roberts G (1996) Why individual vigilance declines as group size increases. Anim Behav 51: 1077–1086.
8. Hwang YT, Larivière S, Messier F (2007) Energetic consequences and ecological significance of heterothermy and social thermoregulation in striped skunks (*Mephitis mephitis*). Physiol Biochem Zool 80: 138–145.
9. Duellmann WE, Trueb L (1986) Biology of amphibians. New York: McGraw-Hill.
10. Kurvers RHJM, Prins HHT, van Wieren SE, van Oers K, Nolet BA, et al. (2010) The effect of personality on social foraging: shy barnacle geese scrounge more. Proc R Soc B 277: 601–608.
11. Terborgh J, Janson CH (1986) The socioecology of primate groups. Annu Rev Ecol Syst 17: 111–136.
12. Chapman CA, Chapman LJ (2000) Determinants of group size in primates: the importance of travel costs. In: Boinski S, Garber P, editors. On the move: how and why animals travel in groups. Chicago: University of Chicago Press. pp. 24–42.
13. Cote IM, Poulin R (1995) Parasitism and group size in social animals: a meta-analysis. Behav Ecol 6: 159–165.
14. Freeland WJ (1976) Pathogens and the evolution of primate sociality. Biotropica 8: 12–24.
15. Alexander RD (1974) The evolution of social behavior. Annu Rev Ecol Syst 5: 325–383.
16. Wells KD (2007) The ecology and behavior of amphibians. Chicago: University of Chicago Press.
17. Wassersug R, Hessler CM (1971) Tadpole behaviour: aggregation in larval *Xenopus laevis*. Anim Behav 19: 386–389.
18. Foster MS, McDiarmid RW (1982) Study of aggregative behavior of *Rhinophrynus dorsalis* tadpoles: design and analysis. Herpetologica 38: 395–404.
19. O'Hara RK, Blaustein AR (1981) An investigation of sibling recognition in *Rana cascadae* tadpoles. Anim Behav 29: 1121–1126.
20. Waldman B, Adler K (1979) Toad tadpoles associate preferentially with siblings. Nature 282: 611–613.
21. Blaustein AR, Waldman B (1992) Kin recognition in anuran amphibians. Anim Behav 44: 207–221.
22. Waldman B (1982) Sibling association among schooling toad tadpoles: field evidence and implications. Anim Behav 30: 700–713.
23. Rödel M-O, Linsenmair KE (1997) Predator-induced swarms in the tadpoles of an African savanna frog, *Phrynomantis microps*. Ethology 103: 902–914.
24. Spieler M (2003) Risk of predation affects aggregation size: a study with tadpoles of *Phrynomantis microps* (Anura: Microhylidae). Anim Behav 65: 179–184.
25. Bazazi S, Pfennig K, Handegard N, Couzin I (2012) Vortex formation and foraging in polyphenic spadefoot toad tadpoles. Behav Ecol Sociobiol 66: 879–889.
26. Sontag C, Wilson DS, Wilcox RS (2006) Social foraging in *Bufo americanus* tadpoles. Anim Behav 72: 1451–1456.
27. Pyke GH, White AW (2001) A review of the biology of the green and golden bell frog *Litoria aurea*. Aust Zool 31: 563–598.
28. Hamer AJ, Lane SJ, Mahony MJ (2002) The role of introduced mosquitofish (*Gambusia holbrooki*) in excluding the native green and golden bell frog (*Litoria aurea*) from original habitats in south-eastern Australia. Oecologia 132: 445–452.
29. Morgan LA, Buttemer WA (1996) Predation by the non-native fish *Gambusia holbrooki* on small *Litoria aurea* and *L. dentata* tadpoles. Aust Zool 30.
30. Pyke GH, White AW (2000) Factors influencing predation on eggs and tadpoles of the endangered Green and Golden Bell Frog *Litoria aurea* by the introduce Plague Minnow *Gambusia holbrooki*. Aust Zool 31: 496–505.
31. Nicieza AG (1999) Context-dependent aggregation in Common Frog *Rana temporaria* tadpoles: influence of developmental stage, predation risk and social environment. Funct Ecol 13: 852–858.
32. Gosner KL (1960) A simplified table for staging anuran embryos and larvae with notes on identification. Herpetologica 16: 183–190.
33. Anstis M (2002) Tadpoles of south-eastern Australia. A guide with keys. Sydney: Reed New Holland.
34. Baugh AT, Ryan MJ (2009) Female túngara frogs vary in commitment to mate choice. Behav Ecol 20: 1153–1159.
35. Hoeffler S, Ariely D (1999) Constructing stable preferences: a look into dimensions of experience and their impact on preference stability. Journal of Consumer Psychology 8: 113–139.
36. Brouwer R, Dekker T, Rolfe J, Windle J (2010) Choice certainty and consistency in repeated choice experiments. Environmental and Resource Economics 46: 93–109.
37. Kennedy M, Gray RD (1993) Can ecological theory predict the distribution of foraging animals - a critical analysis of experiments on the ideal free distribution. Oikos 68: 158–166.
38. Beiswenger RE (1977) Diel patterns of aggregative behavior in tadpoles of *Bufo americanus*, in relation to light and temperature. Ecology 58: 98–108.
39. Black JH (1970) A possible stimulus for the formation of some aggregations in tadpoles of *Scaphiopus bombifrons*. Proc Oklahoma Acad Sci 49: 13–14.
40. Glos J, Erdmann G, Dausmann KH, Linsenmair KE (2007) A comparative study of predator-induced social aggregation of tadpoles in two anuran species from western Madagascar. Herpetological Journal 17: 261–268.
41. Sadler LM, Elgar MA (1994) Cannibalism among amphibian larvae: a case of good taste. Trends Ecol Evol 9: 5–6.
42. Semlitsch RD (1990) Effects of body size, sibship, and tail injury on the susceptibility of tadpoles to dragonfly predation. Can J Zool 68: 1027–1030.

43. Alford RA (1999) Ecology: Resource use, competition and predation. In: McDiarmid RW, Altig R, editors. Tadpoles The biology of anuran larvae. Chicago: University of Chicago Press. pp. 240–278.
44. Brown RM, Taylor DH (1995) Performance and maneuvering behavior through larval ontogeny of the wood frog, *Rana sylvatica*. Copeia 1: 1–7.
45. Grafen A (1990) Do animals really recognize kin? Anim Behav 39: 42–54.

19

Gender, Season and Management Affect Fecal Glucocorticoid Metabolite Concentrations in Captive Goral (*Naemorhedus griseus*) in Thailand

Jaruwan Khonmee[1]*, Janine L. Brown[2], Suvichai Rojanasthien[1], Anurut Aunsusin[3], Dissakul Thumasanukul[4], Adisorn Kongphoemphun[4], Boripat Siriaroonrat[5], Wanlaya Tipkantha[5], Veerasak Punyapornwithaya[1], Chatchote Thitaram[1]*

1 Faculty of Veterinary Medicine, Chiang Mai University, Chiang Mai, Thailand, **2** Center for Species Survival, Smithsonian Conservation Biology Institute, Front Royal, Virginia, United States of America, **3** Chiang Mai Night Safari, Chiang Mai, Thailand, **4** Omkoi Wildlife Sanctuary, Department of National Park, Wildlife and Plant Conservation, Chiang Mai, Thailand, **5** Conservation Research and Education Division, Zoological Park Organization, Bangkok, Thailand

Abstract

Chinese goral (*Naemorhedus griseus*) are a threatened species in Thailand and the focus of captive breeding for possible reintroduction. However, little is known of their biology or what factors in the captive environment affect welfare. Our objective was to determine the impact of gender, season, and management on goral adrenal activity. We hypothesized that differences in fecal glucocorticoid concentrations would be related to animal density. Fecal samples were collected 3 days/ week for 1 year from 63 individuals (n = 32 males, 31 females) at two facilities that house the majority of goral in Thailand: Omkoi Wildlife Sanctuary (Omkoi), an off-exhibit breeding center that houses goral in individual pens (16 pens; n = 8 males, 8 females) and in small family groups (8 pens; n = 8 males, 8 females); and the Chiang Mai Night Safari (NS), a zoo that maintains 31 goral (n = 17 males, 14 females) in one large pen. Glucocorticoid metabolite concentrations were higher in male than female goral at Omkoi throughout the year, and there was a seasonal effect on adrenal activity ($p<0.05$). Goral at Omkoi and NS were used to test the effect of animal density on fecal glucocorticoid excretion of goral housed in similar-sized enclosures. Overall, the highest levels were found at NS (n = 31 adults/pen; 27 m^2 per animal) compared to Omkoi (n = 2 adults/pen; 400 m^2 per animal) ($p<0.05$). Overall findings support our hypothesis that animal density and aspects of the captive environment impact adrenal steroid activity in captive goral. In addition, gender and season also had significant effects on glucocorticoid metabolite production. Potential stressors pertaining to the welfare of this species were identified, which will guide future efforts to improve management and create self-sustaining and healthy populations of this threatened species.

Editor: Cheryl S. Rosenfeld, University of Missouri, United States of America

Funding: This study was funded by National Research Council of Thailand (NRCT). All laboratory analyses and additional lab support was provided by Faculty of Veterinary Medicine, Chiang Mai University. The funders had no role in study design, data collection and analysis, decision to publish, or preparation of the manuscript.

Competing Interests: One or more of the authors are employed by a commercial company (Chiang Mai Night Safari).

* E-mail: jaruwan.khonmee@cmu.ac.th (JK); chatchote.thitaram@cmu.ac.th (CT)

Introduction

Chinese, or grey long-tailed, goral (*Naemorhedus griseus*) are small ungulates with bovid-like features that inhabit mountainous areas of Myanmar, China, India, Thailand, and Vietnam [1–3]. They are agile and easily traverse steep cliffs and rocky crags [3]. Goral are diurnal and live in small family groups of 4–12 individuals; males are territorial and defend home ranges of 25–40 hectares in size [3]. Goral were listed in 1992 as one of 15 protected species under the Wild Animal Reservation and Protection of Thailand [4], and are categorized as Vulnerable by the IUCN Red List [3]. Most goral in Thailand are found within seven protected areas in the northern part of the country, restricted to hills along the Ping River, in the Chiang Mai, Mae Hong Son and Tak Provinces [3,4]. There has been no estimate of the total population size of wild goral, but numbers are declining throughout their range because of habitat loss, over-hunting and disease [3,4]. As a result, there is increasing interest by the Zoological Parks Organization of Thailand to initiate captive breeding programs for goral reintroduction. Today, captive populations are viewed as important "insurance" against environmental or anthropomorphic catastrophe [5,6], so efforts to improve breeding management for species like the goral are warranted.

There are about 100 goral housed among three captive facilities in Thailand, Omkoi Wildlife Sanctuary (Omkoi), Chiang Mai Night Safari (NS) and Chiang Mai Zoo. Omkoi and NS hold all but three of Thailand's captive goral and although there is breeding at both facilities, the populations are not self-sustaining. Goral management at the two facilities differs significantly. Omkoi is not open to the public and houses over 60 goral in small breeding groups (male, female and offspring) in both large- and small-sized enclosures, whereas NS is a tourist attraction and has 31 goral (n = 17 males, 14 females) kept together in one large enclosure. Given that wild goral live in small family groups (herds of four to 12 individuals, usually with one breeding male) within established territories [3], this study examined how these differences in captive housing conditions in Thailand impact individual animal welfare through assessment of adrenal function. There are no published hormonal data, reproductive or adrenal, for goral of either sex. So, there is a need to more fully characterize the biology of this species and identify factors that affect reproduction and welfare to aid propagation and conservation efforts and guide management strategies.

A number of potential stressors exist in captive environments and animal responses can be species-specific. Studies have shown that stress from inadequate housing conditions, inappropriate social interactions or other husbandry factors can lead to heightened glucocorticoid production [7–18]. If a stressor persists or causes consecutive stress responses, chronic glucocorticoid exposure can lead to a number of problems, including abnormal animal behavior, decreased libido, suppressed immune function, poor population performance, and disruption of reproductive hormone secretion [19–23]. One way to monitor welfare is through the analysis of hormonal metabolites excreted in urine and feces [24,25]. Non-invasive glucocorticoid metabolite monitoring is now well established as a valuable tool for understanding adrenal function, and offers significant advantages over blood sampling for assessing stress status [13,22]. In particular, fecal glucocorticoid metabolite analysis techniques have been developed for a number of domestic and wildlife species, which have led to improved *ex situ* management [12,23,26,27]. The ease of fecal collection without animal disturbance and that data reflect pooled values over time makes this a particularly attractive approach for zoo-held species [22,23].

The objective of this study was to use fecal glucocorticoid analyses to determine the influence of gender, season and management on metabolite concentrations in male and female goral. Based on the natural history of goral, we tested the hypotheses that lower fecal glucocorticoid concentrations would be found in goral residing in lower density groups.

Materials and Methods

Ethics Statement

This study was conducted non-invasively, without animal handling. Fecal samples were collected from captive goral. Permission to conduct research at Omkoi Wildlife Sanctuary, a protected forest area in Thailand was granted by Department of National Parks, Wildlife and Plant Conservation (DNP) (Permit Number TS 0907.1/2501). Chiang Mai Night safari permissions were obtained from the staff veterinarian and mammal curator, who also were collaborators on the study. No permits were needed for the fecal sample collection. This study was approved by the Faculty of Veterinary Medicine Chiang Mai University Animal Care and Use Committee (FVM-ACUC) (Permit Number S22/2553).

Seasonal Determination

There are three major seasons in Thailand: summer (February 16 – May 15), rainy (May 16 – October 15) and winter (October 16 – February 15). Information on daily temperature (°C) and rainfall (mm) during the study period at each facility was obtained from The Northern Meteorological Center, Meteorological Department, Ministry of Information and Communication Technology, Chiang Mai, Thailand [31].

Animals and Sample Collection

A total of 63 captive-born goral were used in this study; 32 were housed at Omkoi (17° 48′ 4″ N, 98° 21′ 31″ E) (n = 16 males, 16 females) and 31 at NS (18° 41′ 13″ N, 98° 55′ 8″ E) (n = 17 males, 14 females). Singleton animals were housed in 16 pens (6 m×9 m; Fig. 1A) (n = 8 males, 8 females) at Omkoi (17° 48′ 4″ N, 98° 21′ 31″ E). Another 16 goral were housed at Omkoi in larger pens (30 m×40 m; Fig. 1B), which contained one adult male, one adult female and offspring up to ~5 months of age (8 pens; n = 8 males, 8 females; density = 400 m^2 per animal). At NS, 31 goral (n = 17 males, 14 females) were housed in one large enclosure (35 m×24 m; Fig. 1C) (n = 31; density = 27 m^2 per animal).

Figure 1. Goral pens. Examples of goral housing conditions: (a) as individuals in small pens (6 m×9 m) at Omkoi Wildlife Sanctuary; (b) in family groups in large pens (30 m×40 m) at Omkoi Wildlife Sanctuary; and (c) all animals together in a large (35 m×24 m) at Night Safari.

Fecal samples from dependent offspring were not included in this study.

At Omkoi, the average age was 5.20±0.57 years for males, and 6.11±0.92 years for females. At NS, average ages were 3.43±0.59 and 4.06±1.24 years for males and females, respectively. All animals received natural light, and were fed concentrates (Betagro Company Limited, Thailand; Betagro 009 cattle finisher pellet (12% protein, 2% fat, 13% fiber, 13% moisture) and roughage (Panicum grass; *Brachiaria mutica*) once daily, with unlimited access to fresh water. A mineral block was provided in each enclosure at both facilities. All enclosures at Omkoi had dirt floors, an open shelter, a rock structure for climbing, and several natural trees for shade. The enclosure at NS contained an artificial rock and cliff structure in the middle, with about 13 m of dirt area behind it. Goral stayed primarily on the rock structure, which was ~5 m from the public area, separated by a water mote. NS was open to the public from 1100 to 2200 hours daily.

For health care and status, there was a staff veterinarian at each facility. All animals received annual physical examinations and blood chemistry analyses. They were dewormed every 3 months at Omkoi and every 6 months at NS. Keepers were responsible for noting any changes in health status; all animals were considered in good health during the study period. Identification of feces from individuals was accomplished through keeper observations. At both facilities, old feces were removed every evening, and freshly defecated feces (~30 g) were collected between 0830 and 0930 hours every morning from each goral 3 days/week for 1 year. All samples were stored at -20°C until processing.

Fecal Extraction

All chemicals were obtained from the Sigma Chemical Company (St. Louis, MO) unless otherwise stated. Wet fecal samples were dried using a conventional oven at 60°C for ~24–48 hours and stored at −20°C until extraction. Frozen dried fecal samples were thawed at room temperature, mixed well and 0.1 g (±0.01) of dry powdered feces placed in a glass tube containing 90% ethanol in distilled water. Samples were extracted twice by shaking with a Multi Pulse vortexer (Glas-Col, Terre Haute, IN) set at 70 for 30 min, centrifuging at 2500×g for 20 min and drying the combined supernatants under air in a 50°C water bath. Dried extracts were reconstituted by vortexing for 1 min in 1 ml dilution buffer (0.1 M $NaPO_4$, 0.149 M NaCl, pH 7.0). The extracts were stored at −20°C until further analysis [28]. Extraction efficiency of glucocorticoid metabolites from feces was 89.2% based on the recovery of cortisol added to dried fecal samples before extraction.

High Performance Liquid Chromatography

The numbers and relative proportions of immunoreactive glucocorticoid metabolites in goral fecal extracts were determined using reverse-phase high performance liquid chromatography (HPLC) [29]. Five fecal extracts from five gorals representing different months were combined, air dried, re-suspended in 1 ml methanol, dried again and stored at −20°C until further processing. Extract pools were reconstituted with 0.5 ml in phosphate buffer (0.01 M $NaPO_4$, 0.14 M NaCl, 0.5% bovine serum albumin, pH 5.0) and filtered through a C-18 matrix cartridge (Spice™ Cartridge, VWR, West Chester, PA). The cartridge was washed with 5 ml distilled water and the total steroids eluted with 5 ml 100% methanol, evaporated to dryness, then reconstituted in 300 µl of 100% methanol containing ^{3}H-cortisol and ^{3}H-corticosterone (~3,500 dpm each). Filtered fecal extracts (55 µl) were separated on a Microsorb C-18 column (Reverse Phase Microsorb™ MV 100 C18, 5 µm diameter particle size; Varian Inc., Woburn, MA) using a linear gradient of 20-100% methanol in water over 80 min (1 ml/min flow rate, 1 ml fractions). A subsample of each fraction (100 µl) was counted for radioactivity in a dual-label channel beta scintillation counter (Beckman, Fullerton, CA) to determine the retention times for the radiolabeled reference tracers. The remainder of each fraction (900 µl) was evaporated to dryness, reconstituted in 200 µl assay buffer (0.1 M $NaPO_4$, 0.149 M NaCl, 0.1% bovine serum albumin, pH 7.0) and an aliquot (50 µl) analyzed in singlet in the enzyme immunoassay (EIA).

Enzyme Immunoassay

A single-antibody cortisol EIA was used to quantify glucocorticoid metabolites, which relied on a polyclonal antibody produced in rabbits against cortisol-3-carboxymethyloximine linked to bovine serum albumin (R4866). Horseradish-peroxidase (HRP)-conjugated cortisol served as the label and cortisol was used as the standard. The cortisol R4866 antibody crossreacts with cortisol (100%), prednisolone (9.9%), prednisone (6.3%), cortisone (5.0%), corticosterone (0.7%), 21-deoxycortisone (0.5%), deoxycortisone (0.3%), 11-desoxycortisol (0.2%), progesterone (0.2%), 17α-dihydroxyprogesterone (0.2%), 17α- dihydropregnenolone (0.1%), pregnenolone (0.1%), androstenedione (0.1%), testosterone (0.1%), androsterone (0.1%), dehydroepiandrosterone (0.1%), dehydroisoandrosterone-3-sulfate (0.1%), aldosterone (0.1%), estradiol-17β (0.1%), estrone (0.1%), estriol (0.1%), spironolactone (0.1%) and cholesterol (0.1%) [30]. The EIA was performed in 96-well plates (Nunc Maxisorp, Fisher Scientific, Pittsburgh, PA) coated 16–24 hours previously with cortisol antiserum (50 µl in coating buffer, 0.05 M $NaHCO_3$, pH 9.6; 1:10,000 dilution). Cortisol standards (50 µl, range 3.9–1000 pg/well), diluted in assay buffer and samples (50 µl, 1:2 dilution) were combined with cortisol-HRP (50 µl; 1:15,000 dilution) and incubated at room temperature for 1 hour. Plates were washed five times (Biochrom Anthos Fluido 2 microplate washer, Cambridge, UK) before addition of 100 µl substrate (0.4 mM ABTS) to each well. After incubation for 15–30 min, the absorbance was measured at 405 nM (TECAN Sunrise microplate reader, Salzburg, Austria) until the optical density approached 1.0. The cortisol antibody and HRP were obtained from Coralie Munro (University of California, Davis, CA, USA).

The assay was validated for goral feces by showing that serial dilutions of pooled extracts produced displacement curves parallel to those of the cortisol standard curve. Pearson's correlation coefficient analyses were used to determine the correlation in percent binding between serial dilutions of hormone standards and fecal extract dilutions in the parallelism validation tests ($r = 0.9595$). Addition of unlabeled cortisol standard (Sigma Diagnostics Cat. #H4001) to pooled fecal extracts before extraction resulted in a significant ($p<0.05$) recovery of mass for female ($y = 1.03x-0.10$, $R^2 = 0.99$) and male ($y = 0.97x-0.14$, $R^2 = 0.99$) goral. Physiological validation of the cortisol EIA was demonstrated by showing a significant increase (100–150% increase; $p<0.05$) in concentrations within 24–48 hours after a stressful event (e.g., blood collection, n = 2; semen collection, n = 4). Assay sensitivity was 0.078 ng/ml at 90% binding. Inter-assay CVs were <15% based on binding of high (30%) and low (70%) control samples. Samples were re-analyzed if the duplicate CV was >10%; thus, intra-assay CVs were <10%. Data are expressed as ng/g dry feces.

Data Analysis

Sixteen goral at Omkoi housed in individual pens (n = 8 males, 8 females) were used to study the effect of gender and season on glucocorticoid excretion. Sixteen goral in eight large pens at

Omkoi (n = 8 males, 8 females) and all adult goral at NS (n = 17 males, 14 females) in one large pen were used to study the effect of management at each facility on glucocorticoid production. Goral at NS also were used for the seasonal analysis. Data are reported as the mean ± standard error of mean (SEM). Glucocorticoid metabolite concentrations were averaged by week, followed by calculations of seasonal means. Data were analyzed by fitting a linear model using Generalized Least Squares method with R version 3.0.0 [32] and *nlme* package 3.1-110 [33]. Differences across gender (male vs. female), season (summer vs. rainy vs. winter) and animal density (large at Omkoi vs. large at NS) were analyzed using GLS for repeated measures data followed by a Bonferroni test for multiple comparison analysis. The significance level (α) was set at 0.05.

Results

Analysis of HPLC-purified fecal eluates from male goral revealed the presence of several glucocorticoid metabolites, one of which co-eluted with the cortisol tracer (fraction 39–42) and represented 17.5% of the immunoreactivity (Fig. 2). Two immunoreactive peaks at fractions 25 (4.7%) and 33–37 (18.8%) appeared to be more polar, and five peaks (58.9% total immunoreactivity) were less polar than the tritiated reference tracers.

There was no difference in age between males and females at the two facilities. Fecal glucocorticoid metabolite concentrations were consistently higher ($p<0.05$) in male than female goral at Omkoi (Table 1, Fig. 3). For both sexes, mean glucocorticoid metabolite concentrations differed across seasons and were higher in the rainy season and winter, and lower in the summer ($p<0.05$) (Table 1, Fig. 3). There was no difference in glucocorticoid metabolites between the rainy season and winter for either sex ($p>0.05$) (Table 1).

Data on seasonal average daily temperature and rainfall between Omkoi and NS are shown in Table 2. Average temperature at NS was higher than that at Omkoi in every season ($p<0.05$). Temperatures differed across season for both facilities with the same trend, and were highest during the rainy season and lowest in the winter ($p<0.05$), although the difference between summer and rainy seasons at NS was not significant. The amount of rainfall was similar across facilities for summer and winter ($p>0.05$), but was significantly higher at NS compared to Omkoi. At each facility, rainfall was highest in the rainy season, intermediate in the summer and lowest in the winter ($p<0.05$).

A comparison of glucocorticoid metabolite concentrations between Omkoi and NS is shown in Table 3. Further seasonal analyses revealed a significant facility effect on glucocorticoid concentrations in the summer, being lowest at Omkoi and highest at NS (Table 3, Fig. 4). Within Omkoi, glucocorticoids were lower in the summer ($p<0.05$), with concentrations being similar between the rainy season and winter ($p>0.05$). By contrast, at NS, the highest concentration was observed in summer ($p<0.05$), again with rainy and winter seasons being similar.

Across facilities, animals exhibited significantly higher glucocorticoid metabolite concentrations at NS compared to Omkoi (Table 3, Fig. 4). Because of the within facility difference in summer glucocorticoid responses (being lower at Omkoi and higher at NS), the overall difference between facilities was more than double in that season. By comparison, concentrations at NS were only about a third higher in the rainy and winter seasons compared to Omkoi. Moreover, goral at NS had a lower area per animal; the stocking density at NS was about 14 times greater than that at Omkoi.

Discussion

A cortisol EIA was validated for quantifying fecal glucocorticoids in goral, a threatened species of national importance in Thailand. Findings support the hypothesis that glucocorticoid concentrations are higher in goral housed in animals maintained as one large group at NS compared to smaller, breeding groups at Omkoi. There also was a seasonal effect on glucocorticoid production, although it differed by facility. Specifically, in the summer, concentrations were lowest at Omkoi and highest at NS, possibly due to environmental differences. Higher fecal glucocorticoid metabolite concentrations were observed in male than female goral, irrespective of facility and season. Thus, results suggest that glucocorticoid production in goral is influenced by physiological, environmental and captive conditions, several of which have welfare and management implications.

HPLC analysis found the majority of glucocorticoid immunoreactivity in goral fecal extracts was associated with several peaks, two of which were more polar and five that were less polar than the tritiated reference tracers, indicating the presence of multiple metabolites. A proportion (17.5%) of the immunoreactivity co-

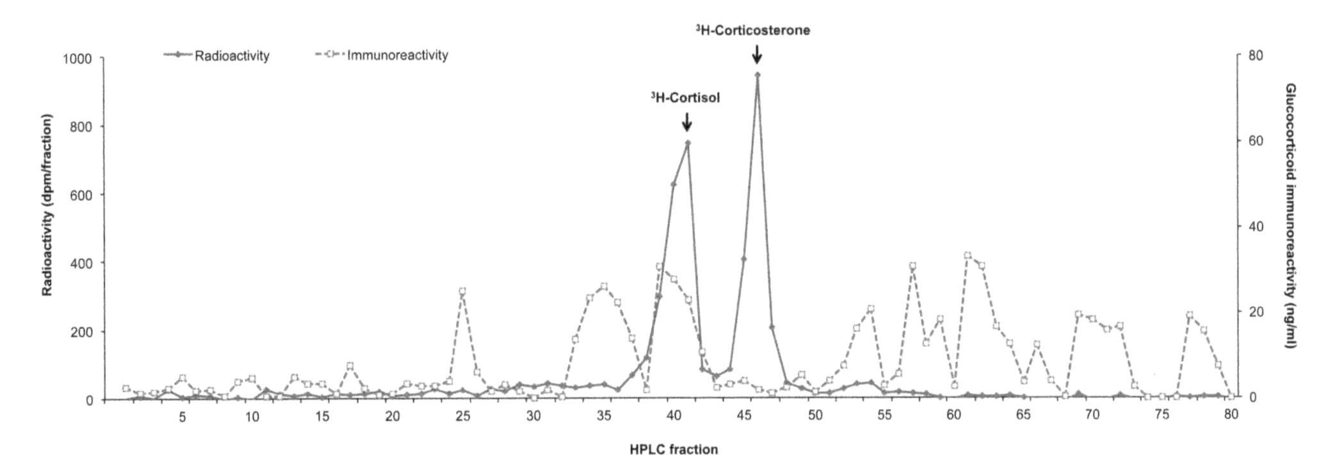

Figure 2. Chromatographic analysis of glucocorticoid metabolite immunoreactivity. Immunoreactivity of glucocorticoid metabolites in fecal extracts of goral was determined by reverse-phase HPLC analysis. Glucocorticoid concentration in each fraction was determined using a cortisol EIA. Elution of ^{3}H-cortisol and ^{3}H-corticosterone reference tracers in HPLC fractions of extracted fecal samples are indicated by the arrows.

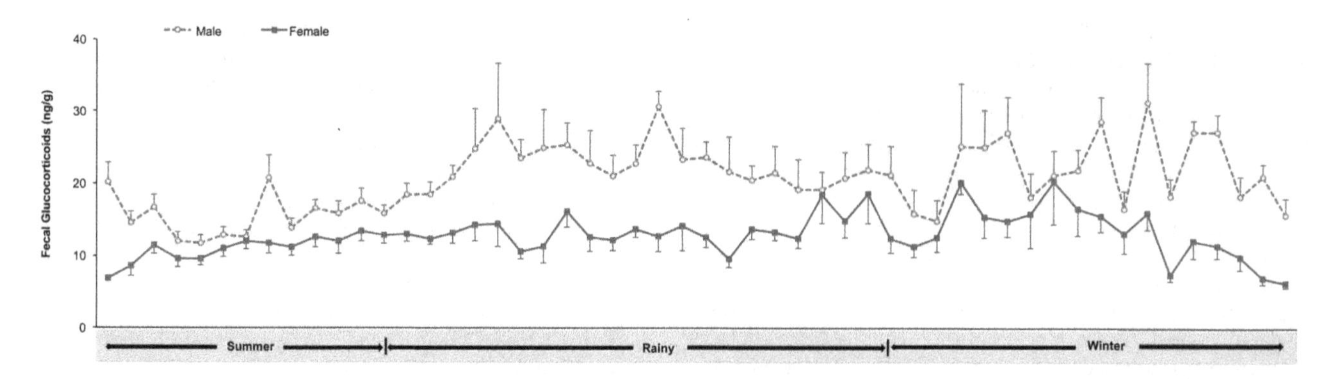

Figure 3. Seasonal pattern of fecal glucocorticoids. Longitudinal mean (± SEM) fecal glucocorticoid metabolite concentrations for male and female gorals were determined by a cortisol EIA. Fecal samples were collected from February 2010 through February 2011, representing the summer (February 16 – May 15), rainy (May 16 – October 15) and winter (October 16 – February 15) seasons.

eluted with radiolabeled cortisol. Several radiometabolism studies have demonstrated the near absence of authentic radiolabeled cortisol and corticosterone in feces; for example, in carnivores [34,35], lagomorphs [36], domestic livestock [37,38] and primates [39]. By contrast, immunoreactive substances in feces of the Himalayan black bear (*Ursus thibetanus*) and clouded leopard (*Neofelis negulosa*) co-eluted with ^{3}H-cortisol, suggesting these species may excrete native cortisol in variable amounts [29], and so it appears that goral do as well.

Gender differences in adrenal activity have been reported previously and may be related to a variety of physiological and behavioral changes within each sex [23]. As with goral, other studies have shown that levels of glucocorticoids (serum or fecal) are higher in males than females, including the laboratory rat [40], marmoset (*Callithrix jacchus*) [41] and spider monkey (*Ateles geoffroyi yucatanensis*) [42]. By contrast, a number of studies have reported higher concentration in females, such as in the domestic dog and cat [35], clouded leopard (*Neofelis nebulosa*) [12], sheep [43,44] and chimpanzee (*Pan troglodytes*) [45]. Still others report no difference between genders; e.g., red deer (*Cervus elaphus*) [13], black (*Diceros bicornis*) and white (*Ceratotherium simum*) rhinoceros [46] and reindeer (*Rangifer tarandus*) [47]. It is not clear if such gender differences are strictly species dependent, or are influenced by other physiological factors. It has been suggested that when glucocorticoids are higher in females, differences may be related to evolutionary adaptations that increase alertness (i.e., increased anticipation of "fight or flight") for protecting and rearing young [45,48] or to avoid aggression from dominant males [43], particularly in species where males are larger and more aggressive [12]. Gender effects may also be due to differences in steroid biosynthesis or metabolism [49]. For example, female rats excrete less hormone into feces presumably because of higher plasma corticosterone-binding capacity [50–51]. There is no size difference between male and female goral, and little aggression is observed within family units. Similarly, infanticide is rare in this species. In the wild, gorals are polygynous and dominant males defend territories and access to females during the breeding season through threatening displays and combat with other males [3]. Thus, males might maintain higher levels of glucocorticoids on average to generate an advantage over competing males, a strategy that persists in captivity.

Glucocorticoid metabolite concentrations in goral varied across seasons, and overall means for both males and females were higher during the rainy season and winter than in the summer. Increased production of glucocorticoids enhances catabolic function during the winter as an adaptation to cold weather [13]. Other seasonal species show fluctuations in fecal glucocorticoid levels related to climate and/or the breeding season. A study of red deer showed a marked increase in fecal glucocorticoid metabolites in December and January, which followed the breeding season in September through November [13]. Deer mice (*Peromyscus maniculatus*) and red-backed voles (*Clethrionomys gapperi*) exhibit increases in fecal glucocorticoids in late August to late September and in mid- to late September, respectively, again following the late summer, early fall breeding seasons [52]. In free-ranging male muriqui monkeys (*Brachyteles arachnoides*), fecal glucocorticoid concentrations are increased during the mating period, which corresponds to the dry season in Brazil [53]. And in African elephants (*Loxodonta africana*), fecal glucocorticoids are higher in the dry season, presumably because of reductions in natural resources [54]. The seasonal increase in glucocorticoid production in goral preceded the purported winter breeding season by several months. Food and

Table 1. Mean (± SEM) fecal glucocorticoid metabolite concentrations (ng/g) between male and female goral at Omkoi Wildlife Sanctuary across the three seasons in Thailand.

Season	Male	Female
Summer	$15.30 \pm 0.49^{a,1}$	$11.22 \pm 0.37^{b,1}$
Rainy	$22.10 \pm 0.73^{a,2}$	$13.37 \pm 0.44^{b,2}$
Winter	$21.98 \pm 0.98^{a,2}$	$13.27 \pm 0.71^{b,2}$

^{a,b}Values differ between male and female gorals, different letters indicate differences ($p<0.05$).
^{1,2}Values differ among seasons, different numbers indicate differences within the same gender ($p<0.05$).

Table 2. Seasonal mean (± SEM) average daily temperature and rainfall at two captive goral facilities in Thailand.

Environmental data	Season	Omkoi	Night Safari
Average temperature (°C)	Summer	$23.47 \pm 2.30^{a,1}$	$27.88 \pm 2.73^{b,1}$
	Rainy	$24.59 \pm 1.85^{a,2}$	$28.21 \pm 2.13^{b,1}$
	Winter	$19.77 \pm 1.69^{a,3}$	$23.88 \pm 2.05^{b,2}$
Rainfall (mm)	Summer	$2.92 \pm 0.29^{a,1}$	$2.34 \pm 0.22^{a,1}$
	Rainy	$5.19 \pm 0.39^{a,2}$	$6.39 \pm 0.48^{b,2}$
	Winter	$1.40 \pm 0.12^{a,3}$	$1.14 \pm 0.10^{a,3}$

^{a,b}Values differ among facilities, different letters indicate differences ($p<0.05$).
^{1,2}Value differ among seasons, different numbers indicate differences within the same facility ($p<0.05$).

water resources were consistent throughout the year for captive goral, eliminating that as a controlling factor. Nor was there a relationship between glucocorticoids and rainfall. However, further analyses revealed a significant facility effect with respect to seasonal glucocorticoid production, especially in the summer months, with concentrations at NS being about double those of Omkoi. Thus, we considered possible explanations for glucocorticoid differences due to both season and location. Animals at Omkoi had more shelter from the sun in the form of natural trees and a shed in each enclosure, whereas animals at NS had no such shade and the overall daily temperatures were higher. As a result, higher glucocorticoids during the summer months at NS could reflect a form of heat stress. High ambient temperatures, direct and indirect solar radiation, and humidity all are environmental stressors that affect animal welfare and can stimulate increased glucocorticoid production, as discussed for various livestock species [55]. A high Temperature-Humidity Index during the rainy season also has been suggested to be a source of stress in tropical species through alterations in the hypothalamo-pituitary-gonadal axis [56]. Secretion of cortisol stimulates physiological changes that allow animals to better cope with a hot environment [57], and for domestic cattle in South Africa, providing shade maintained lower serum cortisol concentrations and rectal temperatures [58]. Thus, the reduced fecal glucocorticoid concentrations across seasons at Omkoi, and especially in the summer, could be the result of a slightly cooler climate and perhaps more importantly, adequate shade being provided to the animals compared to NS. Moreover, Omkoi is located inside a wildlife sanctuary where wild goral live, so animals were exposed to more typical forest cover within their enclosures and a more natural climate.

Besides climate, animals are subjected to a number of other potential factors in the captive environment that can induce stress, such as health problems, limited space, artificial habitats, noise, exposure to the public and unnatural social groupings [15,59,60,61]. For most species, captive facilities are not likely to match the amount of space available to free-ranging individuals, but proper husbandry can enhance welfare and the likelihood for more natural behavior [10,12,14,18]. At both facilities, animals were found to be in good health by staff veterinarians, so that did not appear to be a significant factor in this study. However, the management of gorals at both facilities was different in that at NS animal density was 14 times higher than that at Omkoi. As recently reviewed by Creel et al. (2013), population density is one of the best-documented factors that influences the HPA axis. As far back as the 1950's it was recognized that increased population densities of wild and captive-held species, including mammals, birds, reptiles and amphibians, can result in antagonistic social interactions, suppression of reproduction, increased mortality and heightened adrenal activity (see review [62]). This may be particularly true for territorial species, where conspecific intrusion increases antagonistic encounters. For example, in a study of Peré David deer (*Elaphurus davidianus*), higher fecal glucocorticoids and increased aggression were observed in animals kept at a higher density [17]. Goral family units in the wild generally are under a dozen individuals, and usually include only one male for several females. Thus, a single enclosure containing 31 goral of equivalent gender numbers, such as that at NS, may be perceived as a stressor, and as a result, cause increased adrenal activity. During the breeding season, male goral can become aggressive. Based on keeper records, there was more fighting among the large number

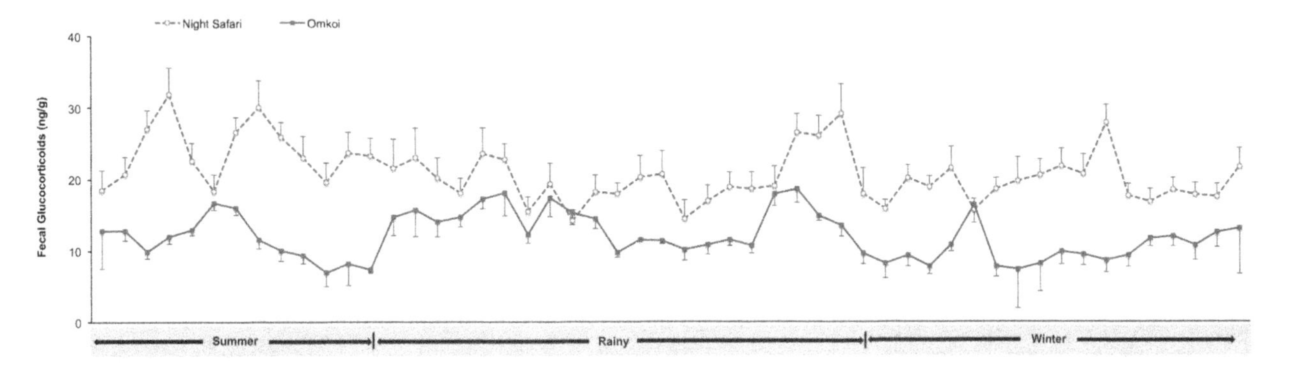

Figure 4. Effect of housing on glucocorticoid production. Longitudinal mean (± SEM) fecal glucocorticoid metabolite concentrations for goral housed at two facilities in Thailand, determined by a cortisol EIA. Fecal samples were collected from February 2010 through February 2011, representing the summer (February 16 – May 15), rainy (May 16 – October 15) and winter (October 16 – February 15) seasons.

Table 3. Facility effect on mean (± SEM) fecal glucocorticoid metabolite concentrations (ng/g) in goral housed in large (30 m×40 m, area 400 m^2 per animal) enclosures at Omkoi Wildlife Sanctuary, and one large enclosure (35 m×24 m, area 27 m^2 per animal) at Night Safari across the three seasons in Thailand.

Season	Omkoi	Night Safari
Summer	$9.94 \pm 0.60^{a,1}$	$23.89 \pm 0.83^{b,1}$
Rainy	$12.54 \pm 0.43^{a,2}$	$20.08 \pm 0.64^{b,2}$
Winter	$12.76 \pm 0.67^{a,2}$	$19.45 \pm 0.54^{b,2}$

^{a,b}Values differ among enclosure sizes, different letters indicate differences ($p<0.05$).
^{1,2}Values differ among seasons, different numbers indicate differences within the same enclosure sizes ($p<0.05$).

of conspecifics, especially males, at NS. By contrast, little aggression was observed among the animals housed in family units at Omkoi. Thus, limited space experienced by goral at NS could be one variable that explains the higher glucocorticoids found at this facility.

Besides more limited space, the animals at NS also were exposed to more noise and the physical presence of humans, which is a zoo and has a high rate of tourist activity. By comparison, Omkoi is a breeding center located in a wildlife sanctuary and not open to the public. The ability of zoo animals to tolerate large numbers of visitors may be species specific; some do well while others do not [63]. However, there are numerous examples of captive-held wildlife being negatively impacted by public exposure. For example, clouded leopards expressed higher fecal glucocorticoids when on display than off [12], and in spider monkey (*Ateles geoffroyii rufiventris*), the number of visitors had a stimulatory effect on the hypothalamic–pituitary–adrenal (HPA) axis [64]. Zoo visitor density also increased fecal glucocorticoid excretion and aggressive behavior in blackbuck (*Antilope cervicapra*) [65], whereas in black rhino, fecal glucocorticoids and mortality rates were correlated positively with the percentage of public visitor access around the enclosure [26], and in honeycreepers heightened glucocorticoid excretion was observed in animals exposed to environmental disturbances caused by humans and equipment [14]. Thus, forced proximity to humans can be harmful to animal well being in captive situations [15]. At NS, not only were goral exposed to tourists for 11 hours per day, but also the public area was only about 5 meters away from the animals. There was only a narrow water mote separating the two, with a rock structure abutting the mote. Goral spent most of the day on the rocks, and so were quite close to the visitors. As reviewed by Tarlow and Blumstein (2007), the distance by which an animal begins to flee from an approaching human is known as the 'flight-initiation distance' (FID), and can be used to define 'set-back distances' or 'buffer zones' when designing facilities. Goral at NS may perceive visitors as being too close, and this could be having an impact on chronic adrenal activity. An analysis of FID at NS could determine if the amount of set-back between goral and the public is adequate [66].

Conclusion

It is undeniable that non-invasive fecal glucocorticoid metabolites monitoring is a valuable tool for advancing our understanding of adrenal function and stress responses in wildlife, and can enhance the *ex situ* management of threatened species. This study validated an EIA for assessing fecal glucocorticoids in goral, a species of high priority in Thailand, and found higher concentrations in males than females, and in animals housed at a higher animal density and exposed to human visitors. As designed, it was not possible to discriminate between the stress caused by a higher animal density or public exposure in this study; there were confounding factors at NS. Thus, additional studies are planned to determine with more certainty what factor(s), exposure to the public, area per animal or stocking density, impact individual animal welfare the most. Nevertheless, we have identified several potential stressors pertaining to the welfare of captive goral. Additionally, we will be relating fecal glucocorticoid metabolites measures with those of reproductive hormone metabolites in the same samples to help unravel how "stress" may be modulating reproductive function/performance and/or success. Together, this information will be crucial for guiding efforts to improve management and create self-sustaining and healthy populations of this nationally important species.

Acknowledgments

The authors wish to thank the staff from Omkoi and Chiang Mai Night Safari for assisting with sample collection. We are also grateful to Nicole Presley (Smithsonian Conservation Biology Institute), Pallop Tankaew (Faculty of Veterinary Medicine, Chiang Mai University), Manisorn Tuantammark and Patharanun Wongchai (Zoological Parks Organization of Thailand) for technical support.

Author Contributions

Conceived and designed the experiments: JK JLB CT VP SR BS. Performed the experiments: JK CT AA AK DT. Analyzed the data: JK JLB VP CT. Contributed reagents/materials/analysis tools: JK JLB CT BS WT. Wrote the paper: JK JLB CT. Collected and extracted the samples: JK AA AK DT.

References

1. Rabinowitz A, Khaing ST (1998) Status of selected mammal species in North Myanmar. Oryx 32: 201–208.
2. Patton ML, Aubrey L, Edwards M, Rieches R, Zuba J, et al. (2000) Successful contraception in a herd of Chinese goral (*Nemorhaedus goral arnouxianus*) with melengestrol acetate. J Zoo Wildl Med 31: 228–230.
3. Duckworth JW, Steinmetz R, Chaiyarat R (2008) *Naemorhedus griseus*. In: IUCN 2012. IUCN Red List of Threatened Species. Version 2012.2. Available: http://www.iucnredlist.org. Accessed 23 April 2013.
4. Chaiyarat R, Laohajinda W, Kutintara U, Nabhitabhata J (1999) Ecology of the goral (*Naemorhedus goral*) in Omkoi Wildlife Sanctuary Thailand. Nat Hist Bull Siam Soc 47: 191–205.
5. Hoffmann M, Hilton-Taylor C, Angulo A, Böhm M, Brooks TM, et al. (2010) The Impact of conservation on the status of the world's vertebrates. Science 330: 1503–1509.
6. Conde DA, Flesness N, Colchero F, Jones OR, Scheuerlein A (2011) An emerging role of zoos to conserve biodiversity. Science 331: 1390–1391.
7. Mellen JD (1991) Factors influencing reproductive success in small captive exotic felids (Felis spp.): A multiple regression analysis. Zoo Biol 10: 95–110.

8. Horton GMJ, Malinowski K, Burgher CC, Palatini DD (1991) The effect of space allowance and sex on blood catecholamines and cortisol, feed consumption and average daily gain in growing lambs. Appl Anim Behav Sci. 32: 197–204.
9. Perkins LA (1992) Variables that influence the activity of captive orangutans. Zoo Biol 11: 177–186.
10. Carlstead K, Shepherdson D (1994) Effects of environmental enrichment on reproduction. Zoo Biol 13: 447–458.
11. Saito TR, Motomura N, Taniguchi K, Hokao R, Arkin A, et al. (1996) Effect of cage size on sexual behavior pattern in male rats. Contemp Top Lab Anim Sci 35: 80–82.
12. Wielebnowski NC, Fletchall N, Carlstead K, Busso JM, Brown JL (2002) Noninvasive assessment of adrenal activity associated with husbandry and behavioral factors in the North American clouded leopard population. Zoo Biol 21: 77–98.
13. Huber S, Palme R, Arnold W (2003) Effects of season, sex, and sample collection on concentrations of fecal cortisol metabolites in red deer (*Cervus elaphus*). Gen Comp Endocrinol 130: 48–54.
14. Shepherdson DJ, Carlstead KC, Wielebnowski N (2004) Cross-institutional assessment of stress responses in zoo animals using longitudinal monitoring of faecal corticoids and behaviour. Anim Welf 13: 105–113.
15. Morgan KN, Tromborg CT (2007) Sources of stress in captivity. Appl Anim Behav Sci 102: 262–302.
16. Moreira N, Brown JL, Moraes W, Swanson WF, Monteiro-Filho ELA (2007) Effect of housing and environmental enrichment on adrenocortical activity, behavior and reproductive cyclicity in the female tigrina (*Leopardus tigrinus*) and margay (*Leopardus wiedii*). Zoo Biol 26: 441–460.
17. Li C, Jiang Z, Tang S, Zeng Y (2007) Influence of enclosure size and animal density on fecal cortisol concentration and aggression in Pere David's deer stags. Gen Comp Endocrinol 151: 202–209.
18. Scarlata CD, Elias BA, Godwin JR, Powell RA, Shepherdson D, et al. (2013) The effect of housing conditions on adrenal activity of pygmy rabbits. Anim Welf 22: 357–368.
19. Liptrap RM (1993) Stress and reproduction in domestic animals. Ann N Y Acad Sci 697: 275–284.
20. Dobson H, Smith RF (1995) Stress and reproduction in farm animals. J Reprod Fertil Suppl 49: 451–461.
21. Möstl E, Palme R (2002) Hormones as indicators of stress. Domest Anim Endocrinol 23: 67–74.
22. Millspaugh JJ, Washburn BE (2004) Use of fecal glucocorticoid metabolite measures in conservation biology research: considerations for application and interpretation. Gen Comp Endocrinol 138: 189–199.
23. Touma C, Palme R (2005) Measuring fecal glucocorticoid metabolites in mammals and birds: the importance of validation. Ann N Y Acad Sci 1046: 54–74.
24. Palme R, Fischer P, Schildorfer H, Ismail MN (1996) Excretion of infused ^{14}C-steroid hormones via faeces and urine in domestic livestock. Anim Reprod Sci 43: 43–63.
25. Schwarzenberger F, Möstl E, Palme R, Bamberg E (1996) Faecal steroid analysis for non-invasive monitoring of reproductive status in farm, wild and zoo animals. Anim Reprod Sci 42: 515–526.
26. Carlstead K, Fraser J, Bennett C, Kleiman DG (1999) Black rhinoceros (*Diceros bicornis*) in U.S. zoos: II. behavior, breeding success, and mortality in relation to housing facilities. Zoo Biol 18: 35–52.
27. Morrow CJ, Kolver ES, Verkerk GA, Matthews LR (2002) Fecal glucocorticoid metabolites as a measure of adrenal activity in dairy cattle. Gen Comp Endocrinol 126: 229–241.
28. Brown JL, Wasser SK, Wildt DE, Graham LH (1994) Comparative aspects of steroid hormone metabolism and ovarian activity in felids, measured noninvasively in feces. Biol Reprod 51: 776–786.
29. Young KM, Walker SL, Lanthier C, Waddell WT, Monfort SL, et al. (2004) Noninvasive monitoring of adrenocortical activity in carnivores by fecal glucocorticoid analyses. Gen Comp Endocrinol 137: 148–165.
30. Young KM, Brown JL, Goodrowe KL (2001) Characterization of reproductive cycles and adrenal activity in the black-footed ferret (*Mustela nigripes*) by fecal hormone analysis. Zoo Biol 20: 517–536.
31. Thai Meteorological Department - Chiang Mai Weather (2013). Available: http://www.tmd.go.th. Accessed 2013 April 24.
32. R Development Core Team (2013) R: A language and environment for statistical computing Vienna: R Foundation for Statistical Computing. Available: http://www.R-project.org/.. Accessed 2013 July 1
33. Pinheiro J, Bates D, DebRoy S, Sarkar D, the R Development Core Team (2013) nlme: Linear and nonlinear mixed effects models. R package version 3.1-110.
34. Graham LH, Brown JL (1996) Cortisol metabolism in the domestic cat and implications for non-invasive monitoring of adrenocortical function in endangered felids. Zoo Biol 15: 71–82.
35. Schatz S, Palme R (2001) Measurement of faecal cortisol metabolites in cats and dogs: a non-invasive method for evaluating adrenocortical function. Vet Res Commun 25: 271–287.
36. Teskey-Gerstl A, Bamberg E, Steineck T, Palme R (2000) Excretion of corticosteroids in urine and faeces of hares (*Lepus europaeus*). J Comp Physiol B 170: 163–168.
37. Palme R, Möstl E (1997) Measurement of cortisol metabolites in faeces of sheep as a parameter of cortisol concentration in blood. Int. J. Mamm. Biol. 62 (suppl. II): 192–197.
38. Möstl E, Messmann S, Bagu E, Robia C, Palme R (1999) Measurement of glucocorticoid metabolite concentrations in faeces of domestic livestock. Zentralbl Veterinarmed A 46: 621–631.
39. Bahr NI, Palme R, Möhle U, Hodges JK, Heistermann M (2000) Comparative aspects of the metabolism and excretion of cortisol in three individual nonhuman primates. Gen Comp Endocrinol 117: 427–438.
40. Cavigelli SA, Monfort SL, Whitney TK, Mechref YS, Novotny M, et al. (2005) Frequent serial fecal corticoid measures from rats reflect circadian and ovarian corticosterone rhythms. J Endocrinol 184: 153–163.
41. Ferreira Raminelli JL, Cordeiro de Sousa MB, Sousa Cunha M, Veloso Barbosa MF (2003) Morning and afternoon patterns of fecal cortisol excretion among reproductive and non-reproductive male and female common marmosets, *Callithrix jacchus*. Biol Rhythm Res 32: 159–167.
42. Rangel-Negrín A, Alfaro JL, Valdez RA, Romano MC, Serio-Silva JC (2009) Stress in Yucatan spider monkeys: effects of environmental conditions on fecal cortisol levels in wild and captive populations. Anim Conserv 12: 496–502.
43. Vandenheede M, Bouissou MF (1993) Sex differences in fear reactions in sheep. Appl Anim Behav Sci 37: 39–55.
44. van Lier E, Pérez-Clariget R, Forsberg M (2003) Sex differences in cortisol secretion after administration of an ACTH analogue in sheep during the breeding and non-breeding season. Anim Reprod Sci 79: 81–92.
45. Buirski P, Plutchik R, Kellerman H (1978) Sex differences, dominance, and personality in the chimpanzee. Anim Behav 26: 123–129.
46. Brown JL, Bellem AC, Fouraker M, Wildt DE, Roth TL (2001) Comparative analysis of gonadal and adrenal activity in the black and white rhinoceros in North America by noninvasive endocrine monitoring. Zoo Biol 20: 463–486.
47. Bubenik GA, Schams D, White RG, Rowell J, Blake J, et al. (1998) Seasonal levels of metabolic hormones and substrates in male and female reindeer (*Rangifer tarandus*). Comp Biochem Physiol C Pharmacol Toxicol Endocrinol 120: 307–315.
48. Gray JA (1987) The psychology of fear and stress. Cambridge: Cambridge University Press. 436 p.
49. Eriksson H, Gustafsson J-Å (1970) Steroids in germfree and conventional rats. Eur J Biochem 15: 132–139.
50. Ottenweller JE, Meier AH, Russo AC, Frenzke ME (1979) Circadian rhythms of plasma corticosterone binding activity in the rat and the mouse. Acta Endocrinol 91: 150–157.
51. Woodward CJ, Hervey GR, Oakey RE, Whitaker EM (1991) The effects of fasting on plasma corticosterone kinetics in rats. Br J Nutr 66: 117–127.
52. Harper JM, Austad SN (2001) Effect of capture and season on fecal glucocorticoid levels in deer mice (*Peromyscus maniculatus*) and red-backed voles (*Clethrionomys gapperi*). Gen Comp Endocrinol 123: 337–344.
53. Strier KB, Ziegler TE, Wittwer DJ (1999) Seasonal and social correlates of fecal testosterone and cortisol levels in wild male muriquis (*Brachyteles arachnoides*). Horm Behav 35: 125–134.
54. Foley C a H, Papageorge S, Wasser SK (2001) Noninvasive stress and reproductive measures of social and ecological pressures in free-ranging African elephants. Conserv Biol 15: 1134–1142.
55. Silanikove N (2000) Effects of heat stress on the welfare of extensively managed domestic ruminants. Livest Prod Sci 67: 1–18.
56. Thitaram C, Brown JL, Pongsopawijit P, Chansitthiwet S, Wongkalasin W, et al. (2008) Seasonal effects on the endocrine pattern of semi-captive female Asian elephants (*Elephas maximus*): Timing of the anovulatory luteinizing hormone surge determines the length of the estrous cycle. Theriogenology 69: 237–244.
57. Christison GI, Johnson HD (1972) Cortisol turnover in heat-stressed cows. J Anim Sci 35: 1005–1010.
58. Muller CJC, Botha JA, Smith WAC and WA (1994) Effect of shade on various parameters of Friesian cows in a Mediterranean climate in South Africa. 2. physiological responses. S Afr J Anim Sci 24: 56–60.
59. Cyr NE, Romero LM (2008) Fecal glucocorticoid metabolites of experimentally stressed captive and free-living starlings: implications for conservation research. Gen Comp Endocrinol 158: 20–28.
60. Poessel SA, Biggins DE, Santymire RM, Livieri TM, Crooks KR, et al. (2011) Environmental enrichment affects adrenocortical stress responses in the endangered black-footed ferret. Gen Comp Endocrinol 172: 526–533
61. Tan HM, Ong SM, Langat G, Bahaman AR, Sharma RSK, et al. (2013) The influence of enclosure design on diurnal activity and stereotypic behaviour in captive Malayan Sun bears (*Helarctos malayanus*). Res Vet Sci 94: 228–239.
62. Creel S, Dantzer B, Goymann W, Rubenstein DR (2013) The ecology of stress: effects of the social environment. Funct Ecol 27: 66–80.
63. Hosey GR (2005) How does the zoo environment affect the behaviour of captive primates? Appl Anim Behav Sci 90: 107–129.
64. Davis N, Schaffner CM, Smith TE (2005) Evidence that zoo visitors influence HPA activity in spider monkeys (*Ateles geoffroyii rufiventris*). Appl Anim Behav Sci 90: 131–141.
65. Rajagopal T, Archunan G, Sekar M (2011) Impact of zoo visitors on the fecal cortisol levels and behavior of an endangered species: Indian blackbuck (*Antelope cervicapra L.*). J Appl Anim Welf Sci 14: 18–32.
66. Tarlow EM, Blumstein DT (2007) Evaluating methods to quantify anthropogenic stressors on wild animals. Appl Anim Behav Sci 102: 429–451.

20

Application of Selection Mapping to Identify Genomic Regions Associated with Dairy Production in Sheep

Beatriz Gutiérrez-Gil[1]*, Juan Jose Arranz[1], Ricardo Pong-Wong[2], Elsa García-Gámez[1], James Kijas[3], Pamela Wiener[2]

1 Dpto. Producción Animal, Universidad de León, León, Spain, **2** The Roslin Institute and R(D)SVS, University of Edinburgh, Roslin, Midlothian, United Kingdom, **3** Animal, Food and Health Sciences, CSIRO, Brisbane, Australia

Abstract

In Europe, especially in Mediterranean areas, the sheep has been traditionally exploited as a dual purpose species, with income from both meat and milk. Modernization of husbandry methods and the establishment of breeding schemes focused on milk production have led to the development of "dairy breeds." This study investigated selective sweeps specifically related to dairy production in sheep by searching for regions commonly identified in different European dairy breeds. With this aim, genotypes from 44,545 SNP markers covering the sheep autosomes were analysed in both European dairy and non-dairy sheep breeds using two approaches: (i) identification of genomic regions showing extreme genetic differentiation between each dairy breed and a closely related non-dairy breed, and (ii) identification of regions with reduced variation (heterozygosity) in the dairy breeds using two methods. Regions detected in at least two breeds (breed pairs) by the two approaches (genetic differentiation and at least one of the heterozygosity-based analyses) were labeled as core candidate convergence regions and further investigated for candidate genes. Following this approach six regions were detected. For some of them, strong candidate genes have been proposed (e.g. *ABCG2, SPP1*), whereas some other genes designated as candidates based on their association with sheep and cattle dairy traits (e.g. *LALBA, DGAT1A*) were not associated with a detectable sweep signal. Few of the identified regions were coincident with QTL previously reported in sheep, although many of them corresponded to orthologous regions in cattle where QTL for dairy traits have been identified. Due to the limited number of QTL studies reported in sheep compared with cattle, the results illustrate the potential value of selection mapping to identify genomic regions associated with dairy traits in sheep.

Editor: Bernhard Kaltenboeck, Auburn University, United States of America

Funding: The authors gratefully acknowledge support from the Spanish Ministry of Economy and Competitiveness (Project AGL2009-07000), Institute Strategic Grant funding from the UK Biotechnology and Biological Sciences Research Council (BBSRC) and the financial support of the European Science Foundation through the GENOMIC-RESOURCES Exchange Grant awarded to Beatriz Gutierrez (EX/3723). BGG is funded through the Spanish "Ramón y Cajal" Programme from the Spanish Ministry of Economy and Competitiveness (State Secretariat for Research, Development and Innovation). The funders had no role in study design, data collection and analysis, decision to publish, or preparation of the manuscript.

Competing Interests: The authors have declared that no competing interests exist.

* E-mail: beatriz.gutierrez@unileon.es

Introduction

Since their domestication 8 000–9 000 years ago (reviewed by [1]), sheep (*Ovis aries*) have been used by humans for the production of wool, meat and milk. Adaptation to very different geographic and climatic conditions and the specialization for specific characteristics have resulted in a phenotypically highly diverse species. The first documented modifications to sheep by human-imposed selection had taken place by the time that illustrations and records first appeared c. 3 000 BC and primarily concerned morphological and coat colour traits with the initial major morphological changes including reduction in the length of the legs, lengthening of the tail and alteration of horn shape [2]. Initially, sheep were kept solely for meat, milk and skins. Archaeological evidence suggests that selection for woolly sheep may have begun around 6000 BC.

Dairy sheep are mainly found in Europe, especially in Mediterranean areas, where they have been traditionally exploited as a dual purpose species, with income from both meat and milk. Sheep milk has a higher solid content than cow or goat milk, which means that it is particularly suited to processing into cheese. Historically, most sheep milk has been produced by multipurpose local breeds with low-to-medium milk yields and raised under traditional husbandry conditions [3]. More recently, modernization of husbandry methods and the establishment of breeding schemes focused on milk production have led to the development of "dairy breeds", facilitated by the implementation of quantitative genetics-based breeding and the use of artificial insemination [2]. The market for sheep milk and sheep dairy products appears to be growing, even in those countries without a history of sheep dairying [4].

Selection sweep mapping strategies, in which regions of the genome are identified that show patterns consistent with positive selection, can be used as a complementary approach to linkage mapping and genome-wide association study (GWAS) analysis to identify regions of the genome that influence important traits in livestock. Various methods have been applied to livestock and other domesticated animals, with the aim of identifying genomic regions with characteristics that reflect the influence of selection: extended low diversity haplotypes [5], overall low heterozygosity (e.g. [6,7]), specific diversity patterns [8], extreme allele frequencies [9] and between-breed differentiation [10,11,12]. Because of their well-documented selection pressures and highly-developed genetic resources, domesticated animal species also provide a valuable resource with the potential to identify the molecular pathways underlying phenotypic traits through the use of selection mapping approaches [10,13].

To perform a search for signatures of selection related to dairy production in sheep, we used genotypes obtained with the *Illumina OvineSNP50 BeadChip* (Illumina Inc., San Diego, CA) for a number of European breeds genotyped within the framework of the Sheep HapMap Project [14]. These breeds include several selected primarily for dairy production and others not used for dairy. In order to specifically target regions under dairy-related selection and not related to other traits that may have been under selection in the sheep populations, only selection signatures commonly identified in different European dairy breeds were considered. We applied two approaches for the detection of selection sweeps: (i) we looked for regions with extreme genetic differentiation between each dairy breed and a closely related non-dairy breed, and (ii) we looked for regions of the genome with reduced heterozygosity in the dairy breeds using two methods. We then searched for candidate genes that could be selection targets within the regions that were identified in multiple breeds and using multiple analysis methods. For these regions we also looked for correspondence with previously reported QTL related to dairy production traits in cattle or sheep. Although the selection history of dairy cattle is quite different from that of dairy sheep, in particular because breeding schemes in sheep are focused on more localized (and in many cases isolated) breeds than the global dairy cattle population, comparison of our results with studies in cattle allowed us to evaluate whether some of the same regions/genes show evidence of selection in both dairy sheep and dairy cattle.

Materials and Methods

Data

Samples. We analysed a subset of the dataset generated in the Ovine HapMap project [14], which included 5 dairy and 5 non-dairy sheep breeds (Table 1).

Genotypes. After an initial quality control procedure described in detail elsewhere [14], this dataset provides the genotypes of 49,034 SNPs (using the *Illumina OvineSNP50 BeadChip*) distributed across the 26 autosomal ovine chromosomes and chromosome X (only one of the markers genotyped belongs to chromosome Y). Markers were filtered to exclude loci assigned to unmapped contigs. The analyses reported here focused on the remaining 44,545 of these SNP located on autosomes. The positions of the markers according to the Sheep Genome Assembly v2.0 (update September 2011) were used for the analyses.

Selection Sweep Mapping Analysis Methods

(i) Genetic differentiation: Pair-wise F_{ST} calculations. In order to search for genomic regions that have been under divergent selection in dairy and non-dairy breeds, we examined genetic differentiation across the genome for five breed pairs. The selection of sheep breeds to serve as non-dairy partners for dairy breeds was based on the shortest divergence time estimates reported by the Sheep HapMap project (based on the extent of haplotype sharing and correlation of linkage disequilibrium values; Supplementary Information Figure S10 and Figure 3 in [14]), and close relationships according to additional Principal Component Analyses (PCA) performed in a selection of breeds (described in detail in File S1).

The following pairs of breeds of European ancestry were considered in the differentiation analysis:

a. Chios (Greek, dairy) *vs* Sakiz (Turkey, non-specialized)
b. Churra (Spanish, dairy) *vs* Ojalada (Spanish, meat)
c. Comisana (Italian, dairy) *vs* Australian Poll Merino (Australian, originated in southwest Europe, wool)
d. East Friesian Brown (highly specialized dairy) *vs* Finnsheep (Finland, primary wool, more recently used as a meat producing breed)
e. Milk Lacaune (French, highly specialized dairy) *vs* Australian Poll Merino (Australian, originated in southwest Europe, wool)
f. Milk Lacaune (French, highly specialized dairy) *vs* Meat Lacaune (French, meat)

For each of these pairs, unbiased estimates of Weir and Cockerham's F_{ST} [15], a measure of genetic differentiation, were calculated as functions of variance components, as detailed in Akey et al. [16]. This type of approach to selection mapping, exploiting between-breed allele frequency differences, has been applied in studies of humans [16] and domesticated animals [10,11,12,17,18] where it has been demonstrated to be effective in identifying genes that are associated with breed differentiation.

(ii) Reduced diversity: Observed heterozygosity. For all the breeds included in the pair-wise F_{ST} calculations, observed heterozygosity (ObsHtz) was calculated for each SNP marker. This approach has previously been applied in selection mapping studies of chickens [6,7], pigs [19] and dogs [20].

(iii) Reduced diversity: Regression analysis for detection of regions with asymptotic heterozygosity patterns. For all the breeds included in the pair-wise F_{ST} calculations, tests of significant asymptotic relationships between heterozygosity and distance from a test position were performed across the genome based on the approach of Wiener and Pong-Wong [8]. This method detects regions with patterns of variation consistent with positive selection: an asymptotic increase in marker variation (heterozygosity; y) with increasing distance (x) from a selected locus $y = A + B\,R^x$ (where R is the asymptotic rate of increase; B is the difference between heterozygosity at the test position and the asymptotic level; A is the asymptotic level of heterozygosity). For each regression (performed in Genstat, [21]), we recorded the parameters of the asymptotic regression, their standard errors, the significance level associated with the regression (p) and the variance explained by the curve. Positive and increasing regressions ($0<R<1$, $B<0$) were considered as being in the direction predicted by positive selection. Analysis of simulated data suggests improved precision of this selection mapping approach compared to an alternative haplotype-based method as well as robustness to demographic influences [8].

Protocols for Selection Mapping Analyses

In order to determine appropriate parameters for the above-mentioned analyses, we investigated their behaviour on a test genomic region encompassing the myostatin (*GDF-8*) gene, which is known to have been under selection in the Texel breed (details in File S2).

Window/bracket sizes. Based on the analysis of the myostatin gene (File S2), window and bracket sizes for the three methods were established. For the differentiation and reduced heterozygosity analyses, F_{ST} and ObsHtz values, respectively, were

Table 1. Breeds included in the present study.

Group	Breed name	Number of samples	Aptitude
Dairy	Chios	23	High milk production
	Churra	96	Double purpose breed (milk and lamb production
	Comisana	24	Highly-specialized dairy breed
	East Friesian Brown	39	Highly-specialized dairy breed
	Milk Lacaune	103	Highly-specialized dairy breed
Non-dairy	Australian Poll Merino	98	Meat production
	Meat Lacaune	78	Meat production
	Ojalada	24	Meat production
	Sakiz	22	Triple-purpose (milk, meat, wool)
	Finnsheep	99	Primary used for wool production; more recently used for meat production.

The classification established into Dairy and Non-dairy groups are presented together with some details about the breed aptitude.

averaged across sliding windows of 9 SNPs (F_{ST}-9SNPW, ObsHtz-9SNPW). For the regression analysis, the test position was moved every 50 Kb across each chromosome and all markers within 10 Mb of this position (10 Mb-bracket size) were considered in the asymptotic regression. A –log(p) value was determined for each test position.

Identification of selection signals by individual methods. Evidence of positive selection was interpreted for window estimates in the extreme of the empirical distributions, as suggested by Akey et al. [10,16] and employed in various subsequent studies (e.g. [11,13]. Specifically, we considered the positions showing signatures of selection as the top 0.5th percent of the distributions for differentiation (F_{ST}) and asymptotic regression (–log(p), for regressions in the predicted direction) or the bottom 0.5th percent for observed heterozygosity. Based on the results of the analysis of the myostatin gene (File S2), a selected "region" was defined as the range of positions within 2 Mb of each other showing evidence of selection by any of the three methods. An additional criterion for selected regions was that they were identified in at least two breed pairs, for F_{ST}, or two dairy breeds, for heterozygosity-based methods (with distances up to 2 Mb allowed between the regions identified for different breeds). For genetic differentiation, we further required that regions of extreme F_{ST} must be detected in at least two different pairs of dairy – non-dairy breeds that did not share a common breed (e.g. top regions found only in the Milk Lacaune-Australian Poll Merino and Comisana-Australian Poll Merino but not in other studied pairs were not included in the list of differentiated regions). By requiring at least two breeds (or breed pairs) for the initial identification of candidate regions for each methodology, this selection mapping strategy will not identify dairy gene variants occurring in only one breed.

Criteria for Identification of Regions with Shared Selection Signals

Based on the selected "regions" identified by the individual methods through the overlapping of at least two breeds or breed pairs, and taking into account that the F_{ST}-based method is expected to specifically target traits relevant for dairy production, whereas signals detected by heterozygosity-based methods may not be specific for dairy-related selection, we defined a "convergent candidate region" (CCR) as one where a signal was identified by the pair-wise F_{ST} comparison and at least one of the reduced heterozygosity methods. Hence, a CCR was labelled where there was overlap between the position ranges of the candidate regions identified by the genetic differentiation methodology and at least one of the two heterozygosity-based methods, such that each CCR was associated with a region identified in at least two breeds (breed pairs) and using at least two different methods.

Identification of Candidate Genes within CCR Regions

We identified the genes mapping to the end of each CCR using the genome browser of the sheep genome reference sequence (v2.0; http://www.livestockgenomics.csiro.au/cgi-bin/gbrowse/oarv2.0/) and identified the corresponding orthologous regions in the bovine genome (Cow (UMD3.1) using Ensembl (http://www.ensembl.org/Bos_taurus/Info/Index). A systematic extraction of all the annotated genes contained within the orthologous genomic ranges in cattle was performed using Biomart (www.biomart.org). Subsequently, an exhaustive search was performed for candidate genes previously linked to cattle dairy traits [22]. In addition, genes not included in this database but reported as candidate genes in the literature in relation to milk production or dairy-related traits were also identified. We also looked for correspondence with genes for which signatures of selection have been reported in studies of dairy cattle [23,24,25] and sheep [14,26].

We evaluated correspondence of the CCR with QTL reported for milk production and other functional traits related to dairy production in sheep (based on the SheepQTL database, http://www.animalgenome.org/cgi-bin/QTLdb/OA/index). We also examined overlap between the CCR and QTL influencing milk-related traits, mastitis and other functional traits related to dairy production in cattle (based on the CattleQTL database; http://www.animalgenome.org/cgi-bin/QTLdb/BT/index), positioned on the bovine genome reference sequence (UMD_version 3.1).

Results

Regions Identified by Individual Methods

Genetic differentiation. The level and range of the top 0.5% of F_{ST} values averaged in sliding windows of 9 SNPs (F_{ST}-9SNPW) varied among the five breed pairs (Figure 1). The lowest genome-wide differentiation within a pair was found, as expected, for the Milk Lacaune-Meat Lacaune pair (0.076), whereas the highest levels of genetic differentiation were found for the East Friesian Brown-Finnsheep pair (0.752, for the 9SNP-window centered on marker OAR3_185527791) (Table 2).

Twenty-eight genomic regions distributed across 15 autosomes were identified in at least two dairy-non-dairy breed pairs (Table S1, where a reference number has been given to each of them: F_{ST}-CandidateRegionX, F_{ST}-CRX). The largest number of F_{ST}-based candidate regions per chromosome was found on OAR3 (5 regions). The length of the F_{ST}-based candidate regions varied from 0.215 Mb (OAR3, F_{ST}-CR8) to 9.211 Mb (OAR6, F_{ST}-CR14).

Reduced observed heterozygosity in dairy breeds. Fifty-five regions showing reduced observed heterozygosity (ObsHtz-CR1–ObsHtz-CR55) in more than one dairy breed were found across 21 of the 26 autosomes (Table S2; where a non-dairy breed showed reduced heterozygosity in the same region, this is also indicated). Eight of the candidate regions found in dairy breeds covered intervals larger than 3 Mb. The largest was that on OAR13 (ObsHtz-CR42; 56.061–63.781 Mb), followed by one on OAR6 (ObsHtz-CR27:34.576–41.863 Mb), while the smallest region was a single window centered on marker on OAR2 (ObsHtz-CR9; 211.205 Mb). A normalized observed heterozygosity (NObsHtz) (based on that introduced by Rubin et al. [6]) was also calculated for all breeds analysed, again averaged in 9-SNP windows. There were no regions in the extreme lower end of the distribution (NObsHtz<-6) in the dairy breeds although the region on OAR6 (*ABCG2* gene region) had a value of −5.99 for the Meat Lacaune breed.

Regression analysis for detection of regions with asymptotic heterozygosity patterns in dairy breeds. Three regions ranging in size from 0.1 to 4.0 Mb were identified with asymptotic heterozygosity patterns (bracket size = 10 Mb) in two or more dairy breeds (RegBrack10-CR1–RegBrack10-CR3) (Table 3, where a non-dairy breed showed reduced heterozygosity in the same region, this is also indicated).

The myostatin analysis suggested that a bracket size of 10 Mb was optimal for identification of selected region. However, because this is a new methodology, the results obtained for the dairy breeds with all three bracket sizes (5-, 10- and 20-Mb) were compared to aid interpretation of results based on this approach. The number of candidate regions identified in at least two dairy breeds decreased with increasing bracket size. For the 5-Mb bracket size, a total of seven candidate regions were observed, whereas only three and one candidate regions were observed for the 10- and 20-Mb bracket sizes, respectively (Table 3). The region commonly identified through the use of all three bracket sizes was located on OAR6 (RegBrack5-CR6, RegBrack10-CR2 and RegBrack20-CR1). The signal for this region was seen in Milk Lacaune (34.875–38.875 Mb, 10-Mb bracket) and Comisana (36.125–38.325 Mb, 10-Mb bracket) breeds. In addition, the Meat Lacaune variety also showed extreme results for this region for all three bracket sizes (34.375–38.175, 10-Mb bracket). Another region on OAR2 (104 Mb) was identified by both of the smaller bracket sizes.

Some of the inconsistencies between bracket sizes were investigated further. In several cases, where regions were not found in the top 0.5% of –log(p) values for a particular bracket size, they did appear in the top 1% of –log(p) values. Regarding the region on OAR20 (~50 Mb) that was identified in two dairy breeds using the 10-Mb bracket size (RegBrack10-CR3, Table 3) but not using the 5-Mb bracket size: for Churra, positions within this region appeared within the top 1st percent of –log(p) values for the smaller bracket size but did not reach the threshold for the top

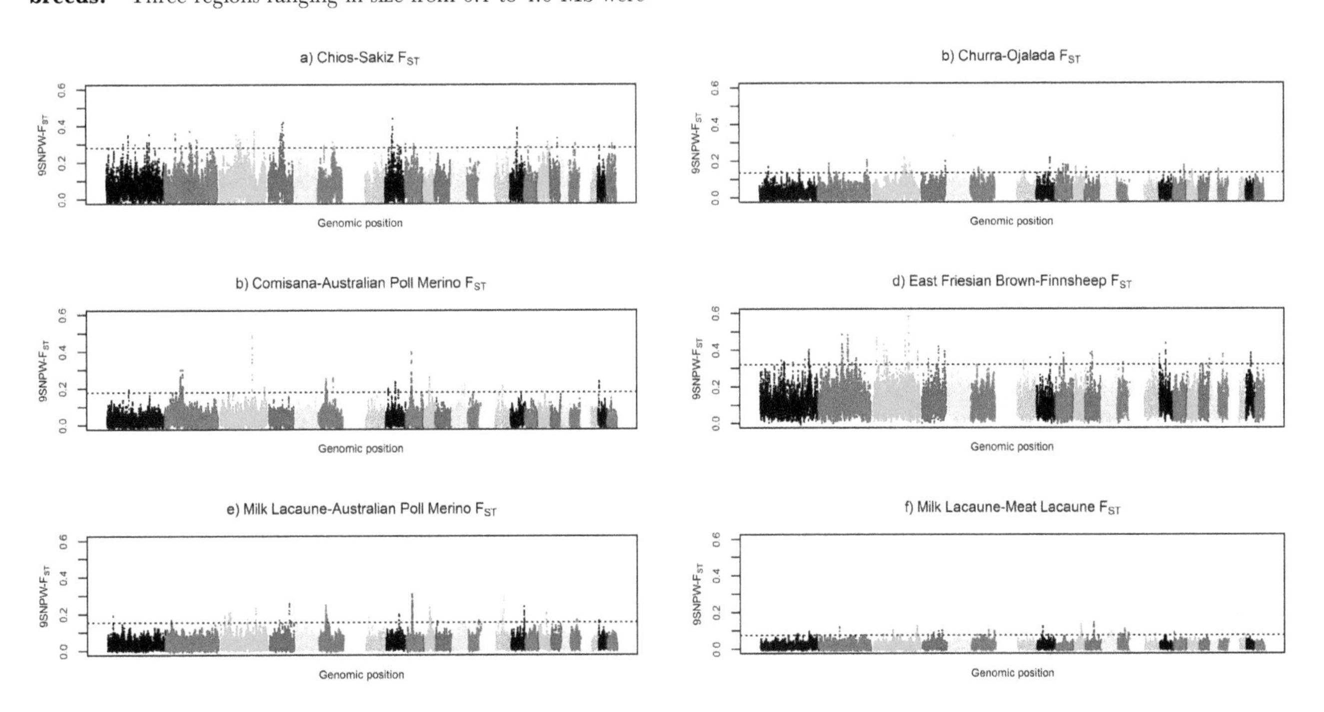

Figure 1. Genome-wide distribution of F_{ST} values for the six analysed breed pairs. The level of genetic differentiation, measured by F_{ST}, was estimated within each dairy – non-dairy breed pair[1], and averaged in sliding windows of 9 SNPs (F_{ST}-9SNPW) across the genome: The horizontal line indicates the top 0.5.th percent threshold considered for the F_{ST}-distributions. These raw results were used to identify F_{ST}-based candidate regions (F_{ST}-CRs) when overlapping significant selection signals (allowing gaps up to 2-Mb) were identified between different pairs. [1]Breed pairs analysed: a) Chios-Sakiz, b) Churra-Ojalada; c) Comisana-Australian Poll Merino; d) East Friesian Brown -Finnsheep, e) Milk Lacaune-Australian Poll Merino f) Milk Lacaune-MeatLacune.

Table 2. Maximum and minimum of the 0.005 top averaged pair-wise F_{ST} values in sliding windows of 9 SNPs (F_{ST}-9SNPW) estimated for the pairs considered in the present work to detect selection signals in dairy sheep.

Breed pair	Min. F_{ST}-9SNPW	Max. F_{ST} -9SNPW
Chios-Sakiz	0.2799	0.4392
Churra-Ojalada	0.1345	0.2193
Comisana-Australian Poll Merino	0.1781	0.4873
East Friesian Brown-Finnsheep	0.3212	0.7515
Milk Lacaune-Australian Poll Merino	0.1547	0.3071
Milk Lacaune-Meat Lacaune	0.0757	0.1449

0.5th percent, whereas for Milk Lacaune, this region was identified using both bracket sizes. Regarding the five regions (Table 3) that were identified in two dairy breeds using 5-Mb bracket size but not 10-Mb, four of the regions were in the top 1st percent of –log(p) values for one or both of the dairy breeds. Two of these regions (RegBrack5-CR1 and RegBrack5-CR3) were found in Chios and Churra, however, while these regions were found for Churra using both the 5- and 10-Mb bracket sizes, for the 10-Mb bracket size, the top –log(p) values for Chios were dominated by regions on OAR13 and OAR16, which did not feature in the top –log(p) values for the other dairy breeds. Thus, these Chios-specific signals may have overwhelmed the more general dairy signals for the larger bracket size in this breed. The region labelled as RegBrack5-CR4, identified at ~75 Mb on OAR3 for Churra and Milk Lacaune using the 5-Mb bracket size, did not feature in the top 1st percent of the –log(p) values for the 10-Mb bracket for either of these breeds. It is worth noting that regions identified using one bracket size but not a smaller one could reflect more recent selection events for which the pattern of heterozygosity with respect to distance from the selected locus appears linear rather than asymptotic in the smaller bracket.

Convergence Candidate Regions (CCR)

Six candidate regions were detected in at least two breed pairs by the pair-wise F_{ST} comparison and in at least two breeds by a heterozygosity-based analysis (Table 4). One of the regions, CCR3 (OAR6:30.367–41.863 Mb), was identified by all three analysis methods. The orthologous bovine genomic regions corresponding to each of the CCR are shown in Table 5. A total of 406 genes (positional candidate genes) were found in these six core regions (Table S3). There were three other regions where an F_{ST}-CR signals was less than 1 Mb from an ObsHtz-CR signal (OAR3:18.648–19.360 Mb, OAR3:167.711–168.959 Mb, and OAR13:95.801–98.865 Mb) but because they did not overlap, they were not considered as CCR.

Among the positional candidate genes extracted from the six CCRs, a search for functional candidates for milk production traits and mastitis was performed by comparison with the genes included in the Ogorevc et al. [22] database of cattle candidate genes for dairy-related traits. A total of 13 genes were common to these two lists (Table 5). The evidence for relationships with milk production traits for these genes was based on the different aspects considered in the Ogorevc et al. [22] database such as gene expression studies related to mammary gland (*TFAP2C, FAM110A, CD82, ABCG2*) or mastitis (*BID, MAFF, AHCY*), mouse model studies in which gene knockouts or expression of transgenes resulted in phenotypes associated with the mammary gland (*FKBP4, MKL1, POFUT1, CHUK*) and association studies of milk production traits (*ABCG2, SPP1, SCD*).

In order to assess whether there was greater overlap between the CCRs and candidate genes than expected by chance, we repeatedly (1 000 000 times) assigned regions of the same length as the CCR at random positions on the bovine genome and checked overlap with all candidate genes from the Ogorevc et al. [22] database that could be positioned on the bovine genome (423 genes). Although we could not do the test with the sheep genome as the annotation is not as complete, the length of the sheep and bovine genomes is very similar and so we expect this test would provide similar results. The number of overlaps between CCR regions and candidate genes based on a model with random positioning of CCR regions was very different from the actual situation: only 8.4% of the replicates contained any overlaps and the maximum number of overlaps was 4.

Some other positional candidate genes not included in the Ogorevc et al. [22] database were identified as possible functional candidates based on their known biological function and an exhaustive literature review of reported signatures of selection in dairy cattle (Table 5). There was also correspondence between the CCR and QTL previously reported in dairy cattle and sheep for milk production traits or functional traits related to dairy production (Table 5), which is discussed below.

Discussion

This study reports the first genome-wide analysis of regions under selection for dairy traits in sheep. For this we have used the valuable information generated in the International Sheep HapMap project [14], through the use of the *Illumina OvineSNP50 BeadChip*, to evaluate a range of European sheep breeds that have been selected for dairy production. With the aim of identifying the signatures of selection specifically due to dairy selection and not related to other traits that may have been selection target in the studied sheep populations (e.g. coat colour), we also included in our study other non-dairy European sheep breeds. Furthermore, because of the difficulties in distinguishing between the effects caused in the genome by genuine selective sweeps rather than demographic events such as population expansion or contraction [16], we used three different analysis methods and only considered for further exploration those six regions identified by the F_{ST}-based method and at least one of the two heterozygosity-based methodologies.

Candidate Dairy Selection Regions

Based on the convergence among the three different analysis methods, six core regions were identified as candidate regions under positive selection in dairy sheep. Based on the comparison

Table 3. Initial candidate regions identified on the basis of the regression analysis performed for detection of regions with asymptotic heterozygosity patterns in at least two of the dairy breeds (top 0.5% results for bracket sizes 5, 10 and 20 Mb).

Analysis	Regression-CR	Chr.	Dairy breed	Start position (Mb)	End position (Mb)	Non-dairy breed	Start position (Mb)	End position (Mb)
Regression top 0.5% bracket 5 Mb	RegBrack5-CR1	2	Churra	51.810	54.110	Ojalada	52.610	53.760
			Chios	52.860	53.410			
	RegBrack5-CR2	2	Milk Lacaune	104.360	104.560	Meat Lacaune	104.360	104.510
			Churra	104.460		Australian Poll Merino	104.410	104.460
	RegBrack5-CR3	2	Churra	122.360	122.910			
			Chios	123.010	123.210			
	RegBrack5-CR4	3	Milk Lacaune	75.192	75.292			
			Churra	75.292				
	RegBrack5-CR5	3	Milk Lacaune	168.742	168.892	Australian Poll Merino	168.692	168.942
			Churra	168.792	168.892	Meat Lacaune	168.792	168.892
	RegBrack5-CR6	6	Milk Lacaune	35.475	36.625	Meat Lacaune	34.725	36.775
			Comisana	36.625	37.325	Australian Poll Merino	35.975	37.175
	RegBrack5-CR7	11	Milk Lacaune	18.380	18.530	Ojalada	18.430	18.530
			Churra	18.430	18.480	Meat Lacaune	18.430	18.480
Regression top 0.5% bracket 10 Mb	RegBrack10-CR1	2	Milk Lacaune	104.410		Ojalada	104.410	104.460
			Churra	104.460	104.510	Meat Lacaune	104.410	104.460
						Finnsheep	104.460	104.510
	RegBrack10-CR2	6	Milk Lacaune	34.875	38.875	Meat Lacaune	34.3747	38.175
			Comisana	36.125	38.325	Australian Poll Merino	35.525	38.225
	RegBrack10-CR3	20	Churra	49.971	50.171			
			Milk Lacaune	50.071				
Regression top 0.5% bracket 20 Mb	RegBrack20-CR1	6	Milk Lacaune	34.825	38.525	Meat Lacaune	34.375	38.175
			Comisana	35.525	38.825	Australian Poll Merino	34.975	38.175

We also indicate if the same signature of selection was also identified in the non-dairy breeds.

Table 4. Convergence candidate regions (CCR) for selection signals identified for dairy sheep.

CCR	Chr.	Method	Individual method candidate region	Start marker*	Start position (Mb)	End marker*	End position (Mb)
CCR1	3	F_{ST}	F_{ST}-CR7	s51772	152.68	OAR3_165450843	154.582
		ObsHtz	ObsHtz-CR17	s26177	153.95	OAR3_165549468_X	154.679
CCR2	3	F_{ST}	F_{ST}-CR9	s34668	209.872	OAR3_234328134_X	215.814
		ObsHtz	ObsHtz-CR21	OAR3_229873996	211.624	s35739	215.403
CCR3	6	F_{ST}	F_{ST}-CR14	OAR6_34086500	30.367	OAR6_44210019	39.577
		Regression	RegBrack10-CR2	OAR6_38919831	34.875	OAR6_38919831	38.875
		ObsHtz	ObsHtz-CR27	OAR6_38585187	34.576	s38254	41.863
CCR4	13	ObsHtz	ObsHtz-CR42	OAR13_60893851	56.061	s63708	63.781
		F_{ST}	F_{ST}-CR24	s48133	62.277	OAR13_71091738	65.811
CCR5	15	F_{ST}	Fst-CR26	s31340	72.774	OAR15_80448054	74.55
		ObsHtz	ObsHtz-CR44	s02793	72.843	s28875	72.948
CCR6	22	ObsHtz	ObsHtz-CR51	OAR22_23392099	19.588	OAR22_24747565	20.991
		F_{ST}	F_{ST}-CR28	OAR22_24682845	20.925	OAR22_26951573	23.157

A CCR region was defined when overlapping selection regions identified by the genetic differentiation analysis (in at least two breed pairs), averaged for a 9-SNP window size (F_{ST}), and by at least one of the two heterozygosity-based analysis methodologies (in at least two breeds): observed heterozygosity, averaged for a 9-SNP window size (ObsHtz), and regression analysis, considering a 10-Mb bracket size (Regression).
*For Regression results, this indicates the closest marker to the Start/End position.

to predicted overlaps for randomly-positioned CCR, these regions were highly enriched for candidate dairy-related loci. We discuss further the CCR regions that meet specific criteria.

Region Identified by all the Three Methods

- **CCR3 (OAR6:30.367–41.863 Mb).** The three analysis methods identified this region of positive selection in the first half of OAR6, which includes the *ABCG2* (ATP-binding cassette, sub-family G (white), member 2) and *SPP1* (osteopontin) genes (at 36.565–36.610 Mb and 36.708–36.720 Mb respectively), and is orthologous to the region of the bovine genome on BTA6 where several QTL for milk production traits have been reported (See Table 5 for QTL identifier number in the CattleQTLdb). This region also includes the *FAM13A* (family with sequence similarity 13, member A) gene, which has been shown to be associated with mastitis in Jersey cows [27]. In dairy cattle, strong selection signals have previously been identified [23,24] in the proximity of the *ABCG2* gene, which harbors one of the few causal mutations or Quantitative Trait Nucleotide (QTN) described in livestock species [28]. In sheep, a selection signal in the *ABCG2* region has also been identified in a work focused on Altamurana sheep, where differences in allele frequencies were compared for animals with high and low milk yields [29].

The identification of a selection signature in this region of OAR6 by the pair-wise F_{ST} comparison (F_{ST}-CR14) was based on four breed pairs. For the Milk Lacaune-Australian Poll Merino and the Comisana-Australian Poll Merino pairs, the signal of genetic differentiation involved the *ABCG2* and *SPP1* genes, whereas for the two other pairs, the identified signal was upstream (Chios-Sakiz; OAR6:30.367–30.380 Mb) or downstream (Churra-Ojalada; OAR6:39.316–39.577 Mb) of these genes. The ObsHtz analysis showed a selection signal (ObsHtz-CR27) for Milk Lacaune, Comisana and Churra dairy breeds, and also for three non-dairy breeds, Australian Poll Merino, Meat Lacaune and Ojalada. Both Lacaune breeds showed low values of ObsHtz extended for long intervals (3.48 and 5.47 Mb for Milk Lacaune and Meat Lacaune, respectively). With regard to the regression-based analysis, this region was the only one detected in multiple breeds for all three bracket sizes (for Milk Lacaune, Comisana, Meat Lacaune and Australian Poll Merino breeds).

Together these results suggest that CCR3 shows selection for dairy traits in several sheep breeds, and that this signal may be related to the documented effects of the *ABCG2* [28] or *SPP1* [30] genes on milk production and lactation regulation, respectively. The selection signal positioned directly at *ABCG2* and *SPP1* was only seen in the highly specialized breeds Milk Lacaune and Comisana (F_{ST}, ObsHtz and Regression). In other dairy breeds for which the selection is more recent and less efficient (e.g. Churra and Chios), selection may not have substantially altered the frequencies of favoured alleles at these loci, which could explain why a strong selection signal directly at these genes was not observed. A previous study in Churra sheep found suggestive associations between the *ABCG2* gene and milk fat percentage and milk yield [31] while no studies to date have tested the effects of these two genes on dairy traits in the Lacaune and Comisana breeds.

The results reported in the current study also suggest that in this region of OAR6 there could be a selection signal related to meat specialized breeds such as Meat Lacaune, Australian Poll Merino and Ojalada. In this regard, it is worth noting that several QTL for growth and carcass traits have been described in the orthologous bovine region [32,33]. Hence, analogous to the observations in the orthologous bovine region, this region of the sheep genome may influence both dairy and meat production traits.

Regions with High F_{ST} in more than Two Breed Pairs

This criterion was used to highlight the CCR regions where the genetic differentiation analysis showed a particularly strong indication of a dairy selection signature, as this is possibly the most effective analysis performed in this study to detect regions specifically affected by dairy selection rather than selection acting on non-dairy-related traits. With the aim of establishing stringent criteria we consider in this section only those regions where more than two breed pairs (none sharing a common breed, as explained above) showed the selection signal. In addition to CCR3 discussed above, this category also includes the following two regions:

- **CCR1 (OAR3:152.680 to 154.679 Mb).** This core region, for which the F_{ST}-selection signals were identified for the Churra-Ojalada, Comisana-Australian Poll Merino and East Friesian Brown-Finnsheep pairs, includes *HMGA2* (high mobility group AT-hook 2), a gene associated with human stature [34]. The identification of this gene as a selection target was also found in an analysis of dogs with divergent stature [10]. The bovine region orthologous to CCR1 includes QTL related to stature (with the *HMGA2* gene suggested as a possible causative locus [35]) and rump length (see Table 5). Hence, the CCR1 signal identified in the present study might indicate selection targeting sheep body conformation traits. This hypothesis would agree with the differences in body size between some of the pairs involved in this selection signal. For example, the adult weight of Australian Poll Merino is significantly higher than that of Lacaune and Comisana; Churra and East Friesian Brown are also generally heavier than their comparison breeds. *HGMA2* has also been suggested as a candidate gene related to ear size and shape in both pigs and dogs [36,37], thus further investigation is required to assess whether there are differences in ear morphology between the sheep breeds showing this selection signal. Although the confidence interval of a QTL for protein percentage reported in Churra sheep [38] (Table 5) overlaps with CCR1, the causal mutation for that QTL was later found in the *LALBA* gene [39], which maps outside of this core region.
- **CCR2 (OAR3:209.872–215.814 Mb).** Four candidate genes in the orthologous bovine region to this CCR (distal end of BTA5) were identified from the Ogorevc et al. [22] database. Two of them were related to mastitis in a disease-induced mouse-model study [40]: *BID* (BH3 interacting domain death agonist), which is a pro-apoptotic induced gene, and *MAFF* (v-maf avian musculoaponeurotic fibrosarcoma oncogene homolog F), which is related to cell proliferation. The identification of two other genes as candidates for dairy traits in this regions, *FKBP4* (FK506 binding protein 4) and *MLK1* (mixed lineage protein kinase), was also based on mouse model studies (http://www.informatics.jax.org/). Furthermore, *FKBP4* is expressed in breast cancer tissue (Genes-to-Systems Breast Cancer database, G2SBC, http://www.itb.cnr.it/breastcancer//index.html) and *MLK1* is expressed in epithelial tumor cell lines of colonic, breast and esophageal origin [41]. QTL effects described in the bovine region orthologous to CCR2 (on BTA5) influence milk production and some conformation traits (Table 5). A previous study in dairy cows found a selection signature in this region [23]. In that case, the gene displaying the strongest evidence of selection was *CD163*, which is involved in the innate immune response and clearance of plasma hemoglobin [42]. This region also includes the gene coding for CSNK1ε (casein-kinase epsilon), which is related to

Table 5. Convergence candidate regions (CCR) for ovine dairy selection sweeps identified in this study.

Convergence candidate regions	Sheep genome range (Mb) (Oar v2.0)	Bovine genome range (Mb) (UMD 3.1)	Functional candidate genes based on Ogorevc et al. [22]	Other candidate genes[1]	QTL described in sheep	QTL described in cattle in relation to milk production and functional dairy traits (CattleQTLdb identifier[2])	Nb. of positional candidates[3]
CCR1	OAR3:152.680–154.679	BTA5:46.720–49.009		*HMGA2*	Milk protein percentage [38]	Somatic cell score (2659), Milk fat yield (4495), Milk yield (2429), Rump length (3422), Stature (16277, 16278), Clinical mastitis (4973)	11
CCR2	OAR3:209.872–215.814	BTA5:106.976–112.636	*BID, MAFF, FKBP4, MKL1*	*CSNK1E*		Milk fat yield (daughter deviation) (9995), Milk protein yield (daughter deviation) (9994), Milk fat percentage (2717), Chest width (4623), Hip height (3420)	100
CCR3	OAR6:30.367–41.863	BTA6:31.710–43.022	*ABCG2, SPP1*	*FAM13A*		Milk protein percentage (EBV) (15002, 15003), Milk fat percentage (1753, 16057), Milk protein percentage (1755, 9913, 16058, 16059), Milk protein yield (daughter deviation) (10145), Milk fat yield (EBV) (11303, 12031), Milk protein yield (EBV) (11304), Milk yield (EBV) (11302), Somatic cell score (EBV) (6165, 6164), Milk fatty acid unsaturated index (11506, 11508, 11509, 11510), Milk myristoleic acid percentage (11507), Milk palmitoleic acid percentage (11505), Teat placement (10285), Udder attachment (10284)	32
CCR4	OAR13:56.061–65.811	BTA13:57.572–67.005	*POFUT1, TFAP2C, FAM110A, AHCY*	*GHRH, ASIP*		Milk protein percentage (2672), Milk protein yield (EBV) (6090), Milk fat yield (2555), Milk protein yield (daughter deviation) (10156), Milk protein yield (2671), Milk yield (2670), Udder attachment (1584), Udder composite index (1589), Udder depth (1588), Udder height (1586), Udder width (1587), Rump angle (3429), Foot angle (1583)	172
CCR5	OAR15:72.774–74.550	BTA15:75.154–76.879	*CD82*			Teat placement (1595), Udder cleft (1600), Udder composite index (1602), Milk fat yield (4503), Milk protein yield (4502), Milk protein yield (EBV), (6103), Milk fat percentage (3452), Milk protein percentage (EBV) (11345), Milk yield (EBV) (11346)	17
CCR6	OAR22:19.587–23.157	BTA26:20.286–24.226	*CHUK, SCD*		Milk fatty acid composition [57,58], Somatic cell score [60]	Milk yield (10452), Milk protein yield (10454), Milk fat yield (3636), Milk protein yield (3638), Milk protein yield (11702), Milk yield (11701), Milk fat yield (10453), Milk fat percentage (2598), Milk fat yield (2572), Milk protein yield (2573), Milk yield (2574), Milk protein percentage (4814), Milk protein percentage (3639), Milk yield (3634), Somatic Cell Count (1503)	74

The interval of each region (in bp) is based on the sheep genome reference sequence v2.0 (http://www.livestockgenomics.csiro.au/cgi-bin/gbrowse/oarv2.0/). The corresponding orthologous bovine genomic intervals are based on the bovine genome reference sequence UMD 3.1 (http://www.ensembl.org/Bos_taurus/Info/Index). The positional candidate genes that map within the bovine candidate range and that are included as candidate genes for milk production and mastitis traits in the database provided by Ogorevc et al. [22] are indicated as functional candidate genes. The affected trait and CattleQTLdb reference for previously reported cattle QTL that map within the bovine candidate genomic regions and that influence milk production traits or other functional traits related to dairy production are also indicated.

[1]Other candidate genes. This category includes two types of genes:

–Those that although are not included in the Ogorevc et al. [22] database may be considered as candidate genes in relation to milk production related traits based on other studies.

–Those that are likely to be related to non-dairy selection signatures such as morphological traits and coat colour features.

[2]CattleQTLdb identifier: Search reference number at http://www.animalgenome.org/cgi-bin/QTLdb/BT/search to find complete details about QTL reported in the orthologous region of the corresponding sheep CCR identified in this study.

[3]Number of positional candidate genes extracted from the orthologous bovine region using BioMart for each of the labeled CCRs (based on Table S3).

circadian rhythms. In a study of the human milk fat globule transcriptome, *CSNK1ε* was identified as one of the nine core "clock" genes that showed differential expression over a 24-hour period time in lactating women [43]. Of particular interest is the finding that this OAR3 region was labelled as a CCR based on the overlap of candidate regions detected by pair-wise-F_{ST} in the pairs including the most highly specialized dairy breeds (Milk Lacaune, Comisana and East Friesian Brown), which may have been under selection for circadian-related adaptation of milk production to intensive milking.

Other Regions

- **CCR4 (OAR13:56.061 to 65.811 Mb).** Several genes included in this core candidate region were also found in the Ogorevc et al. [22] database. The *POFUT1* (protein O-fucosyltransferase 1) gene plays a crucial role in Notch signaling, which regulates mammary stem cell function and luminal cell-fate commitment [44]. *TFAP2C* (transcription factor AP-2 gamma; activating enhancer binding protein 2 gamma) is involved in mammary development, differentiation, and oncogenesis playing a critical role in gene regulation in hormone responsive breast cancer [45], and *AHCY* (adenosylhomocysteinase) has been suggested as potentially involved in mastitis defense based on its disease-associated expression [46]. Another positional candidate gene for this core region is the *GHRH* (growth hormone-releasing hormone) gene. Although the direct relationship of this gene and milk production traits is still not clear [47,48], its link to the somatotropic axis and other functional candidate genes included in the Ogorevc et al. [22] database (*GH, GHR, GHRHR*) suggest a possible influence, directly or indirectly, on dairy traits. In addition to these candidate dairy-related genes, the *ASIP* (Agoutí signaling protein) gene is also located in this region (OAR13:63.028–63.033 Mb). This gene has a major role in metabolic processes [49] and coat colour pigmentation in mammalian species [50]. Based on the known associations between polymorphisms at this gene and coat colour patterns in sheep [51] it is possible that the identified selection signal results from coat colour selection. In their analysis of the complete HapMap dataset, Kijas et al. [14] also identified a selection signal near *ASIP*.
- **CCR5 (OAR15:72.774–74.550 Mb).** This region included the *CD82* (CD82 molecule) gene, which is included in the Ogorevc et al. [22] database based on its expression in the mammary gland. This gene is included in the group of genes that regulate breast cancer metastasis, as a metastasis suppressor [52]. Whereas no studies have reported an association of this gene with dairy related traits, there is a functional relationship between *CD82* and *ERBB3* (Receptor tyrosine-protein kinase erbB-3) [53], which is related to normal mammary development [54].
- **CCR6: (OAR22:19.588–23.157 Mb).** Two functional candidate genes [22] were found in this region: *SCD* (Stearoyl-CoA desaturase) and *CHUK* (conserved helix-loop-helix ubiquitous kinase). The *SCD* gene encodes a multifunctional complex enzyme important in the cellular biosynthesis of fatty acids. Several studies in different populations of dairy cattle have reported associations between polymorphisms at this gene with milk production traits [55] and milk fatty acid composition [56]. In sheep, the *SCD* gene has been suggested as positional and functional candidate gene for a QTL identified on OAR22 in a Sarda × Lacaune back-cross population for the ratio of conjugated linoleic acid to vaccenic acid in sheep milk [57]. A later study in Churra sheep also identified a QTL on OAR22 for the same trait close to the *SCD* position, although various analyses questioned this gene as responsible for the identified effect [58]. The *CHUK* gene is listed in the Ogorevc et al. [22] database because it is expressed in breast cancer tumors and is a regulator of mammary epithelial proliferation [59]. According to the SheepQTL database, this region includes a QTL for somatic cell score described in an Awassi x Merino cross population [60] and it has also been identified as a selection signal by the analysis of allele frequency differences between animals with divergent milk yields reported in Altamurana sheep [29].

The bovine region orthologous to CCR6 (on BTA26), overlaps with a region showing a selection signature in dairy cattle [23], where the *C10ORF76* (chromosome 10 open reading frame 76) gene was associated with the strongest selection signal. Although there is not a reported association of this gene with milk production traits, it is expressed in the mammary gland and it is altered in breast cancer cells, based on the G2SBC database.

Inconsistencies between this Study and Previous QTL and Selection Mapping Studies of Cattle and Sheep

Although all six CCR overlapped with QTL for dairy traits in sheep or cattle (Table 4 and discussed above), our study did not identify a selection signal close to several genes previously associated with dairy traits in sheep and cattle. For example, there were no CCR near the *LALBA* (alpha-lactalbumin) gene (OAR3:137 Mb), where a particular variant has been recently been proposed to explain a QTL for milk protein percentage identified in Churra sheep [39]. The lack of signal near this QTL in the F_{ST} analysis of Churra vs Ojalada is consistent with the fact that the causative mutation is still segregating in Churra, which allowed its identification as QTL.

In addition, in their analysis of the complete Sheep HapMap dataset, Kijas et al. [14] reported positive selection surrounding the *PRLR* gene, which is associated with milk traits in dairy cattle [61]. In our study, although none of the CCRs map to OAR16, where this gene is located (39.250–39.284 Mb), it is worth mentioning that this gene is included in the interval of F_{ST}-CR27 (OAR16:37.347–40.850 Mb), which was identified based on the signals detected in three breed pairs involving the most specialized dairy breeds in this study (East Friesian Brown-Finnsheep, Milk Lacaune-Australian Poll Merino and Comisana-Australian Poll Merino) but was not classified a CCR due to the lack of selection signals from the heterozygosity-based methods. Other regions that were detected by the F_{ST}-pairwise comparison for many breed pairs but that were not supported by the heterozygosity-based methods were found on OAR2 (F_{ST}-CR2:52.346–53.409 Mb) and OAR9 (57.363–60.849 Mb). Whereas the first region does not include any functional candidate gene for dairy traits, the region in OAR9 included three genes related to the metabolism of fatty acids (*FABP4, FABP5* and *FABP9*). *FABP4* and *FABP5* have been shown to be highly expressed in the mammary gland during lactation [62] and significant associations have been found between *FABP4* and fatty acid composition of bovine milk [63]. We acknowledge that one or more of these regions may represent false negatives that were missed by our stringent selection signal criteria. However, because of the difficulty in linking a sweep signal to a given phenotype, we suggest that application of stringent criteria in this type of study is an appropriate option to avoid reporting long lists of candidate regions based on spurious results.

We also did not find evidence of selection on some major candidate genes for milk production for which selection signatures have been observed in cattle (e.g. *DGAT1:* OAR9:13.534–13,543 Mb; *GHR:* OAR16:32.068–32.231). In contrast to our results, the *GHR* gene (BTA20) showed the largest difference in sliding window average allele frequencies in a study of divergent selection between dairy and beef cattle [24], and also showed significant extended haplotype homozygosity [25]. With regard to *DGAT1*, evidence of selection has also been identified when comparing dairy and beef cattle breeds [24].

In their study, Kijas et al. [14] also identified a strong selection signal on OAR10, associated to the presence or absence of horns and close to the gene responsible of the polled phenotype, *RXFP2* (relaxin/insulin-like family peptide receptor 2) gene (OAR10:27.602–27.646 Mb). In our study, a selection signal was identified in this region based on the ObsHtz-based method (ObsHtz-CR33:24.856–27.897 Mb, for the dairy breeds Comisana, Churra and Milk Lacaune) and the F_{ST}-based method (OAR10:25.540–28.983 Mb). However because the F_{ST} signal was due only to the two breed pairs involving the Australian Poll Merino, this was not labelled as F_{ST}-CR.

Apart from the overlap between two CCRs (CCR3 on OAR6 and CCR6 on OAR22) with the selection signals identified in Altamurana sheep for milk yield [29], we did not find evidence of selection near the signals reported for this Italian breed. This lack of correspondence may derive from breed-specific signals reported for Altamurana.

Comparison of the Three Selection Mapping Methods

From our point of view, the analysis method that involved the estimation of pair-wise F_{ST} for pairs of related breeds showing divergent specialization (one for milk production, one not) should be the most powerful analysis in terms of identifying selection specifically related to dairy production. Four out of the 28 candidate regions showing multiple pair-wise F_{ST} signals were detected in four out of the six breed pairs (F_{ST}-CR3, F_{ST}-CR9, F_{ST}-CR14 and F_{ST}-CR18). Of these, F_{ST}-CR3 (OAR2:52.346–53.409 Mb) was not included as a CCR due to the lack of consistency with the ObsHtz or 10-Mb Regression analysis results, although the same region was identified by the 5-Mb Regression analysis (RegBrack5-CR1 in Table 3) in two dairy breeds (Churra and Chios) and one non-dairy breed (Ojalada). Given that no functional candidate genes from the Ogorevc et al. [22] database were found in the orthologous bovine region, it is possible that this region underlies breed differentiation not directly related to dairy traits.

Among the 55 candidate regions identified based on the ObsHtz analysis (ObsHtz-CRs, Table S2), there were only twelve regions showing a signal in dairy but not in non-dairy breeds (ObsHtz-CR3, 7, 8, 10, 12, 25, 28, 29, 31, 41, 45 and 55). Considering that the background genome has been previously selected for meat, maternal characteristics, and other traits, whereas the development of dairy breeds is much more recent, it would be expected that the selection signals specifically related to dairy traits would not be seen in the other breeds (although Meat Lacaune could be an exception). However, as none of these regions showing a reduction of heterozygosity exclusively in dairy breeds were identified by the F_{ST}-based method, they were not identified as final core CCRs (and thus are not present in Table 4). Although the evidence linking these regions to dairy-related selection is weaker than for the CCRs, we performed an additional search for functional candidate genes and dairy-related QTL mapping within these regions, similar to that performed in the eight identified CCRs (see Table S4). A total of 118 genes were extracted from the orthologous bovine regions of these eleven dairy-breed-limited regions of reduced heterozygosity (data not shown). Among them, only the *HSPD1* (Heat shock 60 kDa protein 1; chaperonin) gene is included in the Ogorevc et al. [22] database, due to its expression in the mammary gland. This gene is also included in the G2SBC database although no studies have reported so far its association with milk production traits. Interestingly, among the dairy QTL detected in these regions there is greater overlap with ovine QTL for milk production traits (Table S4) than for the list of core CCRs. Hence, these regions identified exclusively by ObsHtz could include gene variants occurring in individual dairy breeds, as it is the case for many of the QTL described in sheep.

There were eight regions that overlapped between those identified by F_{ST} (including a full set of regions, including those that contained pairs with the same breeds that were removed from Table S1) and ObsHtz (out of 35 and 55, respectively). The explanation for the higher number of regions identified by ObsHz is that the regions identified using F_{ST} were slightly larger (incorporated more windows) than those identified using ObsHtz.

There were far fewer signals identified using the Regression approach than either F_{ST} or ObsHtz. Although the top (or bottom) 0.5th percent results were considered as signals of selection for all methods, the Regression method first filtered out the intervals with non-significant and non-asymptotic regression patterns, and thus the total number of eligible intervals was substantially reduced compared to the other approaches in which the distribution of F_{ST}/ObsHtz values for all markers (with the exception of those on the very ends of the chromosomes) was considered. Thus the implementation of Regression in this study was more stringent than the other methods.

The regions identified by the Regression method showed greater overlap with ObsHtz than F_{ST}, which is not surprising since both Regression and ObsHtz are designed to detect regions with a reduction in diversity. For the 10-Mb bracket size (results considered for the identification of CCR), all three regions identified with the Regression approach overlapped with those identified with ObsHtz while one out of the three, RegBrack10-CR2, overlapped with the regions identified with $F_{ST,}$ and was therefore considered as CCR (CCR3).

Conclusions

The results reported here provide a genome-wide map of selection signatures in the dairy sheep genome. The six core candidate regions identified are likely to influence traits of economic interest in dairy sheep production and can be considered as starting points for future studies aimed at the identification of the causal genetic variation underlying these signals. For some of these regions, strong candidate genes have been proposed (e.g. *ABCG2, SPP1*), whereas some other genes designated as candidates based on their association with sheep and cattle dairy traits (e.g. *LALBA, DGAT1A*) were not associated with a detectable sweep signal. Discrepancies between selection signals in dairy sheep and cattle may be explained either by statistical or biological factors, such as the limited statistical power of the analyses to identify effects of small magnitude or the fact that the genetic architecture of milk production and dairy-related traits substantially differs from sheep to cattle and also between the different breeds of dairy sheep, which have been subjected to different levels of selection pressure. Many of the identified regions corresponded to orthologous regions in cattle where QTL for dairy traits have been identified. Due to the limited number of QTL studies reported in sheep compared with cattle, the results illustrate the potential value of the study of selection signatures to uncover mutations with potential effects on quantitative dairy sheep traits. Additional studies are needed to confirm and refine the results reported here. To this end, the recent availability of the high-density ovine chip (700 K) will provide a valuable tool to perform more powerful and precise selection mapping studies.

Supporting Information

Table S1 Candidate regions for signatures of selection identified on the basis of the pair-wise F_{ST} analysis.

Table S2 Candidate regions identified based on reduced heterozygosity signals identified in at least two of the dairy breeds.

Table S3 List of all genes from the orthologous bovine genome regions corresponding to the six convergence candidate regions (CCR) for dairy selection sweeps identified in this study, extracted using the Biomart tool (http://www.biomart.org/).

Table S4 Candidate regions identified by the analysis based on observed heterozygosity (ObsHtz-CR), averaged in sliding windows of 9 SNPs (ObsHtz-9SNPW), that were exclusively detected in dairy breeds.

File S1 Summary of the criteria for selection of breeds to be included in the study, including the results of a Principal Component Analysis (PCA) performed with the initial set of breeds considered.

File S2 Summary of the results of the analysis performed in this work in relation to the myostatin (*GDF-8*) gene region. These results were evaluated to establish criteria for the analyses performed to detect dairy selection signatures in the dairy breeds.

Acknowledgments

We thank Samantha Wilkinson for providing R scripts.

Author Contributions

Conceived and designed the experiments: BGG PW JJA. Analyzed the data: BGG PW. Contributed reagents/materials/analysis tools: RPW JJA EGG JK. Wrote the paper: BGG PW JJA RPW JK. Conceived the study: BGG PW.

References

1. Legge T (1996) The beginning of caprine domestication. In: Harris DR, editor. The Origins and Spread of Agriculture and Pastoralism in Eurasia. Smithsonian New York: Institution Press. pp. 238–262.
2. Maijala K (1997) Genetic aspects of domestication, common breeds and their origin. In: Piper L, Ruvinsky A, editors. The genetics of sheep. Oxford: CAB Int. pp. 539–564.
3. Barillet F (1997) Genetics of milk production. In: Piper L, Ruvinsky A, editors. The genetics of sheep. Oxford: CAB Int. pp. 539–564.
4. Ida A, Vicovan PG, Radu R, Vicovan A, Cutova N, et al. (2012) Improving the milk production at the breeds and populations of sheep from various geo-climatic zones. Lucrări Ştiinţifice - Seria Zootehnie 57: 17.
5. Gibbs RA, Taylor JF, Van Tassell CP, Barendse W, Eversole KA, et al. (2009) Genome-wide survey ofSNP variation uncovers the genetic structure of cattle breeds. Science 324: 528–532.
6. Rubin CJ, Zody MC, Eriksson J, Meadows JR, Sherwood E, et al. (2010) Whole-genome resequencing reveals loci under selection during chicken domestication. Nature 464: 587–591.
7. Elferink MG, Megens HJ, Vereijken A, Hu X, Crooijmans RP et al. (2012) Signatures of selection in the genomes of commercial and non-commercial chicken breeds. PLoS One 7: ve32720.
8. Wiener P, Pong-Wong R (2011) A regression-based approach to selection mapping. J Hered 102: 294–305.
9. Stella A, Ajmone-Marsan P, Lazzari B, Boettcher P (2010) Identification of selection signatures in cattle breeds selected for dairy production. Genetics 185: 1451–1461.
10. Akey JM, Ruhe AL, Akey DT, Wong AK, Connelly CF, et al. (2010) Tracking footprints of artificial selection in the dog genome. Proc Natl Acad Sci USA 107: 1160–1165.
11. Vaysse A, Ratnakumar A, Derrien T, Axelsson E, Rosengren Pielberg G, et al. (2011) Identification of genomic regions associated with phenotypic variation between dog breeds using selection mapping. PLoS Genet 10: e1002316.
12. Ai H, Huang L, Ren J (2013) Genetic diversity, linkage disequilibrium and selection signatures in chinese and Western pigs revealed by genome-wide SNP markers. PLoS One 8: e56001.
13. Boyko AR, Quignon P, Li L, Schoenebeck JJ, Degenhardt JD, et al. (2010) A simple genetic architecture underlies morphological variation in dogs. PLoS Biol 8: e1000451.
14. Kijas JW, Lenstra JA, Hayes B, Boitard S, Porto Neto LR, et al. (2012a) Genome-wide analysis of the world's sheep breeds reveals high levels of historic mixture and strong recent selection. PLoS Biol 10: e1001258.
15. Weir BS, Cockerham CC (1984) Estimating F-Statistics for the Analysis of Population Structure. Evolution 38: 1358–1370.
16. Akey JM, Zhang G, Zhang K, Jin L, Shriver MD (2002) Interrogating a high-density SNP map for signatures of natural selection. Genome Res 12: 1805–1814.
17. Wilkinson S, Lu ZH, Megens HJ, Archibald AL, Haley C, et al. (2013) Signatures of diversifying selection in European pig breeds. PLoS Genet 9: e1003453.
18. Barendse W, Harrison BE, Bunch RJ, Thomas MB, Turner LB (2009) Genome wide signatures of positive selection: the comparison of independent samples and the identification of regions associated to traits. BMC Genomics 10: 178.
19. Rubin CJ, Megens HJ, Martinez Barrio A, Maqbool K, Sayyab S, et al. (2012) Strong signatures of selection in the domestic pig genome. Proc Natl Acad Sci USA. 109: 19529–19536.
20. Axelsson E, Ratnakumar A, Arendt ML, Maqbool K, Webster MT, et al. (2013) The genomic signature of dog domestication reveals adaptation to a starch-rich diet. Nature 495: 360–364.
21. Payne RW, Murray DA, Harding SA, Baird DB, Soutar DM (2007) GenStat for Windows (10th Edition) Introduction. Hemel Hempstead: VSN. International.
22. Ogorevc J, Kunej T, Razpet A, Dovc P (2009) Database of cattle candidate genes and genetic markers for milk production and mastitis. Anim Genet 40: 832–851.
23. Flori L, Fritz S, Jaffrézic F, Boussaha M, Gut I, et al. (2009) The genome response to artificial selection: a case study in dairy cattle. PLoS One 4: e6595.
24. Hayes BJ, Chamberlain AJ, Maceachern S, Savin K, McPartlan H, et al. (2009) A genome map of divergent artificial selection between Bos taurus dairy cattle and Bos taurus beef cattle. Anim Genet 40: 176–184.
25. Qanbari S, Pimentel EC, Tetens J, Thaller G, Lichtner P, et al. (2010) A genome-wide scan for signatures of recent selection in Holstein cattle. Anim Genet 41: 377–389.
26. Moradi MH, Nejati-Javaremi A, Moradi-Shahrbabak M, Dodds K G, McEwan JC (2012) Genomic scan of selective sweeps in thin and fat tail sheep breeds for identifying of candidate regions associated with fat deposition. BMC Genet 13: 10.
27. Kowalewska-Łuczak I, Kulig H (2013) Polymorphism of the FAM13A, ABCG2, OPN, LAP3, HCAP-G, PPARGC1A genes and somatic cell count of Jersey cows - Preliminary study. Res Vet Sci 94: 252–255.
28. Olsen HG, Nilsen H, Hayes B, Berg PR, Svendsen M, et al. (2007) Genetic support for a quantitative trait nucleotide in the ABCG2 gene affecting milk composition of dairy cattle. BMC Genet 8: 32.
29. Moioli B1, Scatà MC, Steri R, Napolitano F, Catillo G (2013) Signatures of selection identify loci associated with milk yield in sheep. BMC Genet 14: 76.
30. Sheehy PA, Riley LG, Raadsma HW, Williamson P, Wynn PC (2009) A functional genomics approach to evaluate candidate genes located in a QTL interval for milk production traits on BTA6. Anim Genet 40: 492–498.
31. García-Fernández M, Gutiérrez-Gil B, Sánchez JP, Morán JA, García-Gámez E, et al. (2011) The role of bovine causal genes underlying dairy traits in Spanish Churra sheep. Anim Genet 42: 415–420.
32. Eberlein A, Takasuga A, Setoguchi K, Pfuhl R, Flisikowski K, et al. (2009) Dissection of genetic factors modulating fetal growth in cattle indicates a substantial role of the non-SMC condensin I complex, subunit G (NCAPG) gene. Genetics 183: 951–964.
33. Gutiérrez-Gil B, Wiener P, Williams JL, Haley CS (2012) Investigation of the genetic architecture of a bone carcass weight QTL on BTA6. Anim Genet 43: 654–661.
34. Yang TL, Guo Y, Zhang LS, Tian Q, Yan H, et al. (2010) HMGA2 is confirmed to be associated with human adult height. Ann Hum Genet 74: 11–16.
35. Pryce JE, Hayes BJ, Bolormaa S, Goddard ME (2011) Polymorphic regions affecting human height also control stature in cattle. Genetics 187: 981–984.
36. Li P, Xiao S, Wei N, Zhang Z, Huang R, et al. (2012) Fine mapping of a QTL for ear size on porcine chromosome 5 and identification of high mobility group AT-hook 2 (HMGA2) as a positional candidate gene. Genet Sel Evol 44: 6.

37. Boyko AR, Quignon P, Li L, Schoenebeck JJ, Degenhardt JD, et al. (2010) A simple genetic architecture underlies morphological variation in dogs. PLoS Biol 8: e1000451.
38. Gutiérrez-Gil B, El-Zarei MF, Alvarez L, Bayón Y, de la Fuente LF et al. (2009) Quantitative trait loci underlying milk production traits in sheep. Anim Genet 40: 423–434.
39. García-Gámez E, Gutiérrez-Gil B, Sahana G, Sánchez JP, Bayón Y, et al. (2012) GWA analysis for milk production traits in dairy sheep and genetic support for a QTN influencing milk protein percentage in the LALBA gene. PLoS One 7: e47782.
40. Zheng J, Watson AD, Kerr DE (2006) Genome-wide expression analysis of lipopolysaccharide-induced mastitis in a mouse model. Infect Immun 74: 1907–1915.
41. Dorow DS, Devereux L, Dietzsch E, De Kretser T (1993) Identification of a new family of human epithelial protein kinases containing two leucine/isoleucine-zipper domains. Eur J Biochem 213: 701–710.
42. Schaer DJ, Schaer CA, Buehler PW, Boykins RA, Schoedon G, et al. (2006) CD163 is the macrophage scavenger receptor for native and chemically modified hemoglobins in the absence of haptoglobin. Blood 107: 373–380.
43. Maningat PD, Sen P, Rijnkels M, Sunehag AL, Hadsell DL, et al. (2009) Gene expression in the human mammary epithelium during lactation: the milk fat globule transcriptome. Physiol Genomics. 37: 12–22.
44. Bouras T, Pal B, Vaillant F, Harburg G, Asselin-Labat ML, et al. (2008) Notch signaling regulates mammary stem cell function and luminal cell-fate commitment. Cell Stem Cell 3: 429–441.
45. Woodfield GW, Chen Y, Bair TB, Domann FE, Weigel RJ (2010) Identification of primary gene targets of TFAP2C in hormone responsive breast carcinoma cells. Genes Chromosomes Cancer 49: 948–962.
46. Schwerin M, Czernek-Schäfer D, Goldammer T, Kata SR, Womack JE, et al. (2003) Application of disease-associated differentially expressed genes–mining for functional candidate genes for mastitis resistance in cattle. Genet Sel Evol 35 Suppl 1: S19–34.
47. Dybus A, Grzesiak W (2006) GHRH/HaeIII gene polymorphism and its associations with milk production traits in Polish Black-and-White cattle. Arch Tierz Dummerstorf 49: 434–438.
48. Szatkowskaac I, Dybusac A, Grzesiakab W, Jedrzejczakac M, Muszyńskaac M (2009). Association between the growth hormone releasing hormone (GHRH) gene polymorphism and milk production traits of dairy cattle. J Appl Anim Res 36: 119–12.
49. Wolff GL, Roberts DW, Mountjoy KG (1999) Physiological consequences of ectopic agouti gene expression: The yellow obese mouse syndrome. Physiol Genomics 1: 151–163.
50. Bennett DC, Lamoreux ML (2003) The colour loci of mice–A genetic century. Pigment Cell Res 16: 333–344.
51. Norris BJ, Whan VA (2008) A gene duplication affecting expression of the ovine ASIP gene. Genome Res 18: 1282–1293.
52. Debies MT, Welch DR (2001) Genetic basis of human breast cancer metastasis. J Mammary Gland Biol Neoplasia 6: 441–451.
53. Odintsova E, Voortman J, Gilbert E, Berditchevski F (2003) Tetraspanin CD82 regulates compartmentalisation and ligand-induced dimerization of EGFR. J Cell Sci 116: 4557–4566.
54. Lahlou H, Müller T, Sanguin-Gendreau V, Birchmeier C, Muller WJ (2012) Uncoupling of PI3K from ErbB3 impairs mammary gland development but does not impact on ErbB2-induced mammary tumorigenesis. Cancer Res 72: 3080–3090.
55. Alim MA, Fan YP, Wu XP, Xie Y, Zhang Y, et al. (2012) Genetic effects of stearoyl-coenzyme A desaturase (SCD) polymorphism on milk production traits in the Chinese dairy population. Mol Biol Rep 39: 8733–8740.
56. Moioli B, Contarini G, Avalli A, Catillo G, Orru L, et al. (2007) Short communication: effect of stearoyl-coenzyme A desaturase polymorphism on fatty acid composition of milk. J Dairy Res 90: 3553–3558.
57. Carta A, Sechi T, Usai MG, Addis M, M Fiori, et al. (2006) Evidence for a QTL affecting the synthesis of linoleic conjugated acid cis-9, trans-11 from 11-c 18:1 acid on ovine chromosome 22. Proc. 8th World Congress on Genetics Applied to Livestock Production. Belo Horizonte, Brazil. Commun. no. 12–03. Instituto Prociencia, Belo Horizonte, Brazil.
58. García-Fernández M, Gutiérrez-Gil B, García-Gámez E, Sánchez JP, Arranz JJ (2010) Detection of quantitative trait loci affecting the milk fatty acid profile on sheep chromosome 22: role of the stearoyl-CoA desaturase gene in Spanish Churra sheep. J Dairy Sci 93: 348–357.
59. Cao Y, Luo JL, Karin M (2007) IkappaB kinase alpha kinase activity is required for self-renewal of ErbB2/Her2-transformed mammary tumor-initiating cells. Proc Natl Acad Sci USA 104: 15852–15857.
60. Raadsma HW, Jonas E, McGill D, Hobbs M, Lam MK, et al. (2009). Mapping quantitative trait loci (QTL) in sheep. II. Meta-assembly and identification of novel QTL for milk production traits in sheep. Genet Sel Evol 41: 45.
61. Viitala S, Szyda J, Blott S, Schulman N, Lidauer M, et al. (2006) The role of the bovine growth hormone receptor and prolactin receptor genes in milk, fat and protein production in Finnish Ayrshire dairy cattle. Genetics 173: 2151–2164.
62. Sumner-Thomson JM, Vierck JL, McNamara JP (2011) Differential expression of genes in adipose tissue of first-lactation dairy cattle. J Dairy Sci 94: 361–369.
63. Nafikov RA, Schoonmaker JP, Korn KT, Noack K, Garrick DJ, et al. (200 Association of polymorphisms in solute carrier family 27, isoform A6 (SLC27A6) and fatty acid-binding protein-3 and fatty acid-binding protein-4 (FABP3 and FABP4) with fatty acid composition of bovine milk. J Dairy Sci 96: 6007–6021.

21

Genetic Footprints of Iberian Cattle in America 500 Years after the Arrival of Columbus

Amparo M. Martínez[1], Luis T. Gama[2,3], Javier Cañón[4], Catarina Ginja[5], Juan V. Delgado[1], Susana Dunner[4], Vincenzo Landi[1], Inmaculada Martín-Burriel[6], M. Cecilia T. Penedo[7], Clementina Rodellar[6], Jose Luis Vega-Pla[8]*, Atzel Acosta[9], Luz A. Álvarez[10], Esperanza Camacho[11], Oscar Cortés[4], Jose R. Marques[12], Roberto Martínez[13], Ruben D. Martínez[14], Lilia Melucci[15,16], Guillermo Martínez-Velázquez[17], Jaime E. Muñoz[10], Alicia Postiglioni[18], Jorge Quiroz[17], Philip Sponenberg[19], Odalys Uffo[9], Axel Villalobos[20], Delsito Zambrano[21], Pilar Zaragoza[6]

1 Departamento de Genética, Universidad de Córdoba, Córdoba, Spain, **2** L-INIA, Instituto Nacional dos Recursos Biológicos, Fonte Boa, Vale de Santarém, Portugal, **3** CIISA – Faculdade de Medicina Veterinária, Universidade Técnica de Lisboa, Lisboa, Portugal, **4** Departamento de Producción Animal, Facultad de Veterinaria, Universidad Complutense de Madrid, Madrid, Spain, **5** Centre for Environmental Biology, Faculty of Sciences, University of Lisbon & Molecular Biology Group, Instituto Nacional de Recursos Biológicos, INIA, Lisbon, Portugal, **6** Laboratorio de Genética Bioquímica, Facultad de Veterinaria, Universidad de Zaragoza, Zaragoza, Spain, **7** Veterinary Genetics Laboratory, University of California Davis, Davis, California, United States of America, **8** Laboratorio de Investigación Aplicada, Cría Caballar de las Fuerzas Armadas, Córdoba, Spain, **9** Centro Nacional de Sanidad Agropecuaria, San José de las Lajas, La Habana, Cuba, **10** Universidad Nacional de Colombia, Sede Palmira, Valle del Cauca, Colombia, **11** IFAPA, Centro Alameda del Obispo, Córdoba, Spain, **12** EMBRAPA Amazônia Oriental, Belém, Pará, Brazil, **13** Centro Multidisciplinario de Investigaciones Tecnológicas, Dirección General de Investigación Científica y Tecnológica, Universidad Nacional de Asunción, San Lorenzo, Paraguay, **14** Genética Animal, Facultad de Ciencias Agrarias, Universidad Nacional de Lomas de Zamora, Lomas de Zamora, Argentina, **15** Facultad Ciencias Agrarias, Universidad Nacional de Mar del Plata, Balcarce, Argentina, **16** Estación Experimental Agropecuaria Balcarce, Instituto Nacional de Tecnología Agropecuaria, Balcarce, Argentina, **17** Instituto Nacional de Investigaciones Forestales, Agrícolas y Pecuarias, Coyoacán, México, **18** Área Genética, Departamento de Genética y Mejora Animal, Facultad de Veterinaria, Universidad de la República, Montevideo, Uruguay, **19** Virginia-Maryland Regional College of Veterinary Medicine, Virginia Tech, Blacksburg, Virginia, United States of America, **20** Instituto de Investigación Agropecuaria, Estación Experimental El Ejido, Los Santos, Panamá, **21** Universidad Técnica Estatal de Quevedo, Quevedo, Ecuador

Abstract

Background: American Creole cattle presumably descend from animals imported from the Iberian Peninsula during the period of colonization and settlement, through different migration routes, and may have also suffered the influence of cattle directly imported from Africa. The introduction of European cattle, which began in the 18th century, and later of Zebu from India, has threatened the survival of Creole populations, some of which have nearly disappeared or were admixed with exotic breeds. Assessment of the genetic status of Creole cattle is essential for the establishment of conservation programs of these historical resources.

Methodology/Principal Findings: We sampled 27 Creole populations, 39 Iberian, 9 European and 6 Zebu breeds. We used microsatellite markers to assess the origins of Creole cattle, and to investigate the influence of different breeds on their genetic make-up. The major ancestral contributions are from breeds of southern Spain and Portugal, in agreement with the historical ports of departure of ships sailing towards the Western Hemisphere. This Iberian contribution to Creoles may also include some African influence, given the influential role that African cattle have had in the development of Iberian breeds, but the possibility of a direct influence on Creoles of African cattle imported to America can not be discarded. In addition to the Iberian influence, the admixture with other European breeds was minor. The Creoles from tropical areas, especially those from the Caribbean, show clear signs of admixture with Zebu.

Conclusions/Significance: Nearly five centuries since cattle were first brought to the Americas, Creoles still show a strong and predominant signature of their Iberian ancestors. Creole breeds differ widely from each other, both in genetic structure and influences from other breeds. Efforts are needed to avoid their extinction or further genetic erosion, which would compromise centuries of selective adaptation to a wide range of environmental conditions.

Editor: Sergios-Orestis Kolokotronis, Fordham University, United States of America

Funding: This work was partially funded by the Instituto Nacional de Investigación y Tecnología Agraria y Alimentaria (INIA) PET2007-01-C07-04, PET2007-05-C03-03, RZ01-002-C2-1, RZ01-002-C2-2, RZ2004-00009, RZ 2004-00022-00-00, RZ2006-00003-C02-01, RZ 2006-00003-C02-02, RZ 2006-00007-C03-03, RZ2008-00005-C02-02, RZ2008-00006-C02-02 and RZ 2008-00008-00-00 projects. C. Ginja has received funding from the European Union Seventh Framework Programme (FP7/2007–2013) under grant agreement n°PCOFUND-GA-2009-246542 and from the Foundation for Science and Technology of Portugal (DFRH/WIIA/15/2011). This work was also partially funded by the Veterinary Genetics Laboratory, University of California, Davis (USA) and the Diputación de Córdoba (Spain). The funders had no role in study design, data collection and analysis, decision to publish, or preparation of the manuscript.

Competing Interests: The authors have declared that no competing interests exist. The Empresa Brasileira de Pesquisa Agropecuária (EMBRAPA) (Brazilian Enterprise for Agricultural Research) is a state-owned company affiliated with the Brazilian Ministry of Agriculture, which is devoted to pure and applied research on agriculture. EMBRAPA conducts agricultural research on many topics including animal agriculture and crops. EMBRAPA (www.embrapa.br) is not a commercial company therefore the authors think that Jose R. Marques has no conflict of interest.

* E-mail: jvegpla@oc.mde.es

Introduction

"That many breeds of cattle have originated through variation, independently of descent from distinct species, we may infer from what we see in South America, where the genus Bos was not endemic, and where the cattle which now exist in such vast numbers are the descendants of a few imported from Spain and Portugal."

Charles Darwin, in The Variation of Animals and Plants Under Domestication, 1868

Columbus's trip to the Americas was one of the most important events in the history of humanity, as it produced major social and economic changes on both sides of the Atlantic. The Pre-Columbian American civilizations were predominantly agriculturalist but few were livestock keepers. The only domesticated species in the Americas were the dog, turkey, guinea pig and two Andean camelids [1]. One of the major impacts of Columbus's trip was the exchange of plant and animal genetic resources among continents, which revolutionized the way of life and food habits of populations in both Europe and the Americas [2].

Livestock species were brought from the Iberian Peninsula to the Americas since the late 15th century, starting with the second trip of Columbus, which departed from the Spanish city of Cádiz in 1493. In this trip, which had a re-supply in the Canary Islands, Columbus brought horses, cattle, sheep, goats and pigs to the Americas for the first time [3]. Afterwards, many other conquerors and settlers followed, and cattle brought from the Iberian Peninsula, and possibly directly from Africa at a later stage, spread throughout the Americas, adapting to a wide range of environmental conditions and giving origin to the populations currently known as Creole cattle [4]. After nearly 300 years of expansion of Creole cattle in the American continents, and with the development of more intensive production and breeding systems, several other European breeds were introduced into the Americas in the 19th century [5]. By the end of the 19th century, Indian cattle breeds, of the Zebu or *Bos indicus* type, were also introduced and quickly disseminated throughout the Americas, where they were extensively crossed with local populations, especially in tropical regions [6].

For over three centuries, Creole cattle were used as a source of draught power, food and leather, playing a key role in the settlement of human populations and the development of agriculture throughout the Western Hemisphere [7]. However, the successive introduction of different cattle breeds starting in the 19th century resulted in the progressive replacement of many Creole populations, which have completely disappeared in several regions or were displaced to marginal areas, where they still subsist nowadays [8]. Even though these extant populations present high levels of genetic diversity [9] and result from several centuries of adaptation to local environments, it is not clear how much of the ancestral Iberian founder contributions have been retained, or if the successive waves of other cattle introduced over the years have replaced the original contribution of Iberian stock.

The study of genetic diversity within and across breeds provides insight into population structure and relationships, and is essential for the development of conservation and breeding programs. Microsatellite genetic markers have been extensively used to assess between- and within-breed genetic diversity and inbreeding levels, introgression from other genetic groups, genetic differentiation and population structure [10–14]. The phylogeny of cattle has also been investigated with other types of genetic markers, including mtDNA [15], the non-recombining region of the Y chromosome [16] and single nucleotide polymorphisms [17–19] The insight on breed development and introgression provided by the different types of genetic markers is complementary, with neutral genetic markers such as microsatellites essentially reflecting the consequences of genetic drift, founder effects and population admixture. This is particularly important in the case of Creoles where founder effects and genetic drift must have been dramatic considering that the total number of Iberian cattle brought to the Americas was probably less than 1000 [4].

Knowing the genetic history of Creole cattle in the Americas should provide a better understanding of livestock gene flow during the period of discovery and settlement by Iberian colonizers, and the influence that may have resulted from the later introductions of cattle from other European origins and of Zebus from India that begun in the 19th century. In addition, the assessment of genetic diversity and structure of Creole cattle populations is crucial for the development of appropriate management programs aimed at their recognition, conservation and genetic improvement.

The objective of this study was to use neutral genetic markers to retrospectively assess the origins and evolutionary trajectories of American Creole cattle, and investigate the influence that Iberian, European and Zebu breeds may have had on their genetic make-up. The influence of African cattle to the Creole breeds is also discussed, particularly the indirect contribution mediated by their Iberian counterparts. Using a subset of 81 cattle breeds sampled in Europe and the Americas, we show that the majority of the Creole breeds still maintain distinct genetic signatures of Iberian cattle, but some have been admixed with cattle from other geographic regions, mostly of the Zebu type in tropical regions and British and Continental breeds in other parts of the Americas.

Results

Genetic Diversity and Breed Differentiation

A set of 19 microsatellite markers was used to analyze samples of the 81 cattle breeds included in this study (Table S1), which represented the Creole (27 breeds), Iberian (39 breeds), British (5 breeds), Continental European (4 breeds) and Zebu (6 breeds) groups, with the geographical distribution shown in Figure 1.

The microsatellite markers used allowed the detection of a mean number of 6.78±1.88 alleles/locus per breed and 11.93±3.52 alleles/locus per breed group, with global observed and expected heterozygosities of 0.688±0.018 and 0.711±0.025, respectively (Table S2). Taken together, Creole cattle showed the highest mean (14.21±3.74) and effective (4.08±0.57) number of alleles, allelic richness (4.69±0.51), and observed and expected heterozygosities (0.719±0.004 and 0.805±0.014, respectively), when compared with the other breed groups (Table 1).

The average F-statistics and their 95% confidence intervals obtained with 10,000 bootstraps over loci were f = 0.0326 (0.0231–0.0451), F = 0.1360 (0.1250–0.1479) and <theta> = 0.1069 (0.0977–0.1170). The group means for within-breed deficit in heterozygosity were highest for the Spanish and Zebu breeds (nearly 0.048), and lowest for the Continental European breeds (−0.002±0.026). The Portuguese Mertolenga and Brava, the Spanish Negra Andaluza and the Mexican Criollo Poblano had the highest within-breed F_{IS}, with estimates close to 0.11 (Table S2).

Genetic distances among breed pairs, estimated by <theta> values, ranged from 0.01 to 0.33 (results not shown for individual breeds), with a mean distance of Creoles relative to other breed groups as follows: 0.016 for Spanish, 0.018 for Portuguese, 0.023 for Continental European, 0.033 for British and 0.095 for Zebu

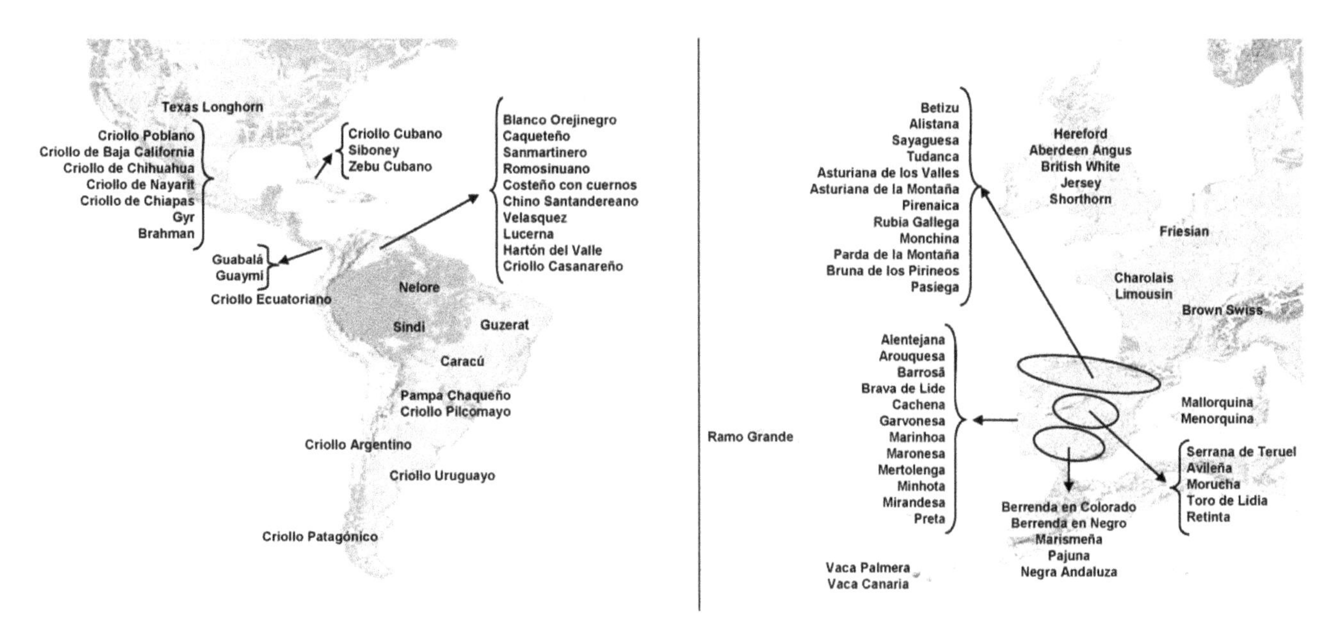

Figure 1. Geographic distribution of the 81 cattle breeds from America and Europe.

breeds (Table 2). The estimated number of migrants, i.e., the number of individuals exchanged between populations per generation that would balance the diversifying effect of genetic drift, was highest for the Creole, Spanish and Portuguese pairs, while the Zebu had the lowest number of migrants relative to all the other groups.

The results from the Factorial Correspondence Analysis (Figure 2) indicate that the first three FCA axes explain <25% of the variability. The first axis accounts for about 16% of the variability and separates *Bos indicus* from the remaining breeds. The second component, which accounts only for 5% of the variability, essentially separates the Iberian and the European breeds, while the third component accounts for a small 4% of the total variability and allows for some splitting among breeds in the same genetic or geographical group. The Creoles occupy a more central position in the graph and, depending on the breed considered, they have a closer proximity to the Iberian, European or Indicine clusters, reflecting the influence that these groups have had in their genetic make-up.

The AMOVA results indicated that the highest percentage of variation among groups (11.4%, P<0.001) was found when breeds deriving from *B. taurus* and *B. indicus* were compared (results not shown). When the genetic differentiation of Creoles relative to Iberian, European and Zebu breeds was considered, the largest amount of variability was found between Creole and Zebu

Table 1. Genetic variability estimated for different groups of cattle breeds.

Breed Group	N	Am ± SD	Ae	Ar	Ho ± SD	He ± SD
Creole	907	14.21±3.74	4.08	4.69	0.719±0.004	0.805±0.014
Spanish	1,199	12.53±3.39	3.85	4.46	0.677±0.003	0.777±0.018
Portuguese	675	10.74±3.59	3.62	4.27	0.677±0.004	0.749±0.025
British	200	8.89±2.21	3.26	4.41	0.653±0.008	0.754±0.015
Continental European	184	9.89±3.45	3.95	4.13	0.720±0.008	0.760±0.020
Zebu	168	11.32±3.16	3.31	4.41	0.654±0.009	0.735±0.026
Mean		*11.93±3.52*	*3.68±0.34*	*4.40±0.19*	*0.683±0.030*	*0.763±0.025*

Number of individuals sampled (N), mean number of alleles (Am), effective number of alleles (Ae), allelic richness (Ar), observed (H_o) and expected (H_e) heterozygosities and their standard deviations (SD). Groups of breeds: ***CREOLE***: Criollo Argentino (CARG), Criollo Patagónico (PAT), Caracú (CAR), Blanco Orejinegro (BON), Caqueteño (CAQ), Criollo Casanareño (CC), Chino Santandereano (CH), Costeño con Cuernos (CCC), Hartón del Valle (HV), Lucerna (LUC), Romosinuano (RMS), Sanmartinero (SM), Velasquez (VEL), Cubano (CUB), Siboney (SIB), Criollo Ecuatoriano (EC), Criollo de Baja California (CBC), Criollo de Chiapas (CHI), Criollo de Chihuahua (CHU), Criollo de Nayarit (CNY), Criollo Poblano (CPO), Guabalá (GUA), Guaymí (GY), Pampa Chaqueño (PA), Criollo Pilcomayo (PIL), Criollo Uruguayo (CUR) and Texas Longhorn (TLH); ***SPANISH***: Alistana (ALS), Asturiana de las Montañas (ASM), Asturiana de los Valles (ASV), Avileña (AVI), Berrenda en Colorado (BC), Berrenda en Negro (BN), Betizu (BET), Bruna de los Pirineos (BRP), Mallorquina (MALL), Menorquina (MEN), Monchina (MON), Morucha (MOR), Marismeña (MAR), Negra Andaluza (NAN), Pajuna (PAJ), Parda de Montaña (PM), Pasiega (PAS), Pirenaica (PIRM), Retinta (RET), Rubia Gallega (RGA), Sayaguesa (SAY), Serrana de Teruel (STE), Toro de Lidia (TL), Tudanca (TUD), Vaca Canaria (VCA) and Vaca Palmera (PAL); ***PORTUGUESE***: Alentejana (ALT), Arouquesa (ARO), Barrosã (BARR), Brava de Lide (BRAV), Cachena (CACH), Garvonesa (GARV), Marinhoa (MARI), Maronesa (MARO), Mertolenga (MERT), Minhota (MINH), Mirandesa (MIRA), Preta (PRET) and Ramo Grande (RG); ***BRITISH***: Aberdeen Angus (AA), British White (BWC), Hereford (HER), Jersey (JER), Shorthorn (SH); ***CONTINENTAL EUROPEAN***: Charolais (CHAR), Friesian (FRI), Limousin (LIM), Brown Swiss (BSW); ***ZEBU***: Brahman (BRH), Gyr (GYR), Guzerat (GUZ), Nelore (NEL), Sindi (SIN), Zebu Cubano (CUZ).

Table 2. Genetic distances among breed groups.

	CRE	SP	PT	BR	EU	ZEB
CRE		0.016	0.018	0.033	0.023	0.095
SP	15.80	-	0.013	0.036	0.020	0.135
PT	13.78	18.51	-	0.041	0.027	0.143
BR	7.43	6.66	5.79	-	0.032	0.159
EU	10.43	11.96	9.12	7.57	-	0.156
ZEB	2.38	1.60	1.50	1.33	1.35	-

Genetic distances estimated by Weir and Cockerham <theta> (above diagonal) and corresponding number of migrants (below diagonal). Breed groups: CRE – Creole; SP – Spanish; PT – Portuguese; BR – British; EU – Continental European; ZEB –Zebu. See Table 1 for the definition of breeds included in each group.

populations (9.15%, P<0.001), and the lowest between the Creole and Iberian breeds (1.09%, P<0.001).

Population Genetic Structure

The Neighbor-net built with the Reynolds distances (Figure 3) supports the existence of two major clusters, corresponding to *B. indicus* and *B. taurus* breeds, with several Creole breeds grouped in the *B. indicus* cluster, which is interpreted as a sign of Zebu influence in their genetic make-up. These included the Creoles from Cuba and Ecuador, the Pilcomayo from Paraguay, the Criollo de Chiapas from Mexico and some Creole breeds from Colombia (Chino Santandereano, Caqueteño and Criollo Casanareño). Among the Creoles showing a residual zebu influence, the Texas Longhorn and the majority of the Mexican Creoles were closely clustered at the centre of the dendrogram, displaying a common origin with the Spanish Marismeña. Another Creole cluster, made-up by the Romosinuano and Costeño con Cuernos from Colombia and, to a lesser extent, the two breeds from Panama, showed a common origin with the breeds from the Canary Islands and the Portuguese Mertolenga. The Creoles from Argentina and Uruguay and the Caracu from Brasil formed an independent cluster at the center of the dendrogram, with a weak relationship with British breeds. On the other hand, the Pampa Chaqueño from Paraguay and the Harton del Valle, Lucerna and Blanco Orejinegro from Colombia showed a clear influence of British breeds.

Among Iberian breeds, several different clusters could be identified, such that nearly all Portuguese breeds grouped together, with the major exception of the Mirandesa, which clustered with breeds with a close geographic distribution, both in Portugal and Spain. Another cluster corresponded to the breeds from the Balearic Islands, which grouped with a few breeds from northern Spain, while the majority of the Spanish breeds clustered together. A distinct cluster corresponded to the breeds from the Canary Islands, which also included the Portuguese Mertolenga. Two Spanish breeds were isolated from the remaining clusters, i.e., the Marismeña and the Berrenda en Negro. The remaining Iberian breeds (Minhota and Ramo Grande from Portugal, Bruna de los Pirineos, Serrana de Teruel and Parda de Montaña from Spain) were close to Continental European breeds, indicating some admixture with these breeds.

The Bayesian clustering model-based method [20] allowed for assessment of the genetic structure and admixture among breeds. When the number of ancestral populations varied from K = 2 to 81, the largest change in the log of the likelihood function (ΔK) was when K = 71 (Figure S1).

The results for K = 2 (Figure 4) indicate a clear separation between *B. indicus* and *B. taurus* breeds. Moreover, these results confirm the admixture of Zebu with some of the Creole breeds, especially Siboney, Criollo Cubano, Criollo Ecuatoriano, Pilcomayo, Casanareño and Velasquez, while other breeds, such as the Creoles from Argentina and Uruguay, and the Romosinuano, Sanmartinero and Blanco Orejinegro from Colombia, show minor signs of Zebu admixture.

When three ancestral populations were inferred, the breeds from Northern Spain and the Portuguese Mirandesa and Marinhoa, separated from the remaining *B. taurus* breeds, whereas the other breeds from Portugal and Southern Spain remained clustered with the Creole breeds. As the number of inferred ancestral populations increased, admixture among breeds became more apparent, but some Creole breeds, such as the two Argentinean and the Uruguayan Creoles, Caracú from Brazil, Texas Longhorn, Creoles of Baja California and Poblano from Mexico, and the Romosinuano and Costeño con Cuernos from Colombia, remained very homogeneous at K = 8. For the 81 cattle breeds analysed, the most likely number of inferred ancestral populations was K = 71 (Figure S2), as assessed by the method of Evanno et al. (2005). The computed individual membership coefficients resulted in about 60–70% of the individuals classified within their source ancestral population, assuming a threshold of $q>0.8$. The Zebu breeds Brahman, Guzerat, Gyr, Nelore and Sindi grouped together in the same cluster with values of q around 0.700 while Cuban Zebu grouped in the same cluster with Criollo Cubano. The Mexican Creoles, with the exception of the Criollo de Chiapas, clustered together, in the same way that the Creoles from Colombia Chino Santandereano, Velasquez, Casanare and Caqueteño formed a unique cluster, although with low q values (Table S3).

Ancestral Genetic Contributions to Creole Cattle

The estimated genetic contributions of each potential ancestral breed group (Iberian, British, Continental European and Zebu) to Creole cattle are shown in Figure 5 and Table S4, as computed by the likelihood estimation of admixture proportions developed by Wang [21], and implemented by the LEADMIX software. The admixture estimates indicate that, for the Creole cattle considered as a single group, Iberian cattle contributed nearly 62% to the genetic pool, Zebu breeds contributed about 17% and Continental European and British breeds about 10% each.

The Neighbor-net indicated the existence of various Creole clusters, which is also supported by the analysis carried-out with STRUCTURE. These clusters likely reflect different contributions from the ancestral genetic groups to the current genetic pool of Creoles. Therefore, a similar analysis of estimated genetic contributions was carried out with LEADMIX for each of the five identified Creole clusters, as shown in Figure 5. These analyses revealed clear differences among the five clusters in the relative contributions of the four parental genetic groups. The Creoles from Panama, Mexico, United States and some Colombian breeds (Clusters 1 and 2) showed the strongest Iberian influence, with nearly 70 to 80% of the genetic pool contributed by Iberian breeds, with the remaining contributions corresponding to Continental and Zebu breeds, in about equal proportions. The Creoles from the southern region of the Americas (Cluster 3) had an important influence of about 60% from Iberian breeds, but also showed influence from British cattle. Cluster 4, which corresponds to Creole breeds widely dispersed in tropical areas, showed an important contribution from Zebu breeds, even though the Iberian contribution was still predominant. The Paraguayan and Colombian breeds included in Cluster 5 show a major influence of

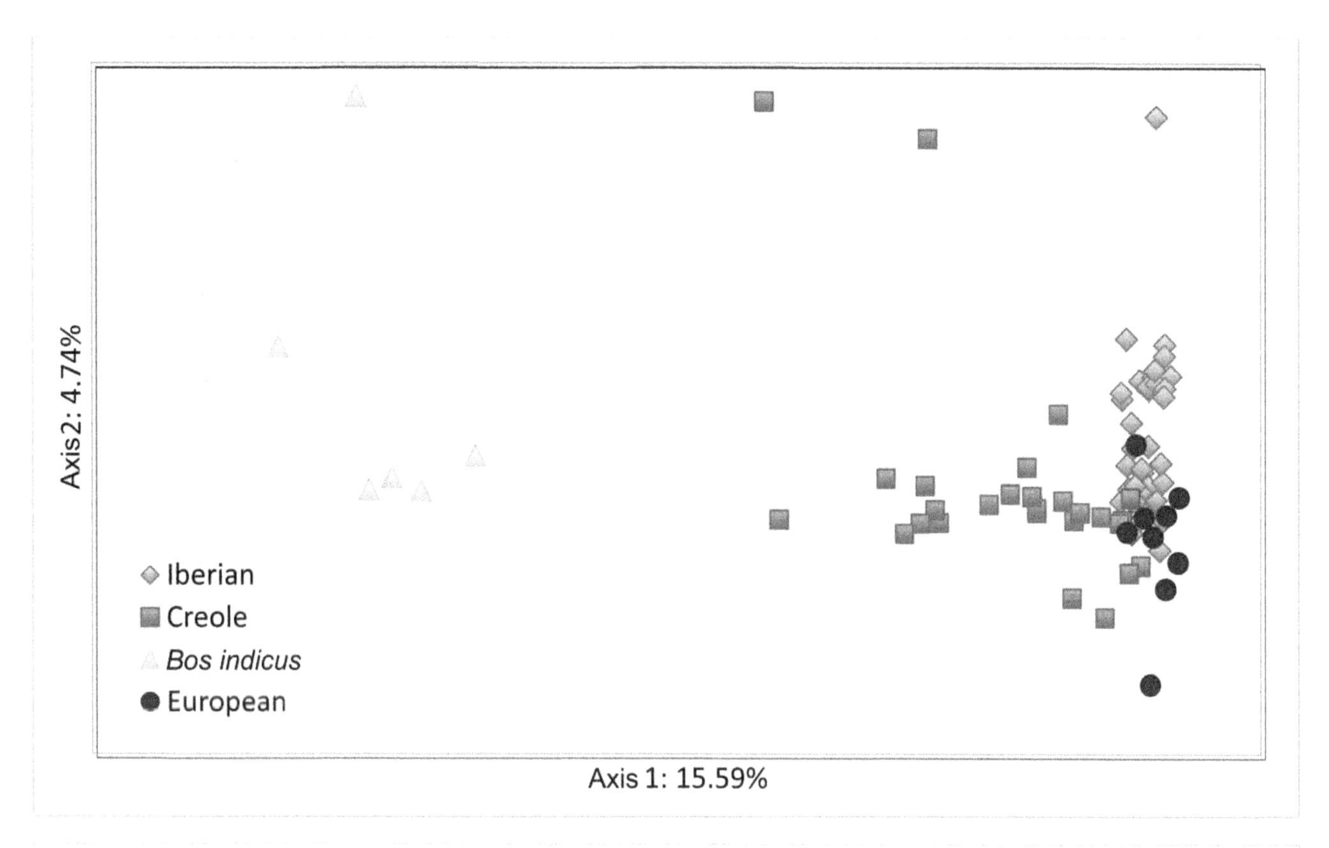

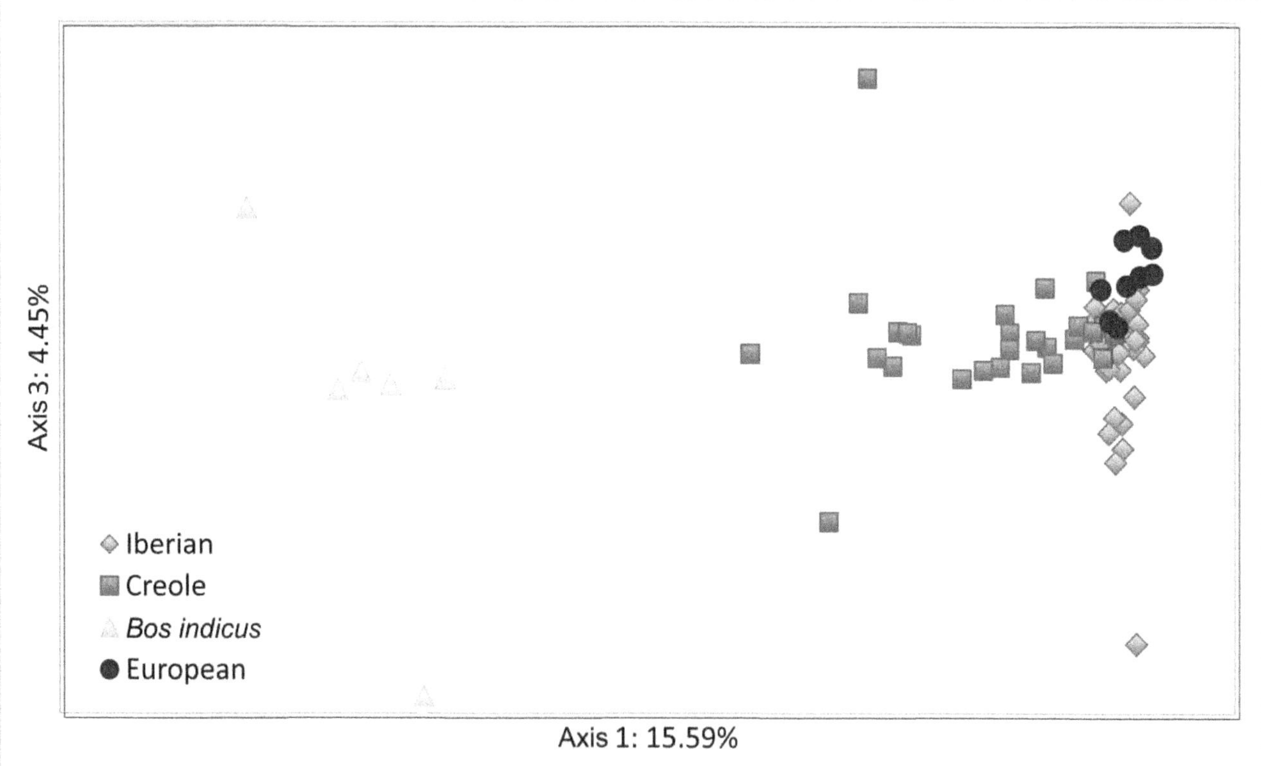

Figure 2. Graphical representation of the three first axes from the factorial correspondence analysis of the 81 cattle breeds from America and Europe.

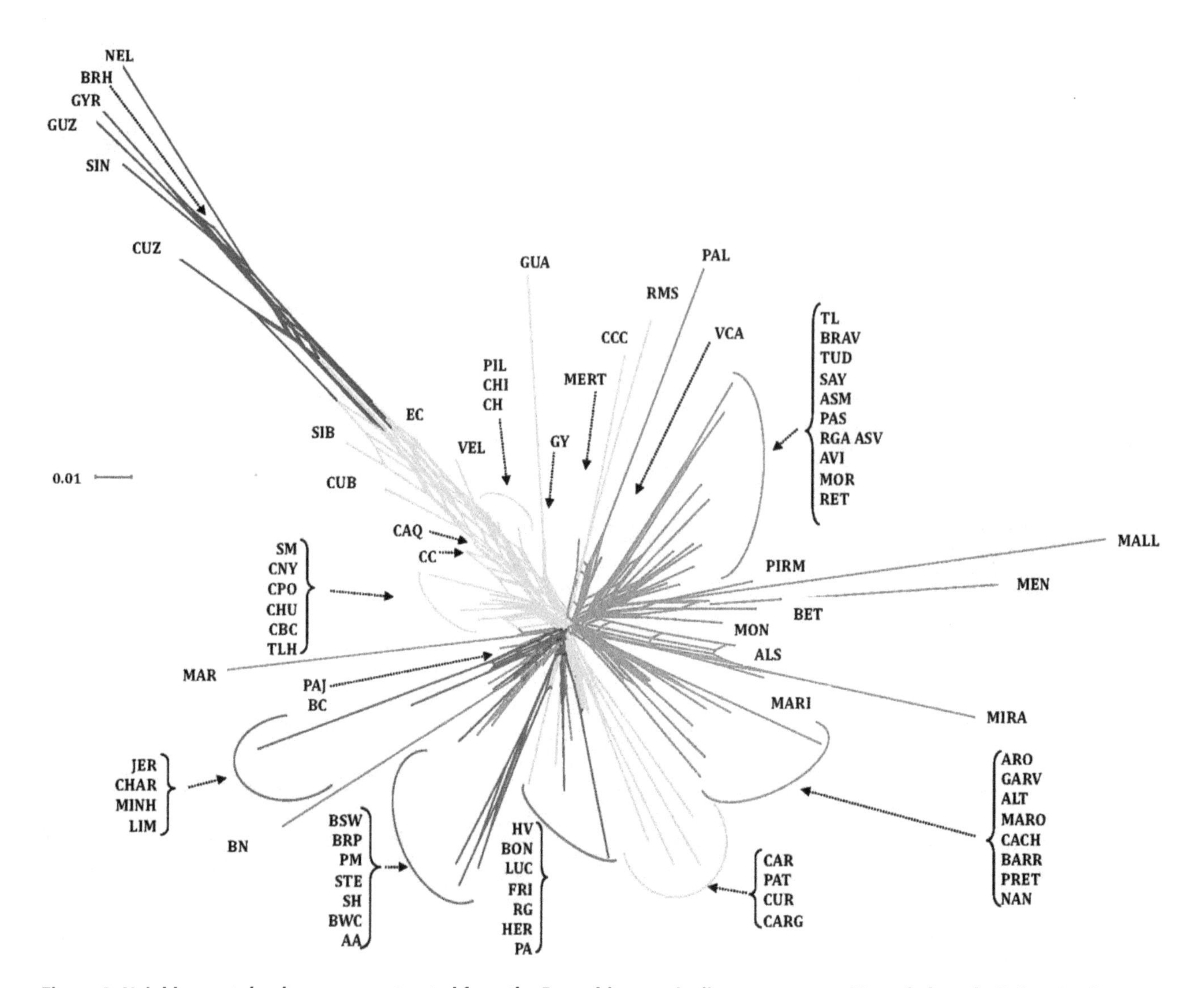

Figure 3. Neighbor-net dendrogram constructed from the Reynolds genetic distances among 81 cattle breeds. Yellow: Creole; Green: Iberian; Pink: British and Continental European; Blue: Indian Zebu. ***SPANISH.*** Betizu (BET), Toro de Lidia (TL), Menorquina (MEN), Alistana (ALS), Sayaguesa (SAY), Tudanca (TUD), Asturiana de los Valles (ASV), Asturiana de las Montañas (ASM), Retinta (RET), Morucha (MOR), Avileña (AVI), Pirenaica (PIRM), Rubia Gallega (RGA), Mallorquina (MALL), Monchina (MON), Serrana de Teruel (STE), Parda de Montaña (PM), Bruna de los Pirineos (BRP), Pasiega (PAS), Berrenda en Colorado (BC), Berrenda en Negro (BN), Marismeña (MAR), Pajuna (PAJ), Negra Andaluza (NAN), Vaca Canaria (VCA), Vaca Palmera (PAL); ***PORTUGUESE.*** Alentejana (ALT), Arouquesa (ARO), Barrosã (BARR), Brava de Lide (BRAV), Cachena (CACH), Garvonesa (GARV), Marinhoa (MARI), Maronesa (MARO), Mertolenga (MERT), Minhota (MINH), Mirandesa (MIRA), Preta (PRET), Ramo Grande (RG); ***CREOLE.*** Guabalá (GUA), Guaymí (GY), Texas Longhorn (TLH), Criollo Poblano (CPO), Criollo de Baja California (CBC), Criollo de Chihuahua (CHU), Criollo de Nayarit (CNY), Criollo de Chiapas (CHI), Blanco Orejinegro (BON), Caqueteño (CAQ), Sanmartinero (SM), Romosinuano (RMS), Costeño con Cuernos (CCC), Chino Santandereano (CH), Velasquez (VEL), Lucerna (LUC), Hartón del Valle (HV), Criollo Casanareño (CC), Criollo Ecuatoriano (EC), Criollo Uruguayo (CUR), Pampa Chaqueño (PA), Criollo Pilcomayo (PIL), Criollo Argentino (CARG), Criollo Patagónico (PAT), Caracú (CAR), Cubano (CUB), Siboney (SIB); ***ZEBU:*** Gyr (GYR), Brahman (BRH), Sindi (SIN), Guzerat (GUZ), Nelore (NEL), Zebu Cubano (CUZ); Other ***EUROPEAN.*** Friesian (FRI), Hereford (HER), Brown Swiss (BSW), Aberdeen Angus (AA), British White (BWC), Charolais (CHAR), Jersey (JER), Limousin (LIM), Shorthorn (SH).

British and Continental breeds, with a smaller but still detectable contribution of Iberian cattle.

Discussion

The genetic relationships between Creole cattle and their presumed ancestral sources remain largely unexplored. Estimates of genetic diversity and population structure have been previously reported for some Creole populations [9,22–26], for Iberian cattle [13,27–30], Zebu breeds [31,32] and European cattle [10,33]. Moreover, mtDNA and Y-chromosome markers were used to investigate the origins of Creole cattle [34–38], but their genetic relationship with other cattle breeds which could have influenced them remained unclear.

Our study combines several data sets that cover a wide range of Creole, European and Indicine cattle populations, thus providing a more comprehensive insight about the genetic influences that Creole breeds received since the arrival of the first Iberian cattle in the American continents in the late 1400's.

Of the total genetic variability, nearly 11% is explained by breed differences, which is slightly higher than what has been reported for other cattle breeds around the world, generally in the range of 7 to 9% [10,28,38]. This could be justified by the inclusion in this study of cattle breeds representing the two well differentiated phylogenetic groups of *B. indicus* and *B. taurus* [39].

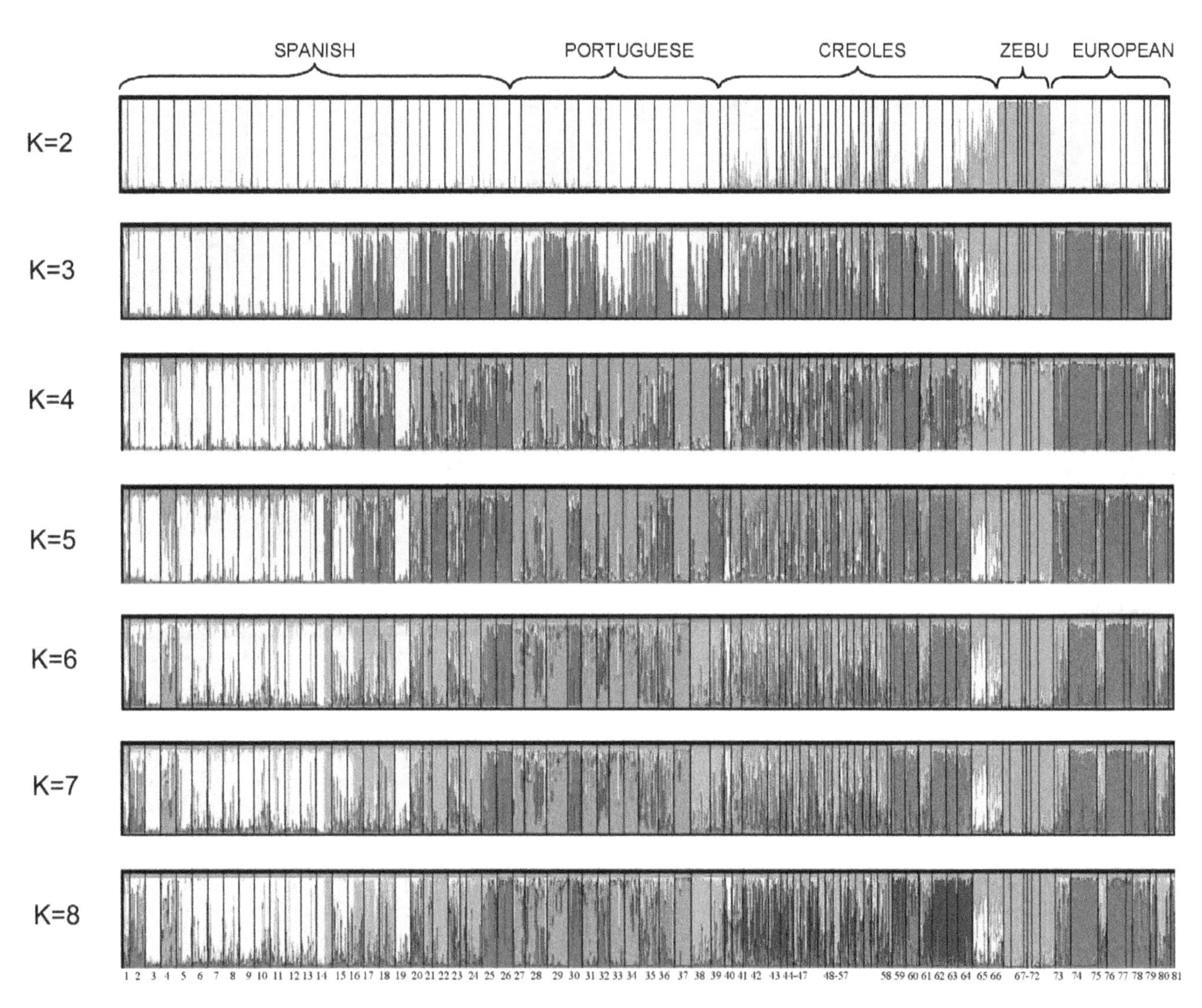

Figure 4. Population structure of 81 cattle breeds based on 19 microsatellite loci using STRUCTURE. Graphical representation of individual genotype membership coefficients (q) when K = 2 to K = 8. Each animal is represented by a single vertical line divided into K colours, where K is the number of clusters assumed and the coloured segment shows the individual's estimated membership proportions in that cluster. Black lines separate the populations. ***SPANISH.*** 1: Betizu (BET), 2: Toro de Lidia (TL), 3: Menorquina (MEN), 4: Alistana (ALS), 5: Sayaguesa (SAY), 6: Tudanca (TUD), 7: Asturiana de los Valles (ASV), 8: Asturiana de las Montañas (ASM), 9: Retinta (RET), 10: Morucha (MOR), 11: Avileña (AVI), 12: Pirenaica (PIRM), 13: Rubia Gallega (RGA), 14: Mallorquina (MALL), 15: Monchina (MON), 16: Serrana de Teruel (STE), 17: Parda de Montaña (PM), 18: Bruna de los Pirineos (BRP), 19: Pasiega (PAS), 20: Berrenda en Colorado (BC), 21: Berrenda en Negro (BN), 22: Marismeña (MAR), 23: Pajuna (PAJ), 24: Negra Andaluza (NAN), 25: Vaca Canaria (VCA), 26: Vaca Palmera (PAL); ***PORTUGUESE.*** 27: Alentejana (ALT), 28: Arouquesa (ARO), 29: Barrosã (BARR), 30: Brava de Lide (BRAV), 31: Cachena (CACH), 32: Garvonesa (GARV), 33: Marinhoa (MARI), 34: Maronesa (MARO), 35: Mertolenga (MERT), 36: Minhota (MINH), 37: Mirandesa (MIRA), 38: Preta (PRET), 39: Ramo Grande (RG); ***CREOLE.*** 40: Guabalá (GUA), 41: Guaymí (GY), 42: Texas Longhorn (TLH), 43: Criollo Poblano (CPO), 44: Criollo de Baja California (CBC), 45: Criollo de Chihuahua (CHU), 46: Criollo de Nayarit (CNY), 47: Criollo de Chiapas (CHI), 48: Blanco Orejinegro (BON), 49: Caqueteño (CAQ), 50: Sanmartinero (SM), 51: Romosinuano (RMS), 52: Costeño con Cuernos (CCC), 53: Chino Santandereano (CH), 54: Velasquez (VEL), 55: Lucerna (LUC), 56: Hartón del Valle (HV), 57: Criollo Casanareño (CC), 58: Criollo Ecuatoriano (EC), 59: Criollo Uruguayo (CUR), 60: Pampa Chaqueño (PA), 61: Criollo Pilcomayo (PIL), 62: Criollo Argentino (CARG), 63: Criollo Patagónico (PAT), 64: Caracú (CAR), 65: Cubano (CUB), 66: Siboney (SIB); ***ZEBU.*** 67: Gyr (GYR), 68: Brahman (BRH), 69: Sindi (SIN), 70: Guzerat (GUZ), 71 Nelore (NEL), 72: Zebu Cubano (CUZ); ***BRITISH AND CONTINENTAL EUROPEAN.*** 73: Friesian (FRI), 74: Hereford (HER), 75: Brown Swiss (BSW), 76: Aberdeen Angus (AA), 77: British White (BWC), 78: Charolais (CHAR), 79: Jersey (JER), 80: Limousin (LIM), 81: Shorthorn (SH).

The high genetic variability found in Creole cattle, even in populations considered as endangered, might reflect recent contributions of cattle from different origins, which are known to have been admixed with some Creole populations over the last century [40,41]. This result is in agreement with the analysis of mtDNA sequences and Y haplotypes, which have shown the genetic heterogeneity of Creole cattle, in which signatures of Iberian, European and Indian cattle are detected, and the direct influence of African cattle has also been claimed [35,37,42].

Recently, Gautier and Naves (2011) [43] used a high-density panel of SNPs to study genetic influences in Creole cattle from Guadeloupe and reported evidence of a direct African ancestry in this breed. In our study, no African samples were included, but some results may be interpreted as indicating a possible African influence on some Creole populations. For example, allele 123 in the BM2113 locus has previously been associated with West African taurine cattle [44], and is present at high frequencies in some Creole populations such as Caqueteño, Sanmartinero and

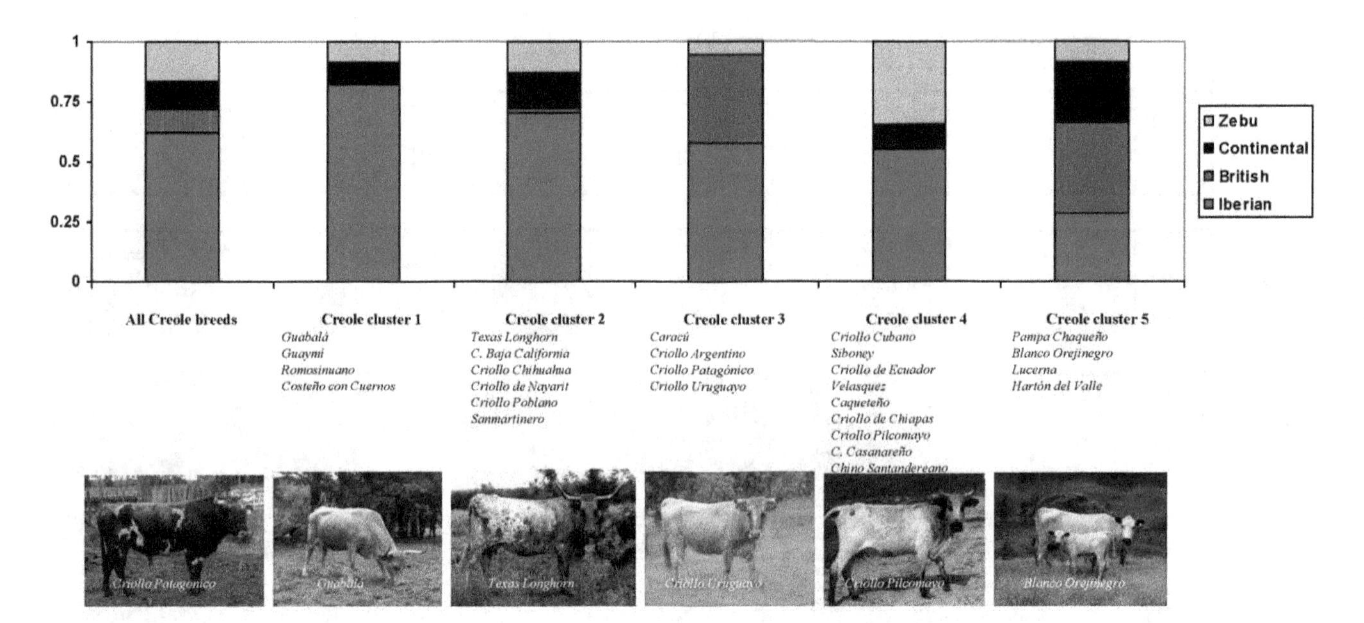

Figure 5. Genetic contributions from Iberian, British, Continental European and Zebu breeds to Creole cattle. Graphical representation of maximum-likelihood estimates of proportional genetic contributions form some groups of breeds to Creole cattle considered as a whole or grouped in five different clusters. The Creole breeds included in each cluster are listed below, and photos of animals representative of each cluster are also shown.

Pilcomayo. On the other hand, allele 143 at the same locus has been considered an indication of African Zebu influence [44], and it is present in Caracú, which could be regarded as further evidence of African influence on Creoles. It is not clear, however, if these African signatures correspond to a direct contribution of African cattle to Creoles, or rather to an indirect influence through Iberian cattle, given that some Iberian breeds in our study also have a high frequency of alleles considered to be African-specific [30]. Also, it has been suggested that the Zebu influence detected in Creoles may be in part due to *B. indicus* cattle imported from Africa during the colonial period [43]. Studies where mtDNA sequence variation was analyzed have confirmed the presence of African matrilines among Creoles [37,42,45], and Y-haplotypes have also revealed a possible West African signature in Creole cattle [35]. Overall, our results provide strong support to the conclusion of an Iberian influence on Creole cattle, but are not so clear in elucidating the possibility of a direct African influence on the different Creole groups. Further studies are necessary, covering a broad sample of African cattle breeds and a combination of different genetic markers, to clarify the African influence on Creoles.

The degree of genetic differentiation among all breeds studied indicates relatively low levels of gene flow and some level of reproductive isolation among most Creole breeds, probably as a result of geographic separation and differentiation. Taken together, Creole breeds differ more from *B. indicus* than from the remaining *B. taurus* breeds, with the lowest levels of differentiation for the Creole-Spanish and Creole-Portuguese pairs. Our results indicate that Creole cattle retain genetic signatures of their Iberian ancestry, in agreement with previous studies based on monoparental genetic markers [35].

The F-statistics, AMOVA and Factorial Correspondence Analysis results confirmed the closer proximity of Creoles to Iberian breeds, and their much larger differentiation from Zebu cattle. Nevertheless, Indian Zebus have had an important influence on some Creole breeds, as is clear from the Reynolds Neighbor-net, the Bayesian approach adopted by STRUCTURE and the maximum-likelihood estimation of genetic contributions from parental populations carried-out with LEADMIX. However, the majority of Creoles seem to have been largely unaffected by the introduction of Zebus into South America in the 19th and 20th centuries. In general, Creole breeds from tropical areas (Siboney, Criollo Cubano, Criollo Ecuatorino, Criollo de Chiapas and some Colombian breeds), showed the highest degree of admixture with Zebu, but this influence extended as far south as the Criollo Pilcomayo from Paraguay.

The present study indicates that the influence of Iberian cattle was mostly due to the breeds from Portugal and Southern Spain, which had a closer relationship with Creoles, as detected in the Bayesian analysis with STRUCTURE. In the period when Central and South America were settled, cattle breeds from Portugal and Spain were probably not very distinct from each other, given that the major period of breed formation started in the late 18th century [46]. Also, Portugal and Spain were united in 1580 under Philip II, King of Spain, and kept together until 1640, and this corresponded to a period of major livestock shipments from the Iberian Peninsula to South America [48]. It is also known that the vast majority of the expeditions to South America departed from Lisbon (central Portugal) or from Cádiz and Seville (Southern Spain), which would explain the closer relationship of Creoles with cattle breeds from these regions. The shipment of cattle directly from Africa to the Americas could have occurred following slave routes and from intermediate ports in the Atlantic Islands.

The establishment of Iberian livestock in the "New World" followed two different migration routes, depending on the predominance of colonizers being Spanish or Portuguese. The first arrival of cattle was in 1493, when Columbus brought animals from Spain to the Caribbean Islands in his second trip, and the presence of hundreds of animals in Cuba was reported four decades later [3]. Cattle from the Caribbean Islands, e.g., Cuban Creole, could thus be considered a remnant of the first cattle brought from Spain, but it is widely recognized that Zebu cattle have been extensively used in crossbreeding with Caribbean cattle over the last century [4,40]; it is, therefore, difficult to identify a specific breed from this area as a good representative of the early Iberian stock.

From the Caribbean Islands, where animals were stocked and bred, cattle were brought to North America through the Mexican port of Veracruz, and from there they expanded throughout Mexico and towards the region corresponding to Texas. The Texas Longhorn and the different Mexican Creole populations could be regarded as representing this path of cattle dispersion.

An additional route of dispersion of animals from the Caribbean was into Central America through the ports of Panama or directly into the northern part of South America, through the port of Santa Marta in Colombia [47]. From these places, they were then distributed to the Northern Peruvian Vice Kingdom (today, Colombia, Ecuador and Peru). The Panamanian and some of the Colombian breeds could be considered as representatives of this path of cattle dispersion.

The Rio de la Plata was an important route of distribution of Iberian cattle in Southern South America, often disseminated by Jesuit missions which had a strong influence in this area [48]. From Rio de la Plata cattle dispersed through most of South America, from Patagonia up to the southern part of the Peruvian Vice Kingdom. In these areas, the Spanish path of dispersion was probably mixed with the Portuguese route, in which cattle were shipped from Portuguese ports into the Cape Verde Archipelago and from there to the Captaincies on the Brazilian coast, mainly to Pernambuco and Bahia in the North, and São Vicente (near Rio de Janeiro) in the South [48]. The southern part of Brazil, Uruguay and Argentina have milder climate, and the Zebu influence was probably less severe than in tropical Central America. Several breeds from Southern South America included in our studies confirm lower levels of Zebu admixture, including the Argentinean, Patagonian and Uruguayan Creoles as well as the Brazilian Caracú, even though the latter may have suffered some Zebu introgression in the past [35]. Nevertheless, it is known that British cattle, especially Hereford and Angus, were introduced and widely expanded in this region in the mid-19th century, and could have had some influence on Creole populations.

Our study provides evidence that Creole breeds still show important influences of Iberian cattle, which contributed with nearly two-thirds of the Creole genetic pool analyzed. Our results further indicate that large genetic differences exist among Creole sub-populations, reflecting the effects of genetic drift as well as the introduction of other breeds through the years after the initial arrival of Iberian cattle. The Neighbor-net representation of the pairwise Reynolds genetic distances generally supports the existence of five Creole clusters, which may correspond to different paths of cattle dispersion into the Americas or, in some cases, perhaps to a more recent admixture of germplasm from other breeds.

The first cluster corresponds to the Panamanian (Guabalá and Guaymi) and some of the Colombian breeds (Costeño con Cuernos and Romosinuano), and could represent the first route of cattle introduction into Central and South America from the Caribbean Islands, as the flow of ships between Panama and Colombia was very important in the 16th century [48]. This cluster presents the highest Iberian and possibly African influences, with minor contributions from Continental and Zebu breeds. Interestingly, a common origin was detected between the Costeño con Cuernos and Romosinuano from Colombia and the breeds from the Canary islands, lending support to the important role played by this archipelago as a point of shipment and reload of animals taken to the Americas [3,48].

The second cluster is represented by most of the Mexican breeds, the Colombian Sanmartinero and the Texas Longhorn, which are near the centre of the radial net, and could correspond to one of the first paths of cattle dispersion into Central and North America. This cluster still shows a major signature of Iberian cattle, with minor influences from Continental and Zebu breeds. A very interesting result in our study was the detection of a common origin shared by this group of breeds and the Spanish Marismeña. It is generally believed that, since Columbian times, the Marismeña has been kept since Colombian times in semi-feral conditions in a natural park near the original point of departure of Spanish sailors [49]. Our results confirm that it could represent a remnant of the animals taken to the Americas in the early period of settlement.

The third cluster contains the Brazilian Caracú and the Argentinean, Patagonian and Uruguayan Creoles. This group is likely a representative of the Rio de la Plata and Brazilian routes of colonization and cattle flow into South America. Even though the Iberian contribution is very clear in this group, the later introduction of British cattle in the region probably resulted in some admixing, which is now detectable in the genetic pool of this cluster, particularly in the Uruguayan Creoles.

The fourth cluster includes breeds with a widely dispersed geographical distribution, such as the Creoles from Cuba and Ecuador, the Velasquez, Caqueteño, Casanareño and Chino Santandereño from Colombia, the Chiapas Creole from Southern Mexico and the Pilcomayo from Paraguay. All these breeds still show a major Iberian contribution but also a strong Zebu influence, confirming the impact of *B. indicus* on the genetic make-up of Creole cattle in many tropical areas during the last century.

The fifth cluster includes three Colombian breeds (Blanco Orejinegro, Lucerna and Harton del Valle) and the Paraguayan Pampa Chaqueño. This cluster shows some proximity with British and Continental breeds, and a minor, but still detectable, representation of the Iberian contribution.

Our results generally support the historical descriptions of cattle introduction and dispersion throughout the Americas [3,48]. Nearly 500 years after their arrival, strong genetic signatures of Iberian cattle are still present in Creoles. The major ancestral contributions are from breeds of Southern Spain and Portugal, in agreement with the historical ports of departure of ships sailing towards the Western Hemisphere. Furthermore, the role of the Canary Islands in the flow of cattle to the Americas was confirmed.

Even though the term "Creole" has been used since early colonial times in Latin America in reference to both people and animals born in the newly-discovered land from parents of Iberian origin [7], it is clear that there is more genetic variability among Creole cattle in comparison to breeds from other geographic regions. This diversity results from differential genetic contributions from several parental populations, genetic drift and some admixing with other breeds over time. Based on the information derived from our study, it is possible to summarize the gene flow that gave origin to the different Creole populations, confirming the influence of Iberian, British, Continental and Zebu breeds.

The major feature that should be retained is the predominant influence of Iberian cattle on Creoles still present today. Signatures of African cattle are also represented in many Creole breeds, and which can result from either direct contributions or indirect influences through Iberian cattle. Evidence of admixing with British breeds is visible in some Colombian and Paraguayan breeds and, to a lesser extent, in Creoles from southern South America. Creoles from tropical areas, especially those from the

Caribbean, show clear admixture with Zebu, which contribute high tolerance to hot and humid climates, and resistance to parasites. Some Creole populations still show a close proximity to their distant Iberian ancestors, and efforts should be made to avoid their extinction or further genetic erosion through admixture with other breeds, which would compromise five centuries of selective adaptation to environmental conditions which range from the deserts of Texas and Mexico, to the mountains of Patagonia.

Overall, our study indicates that: 1) several centuries after the introduction of Iberian cattle into America, Creole breeds still show strong and predominant signatures of Iberian influence; 2) Creole breeds differ widely from each other, both in their genetic structure and in the genetic influences received from other breeds; 3) in some Creole breeds, especially those from tropical regions, the impact of *B. indicus* is very clear, even though the Iberian influence is still prevalent; 4) a few Creole breeds from Colombia and Paraguay have a major influence from British and Continental breeds.

This study provides significant genetic information about cattle populations in the Americas that are remnants of historical colonization. Our findings reveal the evolutionary trajectories of cattle in close association with human dispersal and confirm Creoles as legitimate representatives of cattle from the discoveries. Furthermore, our results provide the means to identify the Creole breeds with different genetic signatures, which will be useful for the development of global and local conservation of cattle genetic diversity.

Materials and Methods

Samples

The study included biological samples of 3,333 animals representing 81 cattle breeds from 12 different countries (Table S1). The origin of the breeds studied (Figure 1) was either Creole (a comprehensive sample of 27 breeds, representing a wide range of Creole cattle, from North America to Patagonia), Iberian (39 native breeds from Portugal and Spain, including 3 breeds from the Atlantic Islands), European breeds (9 *B. taurus* breeds from the British Isles and Continental Europe which have been widely used throughout the world) and Zebu breeds (6 breeds representing the *B. indicus* group).

Semen samples were obtained from germplasm banks. Blood and hair root samples were collected by qualified veterinarians through their routine practice, in the framework of oficial programs aimed at the identification, health control and parentage confirmation of the breeds and populations included in our study. Therefore, the legal restrictions defined in "Spanish Law 32/2007 of November 7, on the care of animals in their husbandry, transportation, testing and sacrifice" do not apply, as they are waved in the case of non-experimental procedures and routine veterinary practices with livestock species, in Article 3d of the above-mentioned Law.

Molecular Markers

Six laboratories were involved in this study (Universidad de Córdoba, Universidad Complutense de Madrid and Universidad de Zaragoza from Spain, Instituto Nacional dos Recursos Biológicos from Portugal, University of California in Davis from the United States of America, and Universidad Nacional de Colombia in Palmira from Colombia).

A common set of 19 microsatellites were selected from a panel of 30 markers recommended for genetic diversity studies by the International Society for Animal Genetics (ISAG) / Food and Agriculture Organization of the United Nations (FAO) working group [50]: *BM1818, BM1824, BM2113, CSRM60, CSSM66, ETH3, ETH10, ETH185, ETH225, HAUT27, HEL9, ILSTS006, INRA032, INRA063, MM12, SPS115, TGLA53, TGLA122* and *TGLA227*.

DNA Amplification, and Genotyping

Genomic DNA was extracted using procedures previously described [13,29,51] The 19 microsatellite markers were amplified in multiplex polymerase chain reactions (PCRs) using fluorescence-labelled primers [13]. PCR products were separated by electrophoresis on ABI instruments (3730, 3130 and 377XL, Applied Biosystems, Foster City, CA) according to manufacturer recommendations and allele sizing was accomplished by using the internal size standards GeneScan™-500 LIZ™ and GeneScan-400HD ROX (Applied Biosystems, Warrington, UK).

Allele nomenclature was standardized following a former European research project on cattle genetic diversity (EU RESGEN CT 98–118, for further details on the project outcome Dr J. A. Lenstra has to be contacted: J.A.Lenstra@uu.nl). To assure compatibility of results from different equipments and laboratories, a total of 30 samples representing the entire allele range for this set of markers was exchanged and genotyped in all laboratories. Allele sizing was standardized across laboratories based on these reference samples. Moreover, reference samples (2) were included in each assay to control for variation between electrophoresis.

Statistical Analysis

Data used in this paper have been archived at Dryad (www.datadryad.org): doi:10.5061/dryad.17 gk0.

Mean number of alleles (Am), observed (Ho) and unbiased expected (He) estimates of gene diversity [52] and their standard deviations were obtained with the MICROSATELLITE TOOLKIT software [53]. Distribution of genetic variability within and between breeds was studied by analysing F-statistics [54] as implemented in GENETIX v4.04 [55]. The within-breed inbreeding coefficient (F_{IS}) was calculated with a 95% confidence interval obtained by 10000 bootstraps across loci. The effective number of alleles (Ae) and allelic richness (Ar) over all loci per breed were calculated with POPGENE [56] and FSTAT v. 2.9.3 [57], respectively. Deviations from Hardy–Weinberg equilibrium (HWE) were assessed with GENEPOP v. 3.4 software [58]. Both global tests across populations and loci as well as tests per locus per breed were carried-out using the method of Guo & Thompson (1992) [59] and the p-values were obtained using a Markov chain of 10000 dememorization steps, 100 batches, and 5000 iterations.

After defining groups of breeds by geographic origin and ancestry (i.e., Creole, Spanish, Portuguese, British, Continental European and Zebu), a hierarchical analysis of variance was performed to partition the total genetic variance into components due to inter-individual and inter-breed differences. Variance components were used to compute fixation indices and their significance was tested using a non-parametric permutation approach [60]. Computations were carried out using the AMOVA (Analysis of Molecular Variance) module of ARLEQUIN 3.01 [61].

Genetic divergence among breeds was estimated by calculating the Reynolds distances [62] with the POPULATIONS software [63]. A Neighbor-net was constructed with the Reynolds distances using SPLITSTREE 4 [64] to graphically represent the relationships between breeds and to depict evidence of admixture.

Factorial Correspondence Analysis [65] was performed using the function "AFC 3D sur populations" of GENETIX v4.04.

The STRUCTURE v.2.1 software [20] was used to investigate the genetic structure of the 81 cattle populations, in order to identify population substructure and admixture, and to assign individuals to populations. Runs of 10^6 iterations after a burn-in period of 300000 iterations were performed for each K to determine the most probable number of clusters, as inferred from the observed genotypic data. Ten independent simulations for K equal to 2 to 81 were performed, and the method of Evanno *et al.* (2005) [66] was used to identify the most probable K, by determining the modal distribution of ΔK. The DISTRUCT v.1.1 software [67] was used to obtain a graphical display of individual membership coefficients in each ancestral population, considering the run with the highest posterior probability of the data at each K value.

In order to assess the relative genetic contributions of breeds from different regions (Iberian, British, Continental European and Zebu) in the development of Creoles, a maximum likelihood estimation of admixture proportions was carried out with the LEADMIX software, following the principles described by Wang (2003) [21]. These analyses were conducted for the full group of Creoles, and for five different Creole clusters, as revealed by the Reynolds genetic distances and the corresponding dendrogram.

Supporting Information

Figure S1 Graphical representation of ΔK values for K = 2 to K = 81. Representation for 81 Cattle breeds based on STRUCTURE results following Evanno criterion.

Figure S2 Population structure of 81 cattle breeds using STRUCTURE when K = 71. Graphical representation of individual genotype membership coefficients (q) when K = 71. Each animal is represented by a single vertical line divided into 71 coloured segments using only 6 colours showing the individual's estimated membership proportions in that cluster. 1: Betizu (BET), 2: Toro de Lidia (TL), 3: Menorquina (MEN), 4: Alistana (ALS), 5: Sayaguesa (SAY), 6: Tudanca (TUD), 7: Asturiana de los Valles (ASV), 8: Asturiana de las Montañas (ASM), 9: Retinta (RET), 10: Morucha (MOR), 11: Avileña (AVI), 12: Pirenaica (PIRM), 13: Rubia Gallega (RGA), 14: Mallorquina (MALL), 15: Monchina (MON), 16: Serrana de Teruel (STE), 17: Parda de Montaña (PM), 18: Bruna de los Pirineos (BRP), 19: Pasiega (PAS), 20: Berrenda en Colorado (BC), 21: Berrenda en Negro (BN), 22: Marismeña (MAR), 23: Pajuna (PAJ), 24: Negra Andaluza (NAN), 25: Vaca Canaria (VCA), 26: Vaca Palmera (PAL), 27: Alentejana (ALT), 28: Arouquesa (ARO), 29: Barrosã (BARR), 30: Brava de Lide (BRAV), 31: Cachena (CACH), 32: Garvonesa (GARV), 33: Marinhoa (MARI), 34: Maronesa (MARO), 35: Mertolenga (MERT), 36: Minhota (MINH), 37: Mirandesa (MIRA), 38: Preta (PRET), 39: Ramo Grande (RG); ***CREOLE***. 40: Guabalá (GUA), 41: Guaymí (GY), 42: Texas Longhorn (TLH), 43: Criollo Poblano (CPO), 44: Criollo de Baja California (CBC), 45: Criollo de Chihuahua (CHU), 46: Criollo de Nayarit (CNY), 47: Criollo de Chiapas (CHI), 48: Blanco Orejinegro (BON), 49: Caqueteño (CAQ), 50: Sanmartinero (SM), 51: Romosinuano (RMS), 52: Costeño con Cuernos (CCC), 53: Chino Santandereano (CH), 54: Velasquez (VEL), 55: Lucerna (LUC), 56: Hartón del Valle (HV), 57: Criollo Casanareño (CC), 58: Criollo Ecuatoriano (EC), 59: Criollo Uruguayo (CUR), 60: Pampa Chaqueño (PA), 61: Criollo Pilcomayo (PIL), 62: Criollo Argentino (CARG), 63: Criollo Patagónico (PAT), 64: Caracú (CAR), 65: Cubano (CUB), 66: Siboney (SIB); ***ZEBU***: 67: Gyr (GYR), 68: Brahman (BRH), 69: Sindi (SIN), 70: Guzerat (GUZ), 71 Nelore (NEL), 72: Zebu Cubano (CUZ), 73: Friesian (FRI), 74: Hereford (HER), 75: Brown Swiss (BSW), 76: Aberdeen Angus (AA), 77: British White (BWC), 78: Charolais (CHAR), 79: Jersey (JER), 80: Limousin (LIM), 81: Shorthorn (SH).

Table S1 Breeds, samples and origins. Breed names, acronyms (Acron.), sample sizes (N), sample type, breed type, genetic group (GG), country of sampling and region of origin (Reg) of the 81 breeds included in this study.

Table S2 Genetic diversity for 81 Cattle breeds. Number of individuals per breed (N), mean number of alleles/locus (Am), mean effective number of alleles/locus (Ae), mean allelic richness per locus corrected for sample size (Ar), mean observed heterozygosity (Ho) and mean expected heterozygosity (He) and their standard deviations, within-breed inbreeding coefficient (FIS) and corresponding confidence interval.

Table S3 Table S3. Estimated membership coefficients in each cluster (q), as inferred by STRUCTURE for K = 71. Contribution of the more important cluster per breed is represented in bold.

Table S4 Genetic contributions from Iberian, British, Continental European and Zebu breeds to Creole cattle. Maximum-likelihood estimates of proportional genetic contributions from Iberian, British, Continental European and Zebu breeds to Creole cattle, considered as a whole or grouped in five different clusters. The SD was obtained from 1000 bootstrapping samples (over loci). **Creole cluster 1**: Guabalá, Guaymí, Romosinuano, Costeño con Cuernos; **Creole cluster 2**: Texas Longhorn, Criollo Baja California, Criollo Chihuahua, Criollo de Nayarit, Criollo Poblano, Sanmartinero; **Creole cluster 3**: Caracú, Criollo Argentino, Criollo Patagónico, Criollo Uruguayo; **Creole cluster 4:** Criollo Cubano, Siboney, Criollo de Ecuador, Velasquez, Caqueteño, Criollo de Chiapas, Criollo Pilcomayo, Criollo Casanareño, Chino Santandereano; **Creole cluster 5:** Pampa Chaqueño, Blanco Orejinegro, Lucerna, Hartón del Valle.

Acknowledgments

The authors gratefully thank the different breeders' associations and research groups who kindly provided the biological samples used in this study and the members of the CYTED XII-H and CONBIAND networks for valuable cooperation over the years.

Author Contributions

Conceived and designed the experiments: AMM CG SD VL IM-B MCTP CR. Performed the experiments: AMM CG SD VL IM-B CR LAA. Analyzed the data: AMM LTG VL. Contributed reagents/materials/analysis tools: AMM JC JVD MCTP CR JLV-P AA LAA EC OC JRM ORM RDM LM GM-V JEM AP JQ PS OU AV DZ PZ. Wrote the paper: AMM LTG JVD. Reviewed and edited the manuscript: JC CG IM-B MCTP JLV-P LAA OC JQ PS PZ.

References

1. Stahl PW (2008) Animal Domestication in South America. In: Handbook of South American archaeology. Springer. 121–130.
2. Crosby AW (1973) The Columbian Exchange: Biological and Cultural Consequences of 1492. Greenwood. 268 p.

3. Rodero E, Rodero A, Delgado JV (1992) Primitive andalusian livestock an their implications in the discovery of America. Arch Zootec 41: 383–400.
4. Rouse JE (1977) The Criollo: Spanish cattle in the Americas. University of Oklahoma Press. 303 p.
5. Willham RL (1982) Genetic improvement of beef cattle in the United States: cattle, people and their interaction. J Anim Sci 54: 659–666.
6. Santiago AA (1978) Evolution of Zebu cattle in Brazil. The Zebu Journal 1: 6.
7. De Alba J (1987) Criollo Cattle of Latinamerica. In: Animal genetic resources. Strategies for improved use and conservation. Available:http://www.fao.org/docrep/010/ah806e/AH806E06.htm. Accessed 5 November 2011.
8. Rischkowsky B, Pilling D, Commission on Genetic Resources for Food and Agriculture (2007) The state of the world's animal genetic resources for food and agriculture. Roma: Food & Agriculture Organization of the United Nations (FAO). 554 p.
9. Delgado JV, Martínez AM, Acosta A, Álvarez LA, Armstrong E, et al. (2012) Genetic characterization of Latin-American Creole cattle using microsatellite markers. Anim Genet 43: 2–10. doi:10.1111/j.1365-2052.2011.02207.x.
10. Cañón J, Alexandrino P, Bessa I, Carleos C, Carretero Y, et al. (2001) Genetic diversity measures of local European beef cattle breeds for conservation purposes. Genet Sel Evol 33: 311–332. doi:10.1051/gse:2001121.
11. García D, Martínez A, Dunner S, Vega-Pla JL, Fernández C, et al. (2006) Estimation of the genetic admixture composition of Iberian dry-cured ham samples using DNA multilocus genotypes. Meat Sci 72: 560–566. doi:10.1016/j.meatsci.2005.09.005.
12. Tapio I, Värv S, Bennewitz J, Maleviciute J, Fimland E, et al. (2006) Prioritization for conservation of northern European cattle breeds based on analysis of microsatellite data. Conserv Biol 20: 1768–1779. doi:10.1111/j.1523-1739.2006.00488.x.
13. Ginja C, Telo Da Gama L, Penedo MCT (2010) Analysis of STR markers reveals high genetic structure in Portuguese native cattle. J Hered 101: 201–210. doi:10.1093/jhered/esp104.
14. Li M-H, Kantanen J (2010) Genetic structure of Eurasian cattle (Bos taurus) based on microsatellites: clarification for their breed classification. Anim Genet 41: 150–158. doi:10.1111/j.1365-2052.2009.01980.x.
15. Beja-Pereira A, Caramelli D, Lalueza-Fox C, Vernesi C, Ferrand N, et al. (2006) The origin of European cattle: Evidence from modern and ancient DNA. Proc Natl Acad Sci USA 103: 8113–8118. doi:10.1073/pnas.0509210103.
16. Edwards CJ, Ginja C, Kantanen J, Pérez-Pardal L, Tresset A, et al. (2011) Dual Origins of Dairy Cattle Farming – Evidence from a Comprehensive Survey of European Y-Chromosomal Variation. PLoS ONE 6: e15922. doi:10.1371/journal.pone.0015922.
17. Gibbs RA, Taylor JF, Van Tassell CP, Barendse W, Eversole KA, et al. (2009) Genome-wide survey of SNP variation uncovers the genetic structure of cattle breeds. Science 324: 528–532. doi:10.1126/science.1167936.
18. Gautier M, Laloë D, Moazami-Goudarzi K (2010) Insights into the genetic history of French cattle from dense SNP data on 47 worldwide breeds. PLoS ONE 5. Available:http://www.ncbi.nlm.nih.gov/pubmed/20927341. Accessed 11 November 2011.
19. Lewis J, Abas Z, Dadousis C, Lykidis D, Paschou P, et al. (2011) Tracing cattle breeds with principal components analysis ancestry informative SNPs. PLoS ONE 6: e18007. doi:10.1371/journal.pone.0018007.
20. Pritchard JK, Stephens M, Donnelly P (2000) Inference of population structure using multilocus genotype data. Genetics 155: 945–959.
21. Wang J (2003) Maximum-likelihood estimation of admixture proportions from genetic data. Genetics 164: 747–765.
22. Egito AA, Paiva SR, Albuquerque M do SM, Mariante AS, Almeida LD, et al. (2007) Microsatellite based genetic diversity and relationships among ten Creole and commercial cattle breeds raised in Brazil. BMC Genet 8: 83. doi:10.1186/1471-2156-8-83.
23. Ulloa-Arvizu R, Gayosso-Vázquez A, Ramos-Kuri M, Estrada FJ, Montaño M, et al. (2008) Genetic analysis of Mexican Criollo cattle populations. J Anim Breed Genet 125: 351–359. doi:10.1111/j.1439-0388.2008.00735.x.
24. Martínez-Correal G, Alvarez LA, Martínez GC (2009) Conservación, Caracterización y Utilización de los bovinos Criollos en Colombia. In: X Simposio Iberoamericano sobre Conservación y Utilización de Recursos Zoogenéticos. Palmira, Colombia.
25. Villalobos Cortés A, Martinez AM, Vega-Pla JL, Delgado JV (2009) Genetic characterization of the Guabala bovine population with microsatellites. Arch Zootec 58: 485–488.
26. Villalobos Cortés AI, Martínez AM, Escobar C, Vega-Pla JL, Delgado JV (2010) Study of genetic diversity of the Guaymi and Guabala bovine populations by means of microsatellites. Livest Sci 131: 45–51. doi:10.1016/j.livsci.2010.02.024.
27. Martín-Burriel I, García-Muro E, Zaragoza P (1999) Genetic diversity analysis of six Spanish native cattle breeds using microsatellites. Anim Genet 30: 177–182.
28. Mateus JC, Penedo MCT, Alves VC, Ramos M, Rangel-Figueiredo T (2004) Genetic diversity and differentiation in Portuguese cattle breeds using microsatellites. Anim Genet 35: 106–113. doi:10.1111/j.1365-2052.2004.01089.x.
29. Martín-Burriel I, Rodellar C, Lenstra JA, Sanz A, Cons C, et al. (2007) Genetic diversity and relationships of endangered Spanish cattle breeds. J Hered 98: 687–691. doi:10.1093/jhered/esm096.
30. Martín-Burriel I, Rodellar C, Cañón J, Cortés O, Dunner S, et al. (2011) Genetic diversity, structure, and breed relationships in Iberian cattle. J Anim Sci 89: 893–906. doi:10.2527/jas.2010-3338.
31. Lara MAC, Contel EPB, Sereno JRB (2005) Genetic characterization of Zebu populations using molecular markers. Arch Zootec 206/207: 295–303.
32. Dani MAC, Heinneman MB, Dani SU (2008) Brazilian Nelore cattle: a melting pot unfolded by molecular genetics. Genet Mol Res 7: 1127–1137.
33. European Cattle Genetic Diversity Consortium (2006) Marker-assisted conservation of European cattle breeds: An evaluation. Anim Genet 37: 475–481. doi:10.1111/j.1365-2052.2006.01511.x.
34. Giovambattista G, Ripoli MV, De Luca JC, Mirol PM, Lirón JP, et al. (2000) Male-mediated introgression of Bos indicus genes into Argentine and Bolivian Creole cattle breeds. Anim Genet 31: 302–305.
35. Ginja C, Penedo MCT, Melucci L, Quiroz J, Martínez López OR, et al. (2010) Origins and genetic diversity of New World Creole cattle: inferences from mitochondrial and Y chromosome polymorphisms. Anim Genet 41: 128–141. doi:10.1111/j.1365-2052.2009.01976.x.
36. Magee DA, Meghen C, Harrison S, Troy CS, Cymbron T, et al. (2002) A partial african ancestry for the creole cattle populations of the Caribbean. J Hered 93: 429–432.
37. Carvajal-Carmona LG, Bermudez N, Olivera-Angel M, Estrada L, Ossa J, et al. (2003) Abundant mtDNA diversity and ancestral admixture in Colombian criollo cattle (Bos taurus). Genetics 165: 1457–1463.
38. Lirón JP, Bravi CM, Mirol PM, Peral-García P, Giovambattista G (2006) African matrilineages in American Creole cattle: evidence of two independent continental sources. Anim Genet 37: 379–382. doi:10.1111/j.1365-2052.2006.01452.x.
39. Achilli A, Bonfiglio S, Olivieri A, Malusà A, Pala M, et al. (2009) The Multifaceted Origin of Taurine Cattle Reflected by the Mitochondrial Genome. PLoS ONE 4: e5753. doi:10.1371/journal.pone.0005753.
40. Uffo O, Martín-Burriel I, Martinez S, Ronda R, Osta R, et al. (2006) Caracterización genética de seis proteínas lácteas en tras razas bovinas cubanas. AGRI 39: 15–24.
41. Martinez AM, Llorente RV, Quiroz J, Martínez RD, Amstrong E, et al. (2007) Estudio de la influencia de la raza bovina Marismeña en la formación de los bovinos Criollos. VIII Simposio Iberoamericano sobre Conservación y utilización de Recursos Zoogenéticos. Quevedo, Ecuador.
42. Miretti MM, Dunner S, Naves M, Contel EP, Ferro JA (2004) Predominant African-derived mtDNA in Caribbean and Brazilian Creole cattle is also found in Spanish cattle (Bos taurus). J Hered 95: 450–453. doi:10.1093/jhered/esh070.
43. Gautier M, Naves M (2011) Footprints of selection in the ancestral admixture of a New World Creole cattle breed. Molecular Ecology 20: 3128–3143. doi:10.1111/j.1365-294X.2011.05163.x.
44. MacHugh DE, Shriver MD, Loftus RT, Cunningham P, Bradley DG (1997) Microsatellite DNA Variation and the Evolution, Domestication and Phylogeography of Taurine and Zebu Cattle (Bos Taurus and Bos Indicus). Genetics 146: 1071–1086.
45. Mirol PM, Giovambattista G, Lir|[oacute]|n JP, Dulout FN (2003) African and European mitochondrial haplotypes in South American Creole cattle. Heredity 91: 248–254. doi:10.1038/sj.hdy.6800312.
46. Lush JL (1943) Animal Breeding Plans. Ames, Iowa, USA: The Iowa State College Press. 457 p.
47. Villalobos Cortés A, Martinez AM, Vega-Pla JL, Delgado JV (2009) History of Panama bovines and their relationships with other Iberoamerican popultions. Arch Zootec 58: 121–129.
48. Primo AT (2004) América: conquista e colonização: a fantástica história dos conquistadores ibéricos e seus animais na era dos descobrimentos. Porto Alegre, Brazil: Movimento. 192 p.
49. Martínez AM, Calderón J, Camacho E, Rico C, Vega-Pla JL, et al. (2005) Genetic characterisation of the Mostrenca cattle with microsatellites. Arch Zootec 206: 357–361.
50. FAO (2004) Secondary Guidelines for Development of National Farm Animal Genetic Resources Management Plans: Management of Small Populations at Risk. Roma: Food & Agriculture Organization of the United Nations (FAO). 225 p.
51. Martínez AM, Delgado JV, Rodero A, Vega-Pla JL (2000) Genetic structure of the Iberian pig breed using microsatellites. Anim Genet 31: 295–301.
52. Nei M (1973) Analysis of gene diversity in subdivided populations. Proc Natl Acad Sci USA 70: 3321–3323.
53. Park SDE (2001) The Excel Microsatellite Toolkit. Trypanotolerance in West African Cattle and the Population Genetic Effects of Selection [Ph.D. thesis]. University of Dublin. Available:http://animalgenomics.ucd.ie/sdepark/ms-toolkit/. Accessed 5 November 2011.
54. Weir BS, Cockerham CC (1984) Estimating F-Statistics for the Analysis of Population Structure. Evolution 38: 1358–1370. doi:10.2307/2408641.
55. Belkhir K, Borsa P, Chikhi L, Raufaste N, Bonhomme F (2004) GENETIX 4.05, logiciel sous Windows TM pour la génétique des populations. Laboratoire Génome, Populations, Interactions, CNRS UMR 5000, Université de Montpellier II, Montpellier (France). GENETIX INTRODUCTION. Available:http://www.genetix.univ-montp2.fr/genetix/intro.htm. Accessed 5 November 2011.
56. Yeh FC, Boyle TJB (1997) Population genetic analysis of co-dominant and dominant markers and quantitative traits. Belg J Bot 129: 157.

57. Goudet J (1995) FSTAT, a program to estimate and test gene diversities and fixation indices. Department of Ecology & Evolution, Biology Building, UNIL, CH-1015 LAUSANNE, Switzerland. Available:http://www2.unil.ch/popgen/softwares/fstat.htm. Accessed 5 November 2011.
58. Raymond M, Rousset F (1995) GENEPOP (Version 1.2): Population Genetics Software for Exact Tests and Ecumenicism. J Hered 86: 248–249.
59. Guo SW, Thompson EA (1992) Performing the exact test of Hardy-Weinberg proportion for multiple alleles. Biometrics 48: 361–372.
60. Excoffier L, Smouse PE, Quattro JM (1992) Analysis of molecular variance inferred from metric distances among DNA haplotypes: application to human mitochondrial DNA restriction data. Genetics 131: 479–491.
61. Excoffier L, Laval G, Schneider S (2005) Arlequin (version 3.0): an integrated software package for population genetics data analysis. Evol Bioinform Online 1: 47–50.
62. Reynolds J, Weir BS, Cockerham CC (1983) Estimation of the coancestry coefficient: basis for a short-term genetic distance. Genetics 105: 767–779.
63. Langella O (1999) Populations 1.2.31 CNRS UPR9034. Available:http://www.bioinformatics.org/~tryphon/populations/. Accessed 5 November 2011.
64. Huson DH, Bryant D (2006) Application of phylogenetic networks in evolutionary studies. Mol Biol Evol 23: 254–267. doi:10.1093/molbev/msj030.
65. Lebart L, Morineau A, Warwick KM (1984) Multivariate descriptive statistical analysis: correspondence analysis and related techniques for large matrices. Wiley. 266 p.
66. Evanno G, Regnaut S, Goudet J (2005) Detecting the number of clusters of individuals using the software STRUCTURE: a simulation study. Mol Ecol 14: 2611–2620. doi:10.1111/j.1365-294X.2005.02553.x.
67. Rosenberg NA (2004) DISTRUCT: a program for the graphical display of population structure. Molecular Ecology Notes 4: 137–138. doi:10.1046/j.1471-8286.2003.00566.x.

Prevalence of Bovine Tuberculosis and Risk Factor Assessment in Cattle in Rural Livestock Areas of Govuro District in the Southeast of Mozambique

Ivânia Moiane[1,2,3], Adelina Machado[3], Nuno Santos[1,2], André Nhambir[3], Osvaldo Inlamea[3], Jan Hattendorf[5], Gunilla Källenius[4], Jakob Zinsstag[5], Margarida Correia-Neves[1,2]*

1 Life and Health Sciences Research Institute (ICVS), School of Health Sciences, University of Minho, Braga, Portugal, 2 ICVS/3B's, PT Government Associate Laboratory, Braga/Guimarães, Portugal, 3 Paraclinic Department, Veterinary Faculty, Eduardo Mondlane University, Maputo, Mozambique, 4 Department of Clinical Science and Education, Södersjukhuset, Karolinska Institutet, Stockholm, Sweden, 5 Swiss Tropical and Public Health Institute, Basel, Switzerland

Abstract

Background: Bovine tuberculosis (bTB), caused by *Mycobacterium bovis*, is an infectious disease of cattle that also affects other domestic animals, free-ranging and farmed wildlife, and also humans. In Mozambique, scattered surveys have reported a wide variation of bTB prevalence rates in cattle from different regions. Due to direct economic repercussions on livestock and indirect consequences for human health and wildlife, knowing the prevalence rates of the disease is essential to define an effective control strategy.

Methodology/Principal findings: A cross-sectional study was conducted in Govuro district to determine bTB prevalence in cattle and identify associated risk factors. A representative sample of the cattle population was defined, stratified by livestock areas (n = 14). A total of 1136 cattle from 289 farmers were tested using the single comparative intradermal tuberculin test. The overall apparent prevalence was estimated at 39.6% (95% CI 36.8–42.5) using a diagnostic threshold cut-off according to the World Organization for Animal Health. bTB reactors were found in 13 livestock areas, with prevalence rates ranging from 8.1 to 65.8%. Age was the main risk factor; animals older than 4 years were more likely to be positive reactors (OR = 3.2, 95% CI: 2.2–4.7). *Landim* local breed showed a lower prevalence than crossbred animals (*Landim* × *Brahman*) (OR = 0.6, 95% CI: 0.4–0.8).

Conclusions/Significance: The findings reveal an urgent need for intervention with effective, area-based, control measures in order to reduce bTB prevalence and prevent its spread to the human population. In addition to the high prevalence, population habits in Govuro, particularly the consumption of raw milk, clearly may potentiate the transmission to humans. Thus, further studies on human tuberculosis and the molecular characterization of the predominant strain lineages that cause bTB in cattle and humans are urgently required to evaluate the impact on human health in the region.

Editor: Pere-Joan Cardona, Fundació Institut d'Investigació en Ciències de la Salut Germans Trias i Pujol, Universitat Autònoma de Barcelona, CIBERES, Spain

Funding: The research leading to these results has received funding from the European Union's Seventh Framework Program (FP7/2007–2013) under grant agreement n° 221948, ICONZ (Integrated Control of Neglected Zoonoses). The contents of this publication are the sole responsibility of the authors and do not necessarily reflect the views of the European Commission. The funders had no role in study design, data collection and analysis, decision to publish, or preparation of the manuscript.

Competing Interests: The authors have declared that no competing interests exist.

* E-mail: mcorreianeves@ecsaude.uminho.pt

These authors contributed equally to this work.

Introduction

Bovine tuberculosis (bTB) is an infectious disease of cattle caused by *Mycobacterium bovis*, a member of the *Mycobacterium tuberculosis* complex. This chronic disease also affects a wide range of other domestic and wildlife animals and may also cause disease in humans [1].

Worldwide, bTB is considered one of the seven most neglected endemic zoonoses, presenting a complex epidemiological pattern and with the highest prevalence rates in cattle found in African countries, part of Asia and of the Americas [2]. In affected countries, the disease has an important socio-economic and public health-related impact, and represents also a serious constraint in the trade of animals and their products [3]. In developed countries, bTB was regarded as one of the major diseases of domestic animals until the 1920s [4], when preventive and control measures based on tuberculin skin test and subsequent slaughter of positive reactors and sanitary surveillance in slaughterhouses, began to be systemically applied [5]. After implementation of the control programs, bTB in cattle populations was greatly reduced or even eradicated [3]. Nevertheless, wildlife species are still considered a significant source of infection and responsible for the failure of the complete eradication of livestock bTB in some developed countries [6]. Unfortunately, a vaccination strategy for animals is not available and present bTB control strategies are expensive and difficult to implement. Consequently, in developing countries, where bTB remains of economic and public health importance [7], these strategies are often not in use or not applied systematically [1,8]. In addition, it is estimated that in Africa 90% of the milk is consumed raw or fermented, increasing the risk of bTB transmission to humans [9].

In order to develop an effective national program for bTB surveillance and control in developing countries, accurate data on bTB prevalence is needed [10]. In Mozambique, data on bTB epidemiology is still scarce and mostly unpublished. However, bTB is estimated to be one of the most important causes of economic losses in cattle production, due to rejection of carcasses at the slaughterhouse and limitations on trade, both intra-community and between districts [11]. Surveillance and control programs based on the tuberculin skin test in cattle at the farm and subsequent slaughter of positive reactors are not applied systematically and do not cover the small holder sector due to the costs with replacement of slaughtered animals [11]. Additionally, there are no effective measures for preventing the transmission of zoonotic diseases and a "bridge" between control programs of bTB and human tuberculosis has not been implemented.

In Mozambique's Govuro district, a great proportion of the population holds livestock animals (especially cattle and goats). According to the findings of positive skin test reactors, associated with lesions compatible with bTB found at slaughterhouses, the Provincial Livestock Services (SPP) considered Govuro positive for bTB [11] but accurate information on the bTB prevalence and its role in human tuberculosis is missing. Control measures are nowadays only based on compulsory test for bTB in cattle to be transferred for breeding or rearing purposes. While slaughter for local consumption is uncommon (only in traditional ceremonies), animals are frequently sold to be consumed in the south of the country. In Govuro there is extensive consumption of untreated milk, direct contact between people and livestock, together with malnutrition and a high prevalence rate of HIV infection, all of which constitute risk factors for this zoonosis. Also still unknown (but crucial to minimize disease propagation) are the main risk factors contributing to the spread of the disease between animals.

Previous studies conducted in Govuro in 2008, using the single intradermal tuberculin test (SITT) in the caudal fold and in the middle neck region of cattle, found prevalence values of 61.94% (n = 268) [11]. This represents the highest recorded prevalence in

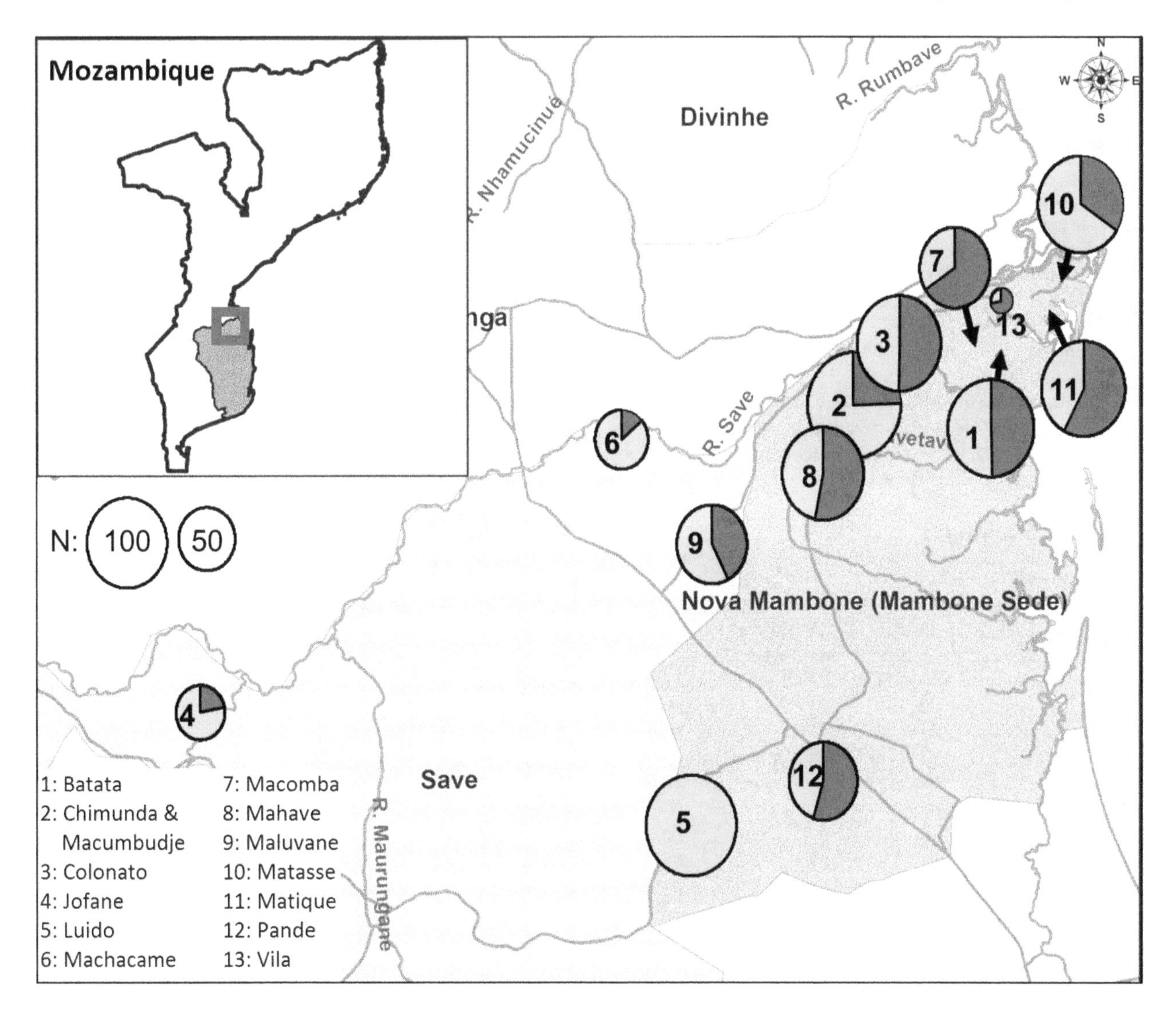

Figure 1. Location of the study district Govuro and spatial distribution of positive reactor cattle. The circle size is proportional to the number of animals tested in each location, and red area denotes the proportion of positive animals.

Table 1. Basic characteristics of the sample.

Characteristics	Classes	n	%
Gender	Female	773	68.0
	Male	362	31.9
	Not recorded	1	0.1
Age	0–1 yr (calf)	38	3.3
	1–4 yrs (steer)	245	21.6
	>4 yrs (bull, cow, ox)	852	75.0
	Not recorded	1	0.1
Breed	*Landim*	468	41.2
	Crossbred (*Landim* × *Brahman*)	534	47.0
	Bonsmara	88	7.7
	Other	11	1.0
	Not recorded	35	3.1
Body condition score	Good	705	62.1
	Reasonable	337	29.7
	Poor	86	7.6
	Very poor	2	0.2
	Not recorded	6	0.5

cattle in the country, however only two livestock areas of the district were analyzed. In order to determine the prevalence rate of bTB in Govuro we conducted a cross-sectional survey covering a representative sample of the cattle population of all livestock areas within the district. The single comparative intradermal tuberculin test (SCITT) was used since its specificity is higher than the one from the SITT [12]. While the SITT is the standard diagnostic test used in the Mozambican Bovine Tuberculosis Control Program, the SCITT is the confirmatory test and can also be used as screening test in herds with a history of cross-reactivity. We also assessed intrinsic determinants of disease associated with SCITT positivity in the study area in order to define strategies suitable to bTB control in cattle in Govuro.

Materials and Methods

Ethics Statement

The purpose of this study was explained to the cattle owners and an informed consent was obtained. While the SITT is the standard diagnostic test used in the Mozambican Bovine Tuberculosis Control Program we used the SCITT due to its higher specificity and to the fact that using this more complete test the data for SITT was also obtained. The Mozambican National Animal Health authority (*Direcção Nacional de Serviços Veterinários*) approved the present study and provided the ethical clearance (Nota 162/ MINAG/DNSV/900/2013).

Study Area

A cross-sectional study was carried out in the Govuro district (Figure 1) located in the northern part of Inhambane province, south-eastern Mozambique. The region is bordered on the north by the Machanga district of the Sofala province (across the Save River), on the east by the Indian Ocean, on the south by Inhassoro district and on the west by the Mabote district. The district covers an area of 3,960 km^2 with an estimated population of 35,500 inhabitants. The climate is tropical dry in the interior and humid close to the coast with an average temperature of 25.5°C (18–33°C). The rainy and dry seasons generally occur around October to March and April to September, respectively [13]. Govuro comprises 2 administrative posts (Nova Mambone and Save), 5 localities, 14 livestock areas and 45 villages containing an estimated number of 773 farmers and 8,760 cattle heads.

Animals and Production Systems Tested for Bovine TB

The animals included in this study were from the small holder sector or from the commercial sector. In the small holder sector most animals were longhorn *Landim* cattle (local breed, mixed *Bos indicus* and *Bos taurus*) and crossbreeds (*Landim* × *Brahman*) and with herds typically comprised of both these cattle breeds. Cattle were kept traditionally in a free-range grazing system using communal grazing grounds (without supplementation) and watering points such as small puddles (formed throughout the grazing area during the rainy season) or the Save River (during the dry season) [11]. Animals received little veterinary assistance, mostly restricted to vaccination. Animals from the commercial sector were mostly of *Simmental* (*Bos taurus*), *Brahman* (*Bos indicus*) and *Bonsmara* (mixed *Bos indicus* and *Bos taurus*) breeds. They were reared under semi-intensive farming with limited grazing areas (maintained in fences separated from cattle of the small holder sector) and with established water sources. Veterinary assistance and supplementation were provided. Whilst in the commercial sector the livestock production was mainly market-oriented, in the small holder farming animals were frequently used to till the ground and to transport material and people. Livestock trading in the small holder sector is restricted to special occasions, essentially when there is scarcity in agriculture production; in case of diseases and medical assistance is needed; to raise money for children's school fees or other essential livelihood assets for the family such as food items, soap and clothes.

Table 2. Apparent prevalence of bTB in Govuro per livestock area.

Livestock area	Animals tested		Bovine PPD SCITT reactors			Bovine PPD SITT reactors			Avian PPD SITT reactors		
	Total	Read	n	%	95% CI	n	%	95% CI	n	%	95% CI
Batata	133	117	58	49.6	40.7–58.5	78	66.7	57.7–74.6	17	14.5	9.3–22.0
Macomba	101	79	52	65.8	54.9–75.3	60	75.9	65.5–84.0	7	8.9	4.4–17.2
Colonato	173	111	56	50.5	42.3–59.6	70	63.1	53.8–71.5	12	10.8	6.3–18.0
Jofane	37	37	8	21.6	11.4–37.2	11	29.7	17.5–45.8	1	2.7	0.5–13.8
Maluvane	84	76	32	42.1	31.7–53.3	39	51.3	40.3–62.2	4	5.3	2.1–12.8
Matasse	123	115	39	33.9	25.9–43.0	53	46.1	37.2–55.2	16	13.9	8.6–21.4
Chimunda	75	62	27	43.5	31.9–56.0	37	59.7	47.2–71.0	16	25.8	16.6–37.9
Mahave	150	105	56	53.3	43.8–62.6	74	70.5	61.2–78.4	15	14.3	8.9–22.2
Matique	142	111	64	57.7	48.4–66.4	83	74.8	66.0–82.0	22	19.8	13.5–28.2
Pande	95	74	41	55.4	44.1–66.2	46	62.2	50.8–72.4	8	10.8	5.6–19.9
Luido	130	125	0	0.0	0.0–3.0	1	0.8	0.1–4.4	8	6.4	3.3–12.1
Vila	43	7	5	71.4	35.9–91.8	5	71.4	35.9–91.8	0	0.0	0.0–35.4
Machacame	45	43	6	14.0	6.6–27.3	9	20.9	11.4–35.2	3	7.0	2.4–18.6
Mucumbudje	90	74	6	8.1	3.8–16.6	20	27.0	18.2–38.1	8	10.8	5.6–19.9
Total	1421	1136	450	39.6	36.8–42.5	586	51.6	48.7–54.5	137	12.1	10.3–14.1

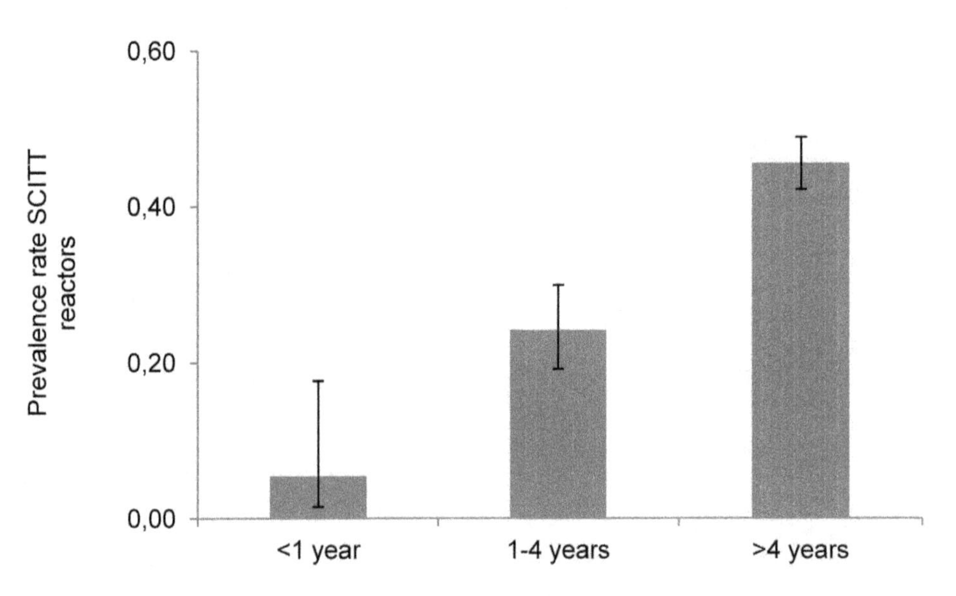

Figure 2. Prevalence rates among SICTT reactors (95% confidence intervals) by age classes.

Sample Size Calculation and Study Animals

To obtain a sample size representative of the Govuro district cattle population, the number of animals to be tested was calculated with Epicalc 2000 (Brixton Books v.1 2), using an expected prevalence of 10% and precision measured as one-half length of the 95% confidence interval of 5%. Sample sizes were calculated for each livestock area and corrected for finite population sizes. The epidemiological unit of this study was the livestock area, which corresponds to the cattle from several owners belonging to the same village. Even belonging to distinct owners, animals have regular direct contact between them and share natural pasture areas and watering sites (grazing groups) or even cowsheds. All cattle of the district (estimated as n = 8,763) was included in the sampling frame. The required sample sizes for the livestock areas ranged from 45 to 139 animals and resulted in a total sample size of n = 1,443 (Details are provided in Appendix S1).

All cattle owners from each livestock area were contacted to participate. The vast majority of the owners brought their animals to pre-defined locations. For the owners that could not bring the animals to the testing place we went to their home place. As a result more than 6,000 cattle participated, out of an estimated population of 8,760. Animals were selected randomly by systematic sampling according to the sample size previously determined and the number of cattle present on the day of the test. They were moved through a cattle chute and every k^{th} animal was selected for sampling, being k the number of animals presented for testing in that livestock area divided by the intended sample size for that same livestock area. At the time of SCITT testing, each animal was identified by a numbered ear tag and individual animal data on age, gender, breed, body condition score (BCS) and owner were registered. Information regarding the age of the animals was provided by the farmers and the breed was determined according to the phenotypic characteristics. The body condition was scored using the guidelines established by Nicholson and Butterworth [14]; all study animals were categorized in four groups: very poor (1), poor (2 to 3), reasonable (4 to 6) and good (score, 7 to 9).

Single Comparative Intradermal Tuberculin Test

The purpose of the study was explained to the owners with the assistance of local veterinary services (SDAE), community leaders, the local prosecutor and trusted intermediaries. SCITT was performed by intradermal injections of both avian and bovine purified protein derivates (PPD) in the middle neck region (usually on the right side) according to the method described by the World Organization for Animal Health standards [2]. Briefly, two sites of about 2 cm^2 diameter, approximately 12 to 15 cm apart, were shaved and the skin thickness was measured using a manual caliper. Aliquots of 0.1 ml containing 20,000 IU/ml of bovine PPD (Bovituber PPD, Synbiotics Europe, Lyon, France) and 0.1 ml with 25,000 IU/ml of avian PPD (Avituber PPD, Synbiotics Europe, Lyon, France) were injected using two different syringes into the dermis in the corresponding shaved area. Palpation of a small grain-like thickening at each site of injection was done to confirm the correct intradermal injection. Three days after injection, the tested animals were brought back for reading. The relative change in skin appearance was classified as swelling or induration followed by measurement of skin thickness at both injection sites. Skin thickness measurements on testing and reading day were performed by the same person to avoid errors related to individual variations in technical procedure.

The SCITT results were analyzed and interpreted according to the recommendations of the World Organization for Animal Health standards [2]. The reaction was considered positive if the increase in skin thickness at the bovine PPD site of injection $(B_{72}-B_0)$ was at least 4 mm greater than the reaction at the avian PPD injection site $(A_{72}-A_0)$. The livestock area was considered positive for bTB if at least one positive reactor was found.

Additionally we determined the SITT results by analyzing the same dataset taking into consideration only the bovine PPD data, using the same cutoff. Also, to assess the prevalence of reactors to other sensitising organisms such as *Mycobacterium avium*, the skin reactions at the injection site of the avian PPD alone were analyzed; animals that reacted to the avian PPD with an increase in skin thickness equal or superior to 4 mm were considered reactors to *M. avium.* Geographical coordinates were registered at the central point of each livestock area by a hand held global positioning system.

Table 3. Risk factors associated with positive reaction to SCITT.

Risk factor	Category	Positive/total	Positive (%)	Univariate analysis		Multivariate analysis		
				OR	95% CI	OR	95% CI	P
Age	≤4 yrs	62/283	22	reference		reference		
	>4 yrs	388/854	45	2.9	2.0–4.1	3.2	2.2–4.7	<0.001
Gender	Female	259/773	38	reference		reference		
	Male	155/362	43	1.1	0.8–1.4	1.2	0.9–1.6	0.300
Breed	*Landim* × *Brahman*	265/534	50	reference		reference		
	Landim	167/468	36	0.7	0.5–0.9	0.6	0.4–0.8	0.002
	Bonsmara[a]	0/88	0	nd		nd		
Body condition score	Good	250/705	35	reference		reference		
	Reasonable	153/337	45	1.0	0.8–1.4	1.0	0.7–1.4	0.820
	Poor and Very poor	44/88	50	1.4	0.8–2.3	1.3	0.7–2.1	0.410

[a]Excluded from the multivariate model to avoid quasi separation.

Data Analysis

All data at individual animal level were entered into a Microsoft Access database. Data analysis was performed in R statistical software (v2.15.1). Prevalence, odds ratios (OR) and their 95% confidence intervals were adjusted for correlation within livestock areas using generalized linear mixed models with binary outcome and livestock area as random effect.

Results

Sample Characteristics

Over the study period a total of 1,419 cattle were injected with PPDs and measurements were obtained from 1,136 animals (80%) belonging to 289 farmers. Table 1 shows the main characteristics of the sample tested. One hundred and twenty five (11%) animals came from the commercial sector (all from the Luido area) and 1,011 (89%) from the small holder sector. About two third of the cattle were female. The age distribution was as follows: 3% of the animals between 0 to 1 years old (calf - "<1 year"); 22% between 1 to 4 years old (steer - "1–4 years"); and 75% were older than 4 years (bull, cow and ox - ">4 years"). Almost half (49%) of the animals were of crossbreeds (*Landim* × *Brahman*), 43% *Landim* and 8% *Bonsmara*. *Simmental*, *Brahman* and *Limousine* were tested only in one herd from the commercial sector in Luido representing 1% of the sample. Sixty-two percent of all animals tested were classified as having good BCS, 30% reasonable BCS, and a small proportion presented poor (8%) and very poor (0.2%) BCS. Characteristic measures were not recorded for a few animals (Table 1).

Cattle bTB Prevalence in Govuro

The results of the SCITT are presented in Table 2 as prevalence per livestock area. The overall apparent prevalence of SCITT positive reactors was 39.6% (95% CI: 36.8–42.5). Except in Vila, where most of the PPD-inoculated animals failed the reading day, representative samples were obtained for each of the other 13 livestock areas. Among them, only in Luido, where the animals were all from the commercial sector, no SCITT positive reactors were detected (Table 2). In addition, data shows that bTB prevalence rates vary remarkably between livestock areas (ranging from undetectable up to 65.8%).

SCITT results showed that 137 (12%; 95% CI: 10.3–14.1) out of 1,136 cattle tested were positive reactors to avian PPD (Table 2). Among the 137 cattle with a positive reaction to avian PPD, 49 (36%; 95% CI: 28.2–44.1) had an overall SCITT test also positive but 24 (18%; 95% CI: 12.1–24.8) showed a stronger response to avian PPD than to bovine PPD.

Risk Factors Associated with Positive Reaction to SCITT

Univariate and multivariate analysis showed that age and breed represented intrinsic risk factors associated with positive reaction to SCITT (Table 3). Animals older than 4 years were more likely to be infected compared to young animals (45.4% *vs* 21.9%; OR = 3.2, 95% CI 2.2–4.7) (Table 3), and this difference was statistically significant even when considering the age classes "<1 year", "1–4 years" and ">4 years" ($\chi^2 = 55.56$; d.f. = 2, P<0.001) (Figure 2). Male animals tended to show higher prevalence rates for bTB (42.7% *vs* 37.2% in females), but there was no statistically significant difference in reactivity to the SCITT test between gender ($\chi^2 = 2.20$; d.f. = 1; P>0.05).

The rates of SCITT bTB reactors were higher in crossbreeds (*Landim* × *Brahman*) when compared to local *Landim* breed. Out of 468 of the *Landim* breed animals tested, 167 were found to be positive for the disease (35.7%; 95% CI 31.5–40.1) whereas 265 out of 534 (49.6%; 95% CI 45.4–53.9) in the crossbred (*Landim* × *Brahman*) cattle were positive reactors (Table 3). These data revealed a statistically significant association between the type of breeds and bTB prevalence, where the *Landim* breed seemed to be at lower risk for infection (OR = 0.6; 95% CI 0.4–0.8). All animals of the breed *Bonsmara* belonged to two private farmers from the livestock area Luido, where no positive reactors were found (0/88).

Discussion

Our results show that bTB is highly prevalent in Govuro district, with an overall prevalence rate of 39.6%. The sample size in each livestock area was slightly lower than targeted and the observed prevalence was closer to 50%, consequently, the precision associated with the prevalence estimates for the single livestock areas was lower than planned in the sample size calculation. However, the overall prevalence and risk factors were associated with high precision and narrow interval estimates. The SCITT has a less than perfect sensitivity, with a range of 52.0–95.5%, dependent on local factors [12]. Adjusting for the relatively low sensitivity of the SCITT, we estimated that the true prevalence in Govuro district is likely to be substantially higher than the apparent prevalence. The Rogan-Gladen estimator yielded a true prevalence of 65%, assuming a test specificity of 0.96 and sensitivity of 0.59, recalculated from data on Chadian cattle [15].

A high prevalence of bTB was observed in almost all livestock areas where small scale farming was practiced, in sharp contrast with what was observed in the commercial sector (only present in Luido), where no SCITT positive animals were detected. While in the commercial sector animals are normally tested for bTB and kept in quarantine before being introduced, trading of animals among breeders in the small holder sector is frequently performed without previous information about bTB status of the animals. In addition, the two tested farms in Luido were established in 2008, only four years before sampling. Interestingly, in the two livestock areas with the lowest bTB prevalence in the small holder sector (Mucumbudje and Machacame), livestock was just recently introduced (years 2007–2008). Our data show that age of the animals was an important intrinsic risk factor, most probably associated with increased exposure to *M. bovis* with lifetime. The type of management system applied in the small holder sector in Govuro, with sharing of water points and grazing areas, and close contact between animals from the same or different herds, promotes the spread of respiratory diseases such as bTB [4,16,17]. Additionally, during vaccination campaigns or external deworming, the animals from different farmers or herds use the same dip tanks. In contrast, animals from the commercial sector are kept inside fences and reared on a rotational grazing system with no contact with cattle from the small holder sector.

A study carried out in 2008 in the same region reported a bTB prevalence rate of 61.9% (95% CI: 55.8–67.8) [11]. Together with the present study, these data suggests that bTB is stable at an extremely high prevalence in the region. In the previous study the covered sample was limited to two livestock areas (Colonato and Vila), whereas in the present study all livestock areas were included. In addition, this study by Macucule et al. [11] made use of the single intradermal tuberculin test (SITT) while in the present study we made use of the SCITT. When our data were analyzed taking in consideration only the bovine PPD result, which corresponds to the SITT, the prevalence rates obtained (63.6%, 95% CI: 0.55–0.72) were similar to what was reported from these two livestock areas in 2008.

The choice of the SCITT instead of the SITT has been shown to be of relevance to differentiate between animals infected with *M. bovis* and those responding to bovine PPD possibly as a result of exposure to other mycobacteria. In fact, in our study the overall prevalence of bTB, taking into consideration only the bovine PPD results, was 51.6% (95% CI: 48.7–54.5), clearly higher than the one determined using the SCITT. This higher rate of positive SITT reactors can be attributed to sensitization with cross-reactive antigens among mycobacterial species and related genera [2].

According to the definitions of positivity, the animals that reacted equally to both PPDs (avian and bovine) were classified as negative reactors to SCITT [2]. Reactivity to the avian PPD may indicate infection or simply exposure to species of the *M. avium* complex or other environmental mycobacteria. This reactivity, however, may indicate a mixed reaction to both agents and hence the classification of bTB negative might also lead to some false negatives. The equal reactivity in both sites of injections (avian and bovine PPD) could be related with a generalized sensitization in which the immune response is not specific to a particular mycobacteria species.

In our study we found 137 (12%) animals that reacted positively to the avian PPD, a finding that has also been previously described. In a cross-sectional study done in Uganda, Inangolet *et al.* [18] attributed the high number of avian reactors in cattle with the existence of large poultry population in the studied areas, where chicken production in a free-ranging system is common. Fecal contamination of the watering sources was indicated as the main route of transmission of *M. avium* to cattle. In our study area, poultry production is a common activity, nevertheless the system where cattle are kept in corrals (mainly during the night), away from residences, do not promote direct and frequent contact between these two species. The reactivity in the avian PPD in Govuro could be associated with the high population density of cattle egret (*Bubulcus ibis*) in the district. This species is usually found along grazing cattle (removing ticks and flies from the animals). In fact, the presence of *M. avium* subsp. *avium* was already found in fecal samples of cattle egret [19] which could constitute a source of spread to the cattle.

The causative agent of avian tuberculosis, *M. avium* subsp. *avium*, was the predominant MAC isolated from tuberculous lesion in cattle [20]. The role of small ruminants (goats and sheep) as vector of *M. avium* subsp. *avium* and *Mycobacterium avium* subsp. *paratuberculosis* has also been identified [3]. In Govuro, the predominant production system is the communal/pastoral system, where the small ruminants graze together with cattle. Sharing of pastures and watering points could represent a potential source of infection of *M. avium* to cattle. However, according to Okuni et al. [21], paratuberculosis in cattle was not reported in Mozambique. Further studies are necessary to clarify the source of the avian PPD reactions found in cattle in Govuro.

In accordance with findings from numerous cross-sectional studies conducted in both developed and developing countries [e.g. 18,22–25], our results show that age was the main individual risk factor. Some authors suggest that it could be related to increased duration of exposure with age, with older cattle being more likely to have been exposed than the younger [24,26]. Out of 38 calves tested only 2 (5.26%) had a positive result on SCITT. The low number of positive cases in young animals may be associated with the predominance of gamma delta ($\gamma\delta$) T cells in calves that have been shown to play a relevant role in antimycobacterial immunity [27]. The positive calves (although in low number) could be due to congenital transmission in utero [28]. In addition, ingestion of contaminated colostrum has already been reported as another route of bTB transmission [29], as well as pseudo-vertical transmission (close contact between cow and its calf) [30].

The analysis of our bTB reactors according to gender showed that, although the reactivity among males was slightly higher (43% *versus* 38%) the difference was not statistically significant. Male cattle were identified as being the group at highest risk in other studies due to their particular longevity in the herd, given their use as draught oxen, facilitating maintenance of the infection in the herds [22]. Higher reactivity of females than males was previously reported in dairy cows [18,25] and associated with their maintenance in the same herd for several years [5]. In Govuro, however, male cattle tend to be maintained for longer periods in the herds since they are commonly used for plowing the land and pull carts for transportation of people and goods.

Most of the cattle included in the present study were crossbreed (*Landim* × *Brahman*) and *Landim* local breed and bTB prevalence rates were found to be significantly higher in this later breed. The cattle recorded as *Simmental* and *Bonsmara* breed, from the two commercial sector farms in Luido, were too few to allow a relevant comparison of susceptibility. Several studies [31,32] have shown a variation in susceptibility to bTB among cattle breeds, with European breeds (*Bos taurus*) being less resistant compared to Zebu cattle (*Bos indicus*). Although crossbred cattle in Ethiopia (local *Bos indicus* breed Arsi × Holstein *Bos taurus* breed) has been suggested to exhibit intermediate levels of susceptibility [31], our data does not support this observation, as animals with more Zebu background (*Landim* × *Brahman*) showed higher bTB prevalence than *Landim* cattle. It should be mentioned that differences in bTB prevalence between breeds - as observed in several studies - can be influenced by different husbandry conditions; however, genetic variations among cattle breeds are also likely to have an influence on susceptibility to infection with *M. bovis*. The genetic variations among cattle breeds have an influence on susceptibility to infection with *M. bovis*. In diverse breeds of British cattle, the genomic regions INRA111 and BMS2753 were strongly associated with bTB infection status [33]. Two others loci have also been linked to susceptibility in Holstein cattle, namely, a variant in the TLR1 gene [34] and BTA 22 [35].

Several studies reported a correlation between body condition and bTB [e.g. 10,36]. In our study animals in reasonable and poor or very poor body condition showed more positive skin test results than animals in good body condition, however this difference did not reach statistical significance. Following recommendations by Humblet *et al.* [5], this parameter should be analyzed carefully, since while a poor BCS might be a cause of disease, it is also extremely influenced by the seasonal climatic changes (rain or dry season) and the consequently more or less availability of pasturage and/or prevalence of intestinal parasites (in the small scale small holder farming of Govuro deworming for internal parasites is uncommon). It was reported previously [10,36] that animals in very poor body condition could be non-responsive to the SCITT due to anergy caused by immune-suppression. Our results do not support this finding. In addition, in cross-sectional studies, the status of the animal before becoming infected is not known, and thereby it is impossible to distinguish if the poor body condition was a risk factor or if it is a consequence of advanced stage of bTB.

This is the first systematic study on bTB prevalence encompassing a representative sampling of all livestock areas of a particular district in Mozambique. The data clearly show that bTB is a serious problem in Govuro district with extremely high prevalence rates being maintained for several years. Our results strengthen the notion that if strong measures were undertaken, as was the case among the commercial sector, the disease might be controlled. It is of relevance to stress that drinking raw milk is a common habit in Mozambique, especially for young children that take in charge the livestock grazing. In addition, due to their stature, children that graze animals may be extremely exposed to *M. bovis* airborne transmission from infected animals. Taking all this into account and the fact that studies on human tuberculosis have not been systematically performed in Govuro (neither for *M. tuberculosis* nor *M. bovis*) our results reinforce the need not just to undertake bTB control measures in the region but also the urgency to investigate the prevalence of tuberculosis in humans, especially in children in Govuro.

Acknowledgments

Disclaimer: The contents of this publication are the sole responsibility of the authors and do not necessarily reflect the views of the European Commission.

The authors thank Yolanda Vaz for critical analysis of the data, Nadine Santos for her careful reading of the manuscript, and important advice and the Provincial Livestock Services of Inhambane Province, District Services of Economic Activities of Govuro district and the field staff for their valuable help and support during fieldwork.

Author Contributions

Conceived and designed the experiments: IM AM JH GK JZ MCN. Performed the experiments: IM AM AN OI MCN. Analyzed the data: IM AM NS JH GK MCN. Contributed reagents/materials/analysis tools: AM JH GK MCN. Wrote the paper: IM AM NS JH GK MCN.

References

1. Etter E, Donado P, Jori F, Caron A, Goutard F, et al. (2006) Risk analysis and bovine tuberculosis, a re-emerging zoonosis. Ann N Y Acad Sci 1018: 61–73.
2. OIE (2009) Manual of diagnostic tests and vaccines for terrestrial animals. World Animal Health Organization. Paris, France. Version adopted by the World Assembly of Delegates of the OIE in May 2009. Available: http://www.oie.int/eng/normes/mmanual/A_summry.htm. Accessed 2013 Jul 12.
3. Biet F, Boschiroli ML, Thorel MF, Guilloteau LA (2005) Zoonotic aspects of *Mycobacterium bovis* and *Mycobacterium avium-intracellulare* complex (MAC). Vet Res 36: 411–436.
4. Cosivi O, Grange JM, Daborn CJ, Raviglione MC, Fujikura T, et al. (1998) Zoonotic tuberculosis due to *Mycobacterium bovis* in developing countries. Emerg Infect Dis 4(1): 59–70.
5. Humblet MF, Gilbert M, Govaerts M, Fauville-Dufaux M, Walravens K, et al. (2009) New assessment of bovine tuberculosis risk factors in Belgium based on nationwide molecular epidemiology. J Clinic Microb 48(8): 2802–2808.
6. Schiller I, Oesch B, Vordermeier M, Palmer M, Harris B, et al. (2010) Bovine tuberculosis: a review of current and emerging diagnostic techniques in view of their relevance for disease control and eradication. Transb Emerg Dis 57(4): 205–220.
7. Awah-Ndukum J, Kudi AC, Bah GS, Bradley G, Tebug SF, et al. (2012) Bovine tuberculosis in cattle in the highlands of Cameroon: seroprevalence estimates and rates of tuberculin skin test reactors at modified cut-offs. Vet Med Int 2012, article ID 798502, doi:10.1155/2012/798502.
8. Ayele WY, Neill SD, Zinsstag J, Weiss MG, Pavlik I (2004) Bovine tuberculosis: an old disease but a new threat to Africa. Int J Tub Lung Dis 8: 924–937.
9. Ibrahim S, Cadmus SIB, Umoh JU, Ajogi I, Farouk UM, et al. (2012) Tuberculosis in humans and cattle in Jigawa state, Nigeria: risk factors analysis. Vet Med Int 2012, article ID 865924, doi:10.1155/2012/865924.
10. Tschopp R, Schelling E, Hattendorf J, Aseffa A, Zinsstag J (2009) Risk factors of bovine tuberculosis in cattle in rural livestock production systems of Ethiopia. Prev Vet Med 89(3–4): 205–211.
11. Macucule BA (2008) Study of the prevalence of bovine tuberculosis in Govuro District, Inhambane Province, Mozambique. MSc Thesis – University of Pretoria, South Africa. 66 p.
12. de la Rua-Domenech R, Goodchild AT, Vordermeier HM, Hewinson RG, Christiansen KH, et al. (2006) Ante mortem diagnosis of tuberculosis in cattle: a review of the tuberculin tests, γ-interferon assay and other ancillary diagnostic techniques. Res Vet Sci 81: 190–210.

13. Ministério da Administração Estatal (2005) Perfil do Distrito de Govuro – Província de Inhambane. Available: http://www.govnet.gov.mz/, 39 p. Accessed 2013 Jul 12.
14. Nicholson MJ, Butterworth MH (1986) A guide to condition scoring in Zebu cattle. Addis Abeba, Ethiopia: International Livestock Centre for Africa. 29 p. Available: http://ftpmirror.your.org/pub/misc/cd3wd/1005/_ag_zebu_cattle_condition_score_ilcaenlp118060.pdf. Accessed 2013 Jul 12.
15. Müller B, Vounatsou P, Ngandolo BNR, Diguimbaye-Djaïbe C, Schiller I, et al. (2009) Bayesian receiver operating characteristic estimation of multiple tests for diagnosis of bovine tuberculosis in Chadian cattle. PLoS ONE 4(12): e8215, doi:10.1371/journal.pone.0008215.
16. Ameni G, Vordermeier M, Firdessa R, Aseffa A, Hewinson G, et al. (2011) *Mycobacterium tuberculosis* infection in grazing cattle in central Ethiopia. Vet J 188(3–4): 359–361.
17. Boukary AR, Thys E, Abatih E, Gamatié D, Ango I, et al. (2011) Bovine tuberculosis prevalence survey on cattle in the rural livestock system of Torodi (Niger). PLoS ONE 6(9): e24629, doi:10.1371/journal.pone.0024629.
18. Inangolet FO, Demelash B, Oloya J, Opuda-Asibo J, Skjerve E (2008) A cross-sectional study of bovine tuberculosis in the transhumant and agro-pastoral cattle herds in the border areas of Katakwi and Moroto districts, Uganda. Trop Anim Healt Prod 40: 501–508.
19. Dvorska L, Matlova L, Ayele WY, Fischer OA, Amemori T, et al. (2007) Avian tuberculosis in naturally infected captive water birds of the Ardeideae and Threskiornithidae families studied by serotyping, IS901 RFLP typing, and virulence for poultry. Vet Microb 119(2–4): 366–374.
20. Pavlik I, Matlova L, Dvorska L, Shitaye JE, Parmova I (2005) Mycobacterial infections in cattle and pigs caused by *Mycobacterium avium* complex members and atypical mycobacteria in the Czech Republic during 2000–2004. Vet Med Cz 50: 281–290.
21. Okuni JB, Dovas CI, Loikopoulos P, Bouzalas IG, Kateete DP, et al. (2012) Isolation of *Mycobacterium avium* subspecies *paratuberculosis* from Ugandan cattle and strain differentiation using optimized DNA typing. BMC Vet Res 8: 99, doi:10.1186/1746-6148-8-99.
22. Kazwala RR, Daborn CJ, Sharp JM, Kambarage DM, Jiwa SFH, et al. (2001) Isolation of *Mycobacterium bovis* from human cases of cervical adenitis in Tanzania: a cause for concern? Int J Tub Lung Dis 5(1): 87–91.
23. Ameni G, Aseffa A, Engers H, Young D, Hewinson G, et al. (2006) Cattle husbandry in Ethiopia is a predominant factor for affecting the pathology of bovine tuberculosis and gamma interferon responses to mycobacterial antigens. Clin Vac Imm 13: 1030–1036.
24. Cleaveland S, Shaw DJ, Mfinanga SG, Shirima G, Kazwala RR, et al. (2007) *Mycobacterium bovis* in rural Tanzania: risk factors for infection in human and cattle populations. Tuberc 87(1): 30–43.
25. Dinka H, Duressa A (2011) Prevalence of bovine tuberculosis in Arsi zone of Oromia, Ethiopia. Afric J Agr Res 6(16): 3853–3858.
26. Cook AJ, Tuchili LM, Buve A, Foster SD, Godfrey-Fausett P, et al. (1996) Human and bovine tuberculosis in the Monze District of Zambia–a cross-sectional study. Brit Vet J 152(1): 37–46.
27. Kennedy HE, Welsh MD, Bryson DG, Cassidy JP, Forster FI, et al. (2001) Modulation of immune responses to *Mycobacterium bovis* in cattle depleted of $WC1^{+}$ $\gamma\delta$ T cells. Inf Imm 70(3): 1488–1500.
28. Ozyigit MO, Senturk S, Akkoc A (2007) Suspected congenital generalized tuberculosis in a newborn calf. Vet Rec 160(9): 307–308.
29. Zanini MS, Moreira E, Lopes MT, Mota P, Salas CE (1998) Detection of *Mycobacterium bovis* in milk by polymerase chain reaction. J Vet Med B 45: 473–479.
30. Phillips CJ, Foster CR, Morris PA, Teverson R (2003) The transmission of *Mycobacterium bovis* infection to cattle. Res Vet Sci 74(1): 1–15.
31. Vordemeier M, Ameni G, Berg S, Bishop R, Robertson B, et al. (2012) The influence of cattle breed on susceptibility to bovine tuberculosis in Ethiopia. Comp Imm Microb Inf Dis 35: 227–232.
32. Ameni G, Aseffa A, Engers H, Young D, Gordon S, et al. (2007) High prevalence and increased severity of pathology of bovine tuberculosis in Holsteins compared to zebu breeds under field cattle husbandry in Central Ethiopia. Clin Vac Imm 14(10): 1356–1361.
33. Driscoll EE, Hoffman JI, Green LE, Medley GF, Amos W (2011) A preliminary study of genetic factors that influence susceptibility to bovine tuberculosis in the British cattle Herd. PLoS ONE 6(4): e18806, doi:10.1371/journal.pone.0018806.
34. Sun L, Song Y, Riaz H, Yang H, Hua G, et al. (2012) Polymorphisms in toll-like receptor 1 and 9 genes and their association with tuberculosis susceptibility in Chinese Holstein cattle. Vet Imm Immunopat 147(3–4): 195–201.
35. Finlay EK, Berry DP, Wickham B, Gormley EP, Bradley DG (2012) A genome wide association scan of bovine tuberculosis susceptibility in Holstein-Friesian dairy cattle. PLoS ONE 7(2): e30545, doi:10.1371/journal.pone.0030545.
36. Nega M, Mazengia H, Mekonen G (2012) Prevalence and zoonotic implications of bovine tuberculosis in Northwest Ethiopia. Int J Med Sci 2(9): 188–192.

23

Short Copy Number Variations Potentially Associated with Tonic Immobility Responses in Newly Hatched Chicks

Hideaki Abe[1,2], Kenji Nagao[3], Miho Inoue-Murayama[1]*

1 Wildlife Research Center, Kyoto University, Kyoto, Japan, **2** Department of Anatomy, University of Otago, Dunedin, New Zealand, **3** Animal Husbandry Research Division, Aichi Agricultural Research Center, Aichi, Japan

Abstract

Introduction: Tonic immobility (TI) is fear-induced freezing that animals may undergo when confronted by a threat. It is principally observed in prey species as defence mechanisms. In our preliminary research, we detected large inter-individual variations in the frequency and duration of freezing behavior among newly hatched domestic chicks (*Gallus gallus*). In this study we aim to identify the copy number variations (CNVs) in the genome of chicks as genetic candidates that underlie the behavioral plasticity to fearful stimuli.

Methods: A total of 110 domestic chicks were used for an association study between TI responses and copy number polymorphisms. Array comparative genomic hybridization (aCGH) was conducted between chicks with high and low TI scores using an Agilent 4×180 custom microarray. We specifically focused on 3 genomic regions (>60 Mb) of chromosome 1 where previous quantitative trait loci (QTL) analysis showed significant F-values for fearful responses.

Results: ACGH successfully detected short CNVs within the regions overlapping 3 QTL peaks. Eleven of these identified loci were validated by real-time quantitative polymerase chain reaction (qPCR) as copy number polymorphisms. Although there wkas no significant *p* value in the correlation analysis between TI scores and the relative copy number within each breed, several CNV loci showed significant differences in the relative copy number between 2 breeds of chicken (White Leghorn and Nagoya) which had different quantitative characteristics of fear-induced responses.

Conclusion: Our data shows the potential CNVs that may be responsible for innate fear response in domestic chicks.

Editor: Zhanjiang Liu, Auburn University, United States of America

Funding: This study was supported financially by the Ministry of Education, Culture, Sports, Science and Technology (MEXT) with a Grant-in-aid for Science Research (#21310150 and 25118005) and the Global Center of Excellence Program "Formation of a Strategie Base for Biodiversity and Evolutionary Research: from Genome to Ecosystem". The funders had no role in study design, data collection and analysis, decision to publish, or preparation of the manuscript.

Competing Interests: The authors have declared that no competing interests exist.

* E-mail: mmurayama@wrc.kyoto-u.ac.jp

Introduction

Tonic immobility (TI) is an unlearned defensive behavior characterized by temporal paralysis [1], and is widely used to measure the extent of fear responses in chickens [2,3] and other animals [4–6]. An individual with fewer induction attempts and longer TI responses is generally considered to be more fearful than those with many induction attempts and shorter TI responses [7]. Indeed, previous studies have demonstrated that the Red Junglefowl (RJF) and domesticated White Leghorn (WL) can be discriminated by different quantitative distributions of TI indices [8,9]. Our previous study also detected significant differences in the TI responses between WL and Nagoya breeds (NG) in newly hatched chicks (days 1–2 after hatching) [10]. These significant levels of interbreed heterogeneity may be attributed to the artificial selection of response insensitivity to human handling during the process of chicken domestication.

Considerable efforts have been taken to understand the molecular basis of anxiety and fear-based responses based on the hypothesis that genetic linkage or pleiotropic gene effects could explain different reactions to fearful stimuli. In chickens, there are 2 major quantitative trait loci (QTL) for individual growth on chromosome 1 (*Growth1* and *Growth2*) [11], and surprisingly, *Growth1* QTL contains several genes which, together, affect personality [8]. Moreover, an important finding has been made regarding genetic links between fear responses and major growth QTLs in an RJF × WL intercross [7]. These findings raise the possibility that the growth QTL may contain genes or genetic regions that influence the extent of fear-related behavior in chickens with far-reaching effects at the molecular and cellular levels.

Another effective and reliable approach for identifying genes or genomic regions responsible for normal behaviors is to perform genome-wide searches for copy number gains and losses. Copy number variation (CNV) is defined as genomic duplications or deletions in relatively long elements (1 kb to several Mb in size). With increasing resolution in the detection of smaller CNVs, this definition has expanded to include short structural variants less than 1 kb, known as short CNVs (sCNVs) [12]. In humans, CNVs have been linked to various behaviors including brain-related disorders [13]. In non-human vertebrates, including chickens, a growing number of studies have focused on the associations between CNVs and observed phenotypic heterogeneities [14], and thus CNV has been recently recognized as an important source of genetic variability that may affect phenotypes because of the rearrangement of the genes or regulatory elements.

The main goal of this study was to identify novel sCNVs between chicks with high and low TI scores by using an array comparative genomic hybridization (aCGH) approach. We targeted 3 different QTL in chromosome 1, for which significant F values had been detected for TI responses in chickens. Our approach provides an efficient way to narrow the number of plausible factors that account for differences in fear-induced behaviors by focusing on the regions containing interesting QTL.

Materials and Methods

Animals Used in this Study

We used 3 breeds/strains of chicken with different selection histories (NG5 [$n=32$], NG7 [$n=39$], and WL [$n=39$]). NG was chosen as the target chicken breed in this study because of the following reason: NG chicks occasionally panic and are hurt when they are frightened by sounds or small stimuli. It is especially important for future management of economically significant breeds to uncover their genetic basis of fear-related behaviors.

The NG breed was originated from a cross between a local chicken from the City of Nagoya and the Chinese Buff Cochin in the early 1880 s. In 1905, this breed was recognized as the first practical breed for poultry farming in Japan, and the NG was formally established in 1919 [15]. Of the various strains, NG5 and NG7 have distinct histories of selection either as a layer-type strain (NG5) or as a meat-type strain (NG7). The details of husbandry of chicks have been described elsewhere [10].

Tonic Immobility Test

A TI test was conducted using male chicks on days 1 and 2 after hatching. We measured TI responses 6 times for each chick (3 times on each of the days 1 and 2), regardless of the success rate of TI induction. We employed the same method used for assessing TI scores in adult chickens [1]; each chick was placed on its back in a V-shaped cradle and was kept there with light pressure on its breast for 5 s. After removing the pressure, chicks were not considered to be in TI status if the bird jumped up or righted itself within 5 s. We carried out the procedure 3 times in succession on each individual. The operator recorded the number of induction attempts required to induce a chick into the TI status as well as the time until righting in each successful TI induction (hereafter expressed as TI_{ind} and TI_{dur}, respectively). If a bird did not enter TI status for all 6 attempts, TI_{ind} was assigned a score of 7. When 10 min had passed since the bird entered TI, the chick was forced to stand up, and TI_{dur} was scored as 600 s. A fixed video camera was also used to ensure that environmental factors such as unexpected noise and changes in light intensity had no influence on chicks' fear-relevant behaviors.

Microarray Design

Blood samples were collected from the 3 chicken breeds/strains of chickens and stored at −20°C until DNA extraction. We isolated genomic DNA from 110 chicks by using either the PUREGENE® DNA Purification Kit (Gentra Systems, Minneapolis, USA) or DNeasy® Blood and Tissue Kit (Qiagen, Tokyo, Japan). DNA concentration of each extract was measured using a NanoDrop spectrophotometer (NanoDrop Technologies, Wilmington, DE). Four individuals with the highest average TI_{dur} were chosen from NG (IDs: NG933 [$TI_{ind}=1$; average $TI_{dur}=248.8$ s], NG4692 [$TI_{ind}=1$; average $TI_{dur}=172.3$ s], and NG3557 [$TI_{ind}=1$; average $TI_{dur}=255.0$ s]) and WL (WL3597 [$TI_{ind}=1$; average $TI_{dur}=199.3$ s]) strains as samples compared in aCGH analysis. One NG chick (NG999 [$TI_{ind}=7$; $TI_{dur}=0$]) that had not been induced the TI status in all 6 attempts was chosen as the reference sample. We used a chicken CGH Microarray 4×180K (Agilent Technologies, Tokyo, Japan), containing 180,000 custom probes of 60-mer, because Agilent's 60-mer offers the highest sensitivity and reproducibility among the currently available commercial platforms [16–18]. We designed these probes by using the eArray software (Available: https://earray.chem.agilent.com/earray/Accessed 2011 May 19). Note that these probes covered a total of 60 Mb in exonic, intronic, and intergenic regions of chromosome 1, where significant F values were detected by previous QTL analysis for fear-related behaviors [7]. Information on QTL for TI attempts (trait ID: 2123) and duration (2124) in the chicken genome was obtained from the QTL database (Available: http://www.genome.iastate.edu/cgi-bin/QTLdb/GG/index. Accessed 2011 Oct 11). The mean probe spacing was 1,029 bp, and the median probe spacing was 264 bp. Our strategy was somewhat analogous to that employed by a previous study [19], which targeted for restricted chromosomal regions in the porcine genome. There were several reasons for targeting sCNVs as a candidate for TI response variability. Although no clear pattern for CNV effect versus CNV-gene distance has been observed, smaller variants less than 1 kb have been found to be more likely to regulate gene transcription than larger variants [12]. Moreover, a recent study suggested that sCNVs tend to originate from the presence of a variable number of tandem repeats, which could provide a source of genetic variability for modifying normal and abnormal human behaviors [20]. All hybridizations were performed using 2 dyes for labeling reference (Cy3) and sample DNA (Cy5). The hybridization and initial data analysis were performed by MACS® Molecular genomic service (Miltenyi Biotec GmbH, Bergisch Gladbach, Germany).

Detection of Copy Number Variations

Statistical analysis for CNV detection was performed using Agilent GENOMIC WORKBENCH Standard Edition 6.5 software (Agilent Technologies). The minimum number of probes present in an aberrant region was 4, and aberrant segments were identified for a CNV locus when the average $\log_2$ ratio was greater than $|\pm 0.4|$. In addition, a less stringent filter was used to infer aberrant regions under the condition that the minimum number of probes present in an aberrant region was 2. In both cases, the statistical analysis of aberrant regions was based on the aberrant detection algorithm ADM-2. The full data set and designs from the oligo aCGH experiments have been submitted to the GEO database [21] under the accession ID GSE38434.

Copy Number Validation by Quantitative Polymerase Chain Reaction

To validate representative aCGH results, quantitative polymerase chain reaction (qPCR) was performed using the chicken β-actin gene (*ACTB*) as a reference for normalization of the real-time PCR experiments. On the basis of the aCGH aberration data, 52 sets of PCR primers for candidate sites were designed using Primer3 software [22]. Each set of primer pair was expected to yield PCR products, ranging from 150 to 200 bp based on the sequence in the chicken genome assembly build 3.1. The primer sequences are listed in Table S1. Prior to real-time PCR, each product was electrophoresed on a 2.0% agarose gel in order to verify the expected product size. Amplification curve and Ct values were generated with the Thermal Cycler Dice® Real Time System (Takara Bio, Shiga, Japan). Each qPCR reaction (15 μL) contained 10 ng of genomic DNA, 2.0 μL of 5 μM forward and reverse primers, 0.3 μL of ROX reference dye, and 7.5 μL of the 2× SYBR® Premix Ex Taq™ II (Takara Bio). The PCR cycle consisted of initial denaturation at 95°C for 30 s, followed by 40 cycles at 95°C for 5 s and 60°C for 31 s. Each sample was run 3 times to obtain accurate qPCR results. For relative amount quantification, Ct value differences were used to quantify the relationship between relative copy number and β-actin. This was calculated as follows: relative copy number = $\log 2^{\Delta Ct}$, where $\Delta Ct = Ct_{\beta\text{-actin}} - Ct_{target}$. The Ct values obtained were in agreement within and between runs.

Statistical Analysis

Since almost all test data showed non-normal frequency distributions that could not be transformed to meet the requirements of parametric statistical analysis, the Kruskal-Wallis test was carried out to examine the difference in TI_{ind} and TI_{dur} scores between breeds/strains. In each sCNV locus, we compared TI scores with the distribution of relative copy number, calculated as $\log 2^{\Delta Ct}$. To compare relative copy number between chicks with high and low TI scores, each chicken cohort was classified into high ($TI_{ind} = 1$; $TI_{dur} \geq 60$ s) and low ($TI_{ind} = 2\sim7$; $TI_{dur} < 60$ s) groups. Biased relationships between copy number and each TI score were examined using F-statistics under the null hypothesis of no association between copy number on the target locus and TI responses.

Ethical Note

All aspects of the study were performed according to the guidelines established by the Ministry of Education, Culture, Sports, Science and Technology in Japan (Notice No. 71). This study fulfilled ethical guidelines of the International Society of Applied Ethology [23]. The protocol was approved by the Committee on the Ethics of Animal Experiments of the Wildlife Research Center of Kyoto University (Permit No. WRC-2012-EC001).

Results

A total of 180,000 unique probes was designed in chicken chromosome 1 targeting 3 identical QTL for fear traits (Figure 1). Table 1 shows the main results from the TI experiment in each cohort of newly hatched chicks. WL chicks were easily induced into TI status according to the number of successful TI inductions to total attempts and TI_{ind} as compared with those of NG ($F_{1,107} = 16.18$; $p<0.001$), whereas TI_{dur} in WL was significantly shorter than that of NG ($F_{1,107} = 4.56$; $p<0.05$). Data from NG5 and NG7 were combined for statistical analysis, since we did not find any difference in their TI scores ($p>0.05$).

A large number of aberrant loci were detected in chicken chromosome1 based on the aCGH analysis between chicks characterized with high and low TI scores. The total number of aberrant segments identified in the 4 comparisons was 202 (average 50.5) in a stringent setting (4 probes) and 477 (average 119.3) in a less stringent setting (2 probes). Of these segments, 288 showed loss variation, and the remaining 391 segments showed gain variation. The duplicated segments (gain) occupied 57.6% in total length aberration. The average length of gain or loss segments was estimated at 3,552 bp (4 probes) and 1,833 bp (2 probes). The 477 CNVs found under the less stringent criteria encompassed 874.4 kb, which accounted for 1.46% of the total target region (60 Mb) in this study. This ratio was similar to the value suggested by whole genome analysis, indicating that CNVs occupied 1.34% of the entire chicken genome [24]. CNVs were not equally distributed throughout the target regions; the distal QTL peak (480–515 cM) of chromosome 1 contained a greater number of sCNVs than the other 2 QTL regions (Fisher's exact test; $p<0.01$). Since a previous study also detected a large number of CNVs in this region [24], this chromosomal region can be regarded as a "CNV hotspot" in which mutations leading to copy number differences between individuals occur more frequently than expected. We further screened these aberrant segments under the following two conditions: the segments were commonly detected in all genomic comparisons with $\log_2$ ratio $|\pm0.4|$, and the segments were detected in at least 2 comparisons with $\log_2$ ratio $|\pm1.2|$. From the whole data set (4 & 2 probes), we extracted 52 loci that satisfied either of these conditions.

Real-time qPCR was performed to validate the aCGH data for 52 candidate loci. For preliminary screening procedure of qPCR validation, 12 DNA samples from NG and WL strains, including samples used for aCGH analysis, were employed as templates. The qPCR analysis displayed different patterns of quantitative variations that could be classified into 5 categories: (1) the same level of ΔCt values was detected in all samples except for the reference sample, whose PCR product was completely absent (described as "deletion" in Table 2); (2) an apparent variation of the ΔCt values was observed in 12 samples including the reference sample (described as "CNV" in Table 2) (3) almost the same level of ΔCt values was observed in all samples including a reference sample, i.e., monomorphic loci; (4) no PCR product was detected in several samples including or excluding the reference sample; and (5) failed PCR amplification in all specimens, probably due to primer mismatch. We found quantitative variations in the following percentage: (1) 11.5%, (2) 9.6%, (3) 63.5%, (4) 3.8%, and (5) 11.5%. We identified 11 loci, belonging to either category (1) or (2), as candidate sCNVs, and sCNVs in the other category were briefly summarized in Table S2.

For the 11 candidate loci, we examined the association between sCNVs and TI_{ind} or TI_{dur} scores in 110 chicks. With regard to both TI indices, we detected significant differences in the copy number distributions between NG and WL in *TIC_3* ($F_{1,107} = 4.10$; $p<0.05$), *TIC_18* ($F_{1,107} = 19.05$; $p<0.001$), and *TIC_42* ($F_{1,107} = 236.59$; $p<0.001$; Figure 2). However, there was no difference in the relative copy number for all target CNVs between chicks with high and low TI scores (TI_{ind} & TI_{dur}) in each NG and WL. Scatter plots of correlation analysis between TI scores and the relative copy number in each locus were shown in the supporting information (Figures S1 and S2).

Discussion

Until now, most studies on chicken domestication have focused on the genetic and behavioral heterogeneities between RJF and WL in order to highlight the alternative histories of domestication. Additionally, our preliminary study [10] detected large differences in the quantitative traits of TI responses between WL and NG populations. WL chicks were prone to be induced into the TI status with fewer attempts (low TI_{ind}), whereas NG had longer TI_{dur} in each successful TI induction. These empirical data strongly support the previous hypothesis that the TI behavior in chickens has a genetic basis with breed- or strain-specific behavioral characteristics [8,25,26].

However, we did not detect a difference in the relative copy number between high and low TI groups within each chicken breed. There are

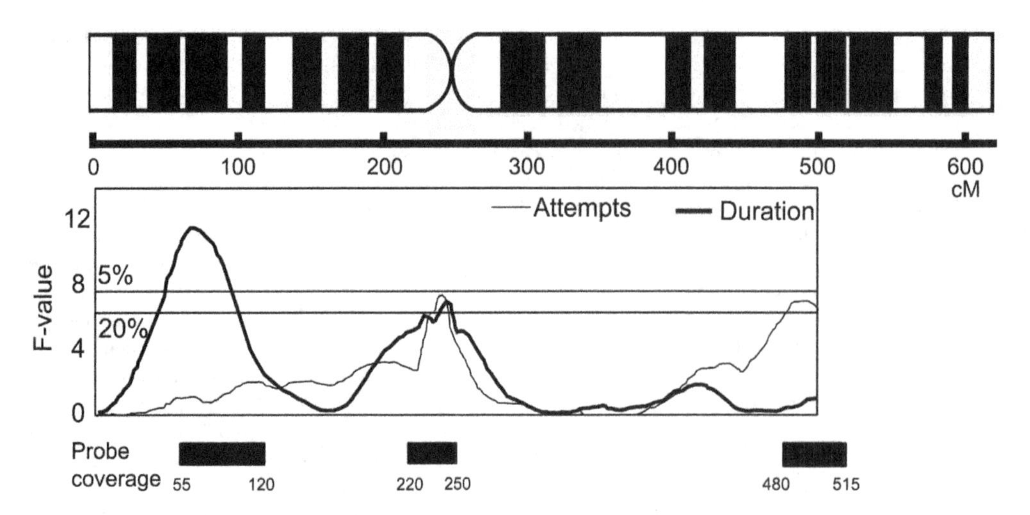

Figure 1. Probe coverage on chicken chromosome 1 for array comparative genomic hybridization. Probes are designed in 3 regions (>60 Mb) where significant *F*-values have been identified by previous quantitative trait loci analysis. Genome-wide *F*-values for tonic immobility duration (thick line) and induction attempts (thin line) are quoted from [7].

several reasons for this, one of which is that the number of chicks used in this study may be insufficient to detect an association between sCNVs and variations in the chicks' sensitivity to fearful stimuli. Considering that a TI response is not simple but has a complicated contextual framework involving cognitive and neural processes, there may be multiple genetic determinants associated with the quantitative traits of TI responses. Therefore, to detect an effect of CNVs on TI, sample sizes would need to be increased more than 20-fold over the current study design, and other genetic and epigenetic factors such as single nucleotide polymorphisms (SNPs) and methylation patterns should be included as possible candidates for affecting fear-induced behaviors in future research. Another explanation for the lack of significant results in this study is the inconsistency in chicken breeds between the reference genomic DNA sequence derived from NG and the genome shotgun sequence of RJF [27] used for primer design in qPCR validation. Based on the patterns of qPCR amplification, more than half (63.5%) of candidate CNV loci showed monomorphism in amplification plots. This is probably due to considerable sequence diversity between NG/WL and RJF, which might lead to an apparent loss of copy number polymorphisms in the validation phase. According to archaeological findings [28], the divergence time of domestic chickens from RJF is estimated to be nearly 8,000 years. Comparison of DNA sequences from 30 introns at 25 nuclear loci revealed that the extent of nucleotide divergence after the split of RJF from their chicken ancestor is as small as 0.01% [29]. However, a more recent study indicated that NG lines were genetically distinct from commercial gene pools, thus making it a unique genetic resource [30]. A way to avoid this complication would be to change the strategy by including target chicken breeds in TI measurements. Finally, we cannot exclude the possibility that innate and learned fear responses are modulated differently by independent neural networks and mechanisms. It should be emphasized that in previous QTL analysis, TI tests were conducted when chickens were 29–30 weeks of age [7], whereas newly hatched chicks were used here for TI testing to preclude secondary social and environmental effects on TI responses. Therefore, if TI scores would fluctuate during the growing stages of chicks and juveniles, as has been suggested by previous studies [31,32], these conflicts may have some impact on the outcome of genome-wide association studies (GWAS) between genetic variants and quantitative TI scores. We therefore consider our aCGH approach for analyzing inter-individual variations in freezing behavior, a useful preliminary analysis capable of generating further hypotheses for future evaluation.

Among the candidate sCNVs, which were excluded by aCGH screening and subsequent qPCR, we detected significant differ-

Table 1. The induction and duration of TI response in each chicken breed/strain.

Strains	*n*	Induction/attempts	Induction	TI duration
NG5	32	51/192 (0.27)	4.4	99.4 ± 15.8
NG7	39	45/234 (0.19)	4.6	114.2 ± 22.5
WL	39	82/234 (0.35)	2.7	74.8 ± 10.2

Note: Standard error of the mean (SEM) are shown with the time until righting (sec).

Table 2. Candidate short Copy Number Variations identified by array Comparative Genomic Hybridization and subsequent qPCR validation.

locus ID	probes	Start	Stop	bp	Gene	qPCR
TIC_03	2	25773437	25773746	309	Non-coding	CNV
TIC_04	2	28079752	28080201	449	Non-coding	CNV
TIC_05	2	28432127	28432740	613	Non-coding	deletion
TIC_15	2,4	87064880	87065977	1097	NME7	deletion
TIC_16	2	170375957	170376395	438	KIAA0564	CNV
TIC_18	2,4	176058363	176059210	847	Non-coding	CNV
TIC_42	2	177404311	177404572	261	NBEA	CNV
TIC_44	2	180954362	180954612	250	CDK8	CNV
TIC_19	2	181380041	181380759	718	NUPL1	CNV
TIC_20	2,4	186702464	186703251	787	Non-coding	CNV
TIC_21	2	188457807	188458430	623	Non-coding	CNV

Note: Only loci displaying quantitative difference in qPCR validation are shown here.

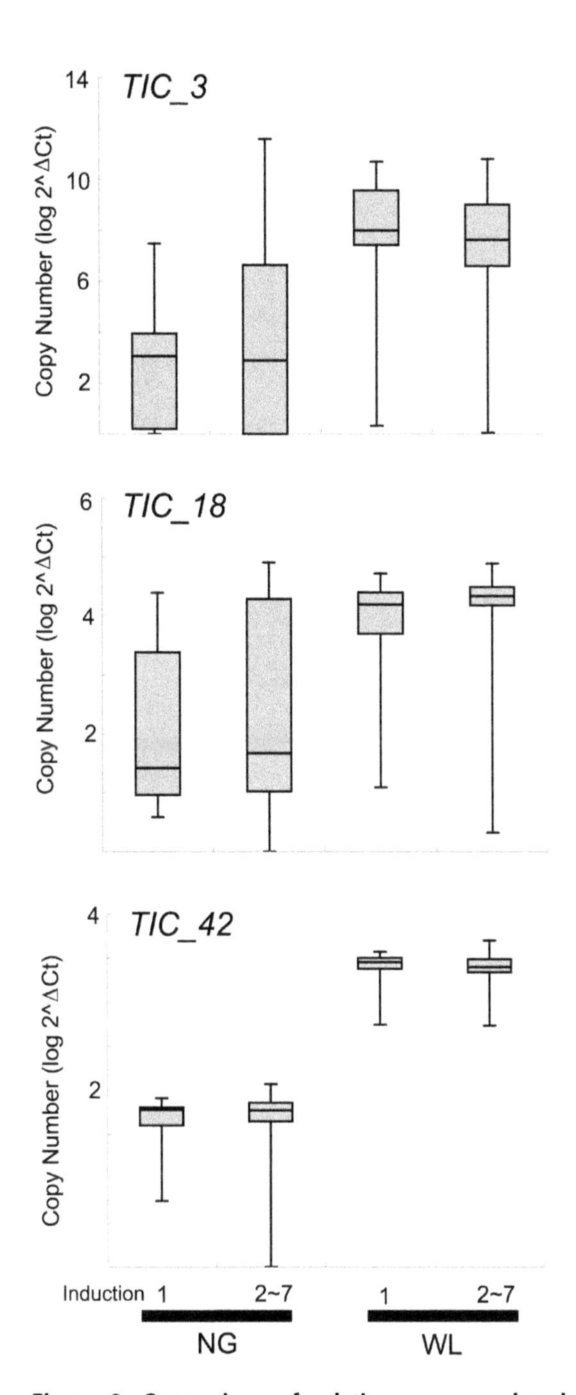

Figure 2. Comparison of relative copy number between chicken breeds with different sensitivity to fear. Relative copy number is calculated as $\log2^{\Delta Ct}$, where $\Delta Ct = Ct_{\beta\text{-actin}} - Ct_{target}$. NG and WL indicate Nagoya and White Leghorn, respectively. The number of samples in each group was; $n = 13$ (NG; TI induction 1), $n = 58$ (NG; TI induction 2~7), $n = 13$ (WL; TI induction 1), and $n = 26$ (WL; TI induction 2~7).

ences in the relative copy number between NG and WL in *TIC_3*, *TIC_18*, and *TIC_42*. The *TIC_3* locus is embedded within *Growth1* QTL, for which the highest *p* value was obtained in a QTL study for TI traits [7]. The fact that this locus was found in a relatively large non-coding region may imply that it affects gene expression through far-reaching cis- or trans-acting mechanisms. The *TIC_18* locus is located upstream of the *TRPC4* gene (*ENSGALG00000017044*), which plays a role in multiple processes, including neurotransmitter release and exocytosis [33]. A recent study suggested that TRPC4 play a pivotal role in regulating dopamine release, which may modulate emotional and cognitive responses in rats [34]. Thus, it is worthwhile investigating the expression of orthologous genes in the chicken brain and further examining the correlation between levels of expression and innate fear responsiveness. The other candidate sCNV (*TIC_42*) was found in the *NBEA* gene (*ENSGALG00000017062*), which has been identified as a putative regulator of membrane protein trafficking associated with the *trans*-Golgi network. [35]. NBEA-deficient mice died immediately after birth apparently from respiratory paralysis [36]. NBEA plays a complex role in the development and functioning of synapses and is believed to play a role in autism spectrum disorders by evoking an excitatory-inhibitory imbalance in synaptic activity [37]. Further information on differential gene expression of these candidate genes in the brain between various chicken selection lines, will provide opportunities to examine their role in shaping behavioral differences.

In conclusion, we identified 11 sCNVs that potentially account for the robust differences observed in TI responses among chicken breeds. None of the genes identified in this study have been directly implicated in TI responses; however, studies on several of these genes, such as *TRPC4* and *NBEA*, have reported brain functions that might be involved in abnormal behaviors in humans and rodents. Future experiments involving gene expression assays and mutagenic approaches are needed to determine whether any of the novel candidate genes identified here play a role in modulating innate fear responses.

Acknowledgments

We thank Akihiro Nakamura, Junichi Ueda, and Hideshi Ohguchi (Aichi Agricultural Research Center) for their technical support. We also thank Azusa Hayano and Miyuki Sato (Kyoto University) for their valuable assistance with laboratory work.

Author Contributions

Conceived and designed the experiments: HA MI-M. Performed the experiments: HA KN. Analyzed the data: HA. Contributed reagents/ materials/analysis tools: KN. Wrote the paper: HA. Conceived and designed molecular experiments: HA. Co-conceived all behavioral and molecular experiments: MI-M. Managed the domestic chicks used in behavioral experiments: KN. Performed molecular experiments: HA. Carried out behavioral experiments and qPCR: KN. Gave final approval of the version to be published: MI-M. Read and approved the final manuscript: MI-M KN HA.

References

1. Jones RB (1986) The tonic immobility reaction of the domestic fowl: A review. World's Poult Sci J 42: 82–96.
2. Maser JD, Gallup GG, Jr. (1974) Tonic immobility in the chicken: Catalepsy potentiation by uncontrollable shock and alleviation by imipramine. Psychosom Med 36: 199–205.
3. Kujiyat SK, Craig JV, Dayton AD (1983) Duration of tonic immobility affected by housing environment in White Leghorn hens. Poult Sci 62: 2280–2282.
4. Webster DG, Lanthorn TH, Dewsbury DA, Meyer ME (1981) Tonic immobility and the dorsal immobility response in twelve species of muroid rodents. Behav Neural Biol 31: 32–41.
5. Cashner FM, Olson RD, Erickson DG, Olson GA (1981) Effects of MIF-I and sex differences on tonic immobility duration in the lizard, Anolis carolinensis. Peptides 2 Suppl 1: 161–165.
6. Erhard HW, Mendl M, Christiansen SB (1999) Individual differences in tonic immobility may reflect behavioural strategies. Appl Anim Behav Sci 64: 31–46.
7. Schtz KE, Kerje S, Jacobsson L, Forkman B, Carlborg Ö, et al. (2004) Major growth QTLs in fowl are related to fearful behavior: possible genetic links between fear responses and production traits in a red junglefowl x white leghorn intercross. Behav Genet 34: 121–130.
8. Wirén A, Jensen P (2010) A growth QTL on chicken chromosome 1 affects emotionality and sociality. Behav Genet 41: 303–311.
9. Campler M, Jöngren M, Jensen P (2009) Fearfulness in red junglefowl and domesticated White Leghorn chickens. Behav Processes 81: 39–43.
10. Abe H, Nagao K, Nakamura A, Inoue-Murayama M (2013) Differences in responses to repeated fear-relevant stimuli between Nagoya and White Leghorn chicks. Behav Process 99: 95–99.
11. Kerje S, Carlborg O, Jacobsson L, Schütz K, Hartmann C, et al. (2003) The twofold difference in adult size between the red junglefowl and White Leghorn chickens is largely explained by a limited number of QTLs. Anim Genet 34: 264–274.
12. Banerjee S, Oldridge D, Poptsova M, Hussain WM, Chakravarty D, et al. (2011) A computational framework discovers new copy number variants with functional importance. PLoS One 6: e17539.
13. Wain LV, Armour JA, Tobin MD (2009) Genomic copy number variation, human health, and disease. Lancet 374: 340–350.
14. Wright D, Boije H, Meadows JR, Bed'hom B, Gourichon D, et al. (2009) Copy number variation in intron 1 of SOX5 causes the Pea-comb phenotype in chickens. PLoS Genet 5: e1000512.
15. Nakamura A, Noda K (2001) Breeding history and genetic characters of Nagoya breed, a Japanese poultry breed of Aichi prefecture. Surv Rep Anim Genet Resour 12: 77–97 (In Japanese).
16. Greshock J, Feng B, Nogueira C, Ivanova E, Perna I, et al. (2007) A comparison of DNA copy number profiling platforms. Cancer Res 67: 10173–10180.
17. Baumbusch LO, Aaroe J, Johansen FE, Hicks J, Sun H, et al. (2008) Comparison of the Agilent, ROMA/NimbleGen and Illumina platforms for classification of copy number alterations in human breast tumors. BMC Genomics 9: 379.
18. Pinto D, Darvishi K, Shi X, Rajan D, Rigler D, et al. (2011) Comprehensive assessment of array-based platforms and calling algorithms for detection of copy number variants. Nat Biotechnol 29: 512–520.
19. Fadista J, Nygaard M, Holm LE, Thomsen B, Bendixen C (2008) A snapshot of CNVs in the pig genome. PLoS One 3: e3916.
20. Conrad DF, Bird C, Blackburne B, Lindsay S, Mamanova L, et al. (2010) Mutation spectrum revealed by breakpoint sequencing of human germline CNVs. Nat Genet 42: 385–391.
21. Barrett T, Troup DB, Wilhite SE, Ledoux P, Rudnev D, et al. (2009) NCBI GEO: archive for high-throughput functional genomic data. Nucleic Acids Res 37: D885–890.
22. Rozen S, Skaletsky H (2000) Primer3 on the WWW for general users and for biologist programmers. Methods Mol Biol 132: 365–386.
23. Sherwin CM, Christiansen SB, Duncan IJ, Erhard HW, Lay DC, et al. (2003) Guidelines for the ethical use of animals in applied ethology studies. Appl Anim Behav Sci 81: 291–305.
24. Wang X, Nahashon S, Feaster TK, Bohannon-Stewart A, Adefope N (2010) An initial map of chromosomal segmental copy number variations in the chicken. BMC Genomics 11: 351.
25. Craig JV, Kujiyat SK, Dayton AD (1984) Tonic immobility responses of white leghorn hens affected by induction techniques and genetic stock differences. Poult Sci 63: 1–10.
26. Albentosa MJ, Kjaer JB, Nicol CJ (2003) Strain and age differences in behaviour, fear response and pecking tendency in laying hens. Br Poult Sci 44: 333–344.
27. Wallis JW, Aerts J, Groenen MA, Crooijmans RP, Layman D, et al. (2004) A physical map of the chicken genome. Nature 432: 761–764.
28. Fumihito A, Miyake T, Sumi S, Takada M, Ohno S, et al. (1994) One subspecies of the red junglefowl (*Gallus gallus gallus*) suffices as the matriarchic ancestor of all domestic breeds. Proc Natl Acad Sci U S A 91: 12505–12509.
29. Sawai H, Kim HL, Kuno K, Suzuki S, Gotoh H, et al. (2010) The origin and genetic variation of domestic chickens with special reference to junglefowls *Gallus g. gallus* and *G. varius*. PLoS One 5: e10639.
30. Tadano R, Nakamura A, Kino K (2012) Analysis of genetic divergence between closely related lines of chickens. Poult Sci 91: 327–333.
31. Heiblum R, Aizenstein O, Gvaryahu G, Voet H, Robinzon B, et al. (1998) Tonic immobility and open field responses in domestic fowl chicks during the first week of life. Appl Anim Behav Sci 60: 347–357.
32. Ghareeb K (2010) Presence of males within laying hens affects tonic immobility response and sociality. Int J Poult Sci 9: 1087–1091.
33. Obukhov AG, Nowycky MC (2002) TRPC4 can be activated by G-protein-coupled receptors and provides sufficient Ca(2+) to trigger exocytosis in neuroendocrine cells. J Biol Chem 277: 16172–16178.
34. Illig KR, Rasmus KC, Varnell AL, Ostertag EM, Klipec WD, et al. (2011) TRPC4 ion channel protein is selectively expressed in a subpopulation of dopamine neurons in the ventral tegmental area. Available from Nature Precedings <http://dx.doi.org/10.1038/npre.2011.6577.1>.
35. Wang X, Herberg FW, Laue MM, Wullner C, Hu B, et al. (2000) Neurobeachin: A protein kinase A-anchoring, beige/Chediak-higashi protein-molog implicated in neuronal membrane traffic. J Neurosci 20: 8551–8565.
36. Medrihan L, Rohlmann A, Fairless R, Andrae J, Döring M, et al. (2009) Neurobeachin, a protein implicated in membrane protein traffic and autism, is required for the formation and functioning of central synapses. J Physiol 587: 5095–5106.
37. Persico AM, Bourgeron T (2006) Searching for ways out of the autism maze: genetic, epigenetic and environmental clues. Trends Neurosci 29: 349–358.

Permissions

All chapters in this book were first published in PLOS ONE, by The Public Library of Science; hereby published with permission under the Creative Commons Attribution License or equivalent. Every chapter published in this book has been scrutinized by our experts. Their significance has been extensively debated. The topics covered herein carry significant findings which will fuel the growth of the discipline. They may even be implemented as practical applications or may be referred to as a beginning point for another development.

The contributors of this book come from diverse backgrounds, making this book a truly international effort. This book will bring forth new frontiers with its revolutionizing research information and detailed analysis of the nascent developments around the world.

We would like to thank all the contributing authors for lending their expertise to make the book truly unique. They have played a crucial role in the development of this book. Without their invaluable contributions this book wouldn't have been possible. They have made vital efforts to compile up to date information on the varied aspects of this subject to make this book a valuable addition to the collection of many professionals and students.

This book was conceptualized with the vision of imparting up-to-date information and advanced data in this field. To ensure the same, a matchless editorial board was set up. Every individual on the board went through rigorous rounds of assessment to prove their worth. After which they invested a large part of their time researching and compiling the most relevant data for our readers.

The editorial board has been involved in producing this book since its inception. They have spent rigorous hours researching and exploring the diverse topics which have resulted in the successful publishing of this book. They have passed on their knowledge of decades through this book. To expedite this challenging task, the publisher supported the team at every step. A small team of assistant editors was also appointed to further simplify the editing procedure and attain best results for the readers.

Apart from the editorial board, the designing team has also invested a significant amount of their time in understanding the subject and creating the most relevant covers. They scrutinized every image to scout for the most suitable representation of the subject and create an appropriate cover for the book.

The publishing team has been an ardent support to the editorial, designing and production team. Their endless efforts to recruit the best for this project, has resulted in the accomplishment of this book. They are a veteran in the field of academics and their pool of knowledge is as vast as their experience in printing. Their expertise and guidance has proved useful at every step. Their uncompromising quality standards have made this book an exceptional effort. Their encouragement from time to time has been an inspiration for everyone.

The publisher and the editorial board hope that this book will prove to be a valuable piece of knowledge for researchers, students, practitioners and scholars across the globe.

List of Contributors

Ylva Telldahl and Jan Storå
Osteoarchaeological Research Laboratory, Department of Archaeology and Classical Studies, Stockholm University, Stockholm, Sweden

Emma Svensson and Anders Götherström
Department of Evolutionary Biology, Uppsala Universitet, Uppsala, Sweden

Yanhong Cao
Institute of Kaschin-beck Disease, Center for Endemic Disease Control, Chinese Center for Disease Control and Prevention, Harbin Medical University; Key Laboratory of Etiologic Epidemiology, Education Bureau of Heilongjiang Province and Ministry of Health (23618104), Harbin, China,
Departments of Orthopaedic Surgery and Biomedical Engineering, University of Tennessee Health Science Center, Memphis, Tennessee, United States of America

Xiaoyun Liu,Yan Jiao, Yonghui Ma, Karen A. Hasty and Weikuan Gu
Departments of Orthopaedic Surgery and Biomedical Engineering, University of Tennessee Health Science Center, Memphis, Tennessee, United States of America

Nan Deng
Department of Medicine, University of Tennessee Health Science Center, Memphis, Tennessee, United States of America

John M. Stuart
Research Service, Veterans Affairs Medical Center, 1030 Jefferson Avenue, Memphis Tennessee, United States of America

Annie Menoud
Institute of Genetics, Vetsuisse Faculty, University of Bern, Bern, Switzerland

Cord Drögemüller
Institute of Genetics, Vetsuisse Faculty, University of Bern, Bern, Switzerland
Dermfocus, Vetsuisse Faculty, University of Bern, Bern, Switzerland

Monika Welle
Institute of Animal Pathology, Vetsuisse Faculty, University of Bern, Bern, Switzerland
DermFocus, Vetsuisse Faculty, University of Bern, Bern, Switzerland

Jens Tetens
Institute for Animal Breeding and Husbandry, Christian-Albrechts-University Kiel, Kiel, Germany

Peter Lichtner
Institute of Human Genetics, Helmholtz Zentrum München - German
Research Center for Environmental Health, Neuherberg, Germany

Rebuma Firdessa, Elena Hailu, Girume Erenso, Teklu Kiros, Lawrence Yamuah and Abraham Aseffa
Armauer Hansen Research Institute, Addis Ababa, Ethiopia

Rea Tschopp
Armauer Hansen Research Institute, Addis Ababa, Ethiopia
Centre for Molecular Microbiology and Infection, Imperial College London, London, United Kingdom
Swiss Tropical and Public Health, Basel, Switzerland

Alehegne Wubete, Melaku Sombo and Mesfin Sahile
National Animal Health Diagnostic and Investigation Center, Sebeta, Addis Ababa, Ethiopia

Stephen V. Gordon
School of Veterinary Medicine, University College Dublin, Dublin, Republic of Ireland

Douglas Young
Centre for Molecular Microbiology and Infection, Imperial College London, London, United Kingdom

Martin Vordermeier, R. Glyn Hewinson and Stefan Berg
Department for Bovine Tuberculosis, Animal Health and Veterinary Laboratories Agency, Weybridge, Surrey, United Kingdom

Jessica L. Petersen, James R. Mickelson and Molly E. McCue
University of Minnesota, College of Veterinary Medicine, St Paul, Minnesota, United States of America

E. Gus Cothran
Texas A&M University, College of Veterinary Medicine and Biomedical Science, College Station, Texas, United States of America

Lisa S. Andersson, Jeanette Axelsson, Gabriella Lindgren and Sofia Mikko
Swedish University of Agricultural Sciences, Department of Animal Breeding and Genetics, Uppsala, Sweden

Ernie Bailey and Kathryn T. Graves
University of Kentucky, Department of Veterinary Science, Lexington, Kentucky, United States of America

Danika Bannasch and M. Cecilia T Penedo
University of California Davis, School of Veterinary Medicine, Davis, California, United States of America

Matthew M. Binns
Equine Analysis, Midway, Kentucky, United States of America

Alexandre S. Borges
University Estadual Paulista, Department of Veterinary Clinical Science, Botucatu-SP, Brazil

Pieter Brama
University College Dublin, School of Veterinary Medicine, Dublin, Ireland

Artur da Câmara Machado and Maria Susana Lopes
University of Azores, Institute for Biotechnology and Bioengineering, Biotechnology Centre of Azores, Angra do Heroı´smo, Portugal

Ottmar Distl
University of Veterinary Medicine Hannover, Institute for Animal Breeding and Genetics, Hannover, Germany

Michela Felicetti and Maurizio Silvestrelli
University of Perugia, Faculty of Veterinary Medicine, Perugia, Italy

Laura Fox-Clipsham and Mark Vaudin
Animal Health Trust, Lanwades Park, Newmarket, Suffolk, United Kingdom

Gérard Guérin
French National Institute for Agricultural Research-Animal Genetics and Integrative Biology Unit, Jouy en Josas, France

Bianca Haase and Claire M.Wade
University of Sydney, Veterinary Science, New South Wales, Australia

Telhisa Hasegawa
Nihon Bioresource College, Koga, Ibaraki, Japan

Karin Hemmann, Hannes Lohi and Marja Raekallio
University of Helsinki, Faculty of Veterinary Medicine, Helsinki, Finland

Emmeline W. Hill and Beatrice A. McGivney
University College Dublin, College of Agriculture, Food Science and Veterinary Medicine, Belfield, Dublin, Ireland

Tosso Leeb
University of Bern, Institute of Genetics, Bern, Switzerland

Nicholas Orr
Institute of Cancer Research, Breakthrough Breast Cancer Research Centre, London, United Kingdom

Richard J. Piercy
Royal Veterinary College, Comparative Neuromuscular Diseases Laboratory, London, United Kingdom

Stefan Rieder
Swiss National Stud Farm, Agroscope Liebefeld-Posieux Research Station, Avenches, Switzerland

Knut H. Røed
Norwegian School of Veterinary Science, Department of Basic Sciences and Aquatic Medicine, Oslo, Norway

June Swinburne
Animal Health Trust, Lanwades Park, Newmarket, Suffolk, United Kingdom
Animal DNA Diagnostics Ltd, Cambridge, United Kingdom

Teruaki Tozaki
Laboratory of Racing Chemistry, Department of Molecular Genetics, Utsunomiya, Tochigi, Japan

Sandrine Hughes, Marilyne Duffraisse and Catherine Hänni
Paléogé nomique et Evolution Moléculaire, Institut de Génomique Fonctionnelle de Lyon, Universitéde Lyon, UniversitéLyon 1, CNRS UMR 5242, INRA, Ecole Normale Supérieure de Lyon, 46 allée d'Italie, 69364 Lyon Cedex 07, France

Helena Fernández, Franc¸ois Pompanon and Pierre Taberlet
Laboratoire d'Ecologie Alpine, CNRS UMR 5553, Université Joseph Fourier, B.P. 53, 38041 Grenoble Cedex 9, France

Jean-Denis Vigne
Centre National de la Recherche Scientifique, UMR 7209, Muséum National d'Histoire Naturelle, «Archéozoologie, Archéobotanique: Sociétés, Pratiqueset Environnements», Département "Ecologie et Gestion de la Biodiversite´ " CP 56, 75005 Paris, France

Thomas Cucchi
Centre National de la Recherche Scientifique, UMR 7209, Muséum National d'Histoire Naturelle, «Archéozoologie, Archéobotanique: Sociétées, Pratiqueset Environnements», Département "Ecologie et Gestion de la Biodiversité" CP 56, 75005 Paris, France
Department of Archaeology, University of Aberdeen, Aberdeen, United Kingdom

François Casabianca
Institut National de la Recherche Agronomique, UR 045 Laboratoire de Recherches sur le Développement de l'Elevage, Quartier Grossetti, 20250 Corte, France

Daniel Istria
Laboratoire d'Archéologie Médiévale Méditerranéenne, CNRS UMR 6572, 5 rue du château de l'Horloge, BP 647, 13094 Aix-en-Provence, France

Marius Warg Næss
Center for International Climate and Environmental Research - Oslo (CICERO), Fram Centre, Tromsø, Norway

Bård-Jørgen Bårdsen
Norwegian Institute for Nature Research (NINA), Arctic Ecology Department, Fram Centre, Tromsø, Norway

Sabrina Briefer Freymond and Rudolf Von Niederhäusern and Iris Bachmann
Agroscope Liebefeld-Posieux Research Station ALP-Haras, Swiss National Stud Farm SNSTF, Les Longs Prés, Avenches, Switzerland

Elodie F. Briefer
Queen Mary University of London, Biological and Experimental Psychology Group, School of Biological and Chemical Sciences, London, United Kingdom

Élia Benito-Gutiérrez, Hermann Weber, Diana Virginia Bryant and Detlev Arendt
Developmental Biology Unit, European Molecular Biology Laboratory (EMBL), Heidelberg, Germany

Ivica Medugorac, Sophie Rothammer and Martin Förster
Ludwig-Maximilians-University Munich, Munich, Germany

Doris Seichter and Ingolf Russ
Tierzuchtforschung e.V. München, Grub, Germany

Alexander Graf and Helmut Blum and Stefan Krebs
Laboratory for Functional Genome Analysis (LAFUGA), Gene Center, Ludwig-Maximilians-University Munich, Munich, Germany

Karl Heinrich Göpel
Goö pel Genetik GmbH, Herleshausen, Germany

Heidi Signer-Hasler and Christine Flury
School of Agricultural, Forest and Food Sciences, Bern University of Applied Sciences, Zollikofen, Switzerland

Tosso Leeb
Institute of Genetics, Vetsuisse Faculty, University of Bern, Bern, Switzerland

Bianca Haase
Institute of Genetics, Vetsuisse Faculty, University of Bern, Bern, Switzerland
Faculty of Veterinary Science, University of Sydney, New South Whales, Australia

Dominik Burger and Stefan Rieder
Agroscope Liebefeld-Posieux Research Station Agroscope
Liebefeld-Posieux (ALP) Haras, Swiss National Stud Farm (SNSTF), Avenches, Switzerland

Henner Simianer
Department of Animal Sciences, Georg-August-Universität, Göttingen, Germany

David M. Wright
School of Biological Sciences, Queen's University Belfast, Belfast, Northern Ireland, United Kingdom

Robin A. Skuce
School of Biological Sciences, Queen's University Belfast, Belfast, Northern Ireland, United Kingdom
Veterinary Sciences Division, Bacteriology Branch, Agri-Food and Biosciences Institute, Belfast, Northern Ireland, United Kingdom

Adrian R. Allen, Thomas R. Mallon and Stanley W. J. McDowell
Veterinary Sciences Division, Bacteriology Branch, Agri-Food and Biosciences Institute, Belfast, Northern Ireland, United Kingdom

Stephen C. Bishop, Elizabeth J. Glass, Mairead L. Bermingham and John A. Woolliams
The Roslin Institute and Royal (Dick) School of Veterinary Studies, University of Edinburgh, Midlothian, Scotland, United Kingdom

Daniel Frynta, Jana Baudyšovaá , Petra Hradcov and, Kateřina Faltusová
Department of Zoology, Faculty of Science, Charles University in Prague,Prague, Czech Republic

Lukáš Kratochví
Department of Ecology, Faculty of Science, Charles University in Prague, Prague, Czech Republic

Laurence Flori
INRA, UMR 1313 GABI, F-78350 Jouy-en-Josas, France
CIRAD, UMR INTERTRYP, Montpellier, France

Sophie Thevenon, Isabelle Chantal and David Berthier
CIRAD, UMR INTERTRYP, Montpellier, France

Mary Isabel Gonzatti, Joar Pinto and Pedro M. Aso
Departamento de Biologı´a Celular, Universidad Simo´n Bolı´var, Caracas, Venezuela

Mathieu Gautier
INRA, UMR CBGP, Montferrier-sur-Lez, France

Marius Warg Næss
Center for International Climate and Environmental Research - Oslo (CICERO), Fram Centre, Tromsø, Norway

Bård-Jørgen Bårdsen
Norwegian Institute for Nature Research (NINA), Arctic Ecology Department, Fram Centre, Tromsø, Norway

Matti Janhunen, Antti Kaus and Harri Vehviläinen
MTT Agrifood Research Finland, Biometrical Genetics, Jokioinen, Finland

Otso Järvisalo
Finnish Game and Fisheries Research Institute, Laukaa, Finland

Barbara Padalino
Department of Veterinary Medicine, University of Bari, Valenzano (Bari), Italy

Lydiane Aubé
Laboratoires d'èthologie animale et humaine EthoS -University of Rennes, Rennes, France

Meriem Fatnassi, Mohamed Hammadi and Touhami Khorchani
Livestock and Wildlife Laboratory, Arid Lands Institute, Médenine, Tunisia

Davide Monaco and Giovanni Michele Lacalandra
Department of Emergency and Organ Transplantation (D.E.T.O.), Veterinary Clinics and Animal Production Section, University of Bari, Valenzano (Bari), Italy

Stephan T. Leu and Michael J. Mahony
School of Environmental and Life Sciences, University of Newcastle, Newcastle, New South Wales, Australia

Martin J. Whiting
Department of Biological Sciences, Macquarie University, Sydney, New South Wales, Australia

Jaruwan Khonmee, Suvichai Rojanasthien, Veerasak Punyapornwithaya and Chatchote Thitaram
Faculty of Veterinary Medicine, Chiang Mai University, Chiang Mai, Thailand

Janine L. Brown
Center for Species Survival, Smithsonian Conservation Biology Institute, Front Royal, Virginia, United States of America

Anurut Aunsusin
Chiang Mai Night Safari, Chiang Mai, Thailand

Dissakul Thumasanukul and Adisorn Kongphoemphun
Omkoi Wildlife Sanctuary, Department of National Park, Wildlife and Plant Conservation, Chiang Mai, Thailand

Boripat Siriaroonrat and Wanlaya Tipkantha
Conservation Research and Education Division, Zoological Park Organization, Bangkok, Thailand

Beatriz Gutiérrez-Gil, Juan Jose Arranz and Elsa García-Gámez
Dpto. Producción Animal, Universidad de León, León, Spain

Ricardo Pong-Wong and Pamela Wiener
The Roslin Institute and R(D)SVS, University of Edinburgh, Roslin, Midlothian, United Kingdom

James Kijas
Animal, Food and Health Sciences, CSIRO, Brisbane, Australia

Amparo M. Martínez, Juan V. Delgado and Vincenzo Landi
Departamento de Geneética, Universidad de Córdoba, Córdoba, Spain

Luis T. Gama
L-INIA, Instituto Nacional dos Recursos Biológicos, Fonte Boa, Vale de Santare´m, Portugal
CIISA - Faculdade de Medicina Veterinária, Universidade Técnica de Lisboa, Lisboa, Portugal

Javier Cañón, Susana Dunner and Oscar Cortés
Departamento de Producción Animal, Facultad de Veterinaria, Universidad Complutense de Madrid, Madrid, Spain

Catarina Ginja
Centre for Environmental Biology, Faculty of Sciences, University of Lisbon & Molecular Biology Group, Instituto Nacional de Recursos Biológicos, INIA, Lisbon, Portugal

Inmaculada Martín-Burriel,, Clementina Rodellar and Pilar Zaragoza
Laboratorio de Genética Bioquímica, Facultad de Veterinaria, Universidad de Zaragoza, Zaragoza, Spain

M. Cecilia T. Penedo
Veterinary Genetics Laboratory, University of California Davis, Davis, California, United States of America

Jose Luis Vega-Pla
Laboratorio de Investigación Aplicada, Cría Caballar de las Fuerzas Armadas, Córdoba, Spain

Atzel Acosta and Odalys Uffo
Centro Nacional de Sanidad Agropecuaria, San Joséde las Lajas, La Habana, Cuba

Luz A. Álvarez and Jaime E. Muñoz
Universidad Nacional de Colombia, Sede Palmira, Valle del Cauca, Colombia

Esperanza Camacho
IFAPA, Centro Alameda del Obispo, Córdoba, Spain, Argentina

Jose R. Marques
EMBRAPA Amazônia Oriental, Belém, Pará , Brazil

Roberto Martínez
Centro Multidisciplinario de Investigaciones Tecnológicas, Direccioón General de Investigación Científica y Tecnológica, Universidad Nacional de Asunción, San Lorenzo, Paraguay

Ruben D. Martínez
Genética Animal, Facultad de Ciencias Agrarias, Universidad Nacional de Lomas de Zamora, Lomas de Zamora, Argentina

Lilia Melucci
Facultad Ciencias Agrarias, Universidad Nacional de Mar del Plata, Balcarce

Estación Experimental Agropecuaria Balcarce, Instituto Nacional de Tecnología Agropecuaria, Balcarce, Argentina

Guillermo Martínez-Velázquez and Jorge Quiroz
Instituto Nacional de InvestigacionesForestales, Agrícolas y Pecuarias, Coyoacán, México

Alicia Pòstiglioni
Área Genética, Departamento de Genética y Mejora Animal, Facultad de Veterinaria, Universidad de la República, Montevideo, Uruguay

Philip Sponenberg
Virginia-Maryland Regional College of Veterinary Medicine, Virginia Tech, Blacksburg, Virginia, United States of America

Axel Villalobos
Instituto de Investigación Agropecuaria, Estación Experimental El Ejido, Los Santos, Panama

Delsito Zambrano
Universidad Te´cnica Estatal de Quevedo, Quevedo, Ecuador

Nuno Santos and Margarida Correia-Neves
Life and Health Sciences Research Institute (ICVS), School of Health Sciences, University of Minho, Braga, Portugal
ICVS/3B's, PT Government Associate Laboratory, Braga/Guimarães, Portugal

Ivânia Moiane
Life and Health Sciences Research Institute (ICVS), School of Health Sciences, University of Minho, Braga, Portugal
ICVS/3B's, PT Government Associate Laboratory, Braga/Guimarães, Portugal
Paraclinic Department, Veterinary Faculty, Eduardo Mondlane University, Maputo, Mozambique

Adelina Machado, AndréNhambir and Osvaldo Inlamea
Paraclinic Department, Veterinary Faculty, Eduardo Mondlane University, Maputo, Mozambique

Gunilla Källenius
Department of Clinical Science and Education, Sö dersjukhuset, Karolinska Institutet, Stockholm, Sweden

Jan Hattendorf and Jakob Zinsstag
Swiss Tropical and Public Health Institute, Basel, Switzerland

Miho Inoue-Murayama
Wildlife Research Center, Kyoto University, Kyoto, Japan

Hideaki Abe
Wildlife Research Center, Kyoto University, Kyoto, Japan
Department of Anatomy, University of Otago, Dunedin, New Zealand

Kenji Nagao
Animal Husbandry Research Division, Aichi Agricultural Research Center, Aichi, Japan

Index